Adam Schwartz
Reinstating the Hoplite

HISTORIA

Zeitschrift für Alte Geschichte
Revue d'histoire ancienne
Journal of Ancient History
Rivista di storia antica

EINZELSCHRIFTEN

Herausgegeben von
Kai Brodersen/Erfurt
Mortimer Chambers/Los Angeles
Martin Jehne/Dresden
François Paschoud/Genève
Aloys Winterling/Berlin

HEFT 207

Adam Schwartz

Reinstating the Hoplite

Arms, Armour and Phalanx Fighting in Archaic
and Classical Greece

Franz Steiner Verlag Stuttgart 2013

Cover illustration:
Departure scene with hoplite wearing a linothorax.
Red-figure stamnos by the Achilles painter, C5m.
British Museum, London (E 448). Reprinted from
Ernst Pfuhl: *Malerei und Zeichnung der Griechen*
(Munich 1923) with kind permission of Stiebner
Verlag GmbH.

Bibliografische Information der Deutschen Nationalbibliothek
Die Deutsche Nationalbibliothek verzeichnet diese
Publikation in der Deutschen Nationalbibliografie;
detaillierte bibliografische Daten sind im Internet über
<http://dnb.d-nb.de> abrufbar.

ISBN 978-3-515-10398-5

Jede Verwertung des Werkes außerhalb der Grenzen
des Urheberrechtsgesetzes ist unzulässig und strafbar.
Dies gilt insbesondere für Übersetzung, Nachdruck,
Mikroverfilmung oder vergleichbare Verfahren sowie
für die Speicherung in Datenverarbeitungsanlagen.
© 2013 Franz Steiner Verlag, Stuttgart
Gedruckt auf säurefreiem, alterungsbeständigem Papier.
Druck: Laupp & Göbel, Nehren
Printed in Germany

CONTENTS

Acknowledgments. 9

1. INTRODUCTION 11
1.1 Conventions 11
1.2 Aims and purposes 11
1.3 Research history 13
1.3.1 The development of the hoplite phalanx. 13
1.3.2 The course of hoplite battles. 17
1.4 Sources and methods. 18
1.4.1 Literary sources. 18
1.4.2 Archaeological evidence. 20
1.4.3 Methods used 22

2. HOPLITE EQUIPMENT AND ITS LIMITATIONS 25
2.1 The hoplite shield 25
2.1.1 Nomenclature 25
2.1.2 Materials and measurements. 27
2.1.3 Characteristics of the hoplite shield 32
2.1.4 'Wielding' the hoplite shield. 38
2.1.5 The hoplite shield in combat. 41
2.1.6 Hoplite race and pyrrhic dance: the case for the light hoplite shield 46
2.1.7 Comparison with a modern combat shield 53
2.2 Headgear 55
2.2.1 Typology 55
2.2.2 Characteristics of the Corinthian helmet. 59
2.2.3 The Corinthian helmet in combat 61
2.3 Body armour. 66
2.3.1 Breastplates. 66
2.3.2 Corslets 70
2.3.3 Cuirasses and corslets in combat 73
2.3.4 Greaves 75
2.3.5 Additional body armour 78

2.3.6	Effectiveness of bronze armour	79
2.4	Offensive weapons	81
2.4.1	The spear	81
2.4.2	Swords and other secondary weapons	85
2.4.3	The spear in combat	86
2.4.4	The sword in combat	92
2.5	Physical limitations – conclusions	95
2.5.1	The weight of armour	95
2.5.2	Physiology	98
3.	THE PHALANX	102
3.1	The development of the phalanx: myth or reality?	102
3.1.1	The development of phalanx fighting: an overview	102
3.1.2	The Homeric poems	105
3.1.3	The *Iliad* and hoplite fighting	108
3.1.4	The phalanx in Archaic poetry	115
3.1.5	The phalanx in Archaic iconography	123
3.1.6	The Amathus bowl	130
3.1.7	The phalanx during the Persian wars	135
3.1.8	The Solonian census classes	141
3.1.9	The development of the phalanx: conclusions	143
3.2	The shield in phalanx fighting	146
3.2.1	Introduction	146
3.2.2	*Rhipsaspia*	147
3.2.3	Comparison with a modern 'phalanx'	155
3.3	Deployment	157
3.3.1	Width of file	157
3.3.2	Phalanx depth	167
3.3.3	The wings	172
3.3.4	Community ties in phalanx organisation	175
3.3.5	Place and function of the general	179
3.4	*Othismos*	183
3.4.1	A vexed question	183
3.4.2	The case against the mass-shove	187
3.4.3	Another purpose of the rear ranks?	194
3.4.4	Maintaining cohesion	195
3.4.5	*Othismos*?	198

4.	DURATION OF HOPLITE BATTLES	201
4.1	The problem of temporal designations	201
4.1.1	Introduction	201
4.1.2	The phraseology of battle duration	202
4.1.3	Hoplite battles described as long	204
4.1.4	Ambiguous points of reference	207
4.2	Contributing factors	208
4.2.1	Necessary phases	208
4.2.2	Different sectors faring differently	213
4.2.3	Pursuit	214
4.2.4	Physical limitations	215
4.3	Unusually long battles	217
4.3.1	All-day battles: a problem of sources	217
4.3.2	All-day fighting: Sphakteria	220
4.4	Excursus: the possible influence of literary conventions	222
5.	CONCLUSION	226
	APPENDIX: BATTLE INVENTORY	235
	BIBLIOGRAPHY	293
	LIST OF ILLUSTRATIONS	303
	INDICES	305
	General index	307
	Index of sources	321
	I. Literary sources	321
	II. Inscriptions and papyri	337

ACKNOWLEDGMENTS

It was Dr Mogens Herman Hansen, director of the Copenhagen Polis Centre, and Gorm Tortzen, MA (Copenhagen and Elsinore), my supervisors, who first drew my attention to the topic of the present dissertation. During my work both have been a constant source of inspiration; and without their valuable suggestions, comments, criticisms and fresh approaches, it is safe to say that this study would not have been possible. I owe them both my deepest thanks.

During the last half of my three years of research especially, Mogens has been absolutely indispensable. I have had the privilege of working with him and having the opportunity to draw on his enormous expertise and discuss problems as they arose on a daily basis, and not once have I come in vain to him for help or advice. Whatever merit this study may possess, I owe it largely to him.

I am also deeply indebted to my colleague Thomas Heine Nielsen, PhD (Copenhagen), for valuable discussions and suggestions. His never-failing support and optimism on my behalf have helped me immeasurably during the creative process. Special thanks are due to Dr Jørgen Mejer (Copenhagen), former director of the Danish Institute at Athens, who kindly and patiently drove me around the Peloponnese on a tour of ancient battlefields (even to the point of securing transportation to Sphakteria by hiring a boat in Pylos). I also wish to single out Prof Nino Luraghi (Princeton) for stimulating discussions and new perspectives, and for permitting me to read his forthcoming article on early Greek mercenaries, as well as the two anonymous referees of the Historia Einzelschriften evaluation committee for many valuable suggestions and acute observations. Furthermore, I am deeply grateful to Fritz Saaby Pedersen, MA (Odense and Copenhagen) for much needed support in times of technical crisis and for general encouragement; to David Bloch, PhD for valuable contributions, corrections and suggestions; to Jean Christensen, MA and my sister Ditte Schwartz, MA, for much-needed and meticulous proofreading and source checking; to Merete Egeskov, DVM for proofreading and assisting in translating from the Dutch, and to Ms Vivi Lund, Ms Lone S. Simonsen and Ms Hanna Lassen, secretaries at the Department of Greek and Latin, University of Copenhagen and the Danish Institute in Athens respectively, whose fluency in modern Greek eased communications with the Archaeological Museum in Olympia immensely.

A large number of Danish and foreign colleagues and friends have helped me greatly during my three years of research. Thanks are due to the following: Gojko Barjamović, PhD (Copenhagen), Dr Adam Bülow-Jacobsen (Copenhagen), Dr Sten Ebbesen (Copenhagen), Pernille Flensted-Jensen, PhD (Copenhagen), Rune Frederiksen, PhD (Oxford), Prof Vincent Gabrielsen (Copenhagen), Dr Erik Hallager (Aarhus and Athens), director of the Danish Institute at Athens, Prof Stephen Hodkinson, Nottingham, The Hoplite Association (UK), Jesper Jensen, MA, vice director of the Danish Institute in Athens, Prof Maria Liston (Waterloo, Ontario), Rune Munk-Jørgensen, MA (Copenhagen), Prof Beat Näf (Zurich), chief inspector Claus Olsen of the Danish police, PhD student Bjørn Paarmann, MA (Fribourg), Prof Kurt R. Spillmann (Zurich), the staff of the collection of classical and Near Eastern antiquities at the Danish National Museum and Dr Martina Stercken (Zurich). Finally, special thanks are due to Stiebner Verlag GmbH, Munich for their very generous permission to reproduce a number of line drawings to which they hold the copyright as well as to the Carlsberg Foundation for enabling me to continue working on the revised edition.

1. INTRODUCTION

1.1 CONVENTIONS

All dates are BC unless otherwise indicated, and are written as C5 (5th century). C5e, f, m, s or l indicate early, first half, middle, second half and late respectively. I have tried to follow the abbreviations for authors used in the *Oxford Classical Dictionary* (1996³), whereas abbreviations for periodicals are those of *L'année philologique*. In transliterating, I have Hellenised rather than Latinised Greek names, although I have preferred the common English transliteration in especially well established names (Aristotle) and where it would interfere with normal pronunciation (Thucydides). This goes for citations of translations in the notes as well, where I have 'normalised' Greek names, rather than keeping their Latinised forms. On the whole, however, I have remained, in the words of Catherine Morgan, "cheerfully, and unapologetically, inconsistent".[1]

During my research, I have worked out an Inventory of 41 major hoplite battles which have served as a 'storehouse' of information and source references. The Inventory has been consulted progressively and formed the basis of the research presented below, and for this reason it has been changed and adapted until the last possible moment. Therefore, a few entries in the Inventory are not discussed in the dissertation; but all entries should be found to contain useful information. The battles are listed alphabetically by battle name, and the information of individual entries has been tabulated under 29 headings. For details, please refer to the introduction to the Inventory.

1.2 AIMS AND PURPOSES

The field of ancient military history has seen a revival of interest in recent years, though the focus of this renaissance has been mainly on the socio-political aspect of warfare. This renewal of interest is hardly surprising, given the fact that war was a fundamental aspect of daily life in antiquity. It has been calculated that Athens in the Classical age was in a state of war no fewer than two out of any three given years in the Classical period, and never experienced ten consecutive years of peace.[2] The Greeks themselves acknowledged this to a large extent. At the beginning of the *Laws,* Plato has the Kretan Kleinias say the following:

---

[1] Morgan (1990) viii.
[2] Garlan (1975) 15.

ἄνοιαν δή μοι δοκεῖ καταγνῶναι τῶν πολλῶν ὡς οὐ μανθανόντων ὅτι πόλεμος ἀεὶ πᾶσιν διὰ βίου συνεχής ἐστι πρὸς ἁπάσας τὰς πόλεις· εἰ δὴ πολέμου γε ὄντος φυλακῆς ἕνεκα δεῖ συσσιτεῖν καί τινας ἄρχοντας καὶ ἀρχομένους διακεκοσμημένους εἶναι φύλακας αὐτῶν, τοῦτο καὶ ἐν εἰρήνῃ δραστέον. ἣν γὰρ καλοῦσιν οἱ πλεῖστοι τῶν ἀνθρώπων εἰρήνην, τοῦτ᾽ εἶναι μόνον ὄνομα, τῷ δ᾽ ἔργῳ πάσαις πρὸς πάσας τὰς πόλεις ἀεὶ πόλεμον ἀκήρυκτον κατὰ φύσιν εἶναι.[3]

It is significant that none of the great Greek philosophers ever questioned war's *raison d'être,* despite their incisive analyses of almost all areas of politics: normally, Greek historians and philosophers are content with discussing the specific causes of this or that war, never war *itself.*[4]

There are urgent cultural idiosyncrasies to explain this phenomenon in part. There is a powerful undercurrent in Greek mentality and culture in the influence from the early epic tradition, and above all the Homeric poems. The *Iliad,* arguably the first great literary work of Greece and Europe, is a mighty epos of war and all its facets, and was known to all Greeks. War, fighting, strife and noble competition are portrayed again and again in the *Iliad* as acceptable ways of achieving social and political recognition, and martial prowess and brave deeds in combat are the standards by which the individual is measured. This, combined with the general agonal aspect of Greek culture, no doubt helped establish war and fighting as legitimate ways of achieving one's goals; and in a civilisation so steeped in competitive mentality as the Greek, it was perhaps inevitable that wars frequently broke out between pocket states that hardly needed much by way of provocation to declare war on each other in and out of season.

Furthermore, Greece was never a predominantly rich and fertile region. Approximately 80% are mountains, and good, arable land is accordingly scarce.[5] Natural resources were therefore always in short supply, and border skirmishes and larger conflicts could easily erupt over matters such as access to pasture land, although quite often such 'territorial' wars were possibly mere pretexts for far more complicated and elaborately codified matters of honour and religion. For these reasons (and many others), war played an absolutely central role in Greek history and culture; and it pervades almost all literature or art in some shape or form.[6]

Central to Greek land warfare throughout Archaic and Classical times was arguably the hoplite, the heavily equipped infantryman armed first and foremost with spear and shield. The primary scope of this dissertation is to assess the military function and fighting style of the Greek hoplite and the hoplite phalanx in the

---

3 Pl. *Leg.* 625e – 626a: "He seems to me to have thought the world foolish in not understanding that all men are always at war with one another; and if in war there ought to be common meals and certain persons regularly appointed under others to protect an army, they should be continued in peace. For what men in general term peace would be said by him to be only a name; in reality every city is in a state of war with every other, not indeed proclaimed by heralds, but everlasting" (trans. Jowett).
4 Cf. Momigliano (1969²) 120–121.
5 80% mountains: Cary (1949) 40. For the hardships of agriculture in Greece, see in general Hanson (1995).
6 See Connor (1988) and Dawson (1996) 47–99 for an analysis of the many levels on which war permeated the Greek society.

period from *c.* 750 to 338.⁷ The year 750 is chosen because of the Argos grave finds, datable to C8l. The grave contained a conical helmet with a high crest-stilt and a precursor of the Archaic 'bell' type bronze cuirass, elements of armour strongly indicative of at least 'proto'-hoplite, phalanx-like tactics. Their wearer can scarcely have been younger than 20–30 years at his death, which pushes the *terminus* back to 750. The other date is furnished by the battle of Chaironeia in 338 (inv. no. 3), in which the Macedonian forces of Philip II swept the last great Greek coalition army off the battlefield, once and for all putting an end to hegemonic *polis* rule and effectively ending the period in which the Classical Greek citizen-soldier, the hoplite, reigned supreme on the battlefield.

The study will focus on the more practical aspects of Greek hoplite warfare and deal specifically and primarily with what was physically feasible and practical under the given circumstances, both for the individual hoplite and for the phalanx as a whole, and on the development of phalanx fighting. It is my hypothesis that the shield above all was what characterised the hoplite and determined his style of fighting, so much space will be devoted to the hoplite shield and its defining characteristics.

1.3 RESEARCH HISTORY

1.3.1 The development of the hoplite phalanx

As mentioned above, warfare in antiquity is a field of research which has seen intensive activity in recent years. Modern scholarship may fairly be said to commence with German scholarship. In 1862, Hermann Köchly and Wilhelm Rüstow's *Geschichte des griechischen Kriegswesens von der ältesten Zeit bis auf Pyrrhos* appeared.⁸ Hans Delbrück's monumental four-volume *Geschichte der Kriegskunst im Rahmen der politischen Geschichte* was published between 1900 and 1920, and 1928 saw another major achievement of German scholarship of that period, Johannes Kromayer and Georg Veith's *Heerwesen und Kriegführung der Griechen und Römer*.⁹ In these the groundwork was laid for much of the later scholarship on the hoplite phalanx, and essentially these works defined the 'canonical' concept of the closed phalanx. They are, however, very much products of their time, and their focus is squarely on such topics as strategics, tactics, logistics and army strengths. In keeping with contemporary scholarship, these scholars regarded the study of warfare in antiquity as an extension of the attempt to understand warfare scientifically, and as a result their analyses are often of a very schematic and rigid nature, despite the fact that they put the sources to good use.

In 1911, Wolfgang Helbig put forward his thesis that the closed phalanx emerged around C7m in Chalkis on Euboia. Helbig regarded the use of javelins and

---

7 All years, unless otherwise stated, are BC.
8 Köchly & Rüstow (1862).
9 Delbrück (1900); Kromayer & Veith (1928²).

light-armed troops, earlier attested in, e.g., Kallinos and Tyrtaios, as inconsistent with a closed phalanx, believing Tyrtaios' phalanx, which he dated to the second Messenian war, to be a transitional phase between wholly open fighting (as seen in the *Iliad*) and the hoplite phalanx. Nevertheless Helbig failed to acknowledge the possibility of auxiliary troops aiding a closed phalanx and, crucially, the fact that the hoplites of the phalanx on the Chigi vase actually carry javelins into battle.[10]

However, the debate over hoplite phalanxes began in earnest in 1947 with Hilda Lorimer's article "The Hoplite Phalanx with Special Reference to the Poems of Archilochus and Tyrtaeus".[11] On the basis of extant Archaic poetry and archaeological finds Lorimer argued that hoplite weapons and phalanx tactics were inseparable, dating the invention and subsequent swift introduction of hoplite arms and armour to C7e.[12] Prior to this, she argued, there were neither hoplites nor phalanxes. The sudden invention of the arms sparked the birth of a new warrior type, who was in turn unable to function outside his chosen type of formation. Lorimer largely rejected iconographical evidence, as this in her opinion was likely influenced by the Homeric poems, while at the same time rejecting the presence of 'hoplite' weapons in them, on the ground that these were interpolations in the 'original' poems.[13] She thus in effect acknowledged the presence of hoplitic elements in the *Iliad*, but assuming a unitarian interpretation of Homer insisted that there were watertight partitions between the poem and the early hoplite phalanx.

This theory was challenged with the Argos grave find, excavated in 1957.[14] Based on stylistic analyses of ceramics in the tomb, the grave was dated to C8l; yet the armour – a bronze cuirass and a conical helmet – bore a strong likeness to hoplite equipment. Anthony Snodgrass countered Lorimer's theory with another approach: basing his arguments on the Argos grave find and the archaeological material, he proposed a longer period of gradual ('piecemeal') development of the armour, which did not immediately bring about a change in tactics.[15] Snodgrass thus maintained that armour and tactics were *not* inseparable: on his interpretation, parts of the equipment were gradually adopted. The next stage was then the adoption of decidedly hoplite tactics. While Snodgrass' assessment of the gradual adaptation is doubtlessly correct, there are certain problems with his theory: what would be the motivation for inventing pieces of equipment (above all the shield) if they were unfit for single combat?[16]

A further analysis of the development of hoplite armour saw the light of day with J.K. Anderson's *Military Theory and Practice in the Age of Xenophon*.[17] Here,

---

10 Helbig (1911).
11 Lorimer (1947).
12 Lorimer (1947) 76, 128–132.
13 See esp. Lorimer (1947) 82 n. 4, 108, 111–114. The 'sudden change' theory has had its defendants: see, e.g., Greenhalgh (1973) 73 and Cartledge (1977) 19–21, correctly stressing the ambiguity of iconographical evidence.
14 For a full excavation report see Courbin (1957). The find has since been corroborated by more finds of a similar type in Argos: *infra* 66.
15 Snodgrass (1964a), (1965) 110–111.
16 Snodgrass (1965) 111 argues, however, that the hoplite shield was adequate in solo fighting.
17 Anderson (1970).

Anderson analysed literary sources and iconograhical evidence from C5l–C4e and convincingly showed that hoplite equipment underwent a notable change towards lightness and less protection in this period: body armour such as cuirasses and greaves are often lacking on vase paintings and funerary reliefs.

Joachim Latacz' pioneering work *Kampfparänese, Kampfdarstellung und Kampfwirklichkeit in der Ilias, bei Kallinos und Tyrtaios* from 1977 opened the discussion of the value of the Homeric poems for an understanding of early massed fighting.[18] Latacz convincingly showed that despite the immediate appearance of duel-based fighting between the heroes in the *Iliad,* there are in fact frequent references to fighting in φάλαγγες or στίχες, i.e. ranks of warriors, arrayed behind each other and led by πρόμαχοι (warriors in the front ranks), thus interpreting the parts identified and rejected by Lorimer as an integral part of the poem. His work demonstrated that the *Iliad* does indeed represent early massed fighting, some of which may actually be hoplite fighting: this is not surprising, since the Homeric poems are ultimately products of an oral tradition, weaving together many layers from different historical periods. Hoplitic elements will at some point have been included in the tradition. Furthermore, Latacz demonstrated that massed fighting is not only present, but is in fact a decisive element in the *Iliad.*[19] It is thus reasonable to assume that hoplite equipment was developed in response to needs perceived in such massed fighting.

Countering this, Hans van Wees has argued that *phalanx* in an *Iliad* context means a more loosely organised group of warriors, comparing the fighting to that found in primitive societies such as those in Papua New Guinea.[20] This, however, ignores the patent references to close ranks and massed fighting which are also on display in the *Iliad,* as demonstrated by Latacz. The two components are essentially different and difficult or impossible to reconcile; but at any rate the presence of both must preclude the notion that the *Iliad* presents a homogenous and consistent image of fighting.

In an important article, Victor Davis Hanson in 1991 stressed the logical causality in matters of weapons development.[21] He noted that while scholars agree that the reduction in armour in C5l–C4e – as shown by Anderson – reflected new strategic needs in infantry employment, "strangely they do not allow for this same phenomenon in reverse chronological order: the preference (well before 700–650 BC) for massing shock troops in close formation led to demands by combatants for new, heavier equipment."[22]

Hans van Wees has presented his view of an extreme 'piecemeal' theory in an article from 2000.[23] According to van Wees, the crucial evidence is iconographical, showing a motley crew of combatants on the battlefield, fighting in no particular

---

18 Latacz (1977).
19 Latacz (1977) 30–31 (citing earlier, but disregarded scholarship – that of Kromayer and Lammert – to the same effect), 46–49.
20 Van Wees (1994) and most recently (2004) 153–158.
21 Hanson (1991), esp. 63–67.
22 Hanson (1991) 64.
23 Van Wees (2000).

order. This development, according to van Wees, possibly did not halt until after the Persian wars, and he maintains that Archaic poetry and even Herodotos show similar signs of loose-order combat. He takes this to be a natural continuation of the loose-order, chaotic fighting which he sees in the *Iliad* and to which he finds parallels in primitive societies. There are several problems with this approach, chief among which the objection that this presupposes a homogenous and consistent Homeric portrayal of society and warfare. Furthermore, it is difficult to argue *chiefly* from iconography, since we cannot always be certain that we can appreciate the artist's intentions and the operative artistic conventions. Very recently, van Wees has also further expounded these views in a monograph with the telling title *Greek Warfare. Myths and Realities,*[24] in which he offers a synthesis of the above-mentioned and a number of other articles.

Most recently, Jon E. Lendon has published a monograph entitled *Soldiers and Ghosts. A History of Battle in Classical Antiquity.*[25] This is an ambitious attempt at analysing the underlying causes of warfare in Greece and Rome. Lendon argues that Greek warfare was above all influenced by two factors: the competitive spirit native to almost all aspects of Greek culture, and more especially as embodied in the Homeric poems. Lendon accordingly argues that the impact of the poems shaped not only the ideals of subsequent Greek warfare, but also its actual practice, to the extent that he more or less ignores such factors as technological advance, socio-political changes and foreign influence. Interesting and refreshingly thought-provoking though it may be, his thesis is somewhat focused on a single cause. His perception of phalanx fighting may illustrate this:

> The phalanx should not be viewed as the submersion of the individual in the mass but as creating in mass combat a simulacrum of individual combat. ... Fighting in the phalanx was hardly a perfect form of individual competition or of competition between states. But it was the best way the Greeks could discover to have men and city compete at the same time in the same way in a form of fighting that worked as a competition in the real world for both.[26]

Strangely, however, Lendon himself hints that if this were the true objective of inter-state 'competing', another outcome would have been more logical: "If the Greeks had wanted a more perfect competition between individuals, they could have surrounded one-on-one fighting with rules and taboos and gone down the road upon which feudal Europe and Japan would travel a good distance."[27]

Sometimes this method leads to putting the proverbial cart before the horse, as when Lendon claims that the cooperation of a phalanx was only "superficially cooperative, for those who fought in the seemingly unheroic phalanx conceived of what they were doing in Homeric terms,"[28] because of 'epic' epitaphs and Homeric heroes in hoplite gear on vases. It is at least as likely, however, that Homeric scenes were portrayed in contemporary garb; and it is hardly surprising that patterns of formal expression should be sought in poetry. Most importantly, however, Lendon

---

24 Van Wees (2004).
25 Lendon (2005).
26 Lendon (2005) 64–65.
27 Lendon (2005) 57.
28 Lendon (2005) 45.

himself admits that the perceived radical influence of the Homeric poems must remain a theory: "in no indiviual case can Homeric inspiration conclusively be proved, but the wider pattern is beyond doubt."[29]

1.3.2 The course of hoplite battles

In 1978, George Cawkwell stirred up controversy by challenging the traditional conception of how hoplites fought.[30] The frequent references to 'shoving' (ὠϑισμός), Cawkwell argued, were misconstrued by scholars who interpreted them as a distinct phase of battle, since this would interfere with the use of weapons. Instead, Cawkwell visualised hoplite battle as essentially consisting of series of weapons duelling between individual hoplites, ending perhaps in a final bout of shoving.[31] One problem with Cawkwell's approach was that, on his interpretation, hoplites would have to open their ranks after marching forward, then join the shields later on for the push, surely impracticable in real life. Nonetheless, Cawkwell's rejection of the bodily shove has been followed by Krentz,[32] Goldsworthy,[33] and, most recently, van Wees.[34] This notion has been countered, above all by Hanson, who in his *The Western Way of War* (1989, second ed. 2000) vividly described the implications of this brutal aspect of hoplite battle.[35]

The question of *othismos* has thus been a bone of contention in recent years. Hanson's *The Western Way of War* offered an interesting analysis of the sources describing the gritty reality of hoplite battle. This study focused on the experience of a hoplite battle from the individual hoplite's point of view, stressing especially the extreme physical exertions and the gruelling, bloody chaos in the front ranks. Particularly important in this respect was his focusing on the amateur aspect of battle between citizen-soldiers. In Hanson's view, hoplite battle was a logical, if chaotic and grim, way of fighting wars between farming *poleis,* since it required no particular technical skill or drill and limited warfare largely to a single day's worth of fighting, and in a way that actually kept casualties on both sides at a minimum.

The individual stages of battle are meticulously analysed by Johann Peter Franz, who, inspired by Snodgrass, Latacz and Anderson, has subdivided his study into chapters dealing with sharply limited periods, assessing the evidence for each separately. In Franz' opinion, this enhances the possibility of determining the development of hoplite arms and armour, but also of the tactics and phases of battle. While this is ostensibly true, it must be said that it is problematic to accept unhesi-

---

29 Lendon (2005) 159.
30 Cawkwell (1978) 150–165, followed up by (1989). The question had been addressed earlier by Fraser (1942), but this article has had little impact.
31 Cawkwell (1978) 152–153.
32 Krentz (1985b), (1994).
33 Goldsworthy (1997).
34 van Wees (2000) 131–132, (2004) 152, 180–181 and esp. 188–191.
35 Hanson (2000²) 28–29, 156–158, 169–178, (1991) 69 n. 18; but see also Holladay (1982) 94–97; Luginbill (1994) 51–61; Lazenby (1991) 97–100; Anderson (1984) 152, (1991) 15–16; Pritchett (1985a) 65–73, 91–92; Franz (2002) 299–308.

tatingly, as Franz does, an historical Homeric society, and furthermore, that it can be safely dated to C8. The evidence for this period must of necessity be limited to Homer, Hesiod and a number of archaeological and iconographical items; and in this respect it is a weakness that there is nothing which decisively links material evidence from C8s with the epic poems. Furthermore, Franz oversteps his own sharply drawn limits time and again, including sources from entirely different periods.[36]

It should also be mentioned that W.K. Pritchett, in this field, as in others, has made considerable contributions, chiefly with his monumental five-volume survey *The Greek City-State at War,* in which he has collected the data on a vast array of pertinent topics.[37]

1.4 SOURCES AND METHODS

1.4.1 Literary sources

The most important sources for this study are literary, and of the Classical period. Literary sources have the great advantage over 'visual' evidence that we can be certain that hoplite activity is actually referred to. It should be obvious that contemporary sources are to be preferred over 'later' sources, i.e. historians and others writing in Hellenistic and later times. This gives natural precedence to authors such as Herodotos, Thucydides and Xenophon. The special importance of Thucydides and Xenophon rests on the fact that they were both military commanders and so doubtless possessed considerable experience in military matters, even when compared with their contemporary audience.[38] In the case of Thucydides, he even claims to have begun his work immediately at the outbreak of the Peloponnesian war, thus ostensibly offering a near-perfect recollection of events. It is a pet criticism of scholars that Xenophon is somewhat naïve and that he displays a 'pro-Spartan' and 'anti-Theban' tendency, but this is highly exaggerated: while biased in his seemingly haphazard selection of events for a number of reasons, it cannot be sufficiently demonstrated that Xenophon actively even disliked Thebes.[39] Certainly Xenophon offers important knowledge about the famously secretive Sparta, which he knew intimately and about leadership of soldiers, a subject that evidently interested him greatly.

Valuable sources are by no means limited to historians. Important information, likely based on first-hand experience, can be found in the great playwrights: Aischylos, Sophokles, Euripides and Aristophanes make frequent allusions to the hoplite experience, which they must have expected a great part of their audience to recognise and understand; and the same applies to numerous fragments of other playwrights. Sophokles served as general twice; and Euripides' tragedies are especially

---

36 See, e.g., Franz (2002) 121, 214 n. 69, 249 n. 37.
37 Pritchett (1971), (1974) (1979) (1985a) and (1991).
38 Thucydides' command: Thuc. 5.26; Xenophon's command: Xen. *An. passim.*
39 See esp. Christensen (2001).

relevant for the warfare of the Peloponnesian war, since they were doubtless influenced by recent events: the horror of war is palpable in many of Euripides' tragedies. It should be noted, however, that care has to be taken in ascertaining the poetic context: since the 'dramatic date' is normally a distant mythical past, elements may occur which were certainly blatant anachronisms in C5; but such details of 'local colour' are normally easily identified: combat details, intended to be recognisable to a contemporary audience, are culled from the shared experience of warfare.[40] C5–C4 logographers and politicians such as Lysias, Demosthenes and Isaios often also offer glimpses into the world of hoplite warfare. Another important element of written sources is the evidence from epigraphy: casualty lists and peace treaties are often preserved on stone, a political decree or a commemoration frozen for posterity.

The above-mentioned sources of course all concern the Classical period. There are, however, also a number of literary sources from the Archaic period, and they should be assessed separately in order to determine whether they reveal change or continuity from the Archaic to the Classical period.

The *Iliad,* for example, contains a great many passages which are surprisingly replete with massed fighting, far more so than is immediately apparent from a glance at the largely duel-based fighting between protagonists of the poem. Many of these contain vivid similes whose *tertium comparationis* is based on the concepts of extreme closeness, contiguity, solidity and powerful forward surges. It is reasonable to assume that these may be connected with phalanx warfare, and even more so since there is a possibility that the Homeric poems were not fixed in writing until perhaps as late as C7.[41]

Elegiac and lyric poets also present an abundance of testimony about hoplite warfare, especially with regard to the Archaic age: poets like Tyrtaios, Kallinos, Mimnermos, Pindar, Archilochos, Alkaios and Simonides are certainly important in this respect. Even when authors such as these are not actually based on first-hand experience, they are, although secondary, in all likelihood at least influenced by actual eye-witness accounts. Epigraphy also plays a role in Archaic sources, as when we have testimonies in the shape of Greek mercenary graffiti from Egypt.

Of decidedly lesser importance are post-Classical sources, of which there is a multitude. Historians such as Polybios, Plutarch and Diodoros lived long after the hoplite era, but discuss much valuable information retrospectively. They may well have preserved relevant information compiled from earlier sources, lost to us. Unless they specify their sources (as is sometimes the case), however, they remain essentially suspect; although the case is somewhat better if they at least are precise with regard to the date of battle in question. When speaking of hoplites and phalanxes, they may do so only in an indirect manner, and actually refer to the Macedonian phalanx, which, for all the similarities, was a different formation, made up

---

40 One example may suffice: in Eur. *Suppl.* 650–730 a messenger reports a battle in which chariots play a predominant role.
41 As the Homeric poems present an especially complex problem in this connexion, they are treated separately below: see chap. 3.1.2 and 3.1.3.

of different warriors altogether.⁴² Diodoros is particularly problematic when it comes to the difficult question of battle duration.⁴³ Plutarch, though very late, is especially interesting because of his *Parallel lives* and *Spartan aphorisms,* in which he collected pithy sayings of the Spartans. The gnomic character of the aphorisms probably testifies to their validity, and the collection itself to their popularity in antiquity.⁴⁴

Even more problematic are the so-called tacticians: Arrian, Ailian and Asklepiodotos all lived in late antiquity, and their writings are suspect for a number of reasons, chief among which is their highly theoretical approach to the subject.⁴⁵ Further exacerbating the problem is the fact that we cannot know for sure whether they deal with the Macedonian phalanx or the Greek hoplite phalanx of earlier times (though the former seems likelier). However, all such later sources must of course be re-evaluated when they refer to events in their own time, or when they describe experiences or phenomena common to or valid at all times.

1.4.2 Archaeological evidence

The archaeological sources may for the present purpose be divided into two main groups: (1) representations of hoplite arms, armour and fighting in works of art, and (2) original weapons or pieces of armour. For an understanding of the weight and size of weapons and armour, original weapons are normally to be preferred, but iconography may assist in making plain the tactics or fighting technique employed.

(1) Warfare is frequently portrayed in Greek art, and many of these images are important for an understanding of the fighting and, to an even higher degree, the equipment. I maintain, however, that iconography is fundamentally difficult to interpret. To put it simply, very early Greek vase painting (C8 and earlier) is often too crude and primitive to determine what is happening with any certainty. As for the painting technique, in many cases it is not until proto- or mature Corinthian vases that the painting technique becomes sufficiently advanced to allow a safe judgment of the contents.

This objection goes only for the representation of objects. The interpretation of tableaux and scenes is even more complicated, though – as with simple objects – identification becomes far easier from C7 onwards. All too often, however, we lack the code or key, as it were, to decode the images. Scenes that may have been perfectly logical to contemporary Greeks are enigmatic to us. We cannot know what conventions were operative, or what elements were simply required, or perceived to be so, in the representation of a particular scene. Worse, we have no way

---

42 See Lazenby (1991) 88.
43 *Infra* 217–218.
44 See Fuhrmann (1988) 137–140 for a review of likely Classical sources and Hammond (1979–80) 108: "In most of the extant classical sources, the *exemplum* [Plut. *Mor.* 241f 16] simply accounted for the Spartan toughness and discipline, that is, it was primarily historical in intention."
45 The tacticians are discussed more fully *infra* 157–159.

of knowing whether battle scenes are intended to show contemporary reality or a mythical battle scene, and this problem is exacerbated exponentially as we go back in time. As a rule of thumb, it may be said that the earlier the representations, the worse the insecurities in interpreting the images 'correctly'. The difficulty remains the same: we cannot determine with any confidence whether an image contains archaising, romanticising or mythologising elements for all the reasons mentioned above, often nor even what the scene is intended to represent.

A single example may suffice to show the sheer amount of ambiguity inherent in interpreting Greek iconography: in Franz' assessment of the source value of vases, he seems to believe as a matter of course that archaising elements are not present in C7 vases: "die Vasen geben ... die Bewaffnung und die Kampfesweise der Zeit, in der sie gemalt wurden, wieder. Archaisierungen oder ähnliche Phänomene, die zumindest ein geringes historisches Verständnis voraussetzen, sind in der für unser Thema relevanten Bildkunst nicht zu erkennen", adding that scholars are generally too quick to discard relevant material "ohne ausreichende Begründung".[46] Yet only two pages later, he claims that vase images are not realistic, but rather portrayed "als heroisches Geschehen".[47] Quite apart from being unsubstantiated, these two principles seem somewhat difficult to reconcile.

The problem with using iconography is exacerbated by the fact that no representations of massed fighting are attested for C5–C4. While hoplites are represented on vases often enough, they are typically portrayed singly or in pairs, and frequently in arming scenes or other non-combat motifs. There are many different potential reasons for this absence; but the fact of the matter is that such scenes do not play any appreciable role in Classical art, thus rendering iconography a very difficult source for a diachronic analysis of phalanx fighting in its later stages.

(2) A fairly large number of ancient weapons have been preserved, chiefly arms and armour from such Panhellenic sites as Delphi and especially Olympia, and these are of course highly important. It was customary to dedicate captured enemy weapons after a victory (a frequent, macabre expression is ἀκροθίνιον ["the best pick of the harvest"]), and consequently we possess a great amount of weapons and especially armour – above all from the Archaic period; but Classical finds, such as, e.g., the Pylos shield excavated in the Athenian agora, are also attested, so that weapons finds actually cover the entire period C8l–C4.[48]

The remains of weapons testify especially to measurements, but can also reveal a great deal about how they were worn or handled in combat. Metal parts of shields have often been partially preserved, including the outer bronze sheathing and the armband, although the organic components – the wooden core, the inner layer of leather and the handlegrip – have long since disappeared. However, there are preserved organic remains of an Etruscan shield from Bomarzo in Italy, both wood, leather and bronze, in the Vatican museum; and the same applies to a Greek shield, found in Sicily and now in Basel. One is an actual hoplite shield, and the other at

---

46 Franz (2002) 16 and n. 85.
47 Franz (2002) 18.
48 See esp. Jackson (1991) 228–232.

least similar in build and structure, and as such they can be used for measurements and assessments of qualities, characteristics and mode of production.

There are also several hundred items of body armour: cuirasses, greaves and supplementary armour such as arm-guards or ankle-guards were frequently dedicated at sanctuaries, presumably because of the value and impressive sight of polished bronze. Again, estimates of weight and measurements of metal items may at least be approached, and such pieces of armour also help establish a relative chronology of the development of weapons. The same goes for helmets: fairly large quantities of helmets have been dedicated at Olympia, and it is thus possible to establish a fairly certain chronology. Furthermore, helmets of the Corinthian type are by far the most frequent, proving its popularity in much of the hoplite era.

With offensive weapons, the conditions are less favourable. By far the most important offensive hoplite weapon was the thrusting spear, and since the shaft was made of wood, we are left with nothing more than iron spear-heads and bronze butt-spikes. However, the diameter of the shaft may be estimated from the sockets, and the length with aid from iconography. Iron swords of several types are also preserved almost intact, if rather corroded. The original weapons and armour are extremely important if we are to understand how hoplites fought and what physical limitations they imposed on their owners. As such, they will be included to a large extent.

1.4.3 Methods used

My method can by now perhaps be guessed from the above. Hoplite weapons did change gradually, as shown by Anderson and Franz; but for a period of some 400 years, there is nonetheless a large degree of consistency within the primary hoplite weapons, chiefly spear and shield. Furthermore, since weapons and tactics are inseparably intertwined, it is assumed that hoplites throughout this period were characterised more by similarities than differences.

In this respect I differ from Franz, who believes that a sharp distinction between more or less arbitrarily defined periods is the only way to achieve precise knowledge.[49] Apart from the problems inherent in using Homer as a historical source, it remains difficult, despite Franz' claims to the contrary, to demonstrate continuity in the development, if we cannot juxtapose sources from two different periods. If, for example, Herodotos cannot be cited to establish anything meaningful about hoplites in the period 479–362, we risk ending up with a lot of *membra disjecta* that cannot be combined to form a whole; and even the analysis of the individual segments suffers. Assuming that hoplites were unable to rally again in this period simply because it is not mentioned directly in the sources while at the same time accepting it for the preceding and following period, is an argument *e silentio*: "Für die Zeit vom 7. bis zum 5. Jh. hatten wir angenommen, daß die Hopliten ihre Schilde wegwarfen, um auf der Flucht schneller laufen zu können und somit ihre Überlebenschancen zu verbessern. Gegen Ende des 5. Jh. konnten sie den Schild

---

49 Franz (2002) 4–7, 11–12.

auf der Flucht offensichtlich wieder mitnehmen." Franz' approach, although unconditionally puristic, is therefore not entirely unproblematic. I maintain that it is possible to regard literary sources from *any* point within these 400 years as valid for an understanding of hoplite tactics and fighting. The following table may help explain the basic approach:

Archaic period	Texts	Weapons	Iconography
Classical period	Texts	Weapons	–

Since there is a definite diachronic typological consistency of the most important hoplite weapons – namely the spear and shield – throughout the Archaic and Classical periods, and since weapon typology and fighting style are arguably interdependent, this provides the basis for an analysis of other types of sources, such as textual evidence and iconography. There can be no doubt that typologically identical hoplite weapons, as attested by weapons finds ranging from C8l to C4, were in use during the entire period, and consequently texts from the entire period are valid sources in the analysis of weapons use. Ultimately, the interpretation of the texts must take place in the light of what can and cannot be done with these weapons; but it should be clear that if it is accepted that there is consistency, there is no contradiction involved in using, e.g., Classical sources to evaluate fighting with typologically similar weapons at any given point of this entire period.

The rather small amount of Archaic texts consists almost exclusively of poetic texts, with different aims and often a more or less fixed vocabulary, sometimes resulting in ambiguity or outright obscurity. Moreover, much of the source material has survived only in fragments and thus often lacks the necessary context for a proper analysis. Nonetheless, the glimpses afforded of hoplite fighting in Archaic poetry are by no means irreconcilable with what Classical sources tell us and are therefore also included.

Conversely, the textual sources from the Classical period (in particular the historians) describe warfare relatively fully and in usually fairly detailed prose, as warfare – to a large extent, hoplite warfare – is the backbone of most historical works and a significant factor in other writings. The fact that Thucydides structured his work by winters and summers[50] – around campaigning seasons – is in itself revealing; and few would dispute the fact that the fullest sources for hoplite fighting are to be found in this period.

---

50 Thuc. 2.1.

Iconography may also be informative in this light but is frequently ambiguous, especially with regard to early images. A major problem here is that there is a tendency to focus on fighting in pairs, possibly even dueling, in Archaic imagery, while apparently there are not even true representations of fighting in larger formation for the Classical period.

Therefore, the weapons themselves, coupled with texts, above all from the Classical period, must form the backbone of the following analysis of hoplite fighting. Iconography and Archaic texts will naturally be discussed as well, but it should be clear that the focus is primarily on the weapons. It is not normally disputed that there is consistency between hoplite weapons and the type of fighting portrayed in Classical texts; but if this is so, and there is a typological similarity between weapons of the Archaic and Classical periods, it follows that Classical texts must also be valid for analysing *earlier* hoplite combat.

2. HOPLITE EQUIPMENT AND ITS LIMITATIONS

2.1 THE HOPLITE SHIELD

2.1.1 Nomenclature

Hoplites were, above all, characterised and defined by the weapons they used. This is clearly reflected in the simple fact that they were named hoplites (ὁπλῖται), long taken to be an ancient denomination derived from the name of their large, concave[51] shield, ὅπλον. The reason for this assumption is in no small degree Diodoros, who in a curious passage asserts just this:

> τῶν γὰρ Ἑλλήνων μεγάλαις ἀσπίσι χρωμένων καὶ διὰ τοῦτο δυσκινήτων ὄντων, συνεῖλε τὰς ἀσπίδας καὶ κατεσκεύασε πέλτας συμμέτρους, [ἐξ] ἀμφοτέρων εὖ στοχασάμενος, τοῦ τε σκέπειν ἱκανῶς τὰ σώματα καὶ τοῦ δύνασθαι τοὺς χρωμένους ταῖς πέλταις διὰ τὴν κουφότητα παντελῶς εὐκινήτους ὑπάρχειν. διὰ δὲ τῆς πείρας τῆς εὐχρηστίας ἀποδοχῆς τυγχανούσης, οἱ [μὲν] πρότερον ἀπὸ τῶν ἀσπίδων ὁπλῖται καλούμενοι τότε [δὲ] ἀπὸ τῆς πέλτης πελτασταὶ μετωνομάσθησαν.[52]

It is immediately apparent that there is some confusion involved here: if peltasts were named after their crescent-shaped shield, the *pelte,* hoplites ought to be named *aspistai.* This has been claimed quite convincingly by J.F. Lazenby & D. Whitehead,[53] who maintain that, in spite of Diodoros, the specific meaning 'shield' of the word *hoplon* cannot be decisively linked to the word *hoplites*. The reason for this is the paucity of ancient sources actually referring to the shield as *hoplon,* as opposed to the relative wealth of sources merely employing the term *aspis*. Nonetheless, as Lazenby and Whitehead's own examples show, *hoplon* sometimes *does* mean precisely 'shield',[54] and no other *specific* weapon or piece of armour, to my knowledge, can claim this.[55] An interesting passage from Xenophon's *Hellenica* to my

---

51 I adopt Hanson's term 'concave' (rather than the usual 'convex') as the default designation, seeing it from the bearer's point of view (Hanson [1991] 68 n. 15); cf. Warner (1979²) 284.
52 Diod. Sic. 15.44.2–3: "[T]he Greeks were using shields which were large and consequently difficult to handle; these he discarded and made small oval ones of moderate size, thus successfully achieving both objects, to furnish the body with an adequate cover and to enable the user of the small shield, on account of its lightness, to be completely free in his movements. After a trial of the new shield its easy manipulation secured its adoption, and the infantry who had formerly been called "hoplites" because of their heavy shield, then had their name changes to "peltasts" from the light *pelta* they carried" (trans. Sherman).
53 Lazenby and Whitehead (1996).
54 E.g. Thuc. 7.75.5 (which I accept); Xen. *Hell.* 2.4.25; Diod. Sic. 17.21.2 (and cf. 17.18.1); Polyaen. 3.8.1, 7.41.
55 Except, perhaps, the puzzling Aristotelian fr. 532 Rose, which seems to refer to a breastplate (normally known as θῶραξ).

mind bears this out, although summarily despatched by Lazenby and Whitehead. It runs thus:

ἀπιόντι [sc. τῷ Κλεομβρότῳ] γε μὴν ἄνεμος αὐτῷ ἐξαίσιος ἐπεγένετο, ὃν καὶ οἰωνίζοντό τινες σημαίνειν πρὸ τῶν μελλόντων. πολλὰ μὲν γὰρ καὶ ἄλλα βίαια ἐποίησεν, ἀτὰρ καὶ ὑπερβάλλοντος αὐτοῦ μετὰ τῆς στρατιᾶς ἐκ τῆς Κρεύσιος τὸ καθῆκον ἐπὶ θάλατταν ὄρος πολλοὺς μὲν ὄνους κατεκρήμνισεν αὐτοῖς σκεύεσι, πάμπολλα δὲ **ὅπλα** ἀφαρπασθέντα ἐξέπεσεν εἰς τὴν θάλατταν. τέλος δὲ πολλοὶ οὐ δυνάμενοι σὺν **τοῖς ὅπλοις** πορεύεσθαι, ἔνθεν καὶ ἔνθεν τοῦ ἄκρου κατέλιπον λίθων ἐμπλήσαντες ὑπτίας **τὰς ἀσπίδας**. καὶ τότε μὲν τῆς Μεγαρικῆς ἐν Αἰγοσθένοις ἐδείπνησαν ὡς ἐδύναντο· τῇ δ' ὑστεραίᾳ ἐκομίσαντο **τὰ ὅπλα**.[56]

The meaning of this passage is clear enough. Kleombrotos' men, on their way over the Kreusis pass, were overtaken by an extremely forceful (ἐξαίσιος) storm which was powerful enough to wrest the shields off their arms and all the way down into the sea below. It is difficult to imagine which other parts of their weaponry would be meant by this. The offensive weapons – spears and swords – will have had too small a surface to risk being swept away (aside from the serious risk inherent in abandoning *all* weapons on enemy territory), and helmets could be worn on the head and secured with a chin-strap. The greaves might be a problem, but were hardly too difficult to handle. This leaves the large, concave shield as the only logical meaning of ὅπλα. The simple solution was to turn the shields upside down and weigh them down by filling them with stones, marching on to relative safety and picking them up later. This generally seems to be the way translators have understood it;[57] and understandably so, since the final sentence leaves us in no doubt: the hoplites left their *aspides* behind, then returned to pick up their *hopla*.[58]

Another important indication that the nomenclature may be less clear-cut than envisaged by Lazenby and Whitehead is furnished by Lysias' speech against Theomnestos (as well as the spurious second speech for the same client, which may be an abstract of the first), in which the prosecutor accuses Theomnestos of slander and, *en passant,* for having thrown away his shield in battle (ῥιψασπία) – a very

---

56 Xen. *Hell.* 5.4.17–18 (key terms emphasised): "However, while he [= Kleombrotos] was on the march back, a most extraordinary wind-storm arose and some people regarded this as an omen of what was going to happen later. There were many instances of the storm's violence; in particular, when he had left Kreusis with the army and was crossing the mountains that slope down to the sea, numbers of pack-asses with their loads were swept down the precipices, and numbers of *hopla* were wrested away from the soldiers and fell into the sea. In the end many of the men were unable to march forward with their *hopla* and left their *aspides* behind scattered about on the heights, lying on the ground with the concave surfaces filled with stones. That day they had whatever sort of an evening meal they could get at Aigosthena in Megarian territory. The next day they collected the *hopla*" (trans. Warner, modified).
57 Warner (1979²) 284; Hatzfeld (1948) vol. II 99.
58 Lazenby & Whitehead (1996) 31 seemingly ignore this: "Another phenomenon is illustrated by *Hell.* 5.4.18: troops unable to handle all their *hopla* (generic) solve the problem by discarding one major item, their *aspides* …". The objection might be made that κατέλιπον should be taken to mean all their weapons, but this is unlikely as the hoplites ventured further down on Megarian territory to prepare their sorry meal: it would be extremely foolhardy to leave spears and swords behind, and would not help much against the wind anyway. Furthermore, Lazenby and Whitehead themselves apparently admit that only the *shields* were discarded – but Xenophon is quite clear that the ὅπλα were retrieved next day.

serious charge, involving possible ἀτιμία. In the course of these two speeches, the legal phrase ἀποβεβληκέναι (or ἐρριφέναι) τὴν ἀσπίδα is used interchangeably with ἀποβεβληκέναι (or ἐρριφέναι) τὰ ὅπλα.[59] The actual text of the law seems to be ἀποβεβληκέναι τὴν ἀσπίδα,[60] which is perfectly unambiguous. Nonetheless, it is demonstrably possible to use ὅπλον in the exact same sense, since there can be no doubt as to the meaning of ὅπλον in the same context and in the very same turn of phrase. We must assume that the language of law is intended to be as unambiguous as at all possible, but this does not deter the speaker from using a synonym for 'shield', and accordingly it cannot have bothered his audience.

Hoplon, then, may mean 'shield' in a very loose everyday sense, as *hopla* was generically applied to nondescript weapons, and, somewhat more particularly, to defensive armour.[61] On Lazenby and Whitehead's interpretation, we do not know of any specific name for the *hoplite* shield type, as opposed to other types. *Aspis* might then be understood as the common term for any shield, the κύριον ὄνομα, almost always used because it precluded any notion of ambiguity; but *hoplon* could – under the right circumstances – mean *hoplite's* shield. *Aspis,* on the other hand, was in fact used generically of *any* kind of shield, as is in evidence in Hellenistic writers.[62] It is thus not wholly inconceivable that the hoplite's shield could be casually spoken of as 'the tool', the piece of equipment which defined him above all.[63] Diodoros may then be taken to mean "those who were called hoplites – after their [type of] shield – had their name changed etc.". This has the additional advantage of restoring meaning to the text of Diodoros, which otherwise seems almost impossibly obtuse.[64]

2.1.2 Materials and measurements

However this may be, there is no denying the fact that hoplites were primarily defined by their main weapons of attack and defence, the spear and the shield, and that it would be pointless to speak of 'hoplites' without the spear and characteristic shield. The shield in particular is what set them apart from any other troop type in

---

59 ἀποβεβληκέναι (or ἐρριφέναι) τὴν ἀσπίδα: Lys. 10.9, 10.12, 10.21, 11.5, 11.7*bis.* ἀποβεβληκέναι (or ἐρριφέναι, or σώσαντες) τὰ ὅπλα: Lys. 10.1, 10.23, 10.25, 10.25, 11.8.
60 Lys. 10.9.
61 Cf. Aesch. *Sept.* 315 and Lorimer (1947) 76 n. 2.
62 Alexander's 3000 élite bodyguards were called the 'Silver Shields' ('Ἀργυράσπιδες): Curt. 4.13.27 and cf. 8.5.4; Just. *Epit.* 12.7.5 – though there were no 'traditional' hoplites in the Macedonian army (except for non-native mercenaries), and consequently no hoplite *aspides.* Menander's fragmentary comedy *The Shield* ('Ἀσπίς) can similarly be dated to C4l, see Jacques (1998) LXXXI–XC. Franz, although seemingly accepting Lazenby and Whitehead's conclusions, acknowledges the problems inherent in this ("das Wort ἀσπίς kann auch andere Schildformen bezeichnen"), and accordingly chooses to use the term *Hoplitenschild* throughout (Franz [2002] 1–2, 121 n. 62 and *passim*).
63 For such euphemisms for instruments of violence in our own culture, cf. *OED s.v.* "tool" 1.b, (b).
64 Lazenby & Whitehead's excellent article has seemingly been universally accepted; cf., e.g., Franz (2002) 1–2 n. 3; van Wees (2004) 47–48 n. 4.

the Greek world; and along with the spear it was also the only part of their equipment which remained completely unaltered all through the hoplite era. Our knowledge about the shield stems especially from two types of sources: (1) extant specimens which have been at least partially preserved – among these the Spartan shield from Pylos, the hoplite shield found in eastern Sicily, the Etruscan hoplite-type shield found at Bomarzo in Italy and numerous shield-covers found at Olympia where they were placed as dedicatory offerings; and (2) enormous numbers of representations in iconography which are frequently quite detailed and revealing, allowing for great accuracy in assessing the measurements.

The *aspis* was 90 cm to 1 m across, circular and noticeably concave, as can be seen from the enormous amount of vase paintings representing hoplites holding their shields in in a variety of ways, offering good views from many different angles.[65] The large diameter would have given it a considerable surface area: between 6362 and 7853 cm², according as the diameter was 90 or 100 cm.[66] The shield consisted primarily of a wooden core,[67] and remains of wood found inside the Bomarzo shield were identified as poplar, while the core of the Basel shield, found in Sicily, was made of willow.[68] This fits nicely with Pliny's description of poplar, along with willow – indeed, all hardwood in the *aquatica* group[69] – as the most suitable wood for production of shields because of their toughness and pliability, which made them difficult to penetrate, yet allowed the wood to contract somewhat after a penetration and so minimise the damage.[70]

65 Several sources attest to the shield's pronounced concavity: Hdt. 4.200.2–3 (in which a shield is used as a sort of giant stethoscope to listen for excavations underground, cf. Aen. Tact. 37.6–7); Thuc. 7.82.3 (where four inverted [ὑπτίας] shields double as dishes for money confiscated from Athenian prisoners); Xen. *Hell.* 5.4.17–18 (where hoplites weigh their shields down to prevent them from blowing away [see n. 56]); Plut. *Mor.* 241f 16, where the shield (again, presumably, inverted) is used as a stretcher (the famous anecdote of a Spartan mother saying to her son ἢ τὰν ἢ ἐπὶ τᾶς); cf. Σ Thuc. 2.39.1; Stob. *Flor.* 3.7.30 and Hammond (1979–80). Several conspicuous landmarks were called 'the *Aspis*'; most notably perhaps a hill or slope in Argos with difficult access (ὀχυρὸς τόπος): [Κλεομένης] νυκτὸς πρὸς τὰ τείχη ἦγε τὸ στράτευμα, καὶ τὸν περὶ τὴν Ἀσπίδα τόπον καταλαβὼν ὑπὲρ τοῦ θεάτρου χαλεπὸν ὄντα καὶ δυσπρόσοδον οὕτως τοὺς ἀνθρώπους ἐξέπληξεν ὥστε μηδένα τράπεσθαι πρὸς ἀλκήν ... ("[Kleomenes] led his army by night up to the walls, occupied the region about the Aspis overlooking the theatre, a region which was rugged and hard to come at, and so terrified the inhabitants that not a man of them thought of defence ..." [trans. Perrin]): Plut. *Cleom.* 17.4–5, cf. 21.3, *Pyrrh.* 32.1–4.

66 If $r = 45$ cm, then $A = \pi \cdot 45^2 = 6362$ cm². The Basel shield measures 91 cm across: Cahn (1989) 15.

67 Blyth (1982) 9–12. A wooden core is also strongly suggested by Brasidas' shield which, when dropped from a ship, drifted ashore (Thuc. 4.12.1; cf. Diod. Sic. 12.62.4).

68 Blyth (1982) 9, 13–14; Rieth (1964) 104–105; Cahn (1989) 15–16. Despite quite good photographs in both Blyth and Rieth's articles, the best illustration of the Bomarzo shield remains that of Connolly (1998²) 53.

69 Franz criticises both Blyth and Hanson for mistaking poplar for hardwood (Franz [2002] 128 n. 118); but this may result merely from a misunderstanding: in English, *hardwood* is regularly used of deciduous trees (and opposed to fir or pine), whereas German *Hartholz* is applied strictly to very hard wood, such as certain tropical types; cf. *OED* s.v. "hardwood"; Duden (1989²) s.v. "Hartholz" ("sehr festes und schweres Holz [Buchsbaum, Ebenholz]").

70 Plin. *NH* 16.209: *Frigidissima quaecumque aquatica, lentissima autem et ideo scutis faciendis*

Fig. 1: A hoplite stands over a fallen centaur. Note the pronounced concavity of the shield bowl and its large size. Inside the shield, the central armband (porpax) *can be glimpsed. The hoplite is wearing a linen corslet (*linothorax*) and beautifully shaped greaves. For a good view of the inside of the shield, with* porpax *and* antilabe *grips, see also figs. 10, 11, 12 and 14. Red-figure kylix by Euphronios (c. 490–480).*

The core of the shield was usually constructed from laths fastened to the rim and maybe interconnected by grooves and tongues fitted together with glue, but quite possibly not fastened together at all.[71] Apparently, some shields had several very thin laths in layers, often so that the grain of the wood crossed at right angles from layer to layer, thus increasing the resilience noticeably – much like modern plywood.[72] The Bomarzo shield had only one layer of wood; and for this reason it was advantageous to affix the handlegrips so that when the shield was held correctly, the grain ran horizontally across the shield, allowing for greater flexibility in case of pressure on both sides of the shield, but not in the middle: this measure will have increased the shield's effectiveness in combat considerably.[73]

The 'bowl' was not equally thick all over: the wood was trimmed (presumably by means of a turning-lathe)[74] so that the central part of the shield (or the bottom of

> *aptissima quorum plaga contrahit se protinus cluditque suum vulnus et ob id contumacius tramittit ferrum, in quo sunt genere ficus, vitex, salix, tilia, betulla, sabucus, populus utraque* ("The trees that have the coldest wood of all are all that grow in water; but the most flexible, and consequently the most suitable for making shields, are those in which an incision draws together at once and closes up its own wound, and which consequently is more obstinate in allowing steel to penetrate; this class contains the vine, agnus castus, willow, lime, birch, elder, and both kinds of poplar" [trans. Rackham]); cf. Franz (2002) 128–129. However, several Greek sources indicate that willow (ἰτέα) was in fact the normal material: Eur. *Heracl.* 375–376, *Supp.* 694–696; *Cyc.* 5–8, *Tro.* 1192–1193; Ar. fr. 650 Kassel-Austin. All these are metonymical uses, in which *itea* is simply used for *aspis*. Both willow and poplar have the added advantage of being much lighter than, e.g., oak or ash.

71 Blyth (1982) 9–13.
72 Cahn (1989) 16; Rieth (1964) 108 (citing Robinson [1941] 444 who merely says "crossing pieces of wood").
73 The situation, as envisaged by Blyth, is not hard to imagine: if the shield-bearer charged (or was charged by) two adversaries, he might as well be squeezed in between them as hit either frontally (Blyth [1982] 17 fig. 6).
74 Judging from the comic compound τορνευτολυρασπιδοπηγοί (Ar. *Av.* 491): "turners-and-fix-

the 'bowl') was thinner: the Bomarzo shield is 10–14 mm thick near the rim, thinning out to a mere 0.5–0.6 mm, although the centre widens out again slightly to 0.8 mm.[75] This is difficult to explain, as the centre of the shield would have been the least curved part of the shield and also the part targeted and struck most often by an adversary; but possibly the original thickness (which can hardly be assessed with any degree of certainty) was deemed sufficient for this part, and the outer parts and the rim needed additional protection to prevent splintering: after all, the rim was most exposed to sword blows striking it at right angles.

The shield was most likely often covered with a bronze sheathing (χάλκωμα), which helped prevent splintering of the wood upon impact by dissipating the force of the blow over a much greater area.[76] The layer of bronze could be paper thin, as is the case with some of the dedicatory shields from Olympia, but the normal thickness is somewhere between 0.3 mm and 0.9 mm.[77] Although it is not possible to determine how much of the metal has corroded, it would seem realistic to assume that the bronze layer was not less than 1 mm.[78]

The inside of the shield was usually also padded with a layer of leather, which was glued on, likely in order to decrease wear and tear, though by Blyth's estimate, the leather inside the Bomarzo shield is too thin to have afforded any protection, and is therefore purely decorative.[79] The shield from Bomarzo serves to illustrate both the thinness of the bronze facing and the remarkable concavity of such shields: "The bronze cover, which is about 0.5 mm thick, forms a shallow bowl about 10 cm deep and between 81.5 cm and 82 cm in diameter, including a rim which projects about 4.5 cm from the wall of the bowl all round."[80] However, many vase images of hoplite shields suggest that the concavity of shields was frequently even more pronounced than this.

Whether the shield had a bronze facing or not, the rim (ἴτυς) was invariably reinforced by a bronze band to protect the vulnerable edge, 'wrapped' around the rim itself and at least in one case covering the rim halfway.[81] It was apparently fas-

    ers-of-lyres-and-shields"; cf. *IG* I³ 476.380, where τορνευτής is listed as a profession, and *Suda* s.v., which explains the word as οἱ ποιοῦντες τὰς λύρας καὶ τὰς ἀσπίδας ("makers of lyres and shields").

75 Blyth (1982) 12.
76 Hdt. 4.200.2–3; Xen. *Lac.* 11.3; Aen. Tact. 37.6–7; Polyaen. 1.45.2, 7.8.1; Cartledge (1977) 12–13. See, however, Snodgrass (1964a) 63–64. The Bomarzo shield also had a bronze sheathing inside which the remains of wood and leather were found, Blyth (1982) 1, 12; Rieth (1964) 101, 106. A bronze facing could be polished quickly, but tarnished only slowly, thus giving the shield an impressive appearance and possibly enabling the bearer to blind the enemy with reflections: Xen. *Lac.* 11.3.
77 E.g. Mallwitz & Herrmann (1980) 106; Franz (2002) 127 n. 110. The Bomarzo shield has a bronze sheathing 0.5 mm thick: Rieth (1964) 102. The sheathing of the Basel shield is 0.3–0.9 mm thick: Cahn (1989) 15.
78 Franz (2002) 127.
79 Blyth (1982) 12; Cahn (1989) 16; Robinson (1941) 444. Another suggestion is that the leather increased comfort for the bearer, both when the shield was hung on the shoulder, and when he pressed against it and cowered inside it during combat.
80 Blyth (1982) 5–6.
81 Rieth (1964) 108. Rieth also mentions the finding of a shield at Olynthos: remains of wood,

tened last of all, since on the Bomarzo and Basel shields it encapsulates the leather layer on the inside.[82] This rim reinforcement was *de rigueur* on all hoplite shields, because otherwise the shield edges would be extremely vulnerable in combat, particularly to sword cuts which might even separate the laths of the core from each other, causing the shield to virtually disintegrate.

The total weight of the Bomarzo shield's component parts is estimated by Blyth at 6.2 kg.[83] Of this, he assesses the weight of grips, straps and fittings at some 0.7 kg, the rest consisting of the wooden core and the bronze sheathing. The weight of the original wood and bronze can of course be no more than an estimate, as the materials have deteriorated and shrunk considerably over the centuries. Assuming, however, that the average shield had a diameter of 90 cm (rather than only 82 in this case), the weight of wood and bronze would be closer to 6.75 kg. Rieth reckons with a very pronounced decay of the wood, and therefore a rather greater mass in the original shield. By Rieth's assessments, then, the shield would have been even heavier than this, possibly as much as 7 or 8 kg.[84] Kunze's listings of the shields dedicated in Olympia show an average diameter of between 80 and 100 cm (though with one specimen actually measuring 120 cm); and the famous Spartan shield captured at Pylos measured 95 by 83 cm. In addition to this, Rieth actually mentions a shield, similar in other respects to the one examined, but measuring no less than 125 cm across.[85]

The only reasonably extant hoplite shields from antiquity are the ones from Bomarzo and Basel, but artistic representations clearly show shields of the hoplite type from early on. A late Geometric amphora (*c.* 760–700) from Eretria shows a procession of warriors, armed with two spears and a large, circular shield, reaching from neck to knee. The shields are decorated with different shield emblems in white paint (an asymmetrical bird emblem or an eight-armed star), a feature typical of hoplite shields, which can be held in one way only.[86] An intact amphora from Attika (*c.* 720–700) shows a similar procession in two different bands; here, at least the topmost warriors carry shields with centrally placed six-armed stars.[87] Lorimer adduces another late Geometric amphora by the Benaki painter, which shows a procession of warriors bearing circular shields of the same, large size and emblazoned with unmistakably figurative emblems such as, e.g., a horse in white paint. On a

circumscribed by a bronze *itys*. See also Cahn (1989) 16.
82 Blyth (1982) 12; Cahn (1989) 16.
83 Blyth (1982) 16.
84 Rieth (1964) 101 and see Donlan & Thompson (1976) 341 n. 4 giving this as the average weight of the shields in Olympia. By way of comparison, the remains alone of the Basel shield weigh 2.95 kg, nearly half of the estimated full weight of a 'new' Bomarzo shield: Cahn (1989) 15.
85 Kunze & Schleif (1938) 70–74; Shear (1937) 347; Rieth (1964) 101. Cf. Paus. 1.15.4.
86 National Museum, Athens (14763): Kourouniotis (1904) 14–23, esp. 22–23 and fig. 7; cf. Boardman (1952) 7 fig. 13a (pl. 3a), where the asymmetry of the figurative avian motif is evident; Coldstream (1968) 55–57. For the one-way interpretation, see Snodgrass (1964) 61–63.
87 Ny Carlsberg Glyptotek (I.N. 3187): Poulsen (1962) 3–9 and figs. 2–3; Coldstream (1968) 55–57.

sherd of a votive shield from Tiryns from *c.* 750, the hindmost part of a warrior figure with an emblazoned shield can also be seen.[88] Renate Tölle has published yet another late Geometric fragment from the Kerameikos showing a warrior carrying a somewhat smaller circular shield emblazoned with what appears to be a horse-head.[89] Finally, the inside of a hoplite shield with *porpax* and *antilabe* is clearly depicted on a late Geometric aryballos from Lechaion, dated to *c.* 690–680.[90]

Although almost no *actual* hoplite shields have been preserved, their unmistakable representation in art thus warrants the assumption that the concept of a circular shield, *c.* 1 m across and decorated with centrally placed, often figurative blazons, was sufficiently well-known on a societal level to have had an impact on iconography from C8l onwards.[91]

2.1.3 Characteristics of the hoplite shield

What above all distinguished the *aspis* from other contemporary types of shields was the unique double-grip system of πόρπαξ and ἀντιλαβή.[92] The shield was fitted with a bronze band at the centre of the shield – the *porpax* – which not infrequently extended from rim to rim in the vertical plane. Usually, the *porpax* ran across the centre of the shield, maximising the control; but at least in one instance it was somewhat eccentric and placed nearer to the 'farther' end of the shield with the other grip. The rim-to-rim extension had the advantage of ensuring a maximum of solidity in the construction, and also made it highly improbable that the band snapped off under the pressure necessarily applied in a combat situation, something that would result in complete loss of control of the shield – and almost certain death.[93] The other grip was an actual handle near the right-hand edge, the *antilabe*. None of these are preserved, and so most likely were no more than leather thongs attached close to the rim with bronze fittings.[94] In order to handle the shield, the bearer's left arm was slipped through the *porpax* until about the elbow and the *antilabe* then gripped with the hand. Consequently, the shield *ideally* was custom-made, ensuring that the shield's radius corresponded at least roughly with the length of the bearer's forearm;[95] although it is of course correct that the *antilabe need* not

---

88 Benaki Museum (7675): Lorimer (1947) 87 and pl. 19. See also 80, 133–138 and pl. 18 Ac; Coldstream (1968) 81.
89 Tölle (1963) 649–650 and fig. 5.
90 Vanderpool (1955) 225 and pl. 68, fig. 10; Snodgrass (1964) 62 and pl. 15b.
91 Snodgrass (1964) 61–67.
92 The technical terms *porpax* and *antilabe* are supplied by Strabo (3.3.6) and apply to contemporary shields, but, as pointed out by Lorimer, there is no reason to doubt that they were used of the hoplite shield as well (Lorimer [1947] 76 n. 1).
93 This is borne out by the fact that *porpages* could be removed: see Ar. *Eq.* 843–859, 1369–1372; Kritias fr. 88 B 37 Diels-Kranz (= Lib. 25.63). A beautiful specimen from C6e can be seen in Ducrey (1996) 50 fig. 29.
94 Lorimer (1947) 76; Snodgrass (1964a) 63.
95 Xen. *Mem.* 3.10.12; Lorimer (1947) 76.

2.1 The hoplite shield

Fig. 2: The painting depicts a running hoplite (possibly participating in a hoplitodromos *race), equipped with shield, helmet and spear. The shield edge is rested on his shoulder, so that the weight of the shield rests on three points – hand, elbow and shoulder – relieving the strain. Notice the concavity of the shield bowl. Terracotta plaque by Euthymides (c. 510–500).*

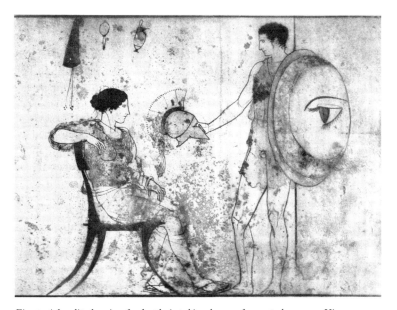

Fig. 3: A hoplite leaving for battle is taking leave of a seated woman. His concave shield is hung on his shoulder to relieve the weight, and he holds his Corinthian helmet in his outstretched arm. Notice the apotropaic eye blazon on the shield surface. White-ground lekythos by the Achilles painter (C5s).

be affixed directly to the rim, as various images of the inside of shields and the enormous size of certain shields attest.[96]

From C5, a puzzling supplement to the shield's protective abilities can be seen in artistic representations. This is the so-called shield apron, a rectangular piece of cloth fastened to the shield and hanging from its lower half.[97] It is normally fringed at the lower end; but for what reasons, other than perhaps those of aesthetics, it is difficult to assess. The most likely material is leather or oxhide, as almost any textile would have been insufficiently stiff. The apron will have given added protection for the lower legs from arrows and other missiles by absorbing much of their kinetic energy. However, as always, the added protection came at a price: the apron added extra weight to the already heavy shield, possibly as much as one kg or even more.[98] Furthermore, an apron will have reduced the shield's mobility further, hampering swift movement and generating even more air resistance. For obvious reasons, no aprons have survived until today.

The double-grip system enabled the hoplite to support the considerable weight of the *aspis* on three points along the arm rather than one, thus relieving the strain on the wrist necessarily generated by a single-grip shield enormously: the shield could be supported simultaneously on the elbow, the wrist and the shoulder – the last point being facilitated by the pronounced 'lip' at the rim.[99] Greek iconography is rife with depictions of just this posture, suggesting that it must have been quite common for the hoplite to assume it whenever he had the opportunity, in or out of actual fighting.[100] This 'hanging' on the shoulder also served as a practical way of holding the cumbersome shield when it was not directly used in combat (e.g. on sentinel duty vel sim.).

In combat, this posture would seem natural to assume, and has decided advantages: it greatly relieves the strain on the arm, and when the rim is rested upon the bearer's shoulder, the shield is carried aslant, its lower rim jutting out before the hoplite, greatly enlarging the zone of protection.[101] Meanwhile, the inclination also served to make spear thrusts glance off the shield, although this would seem a decidedly mixed blessing: as the illustrations show, a shield supported on the shoulder was also directly under the bearer's chin, meaning that a thrust delivered to the top half of his shield might just glance off the polished surface and straight into his face or throat. Worse, a determined kick to the lower edge would cause the shield to

---

96 Greenhalgh (1973) 71.
97 Anderson (1970) 17. For a depiction of a shield apron, see fig. 10 (69).
98 Jarva (1995) 134: "Perhaps the lightest aprons made of some kind of woven pattern were not much above 0.5 kg, but a piece made of medium thick leather could easily exceed 1 kg."
99 Hanson (1991) 68–69, (2000^2) 68; Franz (2002) 132.
100 The examples are very many. I cite but a few by museum and catalogue number: Antikenmuseum und Sammlung Ludwig, Basel (KA424); Fitzwilliam Museum, Cambridge (GR19.1937 and GR5.1930); Martin von Wagner Museum, Universität Würzburg (HA119 and L516); School Museum, Harrow (55); Museo Nazionale Etrusco di Villa Giulia, Rome (26040); Museo Archeologico Nazionale, Naples (81550); Musées Royaux, Brussels (R307); Antikensammlungen, Munich (J421); Musée du Louvre, Paris (ED205; G115).
101 This posture is recognised by van Wees (2000) 128; (2004) 167–169. His sample of illustrations (figs. 3, 4a, 6 and 10) demonstrate quite clearly what is meant.

slam directly into the bearer's vulnerable and sensitive throat region, or at least give him a nasty 'uppercut' on the chin or lower jaw. This would result in immediate, excruciating pain and possible unconsciousness, and thus almost certainly cost him his life in a combat situation. On the other hand, it has been correctly observed that if the shield was held aslant, an incoming spear-head or arrow would have to penetrate a somewhat thicker layer of wood (and possibly also bronze and leather) than if it was merely held vertically.[102] At any rate the illustrations are clear enough, and people who have actually worn replicas of hoplite armour have assured me that this way of handling the shield is not only the logical but indeed the only *possible* way.[103]

This has certain immediate consequences for close combat. A hoplite shield by all lights gave rather good protection against any kind of weapon brought to bear against it, but this protection came at a price. Most scholars agree that a hoplite shield was unusually unwieldy and cumbersome, although this is not universally accepted.[104] Blyth writes, "the total is little more than the weight of a World War II rifle, a weapon which can be handled quite briskly by a trained man ...".[105] It may be that the weight of the Bomarzo shield (which, as indicated above, is in the low end of the range) corresponds to a modern rifle; but this hardly proves anything. A rifle is not, however briskly handled, comparable to a shield. Firstly, to the extent that rifles are handled (as opposed to merely carried), they are so normally with both hands. Secondly, the weight of a rifle is concentrated on a much smaller area, making it easier to deal with. Thirdly, in close combat a shield must be held at the ready at arm's length more or less constantly, or the bearer will risk exposing himself. Fourthly, one will need to move a shield about as quickly as possible (with one arm!) to deflect blows and thrusts from different angles – something that has very little to do with the handling of a rifle in most situations. Finally, a quick examination of typical German, British, French, American and Soviet rifles from World War II reveals that the average weight is in fact closer to $c.$ 4 kg – more than two kg lighter – rendering the comparison altogether useless.[106]

The sheer size and shape of the shield made for unmatched awkwardness in the handling: the mass of the shield is spread over an area of approximately 6400 cm². The greater the area, the more difficulty there will naturally be in handling the object in question; and something the size of a small bridge table is exceedingly unwieldy, whether or not it can be supported on the shoulder of the bearer. Furthermore, the shield had certain drawbacks compared to earlier shields with a single

---

102 Franz (2002) 132–133 and fig. 3.3.
103 I have corresponded with the UK-based re-enactment group 'The Hoplite Association' (http://www.hoplites.co.uk/), whose members have kindly (and patiently) answered my questions. I am also grateful to Nino Luraghi for pointing out the impossibility of holding a modern replica of a hoplite shield except with bent arm and supported on the shoulder.
104 See, e.g., Greenhalgh (1973) 73; Salmon (1977) 85 n.6; Krentz (1985b) 60–61; van Wees (2000) 126; Rawlings (2000) 246–249.
105 Blyth (1982) 17.
106 Mauser Kar98k: 3.92 kg; Lee-Enfield SMLE Mk. III: 3.96 kg; Fusil d'infanterie modèle 1916 M34: 3.56 kg; Garand M1: 4.06 kg; Mosin-Nagant M 1891/30: 3.89 kg. Average weight: 3.87 kg.

central grip. The system dictated that the shield could be wielded with the left arm only, whereas a single-grip shield could be shifted from one hand to the other every so often to ease the strain on the arm and shoulder. The shield might *theoretically* be shifted from the left arm to the right – this could be accomplished by pulling one's left arm out of the twin grips, putting down the shield, turning it upside down and slipping the right arm through them again – but that is hardly something anybody would risk in the heat of battle. Accordingly, the strain on the left arm was very great, and supporting it on the shoulder was a necessity rather than just a convenience. If the shield was to be useful at all, it would have to be held out as far from the body as possible, held aslant in the vertical plane at an angle of *c.* 45°. This increased the angle of deflection and kept possible penetrating weapons as far away from the bearer's body as possible, yet it also increased the strain on the arm considerably. This, in all probability, is what lies behind Aristophanes' comedic character 'Fair Argument's scornful sneer at the hopeless young boys of his own day and age: ὥστε μ' ἀπάγχεσθ' ὅταν ὀρχεῖσθαι Παναθηναίοις δέον αὐτοὺς τὴν ἀσπίδα τῆς κωλῆς προέχων ἀμελῇ τῆς Τριτογενείης.[107] The point is that the boys are unable to even hold out the shield in front of them during the dance, and the result is the disgraceful sight of drooping shields covering the boys' buttocks.[108]

But an even more pressing drawback was the fact, completely overlooked in modern scholarship, that in order to hold a shield with twin grips, the shield can only be held out at *half* an arm's length, since the forearm must of necessity be held at 90° to the upper arm: the zone of protection therefore begins, as it were, at the elbow. The frontal range thus was drastically reduced, since a hoplite shield can only be held out as far from the body as the elbow.

A single-grip shield, on the other hand, could be held out at a full arm's length, i.e. about twice as far as the hoplite shield. This is important, because it means that the shield's surface area must of necessity be much bigger than it needed to be if it could be held with a single hand: a single-grip shield, being held at a full arm's length from the body, could be much smaller, yet offer the same degree of protection – i.e., it decreased the adversary's angle of attack just as well as the large hoplite shield did, simply by being held at twice the distance from the bearer's body. Furthermore, by merely turning the wrist the shield can effectively be rotated to maximise the angle of deflection, even when the bearer reaches across to his own right side. This is due to the single-grip which allows great freedom of movement.

---

107 Ar. *Nub.* 988–989; cf. *Ran.* 726–733: "… and so I choke with rage when they're supposed to be dancing at the Panathenaia and one of them's holding his shield in front of his haunch with no regard for Tritogeneia!" (trans. Henderson).

108 *Pace* Sommerstein (1982) 209, who comments "probably Greek *kole,* lit. 'ham', means 'penis' … though this meaning is not attested elsewhere." The passage is rather better explained by Dover (1968) 219, who says "κωλῆ is the haunch of animal or man … If a man is physically weak, he cannot dance for long holding a heavy shield with his fore-arm at right-angles to his chest or moving it quickly up and down; his arm flops *and the shield covers his side from shoulder to knee*. The exaggeration 'holding it in front of his haunch' is typical drill-sergeant's language" (emphasis mine). It is obvious that the shield, when let go (but still held by the twin grips), would slide sideways and hang in the way described here.

Moreover, incoming attacks could be countered earlier and thus perhaps 'nipped in the bud' by merely parrying with the shield. The hoplite shield could of course also be moved about to a certain degree, but, due to its awkward size and shape, not exactly in a brisk manner; and if it was used actively to parry or block incoming blows and thrusts, this took place too close to the body for comfort: there was very little time and room for secondary measures if a parry came too late or was misdirected. An armour-clad hoplite would doubtless have found it very difficult to reach across his own bronze-covered torso in the attempt to deflect a thrust directed at his right side.

Unlike its predecessors, the hoplite shield had no carrying strap (τελαμών), and accordingly could not be slung around to protect the bearer's back during retreat or flight. This effectively prevented the hoplite from turning his back at any point during battle for whatever reason. The hoplite shield was made for frontal (and partly lateral), not dorsal protection.[109] Even worse than this, if the hoplite was forced to use the *aspis* to deflect a thrust directed at his legs, he would in fact have to stoop in order to actually make the shield reach the attacker's spear or sword, thereby exposing the vulnerable back of his head and neck to a blow from above: possibly the ever-present greaves' inclusion in the 'canonical' hoplite panoply was influenced also by this surely annoying drawback to the double-grip system.[110] A warrior wielding a single-grip shield, on the other hand, might just alter his footing (e.g. by stepping forward) and lower his shield arm. The sheer weight and size of the shield, again, made it exceedingly cumbersome to 'wield': it must be seriously doubted whether anyone, no matter how well trained, was able to sustain its weight, let alone 'wield' it, for any considerable amount of time.

In this connexion, it is interesting that merely standing holding the shield (τὴν ἀσπίδα ἔχων) was considered a sufficient disciplinary punishment in the notoriously tough Spartan army.[111] Franz rejects Hanson's correct interpretation of this passage,[112] apparently because Xenophon goes on to say ὃ δοκεῖ κηλὶς εἶναι τοῖς σπουδαίοις Λακεδαιμονίων ("[which] is regarded by distinguished Spartans as a great disgrace"). It is certainly correct that κηλίς means 'blemish' or 'disgrace';[113]

---

109 Cf. Greenhalgh (1973) 72–73. Greenhalgh correctly observes that it was actually possible to carry the hoplite shield on the back, (since it was apparently often fitted with a cordon on the inside); but that in reality it was too complicated a manoeuvre to disengage the arm from the twin grips, and *then* sling the shield over the shoulder while running in headlong flight.

110 Greenhalgh, however, is excessively sanguine about the manoeuvrability of the *aspis*, and even calls it "the more easily manipulated double-grip shield" (Greenhalgh [1973] 73; *contra* Cartledge [1977] 13 n. 17).

111 Xen. *Hell.* 3.1.9.

112 Franz (2002) 269–270, citing Hanson (2000²) 67. See also Schwertfeger (1982) 263 n. 34: "Daß der schwere Hoplitenschild eine Last war, geht aus dem spartanischen Recht der klassischen Zeit hervor, wo der König im Rahmen seiner Disziplinargewalt gegen ungehorsame Soldaten die Strafe des 'Stehens mit dem Schild' verhängen konnte ..." ("that the heavy hoplite shield was a burden is also evident from Spartan law of the Classical period, where the king, as part of his disciplinary measures, could punish disobedient soldiers with 'standing with the shield' ...").

113 Cf. LSJ⁹ s.v. "κηλίς".

but this does not mean that the dishonourable disciplinary measure was not also intended to be a physical punishment. This interpretation is borne out nicely by a comparison with Plutarch's *Aristeides*, in which king Pausanias' harsh treatment of his allies is discussed: τοῖς τε γὰρ ἄρχουσι τῶν συμμάχων ἀεὶ μετ' ὀργῆς ἐνετύγχανε καὶ τραχέως, τούς τε πολλοὺς ἐκόλαζε πληγαῖς, ἢ σιδηρᾶν ἄγκυραν ἐπιτιθεὶς ἠνάγκαζεν ἑστάναι δι' ὅλης τῆς ἡμέρας.[114] If punishment consisting in holding a shield was merely symbolic, this otherwise similar case is entirely pointless. The point was hardly lost, however, on those unfortunate allies who fell afoul of Pausanias and were forced to stand around holding iron anchors all day, even though they knew nothing of what was considered dishonourable in the Spartan army. The beatings emphasise this even more: clearly, Plutarch sees the anchor-holding as completely parallel to the thrashings, the physical nature of which can hardly be denied. It may have been humiliating or dishonourable for the victim, but it certainly also was a gruelling physical ordeal. Xenophon's remark should be seen in this light: the toughness was obvious, the possible disgrace may have needed an explanation.[115]

2.1.4 'Wielding' the hoplite shield

The above analysis points to the conclusion that the hoplite shield was extremely unfit for single combat of any kind: designwise, every single feature is the opposite of what one would want for this kind of fighting. For this purpose the ideal would be a relatively light, yet sturdy shield, with a single, central grip, and considerably smaller than the hoplite shield. A shield of this type – in effect what might be termed a buckler – could be held at arm's length, could easily be transferred from hand to hand every so often, and above all was easy to manipulate, being light and small. This would also be an even greater necessity, since in solo fighting, attacks may come from any angle. Thrusts and blows could be reasonably easily deflected by relatively small movements of the hand and wrist: if a thrust, cut or blow is met at a sufficient distance, it actually requires very little kinetic energy to deflect the incoming weapon enough that it will not hit one's body: the shield in effect needs to be not so much a cover for the entire body as a sort of 'iron glove' with which to meet and deflect attacks immediately and directly.[116] Diodoros has a pertinent commentary on the use of Iberian targes:

τῶν δ' Ἰβήρων ἀλκιμώτατοι μέν εἰσιν οἱ καλούμενοι Λυσιτανοί, φοροῦσι δ' ἐν τοῖς πολέμοις πέλτας μικρὰς παντελῶς, διαπεπλεγμένας νεύροις καὶ δυναμένας σκέπειν τὸ σῶμα περιττότερον διὰ τὴν στερεότητα· ταύτην δ' ἐν ταῖς μάχαις μεταφέροντες εὐλύτως ἄλλοτε ἄλλως ἀπὸ τοῦ σώματος διακρούονται φιλοτέχνως πᾶν τὸ φερόμενον ἐπ' αὐτοὺς βέλος.[117]

---

114 Plut. *Arist.* 23.2: "The commanders of the allies always met with angry harshness at the hands of Pausanias, and the common men he punished with lashings, or by compelling them to stand all day long with an iron anchor on their shoulders" (trans. Perrin, modified).
115 Greek literature is altogether replete with Spartans either exacting or threatening physical punishment: the sources are collected in Hornblower (2002) 57–60.
116 Cf. Lorimer (1947) 111.
117 Diod. Sic. 5.34.5 ("The bravest among the Iberians are those known as Lusitanians, who carry in war quite small shields which are interwoven with cords of sinew and are able to protect the

Seen from this particular view-point, the hoplite shield would have been a spectacular failure. Nevertheless, it has been assumed repeatedly over the past years that hoplites could function as soloists in a looser formation,[118] most recently and eloquently by Hans van Wees, who believes that the hoplite shield was as good as any when used in single combat. The main thrust of his argument is twofold. (1) He believes that the common view (that the two-grip system on the shield meant that the hoplite's right side was not adequately covered) is wrong, since the correct posture to adopt when fighting with the hoplite shield is a sideways stance, setting the left foot forward and holding the shield more or less at right angles to the body:

> The double grip supposedly meant that the hoplite in effect stood behind the right half of his shield, leaving his right-hand side relatively vulnerable to attack even from the front, and leaving the left half of his shield useless. Useless, that is, unless soldiers stood so close together that they could take cover behind the redundant left halves of their neighbours' shields.[119]

As we have seen, however – and, indeed, as van Wees himself maintains – the correct and even natural posture to assume when using something so clumsy and heavy as a hoplite shield, is to lean it on the left shoulder and turn that shoulder towards the enemy by slightly twisting the torso. In support of this, he cites numerous vase paintings representing hoplites in just this posture.[120] This, according to van Wees, produces an image of hoplite fighting entirely different from the usual, since a hoplite standing with his left shoulder pointing towards the enemy and using a double-grip *aspis* would necessarily "[place] himself behind its *centre,* rather than to one side of it, even if it has a double grip".[121]

Now, it is no doubt true that the natural and ideal position to assume is a slightly sideways one, as is evident from watching almost any martial art: any boxing match or even a Bruce Lee film will suffice to show that this is a basic tenet in almost any kind of close combat. In fact, it makes perfect sense for the fighter to withdraw his centre, keeping his vulnerable head and body as far out of harm's way as at all possible. In this way the fighter will keep his left arm as close as possible to his adversary, allowing him to deflect or block attacks as early on as possible, as we saw above: the farther away from the body the arm is held, the sooner attacks may be parried. Another advantage is that the target of one's body is drastically reduced by turning the side forward. Finally, this stance will keep the fighter's crucial right arm at a convenient distance, thus maximising the potential leverage to be put into thrusts once the fighter decides to abandon his defensive position and strike at the adversary.[122]

---

body unusually well, because they are so tough; and shifting this shield easily as they do in their fighting, now in this direction, now in that, they expertly ward off from the body every blow which comes at them" [trans. Oldfather, modified]).
118 van Wees (2000) 126–130; Krentz (1985b) 53, 61, (1994), (2002) 35–6; Cawkwell (1989) 385–389.
119 van Wees (2000) 127.
120 van Wees (2000) 126–128.
121 van Wees (2000) 129.
122 This last point is also mentioned by van Wees (2000) 128.

It thus seems reasonable enough to assume that this was the hoplite's basic stance for close combat. It is an apt observation, and no doubt a correct one, but it is by no means new. In 1967 it was pointed out by Anthony Snodgrass that "in a frontal attack, the hoplite *could* put his left shoulder forward and draw his unprotected right side back, or, as was more usual in the phalanx, keep so close to his neighbour on the right as to be covered by the overlapping left-hand part of his shield".[123] In his description Snodgrass correctly hints at the constant flux inherent in any combat situation: it does not follow from the recognition of the importance of the stance that a hoplite would always stand with his torso at right angles to his adversary, and van Wees' conclusion, that "surely this was how hoplites in reality ... stood in combat and handled their shields", perhaps seems slightly premature.[124] It is no less inconceivable that the hoplite would stand sideways with twisted torso *constantly* during combat than that he would at all times stand facing forward, his shield held out directly in front of him. Even if this were the case, he would have to abandon the defensive position again and again: every time he decided to strike a blow at his opponent, he would be forced to twist his torso forwards in order to obtain the necessary leverage on his blow: if the sideways-on stance is in fact natural because it allows for greater force in striking, it must be precisely because the blow can be 'followed through', as the golf (or tennis) term is for a somewhat similar phenomenon. It would not be possible for a hoplite to stay safely behind the shield continuously if he were to actually take part in the fighting.

(2) The second part of van Wees' argument is that hoplite shields were not worse than single-grip shields in any crucial respect, since "the shield ... at most tended to slow down movement on the battlefield: it did not in itself impose a static form of combat". Thus, the loss of manoeuvrability which the hoplite shield entailed "should not be exaggerated", since no type of shield can be brought very far to the right "without badly impeding the use of weapons".[125] It is of course true that any shield – or, for that matter, just one's left arm – would seriously impede the functioning of a spear or sword. It is nearly impossible to deflect or parry a blow or thrust with the shield in one hand and strike or thrust with an offensive weapon in the other *simultaneously*. But then that is not a realistic conception of fighting with shield and spear (or with any kind of weapon for that matter).

The required technique – as indeed in any martial art – was to ward off immediate danger by parrying, blocking or deflecting an incoming attack (delivered to, say, the defender's right side), and only *after* that going into the offensive. By this time the hoplite's shield would have been retransferred to its usual position to the front and slightly to the left of the bearer, enabling him to strike with much greater force. This again points to the constant flow of combat, as hinted at by Snodgrass. The hoplite would have been forced to turn his shield this way and that, as the needs of combat dictated, but this will actually have been in accordance with the continuous change: once he had succeeded in parrying a blow, he will have been in

---

123 Snodgrass (1967) 54 (emphasis mine). The point is discussed also by Greenhalgh (1973) 72, by Cawkwell (1989) 384–385 (briefly) and by Luginbill (1994) 53–54.
124 van Wees (2000) 128.
125 van Wees (2000) 127; *contra* Schwartz (2002) 35–38.

a good position to strike a blow himself. This dictated the natural rhythm of all close combat: parry-attack, parry-attack.[126]

In comparison with this, a fighter can much more easily 'reach across' to his right side and back again with a smaller, lighter shield: it requires no more than a slight twist of the torso and a bending of the arm, whereas, with a hoplite shield, the bearer's arm is already bent, since this is the only possible way to hold the shield. The required movement therefore is a strong twisting of the torso, coupled perhaps with an attempt at stretching the arm to enhance the reach with the shield. Furthermore, it is worth remembering that the hoplite will often have worn a bronze breastplate or other body armour; and this will have made reaching across his own torso considerably more difficult, increasingly so as we go back in time towards the clumsy bell-shaped breastplate.

Arguing along much the same lines as van Wees, Louis Rawlings has suggested that hoplites were quite agile: they are, he maintains, often mentioned in the sources as performing various feats of fighting and working which, according to orthodoxy, should be impossible for hoplites. On the whole, Rawlings succeeds in demonstrating that hoplites were capable of carrying out many different tasks, but he is considerably less successful with regard to the second part of his argument. Hoplites may have been versatile enough, particularly in the late hoplite era (C5l – C4), but not *qua* hoplites: the vast majority of Rawlings' examples are either soldiers who are merely assumed to be hoplites (e.g., on garrison duty), or it cannot be shown in any way that they are actually equipped as hoplites when they perform other duties. His conclusion – that hoplite equipment was not much of an encumbrance – therefore does not necessarily follow.[127]

2.1.5 The hoplite shield in combat

The sources on how the shield was actually handled singly in a combat situation are few, and to the extent that they treat of the matter at all, they do so in a rather roundabout way. However, a central passage is found in Euripides' *Phoinissai*, where Eteokles and Polyneikes are duelling. The 'Thessalian feint' employed by Eteokles is described in great and vivid detail:

> ἔνθεν δὲ κώπας ἁρπάσαντε φασγάνων
> ἐς ταὐτὸν ἧκον, συμβαλόντε δ' ἀσπίδας
> πολὺν ταραγμὸν ἀμφιβάντ' εἶχον μάχης.
> καί πως νοήσας Ἐτεοκλῆς τὸ Θεσσαλὸν
> ἐσήγαγεν σόφισμ' ὁμιλίᾳ χϑονός·
> ἐξαλλαγεὶς γὰρ τοῦ παρεστῶτος πόνου,
> λαιὸν μὲν ἐς τοὔπισϑεν ἀμφέρει πόδα,
> πρόσω τὰ κοῖλα γαστρὸς εὐλαβούμενος,

---

126 Schwartz (2002) 34–36.
127 Rawlings (2000).

προβὰς δὲ κῶλον δεξιὸν δι' ὀμφαλοῦ
καθῆκεν ἔγχος σφονδύλοις τ' ἐνήρμοσεν.
ὁμοῦ δὲ κάμψας πλευρὰ καὶ νηδὺν τάλας
σὺν αἱματηραῖς σταγόσι Πολυνείκης πίτνει.[128]

It must be said here that the language and the mythical context clearly are meant to evoke those of the epic genre; and this is also reflected in the fact that the two brothers face off in a duel-like combat. All the same, the situation is probably meant to evoke a duel between warriors equipped as hoplites. This can be seen from the description and also follows naturally, since an inclusion of contemporary equipment was the logical way to make the audience relate directly to the action. The hoplites bash their shields together, each trying to press his adversary backwards and perhaps topple him or at least throw him off balance. This is a natural way to fight with a large shield, since, as we have seen, the hoplite shield could hardly be wielded quickly or adroitly enough to be used in a free-fighting style. The ideal position to adopt was to pull the shield as close to the body as possible and hide the upper body *inside* the concave shield.[129] The hoplite could duck and pull his head down reasonably quickly while at the same time raising the shield a little and thus retract his head down below the rim of the shield, and out of reach of a sword blow. There is even a striking description in Greek literature of the stain of sweat impressed on the *itys* from the hoplite's constant leaning his face there during combat, suggesting that this was a well-known phenomenon.[130] From the relative safety of this position, the hoplite would then try to get at his opponent, more or less blindly, hoping to get a fatal blow in without exposing himself unnecessarily.

The 'Thessalian feint' relies on just this locked type of shoving competition. If both fighters were applying force to a somewhat similar degree, the easiest way of throwing the opponent off balance was to suddenly relieve the pressure: Eteokles swiftly pulled back his left foot (which would be forward in this situation) while keeping a steady grip on the shield, keeping it close to the body. By pulling his left foot backwards, he took a half-step to the rear, bringing his right side forward and making a half-turn with his upper body so that his sword-arm was now in advance. It is therefore not so much a matter of stepping forward on the right foot (despite

---

128 Eur. *Phoen.* 1404–1415 and Σ: "Then, both seizing sword's hilts, / they came together and clashing shields / they made a loud battle-sound, close-locked. / Eteokles, somehow knowing the Thessalian / feint, through visiting that land, brought it to bear. / Freed from his current bout, / he drew his left foot back / guarding his belly's hollow in front, / thus advancing his right foot, through the navel / he plunged his sword and brought it to the backbone. / Doubling chest close to stomach, wretched / Polyneikes falls with blood streaming" (trans. Craik, modified).

129 Polybios later stated that a serious flaw of the oblong Celtic shield (θυρεός) gave it a decided disadvantage against the better Roman legionary's shield (*scutum*), because it did not cover the body (τὸν ἄνδρα περισκέπειν). The prefix περι- illustrates quite well that a hollow shield could be used for cowering inside: Polyb. 2.30.3. A similar mode of fighting is discernible in the famous story of Titus Manlius Torquatus killing his Gaulish adversary mainly relying on his shield: Claudius Quadrigarius fr. 10b Peter *apud* Aul. Gell. 9.13.7–19.

130 Eur. *Tro.* 1196–1199.

what the scholiast says[131]) as withdrawing the left foot behind the other one, turning 180° in the process. By this manoeuvre the pressure on the enemy's shield was released so suddenly that he lost balance and stumbled forward, the difference being that *he* did not alter his footing: his left side and shield was still in advance. Besides, the sudden release of pressure meant that the opponent's shield would be projected away from the body, and the weight of the shield would in turn assist in bringing him out of balance. A gap would open between the tricked hoplite's shield and body, and his adversary would be in an excellent position to drive home his weapon, because he now had his sword-arm in advance, right where the gap opened. The force of the sword thrust would be increased by the momentum of the hoplite rushing headlong onto the point of the sword (or spear). The effect of Eteokles' sword thrust was terrible: "through the navel / he plunged his sword and brought it to the backbone".[132]

Euripides deemed it necessary to explain to his audience just what a Thessalian feint was, and the reason for this may either be that it should be made understandable to the non-hoplites in the audience, or that tricks as these were not commonly known and used in the rank and file – or perhaps simply that the *name* might be the only thing new about it. (Nevertheless, it is quite plain that the bone-grating realism of the actual wounding is drawn from personal experience, possibly firsthand: this must have been recognisable to a large part of the audience.) Furthermore, Thessalians were proverbial for untrustworthiness,[133] and Euripides may merely be implying here that Eteokles resorted to a 'dirty trick' to win the fight.[134] At any rate it is a depiction of hoplite in-fighting which does not seem at all irreconcilable with what we know about what could and could not be done with a hoplite shield. The only matter for speculation is whether there would be room enough for such tricks in the thick of phalanx fighting – i.e., would a hoplite in fact have room enough behind

---

131 Σ Eur. *Phoen.* 1408: τὸν μὲν ἀριστερὸν πόδα λάθρα εἰς τοὐπίσω ἀνάγει τὸ πρόσω τῆς γαστρὸς φυλάττων, τὸν δὲ δέξιον προβάς ... ("he stealthily draws his left foot back, protecting his stomach, and thus putting the right foot in front ..."). All commentators have focused on the right foot being projected: Borthwick (1970b) 19; Pritchett (1985a) 64; Craik (1988) 253–254; Mastronarde (1998) 542–543. I would suggest, however, that the προβάς here was no more than the result of retracting the left foot, not an active move in itself. Not only would this take too long, spoiling the element of surprise, it would also result in a too long forward step, which runs a serious risk of throwing the attacker off balance. It is surely much safer (and more economic as well) to complete the half-turn by retracting the left foot far enough. Mastronarde aptly compares this altering to a similar move in Oriental martial arts: "In karate a sudden 180-degree shift of stance is supposed to upset an opponent" (Mastronarde [1994] 542). The value of such tricks is actually well known in our own combat sports as well: the thought of a 'southpaw' opponent is unsettling to most right–handed boxers (cf. Borthwick [1970b] 18 and n. 16).
132 Trans. Craik (1988) 141.
133 Σ Eur. *Phoen.* 1408: παροιμία τὸ Θεσσαλὸν σόφισμα τάττεται δὲ ἐπὶ τοῦ παραλογίζεσθαι καὶ ἀπατᾶν· ποικίλοι γὰρ τὰ ἤθη οἱ Θεσσαλοὶ καὶ οὐκ ὀρθοὶ τὴν γνώμην ("The 'Thessalian feint' is a byword; it is applicable to misleading and deceiving; for the Thessalians have a wily character and dishonest way of thinking") cf. Dem. 1.22; Zen. 4.29.
134 Craik (1988) 254; Mastronarde (1994) 542–543. Mastronarde argues that the definite article rules out any possibility of an "ethnic slur": it must be a specific manoeuvre.

him for the half-step backwards?[135] This question is better left unanswered at present and examined along with other problems concerning the tactics and fighting methods of the phalanx.

Another way in which the shield may have been used offensively is a staple of most shield fighting, namely that of hooking the opponent's shield with one's own. In hoplite fighting, it could be carried out by thrusting one's own shield against that of the adversary, then sliding it to the right while still keeping up the pressure. One's own rim would now slide off the enemy's rim, at which point it could be hooked sharply and pulled violently to the left, thereby at once trapping the adversary's right arm, effectively preventing him from using it, and jamming his left arm with the shield across the chest, exposing his left side and armpit. Here, one's own weapon, be it sword or spear, could quickly and effectively be thrust in – it is a very short way to the heart through the armpit or between the ribs. Alternatively, one could just pull the enemy's shield vigorously toward oneself, rather than across his body: this would make him lose his balance and create a situation in many ways similar to the above-mentioned Thessalian feint. The only difference is that whereas the former relied on a sudden and unexpected release of pressure, the latter employed just the opposite, an active pull towards oneself. The result was much the same: the adversary's shield was suddenly and violently jerked away from his body, thus exposing his left side and his stomach.

The element of surprise played a great part in both movements and was further enhanced by the blow or thrust being delivered to the 'wrong' side of the body: in all likelihood hoplites did not expect their better protected left sides to be attacked, and consequently probably concentrated on their right, relatively unprotected sides. To effectively perform the attack, however, the hoplite would have to take a step forward, bringing his right-hand side forward, much like the 180° turn in the Thessalian feint. If not, it is difficult to see how the hoplite could effectively reach across his own torso and the large shield to get at his opponent's side. Much like with the Thessalian feint, it is an open question whether this technique could be used within the narrow confines of the phalanx. To be sure, it probably requires less space to the rear than the former, since there is no step backwards here; but there is the problem of stepping forward and of performing the turn.

As we have seen, the shield was often, perhaps even most of the time, supported on the shoulder as well as on the wrist and elbow, which resulted in the lower edge of the shield jutting out noticeably, as represented so often on vase images. This position has formed the basis of another conception of how the shield might be used offensively:

> Tilted back against the shoulder, the shield's bottom rim stuck out a couple of feet in front of its bearer, so that in very close combat the lower edges of opponents' shields were liable to touch ... 'Pushing' would thus consist of shoving the protruding lower part of one's shield against the corresponding part of the enemy's shield – with the aim, no doubt, of driving him back, disturbing his balance, or at least breaking his cover[136]

---

135 Lazenby (1991) 94; cf. Anderson (1991) 35.
136 van Wees (2000) 131 and figs. 3 and 4, four vase paintings supposedly demonstrating the truth of this view. One, however, depicts a hoplite crouching for cover and as such can hardly be

This representation of offensive shield use is not very realistic. Jabbing against the smooth and quite convex surface of the enemy's shield would make one's own shield rim glance off to an even higher degree than a spear thrust; and the thrust might very well end up wrong-footing oneself because of the rather greater momentum in a shield weighing some seven or eight kg. Even if the rim did not glance off, however, it is not likely that such a method of 'pushing' would prove terribly effective: again, because of the shield's weight, it cannot be used to strike quickly with one arm; nor should this topple an adversary wearing some 20 kg worth of armour and assuming a robust sideways-on stance similar to one's own. What is more, if two shield bottoms are pushed against each other with even a moderate amount of force, would it not rather result in the two shields being pressed slowly together in a vertical position?

A more realistic and logical way of exploiting the peculiar slant of the hoplite shield would be to try to get one's own shield tucked under the lower edge of the opponent's. In this way, it would be fairly easy to suddenly bring one's own shield upwards, lifting that of the opponent, who would, in all probability, be concentrating on defending against attacks leveled at him frontally or laterally. With any luck, he would thus be pressing forwards and be caught off guard when his own shield was suddenly lifted up above his head to a position where it could not protect anything.

The attack could be completed simply by taking a step forward and driving home the spear or sword, as the case may be, as the surprised opponent would likely be unable to fend off the attack. This interpretation of the slanting shield is superior on at least two points: firstly, the trick could be used without changing one's own position backwards or to the sides: all that was required was a small step forward once the lift was carried out successfully. Secondly, it could be done with little or no disadvantage to the attacker, even if it did not work out as planned, since the attacker could cower inside his own shield during the entire operation. Furthermore, it is entirely reconcilable with the great number of vase images that depict just this position.

If the shield were to be used for pushing directly against that of an opponent, the bearer would do wisely to keep the lower edge of the shield rather closer to the body or risk a sudden lifting of the shield, which meant fatal exposure. On that account, it is scarcely possible to interpret the iconographic representations of slanting shields as something involving a push: the mutual pressure on the shields would make the slant impossible, and the possibility of the shield lift would make it a quite dangerous technique in close combat.

---

used to show this particular combat situation. Another shows a man running with hoplite shield, spear and helmet, and the same must apply here. The interpretation is repeated in van Wees (2004) 190.

2.1.6 Hoplite race and pyrrhic dance: the case for the light hoplite shield

To support the view that the shield was highly manoeuvrable, both Rawlings and van Wees further argue that the shield was regularly included in such festive events as the hoplite race and the pyrrhic dances, and consequently cannot have been heavy or unwieldy.[137] While it is correct that the shield featured prominently in these events, there are several features about them which are not considered by either Rawlings or van Wees.

(1) The hoplite race was simply a *diaulos* race[138] (in Olympia, probably the longest stadium, this corresponds to *c.* 400 m[139]), run in hoplite armour; and this, it would seem, supports the view that hoplite armour (and especially the shield) were easily carried and manipulated.

Firstly, however, the apparent intimate connexion between war and the hoplite race is somewhat disrupted by the fact that it was introduced into the games rather late: the *terminus post quem* is 665/4,[140] and we know that at the Plataian games the hoplite race was not introduced until after the Persian Wars.[141] Since the Olympic games date from 776, there is an interim of at least some 110 years; and as for Olympia, Pausanias claims that the *hoplitodromos* was introduced only in "the sixty-fifth Olympiad" or 520 – no less than 256 years after the first Olympic games.[142]

Moreover, it has been suggested that the hoplite race originated at the games in Argos: the prize here was a bronze shield, and the games themselves might be referred to simply as the 'bronze in Argos'.[143] The shield always played an important role in Argive tradition (hence possibly the name ἀσπὶς Ἀργολική), as the somewhat confused tradition reveals: a number of different aetiological explanations appear in quite a few ancient authors).[144] The centrality of the shield in Argive culture may even have its roots in religious cult, as many archaeological finds in the vicinity of

---

137 van Wees (2000) 127 n. 6, citing Rawlings (2000).
138 Paus. 2.11.8, 10.34.5. The word δίαυλος could mean either the 'normal', naked race of that length or the *hoplitodromos* (whether defined with the qualifier μετὰ τῆς ἀσπίδος or not): Ar. *Av.* 291–292. The name could also be ἐνόπλιος δρόμος or simply ὁπλίτης: cf. Moretti (1953) no. 59 (*passim*).
139 Lee (2001) 56.
140 Paus. 3.14.3 describes the victories of the Spartan runner Chionis this year: ἐνταῦθα δὲ ἑπτὰ ἐγένοντό οἱ νῖκαι, τέσσαρες μὲν σταδίου, διαύλου δὲ αἱ λοιπαί· τὸν δὲ σὺν τῇ ἀσπίδι δρόμον ἐπὶ ἀγῶνι λήγοντι οὐ συνέβαινεν εἶναί πω ("The Olympian victories were seven, four in the single-stade race and three in the double-stade race. The race with the shield, that takes place at the end of the contest, was not at the time a part of the events" [trans. Jones & Ormerod]). Accordingly, the hoplite race was scarcely so 'ancient' as Philostratos would have us believe (Philostr. *Gymn.* 7).
141 Philostr. *Gymn.* 8.
142 Paus. 5.8.10.
143 Pind. *Nem.* 10.19–24, *Ol.* 7.81–87 and Σ; Ringwood Arnold (1937). The name 'Shield of Argos' was common in Hellenistic and imperial times; Amandry (1983) 627–628.
144 Callim. *Hymn* 5.35–41; Paus. 2.17.3, 2.25.7, cf. Apollod. *Bibl.* 2.2.1; Hor. *Carm.* 1.28.9–15 (cf. Diog. Laert. 8.1.4–5; Max. Tyr. 10.2; Iambl. *VP* 14.63; Porph. *VP* 26–27); Hyg. *Fab.* 170.9–11, 273.1–3.

Argos seems to attest.¹⁴⁵ If this is true, then the hoplite race may in fact have its origins in cultic practice at the Argive games.

On the whole it may be safer to interpret the hoplite race as a sort of cultic or festive and quasi-symbolic event at the end of the games, quite effectively and bluntly signalling the transition from the relatively peaceful state of ἐκεχειρία to normal conditions. This is certainly how Philostratos takes it to be, and there is no lack of sources testifying that the *hoplitodromos* was always scheduled as the last event.¹⁴⁶ The hoplite race is probably more realistically interpreted as part and parcel of the general agonistic Greek spirit, which had a strong penchant for making competitions out of almost everything.¹⁴⁷

Secondly, Pausanias informs us that greaves and helmet were at some point removed from the race, so that the only part of the panoply that remained was the shield.¹⁴⁸ It is puzzling why these other very traditional items of hoplite equipment were discarded, if it is not because they were simply deemed too heavy and cumbersome. In an interesting passage, Philostratos remarks on the physical requirements for *hoplitodromos* runners:

ὁπλίτου δὲ καὶ σταδίου ἀγωνιστὴν καὶ διαύλου διακρίνει μὲν οὐδεὶς ἔτι ἐκ χρόνων, οὓς Λεωνίδας ὁ Ῥόδιος ἐπ' ὀλυμπιάδας τέτταρας ἐνίκα τὴν τριττὺν ταύτην, διακριτέοι δ' ὅμως οἵ τε καθ' ἓν ἀγωνιούμενοι ταῦτα καὶ ὁμοῦ πάντα. τὸν μὲν ὁπλιτεύσοντα πλευρά τε εὐμήκη παραπεμπέτω ὦμος <τ'> εὐτραφὴς καὶ σιμὴ ἐπιγουνίς, ἵν' εὖ φοροῖτο ἡ ἀσπὶς ἀνεχόντων αὐτὴν τούτων. ... οἱ δὲ τῶν τριῶν ἀγωνισταὶ δρόμων ἀριστίνδην συντετάχθων συγκείμενοι ἐκ πλεονεκτημάτων, ὧν οὗτοι κατὰ ἕνα. τουτὶ δὲ μὴ τῶν ἀπόρων ἡγείσθω τις, δρομεῖς γὰρ δὴ καὶ ἐφ' ἡμῶν τοιοῦτοι ἐγένοντο.¹⁴⁹

It would seem that an average, slender build was sufficient to compete in both long-distance running, sprints and *hoplitodromos*. Common sense indeed suggests that a slender, athletic build would be ideal for such multi-purpose athletics – but Philostratos says that those concentrating only on *hoplitodromos* ideally possess the heaviest build of the three types. It is not so clear what is meant by σιμὴ ἐπιγουνίς. Although *epigounis* or *epigonatis* properly means 'patella',¹⁵⁰ it may be taken to simply mean 'knee', as can be seen in Homer.¹⁵¹ No matter how this should be un-

---

145 Ringwood Arnold (1937) 438–439.
146 Paus. 3.14.3; Plut. *Mor.* 639d–e; Artem. 1.63; Philostr. *Gymn.* 7.
147 The examples really are legion, and fall outside the scope of this dissertation, but among the more curious may be mentioned that there is recorded an *agon*, or competition of skill, between potters in Athens in C4: see *IG* II² 6320.
148 Paus. 6.10.4. Apparently at the very earliest times a breastplate was also part of the equipment for *hoplitodromos;* cf. Lee (2001) 66.
149 Philostr. *Gymn.* 33: "There is no longer any difference between participants in *hoplitodromos*, stade-race and double-stade-race, since Leonidas of Rhodes won all three disciplines over a span of four olympiads. Nevertheless, one must distinguish between athletes who compete in all three disciplines, and those who concentrate on one only. The *hoplitodromos*-runner must have a slender waist, a well-built shoulder and a curved knee, so that the shield can be easily borne and supported on these three points. ... However, participants in all three disciplines should possess all the best bodily qualities of other runner types. This should not be considered impossible; such runners have existed also in my time."
150 [Gal.] 14.724; Gal. 2.303, 2.775, 3.253, 18¹.737 Kühn; Poll. *Onom.* 2.188–189.
151 *Od.* 18.74.

derstood, it seems that the knee assisted in supporting the shield. It is impossible to ascertain whether – or, indeed, how – this was to be accomplished during the race itself. If the shield was really supported on the knee while running, it is difficult to see what kind of running can have been involved: it must have been a sort of limping or hobbling, hardly worthy of the name running. Philostratos may of course have meant that the knee should be able to support the shield *prior to* running (i.e., in the 'ready' position); but this seems to be ruled out by his insistence that all three factors – σιμή knee(cap), robust shoulders and a long waist – were required to ensure that the shield could be carried properly.

Be that as it may, it should still be apparent that like other athletic contests, what was admired in the *hoplitodromos* was the larger-than-life feat: this was spectacular and impressive precisely *because* it displayed a combination of strength and agility quite out of the usual.[152] This impression is reinforced by the information that the different hoplite races were not considered equal: the one in Plataiai was considered far tougher than the others due to the length of the race and the heaviness of the equipment.[153] This differentiation between the games with regard to basic elements such as length of race and toughness of the ordeal is perhaps akin to something like modern triathlon, which has several varying distances, commanding corresponding degrees of respect and admiration, and which is also often regarded as something requiring almost superhuman abilities and extreme physical fitness, as is aptly demonstrated by denominations such as 'Iron Man'.[154] Similarly, the *hoplitodromos* was likely recognised as the toughest of races, the very paradox of running long distances in heavy armour being impressive precisely *because* it was so spectacular a feat.[155]

---

152 Philostr. *Gymn.* 8. This is confirmed by Donlan and Thompson's experiments, proving that running some 200 m with 'hoplite equipment' weighing only 4.9 kg exhausted even the fittest test subjects: "none were able to maintain the chest-high position after 75 yards [= 82.02 m] ... and only one of the subjects (a varsity long-distance runner) completed the run" (Donlan & Thompson [1976] 339–341).

153 Philostr. *Gymn.* 8. Philostratos' phrasing is interesting in this connexion: ἄριστος δὲ ὁ κατὰ Βοιωτίαν καὶ Πλάταιαν ὁπλίτης ἐνομίζετο διά τε τὸ μῆκος τοῦ δρόμου διά τε τὴν ὅπλισιν ποδήρη οὖσαν καὶ σκεπάζουσαν τὸν ἀθλητήν, ὡς ἂν εἰ καὶ μάχοιτο ... ("The *hoplitodromos* of Plataiai in Boiotia was considered the most distinguished because of the length of the race, and because of the runners' equipment which reached to their feet and covered them, as if one were to actually fight"). It seems that hoplite races at other games featured equipment that was not easily mistaken for full combat gear – another indication that the connexion between warfare and sports is perhaps not so certain. Moreover, the winner of the hoplite race at Plataiai received the title 'best among the Greeks' (ἄριστος Ἑλλήνων).

154 'Iron Man': 3.8 km swimming in open sea, 180 km bicycling, 42 km running (or a Marathon run).

155 It is refreshing to read E.N. Gardiner's thoughts on the subject of hoplite race: "The armed race belongs to what we may call mixed athletics, that is to say competitions conducted under fancy conditions, such as obstacle races, races in uniform, swimming races in clothes and all the many events which make up a modern Gymkhana meeting. Such events are popular in character: they are not intended for the specially trained athlete. ... Signs are not wanting that the armed race belonged to this class. The entries were apparently large. Twenty-five shields were kept at Olympia for use in the race, though the starting lines only provided separate places for

A curious parallel to the *hoplitodromos* may perhaps be found in the strange event of the Panathenaia in C5 and C4 (and also at the Amphiaria at Oropos) known as ἀποβάτης, in which hoplites, wearing full or partial armour, apparently jumped on and off chariots (driven by a charioteer). It is difficult to assess the exact nature of the event, but it seems that the hoplites would dismount at fixed times and run along, then jump up again. Again, it is difficult to see precisely how this would reflect contemporary warfare, or be practical training for combat, except perhaps in the broadest possible sense.[156]

Finally, an interesting passage in Euripides' *Autolykos* contains a scathing criticism of the value of sports for warfare: nobody, it is stated, has ever defended their *polis* by wrestling, boxing or running; and nobody has ever fought the enemy holding a *diskos* or driven them off by punching a fist through their line of shields (δι' ἀσπίδων).[157]

(2) The *pyrrhiche,* or war dance, also included exotic and exaggerated movements with at least the shield, such as twirling it and waving it about. That a hoplite shield was normally employed can be seen chiefly from the vase images.[158] To the Greek mind, dancing and military prowess were often connected, as is perhaps most clearly seen from Epameinondas' chilling remark that Boiotia – level and therefore well suited for hoplite battle – was "the dancing-floor of War" (πολέμου ὀρχήστραν).[159] It needs to be asked, however, whether the dances were in fact so directly related to warfare that they were employed specifically as training for real life combat. Spartan children, we are told in later literature, were taught to dance the *pyrrhiche* from the tender age of five, and something similar applies to Crete.[160] It is odd, however, that an armed dance suitable as a preparation for phalanx fighting should have been practised on Crete, which was notable for never truly adopting hoplite armament and tactics; and equally odd that the Spartan youth in Luki-

twenty runners. In such races the more competitors the better. Again the armed race was the last event on the programme at Olympia, and elsewhere, and the last event is often of a less serious character than those that have gone before. In moderns sports we often end with a sack race, or an obstacle race, and we find the same motive on the Greek stage, where the tragic trilogy was followed by a satyric drama by way of relief. ... Again there is always something incongruous and comic in the sight of a person running fast in inappropriate costume, a gentleman in a top hat and frock coat with an umbrella in his hand. ... There must have been something comic in a race of Greek hoplites with shields and high crested helmets ..." (Gardiner [1903] 281).

156 [Dem.] 61.23–29; Plut. *Phoc.* 20.1; Kyle (1992) 89–91; Miller (2004) 142–143; Boegehold (1996) 97. *Apobates* at Oropos: [Dem.] 61.25.
157 Eur. fr. 282.16–21 Nauck; cf. Poliakoff (1987) 99 n. 34. Tyrtaios displays the same hostile attitude to athletics: Tyrt. fr 12.1–14 West, and cf. Xenoph. fr. 11 B 2.14–22 Diels-Kranz.
158 See, e.g., Goulaki-Voutira (1996).
159 Plut. *Mor.* 193e 18; cf. Ath. 14.628f; Lucian *Salt.* 14; Philostr. *Gymn.* 19 and Wheeler (1982) 223. See also Anderson (1991) 29–30.
160 Ath. 14.631a; Strabo 10.4.16; Lucian *Salt.* 8, 10. Pritchett's contention that *paidotribai* taught the *pyrrhiche* in Athens is not proved by his source, Plut. *Mor.* 747b 1: Pritchett (1985a) 63 n. 192; but it is well attested that it was known and used at the Panathenaic games: Is. 5.36; Lys. 21.2, 21.4; *IG* II² 2311.72–74. Further places where the *pyrrhiche* is attested: Megara (*IG* VII 190), Aphrodisias (*CIG* 2758 G [= Roueché 52.IV], *CIG* 2759 [= Roueché 53]).

anos' time – long past the demise of the hoplite phalanx – were apparently *still* being taught the dance οὐ μεῖον ... ἢ ὁπλομαχεῖν.[161] Surely by this time, no one was taught in earnest how to fight with a hoplite shield.

An examination of the sources yields no proof that weapon dancing was intended for any other purpose than general fitness and the impressive sight of arms being handled in a spectacular and impressive fashion. One of the most detailed accounts of dancing in arms is supplied by Xenophon, who witnessed it at a party in Thrace. The description clearly implies that the movements of the dancing was of an exaggerated, acrobatic nature:

> μετὰ τοῦτο Μυσὸς εἰσῆλθεν ἐν ἑκατέρᾳ τῇ χειρὶ ἔχων πέλτην, καὶ τοτὲ μὲν ὡς δύο ἀντιταττομένων μιμούμενος ὠρχεῖτο, τοτὲ δὲ ὡς πρὸς ἕνα ἐχρῆτο ταῖς πέλταις, τοτὲ δὲ ἐδινεῖτο καὶ ἐξεκύβιστα ἔχων τὰς πέλτας, ὥστε ὄψιν καλὴν φαίνεσθαι. τέλος δὲ τὸ περσικὸν ὠρχεῖτο κρούων τὰς πέλτας καὶ ὤκλαζε καὶ ἐξανίστατο· καὶ ταῦτα πάντα ἐν ῥυθμῷ ἐποίει πρὸς τὸν αὐλόν.[162]

The Mysian in question impressed his audience of ethnically diverse party-goers by his outlandish dancing, which incorporated a wide variety of moves – feinting, jumping, whirling of the twin *pelte* shields, and even turning somersaults or cartwheels *with* the shields – that are a mixture of pure acrobatics, and steps and figures supposed to mimic combat. The artistic and impressive nature of the show is underlined by Xenophon's almost disbelieving remark "and he did all this keeping time to the flute". Xenophon's tone is reminiscent of someone describing a particularly spectacular trapeze act performed in a circus.

More to the point, such mixtures of martial arts moves and impressive acrobatics with no discernible fighting purpose in themselves are part and parcel of many cultures the world over. The East Asian martial arts in particular are heavily based on such sequences of movements (known in Japanese martial arts as *kata* (型 or 形), in Chinese as *tao lu,* 套 路), meant to emulate defending against and attacking one or more opponents. The purpose of such exercises is to improve the performer's agility and suppleness, as well as his general physical fitness while he is perfecting his punches and kicks. At the same time, they should ideally be performed with seemingly perfect ease in order to produce a beautiful overall impression on the spectator. Certainly the Chinese national sport *wushu,* which is a strongly traditional sport-like art form of a highly gymnastic nature, regularly includes rather exaggerated movements – such as cartwheels – whose martial value are, at best, doubtful. While *wushu* certainly has roots in self defence, it cannot be described as direct and pure combat training: for this purpose, there are several better options, no-nonsense close combat techniques (usually developed in modern times), which

---

161 Pl. *Leg.* 625d; Lucian *Salt.* 10 ("no less than weapons fighting").
162 Xen. *An.* 6.1.9–10 ("After this a Mysian came forward with a light shield in each hand and danced. He made himself look sometimes as though there were two people attacking him, and sometimes he used the shields as though he was fighting against one other person; and sometimes he would whirl round and go head over heels, still holding the shields, and giving a very fine show. Finally he danced the Persian dance, clashing the shields together and bending his knees and then leaping up again; and he did all this keeping in time to the flute" [trans. Warner]).

are taught and used by military and police special forces; but which, though highly effective, are normally brutal and unpleasant to look at. It seems that the Mysian's weapon dance in this case was something very much akin to the former: a showpiece containing elements of both realistic fight simulation and impressive acrobatics with no external purpose. In this respect it is interesting enough that Plato informs us that there were two basic types of dance: one which imitated defensive fighting (evasive manoeuvres such as dodging, jumping, ducking, retreating, parrying), and one which emulated attacking (blows, feints, counterthrusts etc.).[163]

Now, it may be argued that this was not in fact a pyrrhic dance. However, on the same occasion there was also a more traditional weapon dance, performed by a professional female dancer (ὀρχηστρίς) equipped with a light shield.[164] In Xenophon's case, this was a source of much amusement to the men present at the symposium; but women dancing the *pyrrhiche* are frequently attested on a number of C5 Attic vases, often seemingly in festive surroundings, such as symposia. The women are sometimes dressed to look as Athena, but often quite naked apart from the required weapons: shield, lance and helmet. They may be *hetairai*, performing as part of the evening's entertainment at symposia.[165] The fact that women are depicted so often, both performing and training, and that children were taught the dance as well, suggests that the *pyrrhiche* was perhaps used not so much as a direct exercise for fighting in the phalanx: rather, it was thought of as a fitness exercise or a beautiful dance, including (although not exclusively) moves vaguely mimicking fighting of a rather generic kind.[166] This is borne out by Plato, who in the *Laws* recommends the pyrrhic dance for its overall positive effect on the performer's mind and body: τό τε ὀρθὸν ἐν τούτοις καὶ τὸ εὔτονον, τῶν ἀγαθῶν σωμάτων καὶ ψυχῶν ὁπόταν γιγνήται μίμημα, εὐθυφερὲς ὡς τὸ πολὺ τῶν τοῦ σώματος μελῶν γιγνόμενον, ὀρθὸν μὲν τὸ τοιοῦτον, τὸ δὲ τούτοις τοὐναντίον οὐκ ὀρθὸν ἀποδεχόμενον.[167]

---

163 Pl. *Leg.* 815a. Xenophon also tells of a Thracian dance with two participants, emulating armed hand-to-hand combat, which was so realistic that the spectators actually believed one of them had been killed or wounded (Xen. *An.* 6.1.5). Borthwick (1970a) carefully examines a number of movements included in the pyrrhic dance, such as reclining of the head. See also Anderson (1991) 29–30.
164 Xen. *An.* 6.1.12.
165 Goulaki-Voutira (1996) 5–7. This aspect of eroticism in the dance is further evident in *Anth. Lat.* 104 Shackleton-Bailey. In Xenophon's *Symposium* a highly erotically charged dance is performed by a girl leaping and turning somersaults and cartwheels between upright swords with the hilt stuck into a wooden κύκλος, Xen. *Symp.* 2.11.
166 See also Ath. 14.628f: dancing in Sparta was an ἐπίδειξις οὐ μόνον τῆς λοιπῆς εὐταξίας, ἀλλὰ καὶ τῆς τῶν σωμάτων ἐπιμελείας ("a display, not merely of discipline in general, but also of care taken for the body" [trans. Gulick]).
167 Pl. *Leg.* 815a-b ("And when the imitation is of brave bodies and souls, and the action is direct and muscular, giving for the most part a straight movement to the limbs of the body – that, I say, is the true sort; but the opposite is not right"); cf. 816d and Goulaki-Voutira (1996) 3 (but see van Wees [2004] 189). Galenos agrees completely with Plato on this point: Gal. 6.155 Kühn and cf. Borthwick (1967) 20 n. 17. Plato is generally critical of athletics as preparation for civic duties: see esp. Poliakoff (1987) 99–100.

A further testimony to the inaptitude of the stylised movements of the pyrrhic dance in actual hoplite fighting is found in an interesting passage in Euripides' *Andromache*. Here, Neoptolemos defends himself against the attacking Delphians:

> τῶν δ' οὐδὲν οὐδεὶς μυρίων ὄντων πέλας
> ἐφθέγξατ', ἀλλ' ἔβαλλον ἐκ χειρῶν πέτροις.
> πυκνῇ δὲ νιφάδι πάντοθεν σποδούμενος
> προύτεινε τεύχη κἀφυλάσσετ' ἐμβολὰς
> ἐκεῖσε κἀκεῖσ' ἀσπίδ' ἐκτείνων χερί.
> ἀλλ' οὐδὲν ἧνον, ἀλλὰ πόλλ' ὁμοῦ βέλη,
> οἰστοί, μεσάγκυλ' ἔκλυτοί τ' ἀμφώβολοι
> σφαγῆς ἐχώρουν βουπόροι ποδῶν πάρος.
> δεινὰς δ' ἂν εἶδες πυρρίχας φρουρουμένου
> βέλεμνα παιδός.[168]

Euripides' careful phrasing in the description of this situation is revealing. Neoptolemos finds himself in an unusual situation: he is defending himself alone against large numbers of enemies pelting him with huge quantities of missiles of all kinds, and from all sides. In this situation, he is naturally forced to point his shield this way and that, according as he is being targeted.[169] These unorthodox movements are compared to a *pyrrhiche*; and the fact that the comparison begins with ἂν εἶδες, a 'supposition contrary to fact' (ἄν with a secondary tense), emphasises the point: it is almost *as if* one *were* watching a "grim [= deadly serious] *pyrrhiche*".[170] The phrasing indicates that this is something one would not expect to see under normal combat circumstances.

A final argument against the significance of the *pyrrhiche* in training specifically for hoplite needs is the very old age of the dance. There are a number of different aetiologies for the dance, suggesting its great antiquity.[171] Certainly it was included in the programme at the Panathenaic games, and in this connexion was believed to have been invented by the goddess herself.[172] If it does in fact go back to the origins of the Panathenaic games, the cultic and symbolic element of the *pyrrhiche* must predate hoplite fighting by a good margin; and Everett L. Wheeler has plausibly argued that the martial elements in the dance are traditional atavistic fea-

---

168 Eur. *Andr.* 1127–1136 ("Not one uttered a word. They picked up stones, and began / to throw them. A thick hail of missiles pelted him / from all directions. He had a shield on his left arm, / and held it before him, now on this side, now on that, / to protect himself. It was no use. Weapons came at him / flying in a mass – arrows, spits, sacrificial knives – / and fell at his feet. As he avoided each attack / he looked like one performing a ghastly Pyrrhic dance" [trans. Vellacott]).

169 Something similar is said of Athena's dancing the *pyrrhiche* immediately after springing from Zeus' forehead: ἡ δὲ πηδᾷ καὶ πυρριχίζει καὶ τὴν ἀσπίδα τινάσσει καὶ τὸ δόρυ πάλλει ... ("She's leaping up and down in a Pyrrhic dance, shaking her shield and poising her spear ..." [trans. MacLeod, modified]) Lucian *Dial. D.* 13.

170 Cf. [Longinus] *Subl.* 26: this particular rhetorical device is supposed to hold the attention of every single member of the audience.

171 Archil. fr. 190 Bergk; Arist. fr. 519 Rose; Dion. Hal. *Ant. Rom.* 7.72.7, cf. Strabo 10.4.16; Lucian *Dial. D.* 13, *Salt.* 9.

172 Borthwick (1970a) 318–319. Wheeler (1982) 231 n. 43 cites Paolo Scarpi, who believes that the pyrrhic dance may be a ritual initiation dance much older than the Panathenaia.

tures from much older times, which had nothing to do with fighting in the phalanx.[173]

Rather, it seems that the pyrrhic dance was a centuries-old tradition containing elements of basic combat moves of both defensive and offensive nature; a dazzling, pyrotechnic display of agility and weapons handling, conducive certainly to overall fitness and suppleness, and showing off a number of martial moves to good effect. In this respect it has close affinities to both Oriental martial arts characterised in no small degree by showmanship and acrobatics, and, perhaps, to something like the sights that can be seen at a modern military tattoo. It is therefore somewhat rash to conclude that we "get some clue to the nature of Greek hand-to-hand fighting" by studying the pyrrhic dance:[174] ultimately the question of the value of dancing for hoplite combat cannot be assessed with any certainty; but so far as can be judged from the textual sources, it contributed only indirectly and insignificantly to the hoplite's weapon skills, but probably to a great degree to overall suppleness, agility and bodily strength.

2.1.7 Comparison with a modern combat shield

A major problem facing scholars trying to assess the combat aptitude of ancient weapons is naturally the scarcity of possibilities to try to handle them, and much less under anything resembling actual fighting situations. Accordingly, it would be the next best thing if it were possible to obtain this much needed information from somebody who has actually tried it.

Police forces around the world have regularly used shields against rioters throwing stones, bottles or even Molotov cocktails; and Danish police has often seen action, particularly against squatters in the 1980s, but on many other occasions also. The police must be said to be among the very few today who have any experience with the handling of a shield, and the theory and practice of shield fighting employed by them is therefore very relevant in a discussion of what can and cannot be done with a shield. For this reason, I contacted the riot squad section of the Danish police academy in Copenhagen. The following is the distilled result of a long interview with chief inspector Claus Olsen of the Danish police, who supervises the police training section and has taught riot control for many years, including the use of double-grip shields in phalanx-like formations. He is thus among the few today who have any practical experience of handling a shield in actual combat.

Danish police riot control forces regularly used shields from the 1970s until quite recently, when it was almost completely abandoned in favour of more mobile and offensive tactics. The shield in question is rectangular with rounded corners and made of Plexiglas, and so its shape is possibly more reminiscent of a Roman legionary *scutum*. The shield is fitted with a double-grip carrying system which allows for ambidexterity, placed in the middle; and the twin *'porpax'* and *'antilabe'*

---

173 Wheeler (1982) 230–232; *contra* Rawlings (2000) 248–249; Pritchett (1985a) 61–63.
174 Pritchett (1985a) 61 n. 189; cf. Rawlings (2000) 249.

are c. 45° to the vertical edge, so that the arm is inserted at an oblique angle. The shield measures 95 × 60 cm, or 5700 cm²; not incomparable with the surface area of a hoplite shield (which, as we saw, would typically be between 6362 and 7853 cm²), though of course somewhat smaller. The weight is nonetheless kept down to a mere 2.74 kg. Despite the shield's weighing no more than between 34% and 39% of a hoplite shield, however, it was considered a weapon suitable only for defensive fighting: policemen would typically form a line, advance to the combat zone and keep their position. They would form defensive lines ('chains') and stand so close that the edges of their shields actually touched. The defensive character of these formations was underlined by the fact that policemen in combat gear would also be equipped with helmets (with visors), greaves, bullet-proof vests and thick, padded gloves.

According to chief inspector Olsen, the shield was deemed too heavy, large and awkward to be wielded freely and to be put to offensive use; so much so in fact that a provisional concept was devised for offensive action. The stationary shield line might under certain circumstances be supported by hastily summoned plain-clothes policemen, who would be equipped merely with modified standard shields. The modified shield is identical to the normal type, but is simply sawn off just above the middle, so that a little less than half the shield remains, just enough that the grips are still attached. Much like a buckler or targe, this much lighter shield can be swung around with relative ease; and unlike the large shield the adapted version could therefore be used offensively, combined with little or no body armour to ensure crucial mobility. These policemen, cowering behind the wall of shields held by the front line in full combat gear, would then be able to dart forward and close with rioters who had ventured too close to the defensive police line.

The defensive, immobile police line thus benefited from the unarmoured, lighter troops, who ensured that missile-throwing rioters could not come close to their position with impunity, otherwise a distinct possibility. In other words, policemen with shields and in combat gear were unable to fend off this particular threat themselves, though they were very well suited indeed to braving barrages of thrown cobblestones and bottles. Individually, however, they could not do much more than that, and the practice of leaving the line to pursue rioters was discouraged for two reasons: firstly, this would threaten to disrupt the shield wall, thereby endangering the entire position and their colleagues; secondly, although well protected, policemen were unfit for single combat due especially to the large, heavy and unwieldy shield.

The standard shield, deemed too heavy and clumsy by well-trained and physically fit riot squad policemen, it is worth remembering, weighed not much more than a third of a typical hoplite shield. It seems unlikely that hoplites in bronze armour would have been able to do what larger, fit and trained policemen cannot or at least deem hopeless; namely to fight individually in *monomachiai,* wielding their *three times heavier* shields with ease against attacks from all corners.

2.2 HEADGEAR

2.2.1 Typology

The hoplite's torso was well protected by shield and breastplate, which left the head exposed. This was remedied by the wearing of a bronze helmet (κόρυς, κράνος, κυνέη). At a very general level, Greek helmets may be divided roughly into two distinct 'families': the *Kegelhelm* and the 'Illyrian' type on one side, and the 'Corinthian' type, comprising the Corinthian, Attic and Chalcidian helmet on the other.[175] Without doubt, the simplest and earliest type is the *Kegelhelm*, which consists in little more than a conical bronze cap with separate cheek-pieces riveted onto the helmet.[176] The riveted cheek-guards set it apart from virtually all other types of early Greek helmets, which were beaten out of a single sheet of bronze and had the necessary orifices cut out afterwards. *Kegelhelme* were joined together of five pieces and often a very high crest-stilt was affixed to the apex of the cap. The chronological priority of the type is clearly suggested by its inclusion in the Argos grave panoply, which can be dated to *c.* 725.[177]

The 'Illyrian'[178] helmet was a slight improvement over this,[179] as its cheek-pieces were an integral part of the cap, which was rounded and thus fitted the wearer significantly better than the conical *Kegelhelm*.[180] The Illyrian helmet facilitated manufacture by reducing the five necessary components of the *Kegelhelm* to only two, being joined transversally along the crest ridge, and later still to just one, being beaten out of a single sheet of bronze. Two ridges or grooves along the crown of the helmet for attaching a crest are invariably part of the design. Some variants of the type display a prolonged neck-guard or cheek-guards or holes cut out for the ears.[181] Unlike the *Kegelhelm*, which disappeared towards the end of C7, the Illyrian helmet continued to be used right through to C5, although it was soon ousted, to the point of insignificance, by a far superior and much more popular type.

The 'Corinthian' helmet was developed early in C7 and owes its modern name to Herodotos, who mentions a helmet which, by all appearances, is similar to the one called by that name in modern terminology.[182] Apparently, Corinth was either

---

175 An excellent overview of the typology and the relative chronology of the several types is supplied by a lucid diagram in Connolly (1998²) 60–61.
176 Snodgrass (1964a) 13–16.
177 Courbin (1957) 333–340; Snodgrass (1967) 41–43, (1965) 110–112; Hanson (1991) 64–65. See also fig. 9 (67) for a picture of the complete Argos panoply.
178 As Kunze and Snodgrass show, the name 'Illyrian' is a misnomer: the overwhelming majority of finds are from Olympia, suggesting the Peloponnese as its likely source (Snodgrass [1964a] 18–19, citing Kunze; and see Kunze [1958] 125 and [1967] 116).
179 Kunze (1958) 125–126 and pl. 39–45.
180 Snodgrass (1964a) 18.
181 Kunze (1967) 121 fig. 41.
182 Hdt. 1.60.4–5. The helmet is placed on the head of the peasant girl Phye by Peisistratos in order to make her look like the goddess Athena, who is almost always portrayed with a helmet of this type pushed back on her head to symbolise her status as the goddess of warfare. The girl was also dressed πανοπλίῃ Ἑλληνικῇ.

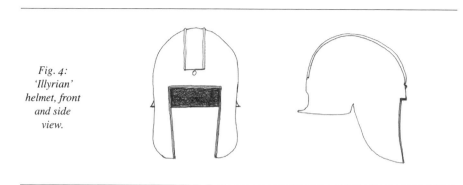

Fig. 4: 'Illyrian' helmet, front and side view.

the place of origin of the type or the chief place of production, or both. Soon after its introduction, the Corinthian helmet quickly rose to a position of predominance. The vast majority of extant helmets from at least the Archaic age are of this type, and this is confirmed by its overwhelming preponderance in contemporary Greek art, due no doubt also in part to its emblematic design.[183]

The Corinthian helmet was in fact a quite refined design. It was hammered out of a single sheet of bronze (although early specimens were something like the Illyrian helmet, made from two pieces joined at the crest). Early helmet specimens had a very round appearance, with a completely straight lower edge and were crudely shaped with a characteristic inward slant at the back and sides – much, in fact, like a modern motorcycle helmet – making the helmet look not entirely unlike a cowbell. The only parts of the face left uncovered by a T-shaped slit were the eyes and the mouth and chin, giving the Corinthian helmet its uncannily enigmatic and forbidding appearance.

Like other types of helmet, the Corinthian underwent a long period of increased refinement: the back and sides of the helmet were 'straightened out', and it was supplied with a neck-guard jutting out at the lower back edge to protect the vulnerable nape of the neck. Furthermore, the nose-guard, which in the earliest times was hardly more than implied,[184] was extended and made to protrude noticeably, enhancing nose protection greatly, sometimes seemingly threatening to shut off the narrow slit between the forward-sweeping cheek-guards entirely. Later helmets had even more elongated cheek-guards pointing downwards, so as to encircle and protect the neck and throat of the wearer. These prolonged cheek-guards tapered from about the inferior maxillary to a point near the clavicle, giving the lower part of the helmet a characteristically pointed appearance and enclosing the wearer's head and face completely, except for the narrow, almond-shaped slits for the eyes and the narrow gap between the cheek-guards. The gap no doubt facilitated breathing and possibly, to a certain extent, communicating.

183 Out of 350 Archaic helmets excavated at Olympia, 31 were *Kegelhelme*, 41 were Illyrian helmets, "at least" 31 were Chalcidian helmets. Some 250 helmets – or 71% – were Corinthian helmets: Jarva (1995) 111–112.
184 Kunze (1961) 64.

As the helmet found favour, further refinements caught on: the foremost part of the helmet, covering the forehead, and the nose-guard, were strengthened while the rest of the helmet was kept relatively light and rigid: the difference could be 1 mm for the crown and as much as 3 mm for the reinforced parts.[185] These areas were in need of more protection because they were targets of choice; and the need for *tripled* frontal protection also seems to indicate that phalanx fighting was predominantly defensive. The characteristic rounded crown of a Corinthian helmet was not only a thing of beauty, it also ensured that the helmet fitted fairly snugly on the wearer's head and helped deflect spear-thrusts and sword blows directed at the hoplite from above.[186]

A late type of helmet (late C5) also deserves mention. This is the so-called πῖλος type, a slightly conical cap-type covering only the crown of the head. No doubt it was so named for its similarity to the civilian skull-cap made of felt (πῖλος) worn by shepherds and the like, but there is some doubt as to the material chosen for the military *pilos* helmet. Certain sources seem to indicate that the helmet was in fact also made of felt: firstly, Thucydides' observation that the Spartans' *piloi* did nothing to protect them from the Athenian arrows and javelins on Sphakteria in 425 seemingly rules out bronze, since this material should be able to keep out arrows with relative ease.[187] Secondly, another revealing indication appears on a tombstone from Megara *c*. 420–10, in which a hoplite is holding a *pilos* in his hand, of which Beazley observed: "The hand sinks into the hat and presses it out of shape … I have no doubt that the material is felt and that felt piloi were used in war as in peace."[188] The oppressive heat of Greece in summer may also have played a part in the adoption of lighter, though still protective headgear.[189]

However, bronze *piloi* have been found; and the amusing passage in Aristophanes' *Lysistrate* from 411 in which a soldier is shopping for porridge in the *agora* and stuffing it in his *pilos* seems to rule out the possibility of a felt hat here (unless perhaps Aristophanes, not entirely unlikely, was aiming for a particularly disgusting comic effect).[190] The best solution probably is that the word *pilos* simply

---

185 Jarva (1995) 141 and n. 989 asserts that nose-guards could be as thick as 6–8 mm.
186 Franz (2002) 135.
187 Thuc. 4.34.3 and Σ (inv. no. 34). Thucydides, ever the conservative, may be warning his Athenian readers against the drawbacks of the 'new' headgear: see Anderson (1970) 30. The scholiast explains the word thus: οἱ πῖλοί εἰσι τὰ ἐξ ἐρίου πηκτὰ ἐνδύματα, ὥσπερ θωρακιά τινα ὑπὸ τὰ στήθη, ἃ ἐνδυόμεθα· οἱ δὲ τὰ ἐπικείμενα ταῖς περικεφαλαίαις ("*Piloi* are compressed wool garments, such as the small *thorakes* worn under the chest, or according to some, lining for helmets").
188 Quoted in Anderson (1970) 31. Chrimes (1949) 362–363 strangely assumes that the *pilos* was in fact made of leather.
189 Cornish miners in the 19th century AD wore headgear apparently strikingly similar to the *pilos:* "In such temperatures miners often worked virtually naked. Flannel trousers, heavy boots without socks and a strong, resin-impregnated felt hat with a convex crown onto which was stuck a lump of clay to secure a candle, was all that most could suffer to wear." http://www.cornish-mining.org.uk/story/work.htm.
190 Ar. *Lys.* 562.

Fig. 5: Early Corinthian helmet, front and side view. The nose-guard is barely noticeable, and the form of the helmet itself is primitive compared with later Corinthian helmets.

Fig. 6: Corinthian helmet, front and side view. The outline of the helmet is more anatomical in its design, while the eye slits are as yet relatively large.

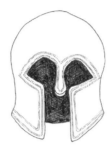 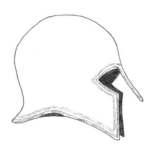

Fig. 7: Late Corinthian helmet, front and side view. The shape is much more advanced, and the aperture between the elongated cheek-guards is almost non-existent. The eye slits are very narrow.

Fig. 8: Pilos helmet (front and side view identical).

means 'a helmet of *pilos* shape', but that *piloi* of both materials were used for protective headwear.[191]

Be that as it may, it is certain that the introduction of the *pilos* helmet reflects the desire for lighter, less oppressive headgear: it is noteworthy that the helmet is devoid of any protection for ears and eyes. Like the Corinthian helmet, the *pilos* might be fitted with a crest of horsehair, but there is scarcely any doubt that the construction as a whole was rather lighter.[192] It is almost certain that the design is indicative of considerable dissatisfaction with the Corinthian and other, similar types of closed design: the decision to sacrifice the excellent protection of the entire head and neck which the Corinthian helmet offered must mean that the Corinthian helmet was finally perceived as too obtrusive and awkward, despite its protective advantages.

2.2.2 Characteristics of the Corinthian helmet

Like its Illyrian cousin, the Corinthian helmet might be fitted with grooves or ridges for attaching a crest, and from early on Corinthian helmets were often lavishly decorated with geometric patterns around the edges of the slits for eyes and nose, or even inlaid with contrasting metals or ivory.[193] The helmet could be further embellished by fixing a crest directly to the crown of the helmet, or, in early helmets, to have the horsehair crest raised on a stilt and placed in a metal holder, thereby moving the centre of gravity even higher. The crest seems to have been more or less *de rigueur* from the earliest times. Crests and other high headgear (whether fixed on helmets or otherwise) is a common martial phenomenon in almost any culture, as is plain from even a cursory glance at, e.g., Zulu warriors, Japanese samurai and the impressive array of enormous headgear common in European armies at the time of the Napoleonic wars. They make the wearer appear taller and therefore more frightening and at the same time are indicative of a desire for appearing more conspicuous or impressive, satisfying "demands that [are] at least partly psychological or aesthetic".[194] Although no crests are preserved, it would seem from artistic representations that a high, narrow crest of dyed horsehair[195] was very much the norm with Corinthian helmets, making the hoplite appear taller and more intimidating.[196] This is the basis for the celebrated, moving scene in the *Iliad,* where Hektor tries to pick up his infant son Astyanax while wearing his plumed helmet: ἂψ δ' ὁ πάϊς πρὸς κόλπον εὐζώνοιο τιθήνης / ἐκλίνθη ἰάχων, πατρὸς φίλου ὄψιν ἀτυχθείς / ταρβήσας χαλκόν τε ἰδὲ λόφον ἱππιοχαίτην, / δεινὸν ἀπ' ἀκροτάτης κόρυθος νεύοντα νοήσας.[197]

---

191 On *piloi,* see Sekunda (1994) 175–178, with a selection of excellent plates.
192 Franz (2002) 277, 344 gives the weight as just below one kg.
193 For a sample of the decorative patterns, see, e.g., Kunze (1961) 69, 73, 98 and 93 fig. 43.
194 Cartledge (1977) 14; cf. Jackson (1991) 230.
195 Dyed white in Alc. fr. 140.4–8 Voigt: κὰτ τᾶν λεῦκοι κατέπερθεν ἴππιοι λόφοι / νεύοισιν, κεφάλαισιν ἄνδρων ἀγάλματα ("down from which nod white horse-hear plumes, adornments for men's heads" [trans. Campbell]).
196 Crested helmets are specifically mentioned by, e.g., Tyrt. fr. 11.26, 32 West.
197 *Il.* 6.467–470 ("Back into the bosom of his fair-belted nurse shrank the child, crying, fright-

That hoplites were very well aware of the potential psychological effect of the crest can be clearly seen from Tyrtaios, who encourages the Spartan youth to strike fear into the enemy by "shaking the terrible plume on top of his head".[198] For ease of storage and transport, the plume could be removed altogether.[199]

Corinthian helmets were presumably padded on the inside to prevent head trauma if the wearer was hit, as is suggested by minuscule holes around the edges of extant helmets: probably the padding, whatever it was, was attached here. Alternatively, a skull-cap of some sort might be worn loosely under the helmet: there is a beautiful rendering of a hoplite wearing such a cap *without* his helmet.[200] Aristotle mentions a certain, very rare kind of sponge (σπόγγος 'Αχίλλειος), which was used to polster both helmets and greaves.[201] Apparently, this particular sponge was extremely light, compact and tough (λεπτότατος καὶ πυκνότατος καὶ ἰσχυρότατος), which made it suitable for this purpose, although it has been argued that the density of the sponge prevented it from absorbing moisture from the wearer's scalp.[202] Another possible material was felt, which is mentioned as padding material in the *Iliad*.[203] The capacity for absorption in the case of the sponge may be a matter for speculation, but felt certainly is very absorbent. However, it is an open question whether a thoroughly soaked helmet padding or rivulets of sweat constantly trickling down one's neck would be preferable.

The padding, apart from damping the sound of blows and enhancing comfort for the wearer, had the much more important function of absorbing the force of blows. The protective principle of any helmet is that it dissipates the impact over a much larger area, thereby helping absorb the energy. Nonetheless, the energy of the impact may still be such that, though deflected or absorbed to a certain degree, it can cause injury to the skull underneath the helmet shell. The bronze of the helmet may normally have been sufficient to prevent outright perforation,[204] but was hardly immune to being dented. Naturally, a helmet that is only dented is in a sense intact, but the dented area itself might easily reach the hoplite's head and cause severe in-

---

ened at the sight of his dear father, and seized with fear of the bronze and the crest of horsehair, as he caught sight of it waving terribly from the top of the helmet" [trans. Murray]); cf. 13.132–133, 16.137–138. Quite likely in this case the poet is not thinking of a Corinthian helmet; but the argument works equally well with any type of helmet with a nodding horse-hair plume.

198 Tyrt. fr. 11.26 West (κινείτω δὲ λόφον δεινὸν ὑπὲρ κεφαλῆς).
199 Ar. *Ran.* 1037–1038; *Ach.* 1180–1183.
200 On a wine cup by the Sosias painter (Berlin, Antikensammlung [F2278]; or see Salazar [2000] fig. 1 or Majno [1975] 149 fig. 4.9).
201 Arist. *Hist. an.* 548a 31 – b 4.
202 Franz (2002) 137. The suitability of sponges is confirmed by the ubiquitousness of synthetic sponge material as padding in a host of modern products such as mattresses, seat cushions and carpet pads.
203 *Il.* 10.265.
204 Gabriel & Metz (1991) 57–58. There seems, however, to be at least one instance where the helmet was perforated: see Salazar (2000) 231–232 who estimates genuine battle damage to a helmet in the British Museum. Salazar believes that helmets may occasionally have been perforated by unusually forceful blows, such as those from sling projectiles: Salazar (2000) 13–15 (with sources); cf. Jarva (1995) 142 and n. 999.

jury. The padding will naturally have helped absorb the force of such blows; but what was really lacking was a form of interior suspension, such as that employed in, e.g., modern military (steel) helmets, which prevents physical contact between the area of impact and the wearer's skull.[205] The consequence of this is of course that the hoplite might otherwise easily be knocked unconscious by a severe blow to the head (perhaps causing internal haemorrhaging or contusion), or possibly suffer an equally disabling 'whiplash' injury to the cervical vertebrae.[206]

2.2.3 The Corinthian helmet in combat

What advantages in increased protection the Corinthian helmet gained by enclosing the entire head and face was certainly made up for by the evident drawbacks to the design. First of all, the very fact that it encased the face completely must have impaired the bearer's range of vision sharply. This is contested by Franz, who has been given access to helmets in the Sammlung Guttmann and claims that the range of vision afforded by the helmet is quite good: he was apparently able to see to both sides out of the corner of his eye without having to turn his head, because "die Großen, der Augenform angepaßten Augenausschnitte eine sehr gute Sicht zur Seite zulassen".[207]

I carried out an experiment myself in order to verify this. Obtaining the correct measurements from the Archaeological Museum in Olympia, I made a model to scale of a reasonably average Corinthian helmet in the collection.[208] Wearing the helmet, the test subject could relatively clearly discern objects at height of head one metre to the front of her and within a radius of *c.* one metre to either side. Beyond this line objects quickly blurred and disappeared out of sight. At 0.5 m distance, only objects within 0.5 m to either side were visible. Without the helmet, the range of vision at a distance of one metre was more than 1.5 m to either side, and as much as one metre at 0.5 m distance. It is difficult to achieve complete accuracy; but it is certainly safe to say that a helmet of Corinthian type does restrict vision.

---

205 Hanson is misrepresented by Franz (2002) 136: "Diese Aussage beruht allerdings auf Hansons Annahme, daß der Helm kein Innenfutter besessen habe und einfach auf die Haare des Trägers aufgesetzt wurde". This *Aussage,* however, is not to be found in Hanson. What Hanson *does* say is that helmets were lacking any kind of suspension system and therefore rested – padding and all – directly on the wearer's head. See Hanson (2000²) 73: "… punched holes around the perimeter of extant helmets suggest that there was some type of interior felt or leather padding attached to the bronze surface …", 79.
206 Hanson (2000²) 213; cf. Franz (2002) 136–137. See, however, Gabriel & Metz (1991) 57–58.
207 Franz (2002) 135 ("the large eye apertures, shaped after the form of the eye, provide very good side view"). Curiously, the good range of vision inside a Corinthian helmet is contrasted favourably with that of an Illyrian helmet, which, according to Franz, actually restricted vision. This is rather surprising, as one would suppose than an entirely open-faced helmet would present no obstacles in this respect: Franz (2002) 135, 52–53. Apparently, the cheek-guards are causing the problem; and it is certainly premature to make this judgement based on trying on a single Ilyrian helmet (Franz [2002] 52 n. 208).
208 The helmet has catalogue no. M 24 and can be found in Ducrey (1986) 58–59 fig. 41b (given there as B 4198).

Moreover, as anyone who has tried on a domino mask will admit, eye-shaped and -sized apertures can easily block or obscure the vision precisely *because* they are shaped after the eye: unless it is very securely fastened to the face, it takes only the tiniest of shifts for a domino mask to become dislodged and obscure vision or cut it off entirely. This must have been the case to an even higher degree for a Corinthian helmet, since the helmet was in essence a closed bronze bell, save for the eye-shaped openings: if the helmet was dislodged in the least, the wearer's vision will have been instantly obscured or completely blocked. That this was a reality for hoplites seems to be demonstrated by the fact that they were fastened with chin-straps, as is casually hinted at by Aristophanes: πρώην γοῦν, ἡνίκ' ἔπεμπεν, τὸ κράνος πρῶτον περιδησάμενος τὸν λόφον ἤμελλ' ἐπιδήσειν.[209] This is of course a rather inane way of donning the helmet; but the passage illustrates well enough the necessity of a chin-strap. It is certainly intended to get a laugh from the audience, which must therefore have understood quite well what Aristophanes meant.[210]

The chin-strap is an indication that helmets usually did not fit the wearer perfectly. It might be possible for armourers to 'tailor' the helmets to specification and so achieve a near-perfect fit; but it would be exceedingly difficult, if not impossible, to make it fit with such a degree of perfection that it would not shift *at all* when the hoplite moved – or was struck on the head. Moreover, even if helmets were made to specification, they were certainly very costly and not just thrown on the scrap-heap simply because the first owner died or no longer was required to serve as a hoplite. We know that shields were passed on from father to son over generations and proudly displayed above the fireplace.[211] The same must apply to other costly items of the hoplite panoply, so it is a fair guess that a minority of helmets in use at any given time were actually tailor-made to suit the men currently wearing them.

The second problem with a Corinthian helmet was that it severely restricted hearing.[212] The helmet's closed construction ensured that the wearer's ears were completely encased in bronze, and it very nearly cut off his sense of hearing entirely, since there were no orifices for the ears. This will have impeded hearing badly: anyone who has ever tried on a crash helmet may have an idea of the 'sound level' inside such a bronze bell. Franz is of the opinion that hearing was not too difficult inside the helmet, however. He summarily dismisses the problems in two sentences: "Auch die Hörfähigkeit kann durch den korinthischen Helm nicht übermäßig eingeschränkt worden sein. Denn die Krieger sind offensichtlich in der Lage gewesen, die Signale von Aulosbläsern zu hören."[213]

This argument is tenuous. It is attested nowhere in the sources that *auloi* were ever used to sound signals during combat, although van Wees has come to the same

---

209 Ar. *Ran.* 1037–1038 ("The other day ... when he was in a procession, he was trying to fasten the crest to his helmet *after* first putting the helmet on his head" [trans. Sommerstein]).
210 Again, Franz (2002) 138 strangely imputes something to Hanson which is stated nowhere: nothing is said in Hanson (2000²) about the lack of chin-straps.
211 Ar. *Ach.* 277–279 and cf. 57; Plut. *Mor.* 241 f. Alkaios fr. 140.1–8 Voigt speaks with palpable enjoyment of a room with its ceiling adorned with helmets.
212 Hanson (2000²) 71–73.
213 Franz (2002) 134 n. 151, citing the Chigi jug as evidence.

conclusion regarding the *aulos* player on the Chigi jug. There is no good reason to assume that he is sounding "a call to arms, as trumpeters do elsewhere", however: this would require precisely a trumpeter.[214] The piper rather seems to be setting a marching rhythm for the hoplites on the left-hand side. Thucydides, no novice to military matters, is quite explicit that this was the function the Spartan *auloi* fulfilled,[215] and Greek literature is littered with references to signals, if necessary, being given with the *salpinx* as a matter of course.[216] The marching rhythm may have been set by the oboe-like *auloi*; but it is difficult to imagine that anyone except the nearest could hear them in a phalanx on the march; and certainly they could not all hear *aulos* signals, especially in the tumult of fighting.[217]

It is noteworthy, too, that later versions of the Corinthian helmet, as well as its late derivations (Chalcidian, Attic) all have cut-outs for the ears in some shape or form, suggesting that the original design left quite a lot to be desired with regard to hearing.[218] Simply put, if hearing in a Corinthian helmet was not a problem, why abandon much better and much needed protection and solidity at the helmet's sides? Why were hoplite helmets with orifices for the ears ever devised?

Another problem with the helmet was that it was bound to be rather heavy compared to a more open type or a simple 'cap'-type covering only the top of the head. Extant examples weigh approximately 1.6 kg on average (with extremes at 1.1 and 1.9 kg);[219] but as with the Bomarzo shield a certain amount of corrosion has to be taken into account. Furthermore, organic materials fitted to the metal shell, such as padding, chin-strap and crest, naturally have not survived. Franz estimates an additional 25% to compensate for deterioration and loss of organic material, but gives no specific arguments. Accepting this argument, the 'average' helmet will thus have weighed close to 2 kg. Hanson has estimated the average helmet at 2.27 kg (= 5 lbs.) but is unfortunately even less helpful as to how the result has been obtained. Franz criticises Hanson for terming the helmet 'heavy'; but the very small difference between their respective assessments – not quite 300 g – is scarcely sufficient to warrant a complete re-appreciation of the helmet as "nicht besonders schwer".[220] For a helmet of this construction principle, the Corinthian helmet may have been relatively light; but it measures unfavourably against most modern helmets manu-

---

214 That this should be the reason why we see but one piper and that he is "evidently blowing at the top of his lungs" (van Wees [2000] 139) seems to me unlikely. In that case, this is the only known instance of a piper acting as a trumpeter in battle. One can only wonder how many would be able to hear the oboe-like *aulos* over the din. See also Schwartz (2002) 53; Hurwit (2002) 15; Helbig (1911) 39.
215 Thuc. 5.70; cf. Aul. Gell. 1.11.1–10.
216 The sources are conveniently collected in Krentz (1991).
217 Arr. *Tact.* 27.3–4 has a vivid and realistic description the noise on a battlefield, which, although late, is probably valid for any ancient battles. Onas. 26 recommends that watchwords and orders be accompanied by gestures with hands or weapons.
218 Snodgrass (1967) 69–70, 93–94; Connolly (1998²) 60–63.
219 These assessments by Franz (2002) 134, who has weighed an unspecified number of Corinthian helmets in the Sammlung Guttmann. See also Jarva (1995) 134, who cites Kukahn's calculations between 1.2 and 1.5 kg.
220 Franz (2002) 134.

factured in steel. A sample of three standard issue army steel helmets from the last century show an average weight of a mere 1.08 kg.[221] These helmets were certainly considered heavy and bothersome enough by the infantry troops wearing them.[222] Accordingly, it is hardly wrong to say that the Corinthian helmet was in fact heavy, by almost any definition of the word. It was surely heavy enough to be awkward to wear for a longer period of time, as even lighter modern helmets are sufficient to make the neck muscles sore after a relatively short while.

If iconographic evidence is anything to go by, the helmet was normally pushed to the back of the head when not used directly in combat. The famous portrait bust of Perikles is an example of this; and while it is an artistic convention when hoplites are depicted *fighting* with their helmets pushed back, there is hardly any doubt that any hoplite would push it back given half a chance, if he wore it at all. This naturally seems to underscore the fact that the helmet severely restricted vision and hearing; but also the fact that the bronze, like most metal, conducted heat all too well. It is difficult to imagine how hot the helmet became on a summer day in the campaigning period: it would quickly get unbearably stuffy inside the closed helmet, especially if the hoplite did not wear his hair and beard rather closely cropped.[223]

---

221 British Mk I 'Brodie': 1.14 kg; American M-1: 1.02 kg; German M 16/35/42: 0.98–1.180 kg (average 1.08 kg). See, e.g., http://www.thevietnam-database.co.uk/usmc/mhelmets.htm; Baer (1994) vol. I 172. Incidentally, the German M 16/35/42 was the first military helmet for centuries to take up the Corinthian helmet's exocranium construction principle; a timely reminder of its tremendous protective effectiveness. Even more intriguingly, Friedrich Schwerd's original M 16 design is echoed in and further enhanced by the modern PASGT kevlar army helmet used by most NATO forces.

222 The German *Armeeabteilung* 'Gaede' was among the first to experiment with a sort of leather helmet, and also among the trial units who were supplied with the first 30,000 M16 helmets manufactured in early 1916. An excerpt from the subsequent report to the war ministry fully reflects the perceived heaviness of the new headgear: "Die Stahlhelme können den Lederhelm nicht ersetzen. Sie sind als Marschhelme zu schwer. ... Der Stahlhelm eignet sich gut zum Schutz einzelner gefährdeter Posten, aber nicht für allgemeine Einführung" ("The steel helmet cannot replace the leather helmet. They are too heavy for marching with. ... The steel helmet is suitable for protection on certain exposed posts, but should not be generally introduced"). The final report from 'Gaede' to the war ministry included complaints from the rank and file: "Nach einigen Truppenurteilen steht eine Beeinträchtigung der Marschleistung, selbst wenn sich die Truppen an den Helm gewöhnt haben, bei Ersatz des Lederhelmes durch den Stahlhelm zu erwarten. Für diesen Fall wäre zu erwägen, ob nicht die Truppen mit einer geeigneten Feldmütze zu versehen sein werden" ("According to some troop reports a limitation of marching capacity should be expected if the leather helmet is replaced by the steel helmet, even if the troops have grown used to it. For such cases supplying the troops with some sort of cap should be considered") Quoted in Baer [1994] vol. I 50–53).

223 Hanson (2000²) 72–73; Franz (2002) 135. Franz again misrepresents Hanson: "Viel wichtiger ist die nicht von Hanson erwähnte Einschränkung der Luftzirkulation ..." ("even more important is the limitation of air circulation, not mentioned by Hanson ..."). The word 'circulation' may not appear in Hanson's text; but it is difficult to see how anything else could be meant: "the helmet was uncomfortable ... because of the heat it generated around the eyes, mouth, nose, and ears ... there could not have been a more stifling type of headgear ... the beard and hair ... could only make the helmet more uncomfortable and stuffy".

This was not normally the case, because close-cropped hair was commonly regarded as a mark of slavery and thus unfitting for a free man.

The question needs to be asked whether the protective advantages of this helmet design outweighed the considerable drawbacks that restricted vision and hearing meant in a combat situation. As we have seen, the closed helmet must have been singularly uncomfortable and stuffy; and the more so the better the fit. Breathing freely will almost certainly have been difficult, and the air inside the helmet must rapidly have become hot, humid and stale owing to the poor ventilation. The scorching sun in normal Greek summer conditions will have exacerbated the situation considerably. Franz claims that the weight of a hoplite helmet is comparable to a 'Montefortino' helmet of a Roman legionary of the Augustan age, i.e. *c.* 2.1 kg.[224] It may very well be that the helmets are comparable with regard to weight alone, but otherwise there are not many points of contact between the two types. A Montefortino helmet is of another type altogether. It is basically a helmet of the 'cap'-type with two cheek-guards placed on hinges and a quite pronounced neck-guard which extends to roughly halfway between the nape of the neck and the shoulders, providing very good protection for the back of the head and neck. Nevertheless, it leaves the face and ears completely free, and the cheek-guards, even though hinged, still have small incisions to provide better range of vision and facilitate hearing. The type frequently has a plain rounded crown and likely no crest or other embellishments to shift the centre of gravity upwards.[225] While probably just as heavy, the Montefortino helmet thus cannot have been anywhere near as stuffy and uncomfortable as the Corinthian helmet, and it certainly did not interfere with vision and hearing to the same degree.

Consequently, the comparison should take this into account in an assessment of the burden on the wearer. Although quite similar in terms of mere weight, the Montefortino helmet was far lighter in practice, and its aptitude in combat is instantly recognisable. The Corinthian helmet looks less than ideal in comparison, and *a priori* it is difficult to understand why anyone would want to wear in combat a helmet apparently so ill-suited to the purpose. It was heavy – perhaps not more so than that of other types of warrior at other times, but certainly just as heavy – and the design did nothing to alleviate the problem. The rather small eye slits will have permitted a range of vision that, even if adequate, could all too easily be restricted severely or cut off altogether, when the helmet shifted on the wearer's head, either from his own movements or when struck by a weapon. That it could and did shift is suggested by the chin-strap, which was there to keep the helmet in place as much as possible. The helmet was rather top-heavy, not least because of the ever-present crest, and even more so if the crest was placed in a crest-holder: this made it even more unstable. Hearing must have been cut off effectively if the helmet was to fit even tolerably closely on the sides of the head; and the padding, hair and beard will have muffled sound even more. Quite possibly his own laboured breathing was what the hoplite heard most of all inside the bell-shaped helmet. Lastly, the closed

---

224 Franz (2002) 134, citing Junkelmann.
225 Connolly (1998²) 228; cf. 308 fig. 8.

design ensured that breathing was indeed troubled, causing bad ventilation and hot, humid, fetid air on the inside. This is hardly descriptive of a helmet with a reasonable balance between protection and comfort and, indeed, aptitude to combat. One cannot help but feel that the Roman legionary in his open-faced Montefortino helmet was far better equipped for close combat than was the Greek hoplite of the Archaic and Classical periods.

2.3 BODY ARMOUR

2.3.1 Breastplates

Hoplite breastplates may conveniently be divided into two main groups: those beaten out of bronze sheet, and 'composite corslets', made from organic material (leather, linen). The bronze breastplates may in turn be subdivided into bell corslets and the later muscular breastplates.

The 'bell cuirass' is in evidence since the earliest extant 'proto'-specimen, arguably the one excavated in the Argos grave in 1953 and found together with a *Kegelhelm*. On the basis of the proto-Corinthian ceramic material found in the grave, Paul Courbin convincingly dated the grave to the C8l.[226] The skeletal remains of the interred warrior revealed his age to be between 25 and 30 ("grâce à une dent de sagesse et à l'état des sutures craniennes"),[227] so the armour can scarcely be older than 750, given that the armour must have been in use for some time before its owner died. This is therefore the *terminus ante quem* for corslets of the 'bell' type, and the Argos cuirass is in fact the earliest 'panoply' yet discovered. Before it there is only the Dendra armour, some 700 years its senior.[228]

This is corroborated by two additional contemporary finds of a similar type, made in the 1970s, also at Argos: the first grave contained a helmet of the same type, and seemingly made at the same workshop (lacking only the bronze crest of the 1957 find), and also traces of a decayed bronze cuirass and an iron sword hilt.[229] The second grave contained a warrior equipped with two spears grasped in the right hand (the spear-heads are preserved), and a different type of helmet without cheek-pieces.[230]

The name 'bell cuirass' (*Glockenpanzer*) is apt, since the corslet curves inward at the waist and juts out at the bottom, giving it a vague bell-shape. The cuirass has openings for arms and head, and it is joined together from two halves (γύαλα) at the sides and shoulders, with hinges at one side and holes for lacing or straps at the other. Fragments of other bell cuirasses indicate that the breastplate and the back-

---

226 Courbin (1957) 333–340.
227 Courbin (1957) 326.
228 Snodgrass (1964a) 73, (1965) 110–112; Connolly (1998²) 54.
229 *ArchDelt* 26 ([1971] 1973) *Chron.* 81–82, n. 22 and plate 68.
230 *ArchDelt* 28 ([1973] 1977) *Chron.* 99, plate 95 ε.

plate were pulled together by means of a strap passing through a ring on either half, and a pin was inserted into and through the hinge rings.[231]

The neck was encased and protected by a sort of 'stand collar', presumably meant to catch upward thrusts along the corslet before they hit the throat. The 'skirt' below served at least two purposes: it deflected downward thrusts,[232] much like the collar, but it also no doubt made it possible for the hoplite to sit and squat while wearing the rigid corslet. Contrariwise, the outward-turned hem – more or less like a funnel – also virtually guided home random thrusts delivered underhand or upwards towards the vulnerable and extremely sensitive groin area. Without doubt, this was a considerable drawback to the overall usefulness of the armour.

From the Argos cuirass onwards, stylised representations of anatomy can be seen in the cuirasses: the pectoral muscles stand out perceptibly and are further indicated by two semicircular curves (sometimes spiralling inwards), and the higher abdominal muscles are represented by what looks like a pointed bell curve (the 'omega' curve). Also, the shoulder blades are hinted at on the back plate, and along the spine there is typically a groove in the metal. All these features become steadily more naturalistic over time. The height of extant bell cuirasses falls between 40 and 50 cm with plate thickness from 0.5 to 3 mm.[233] The bell cuirass was used at least throughout the Archaic period, with an apparent peak in popularity about 640 to 575.[234]

The weight of the Argos cuirass in its current state is 3.360 kg, but the original weight was of course rather greater, as the cuirass has corroded. Connolly loosely estimates approximately 6 kg for the original armour, with lining.[235] Franz has criticised Hanson for assessing the weight of bronze cuirasses based on the Vergina finds of Macedonian cavalry armour made of iron; but in fact Hanson's data are derived from W. Donlan and J. Thompson, who give the total weight of "breastplate, backplate, helmet and sword to be slightly under 35 lbs. [15.91 kg]" on the authority of the curator of the John Woodman Higgins Armory Museum.[236] In comparison with

Fig. 9: The Argos panoply. The vague bell shape of the skirted bronze cuirass is already in evidence in this specimen, complete with hints at the curvature of the pectoral muscles and the 'stand collar'. Notice the clumsy and top-heavy Kegelhelm *with its riveted cheek-guards and very high crest-stilt.*

231 Connolly (1998²) 54–55.
232 Jarva (1995) 24. Franz (2002) 58–59 n. 254 misinterprets Jarva and rejects "von unten geführte Stöße" on the basis of fatal wounds of this kind in Homer; but Jarva clearly suggests that the hem protected the wearer against thrusts from *above*.
233 Jarva (1995) 24.
234 Jarva (1995) 27.
235 Courbin (1957) 350; Connolly (1998²) 58.
236 Franz (2002) 139; Hanson (2000²) 78; Donlan & Thompson (1976) 341 n. 4.

Connolly's findings, this is rather heavier, even if we allow for as much as 6 kg to helmet and sword. A specimen in the Sammlung Guttmann weighed as little as 1.83 kg, and Franz tentatively sets the original weight at 2.3 kg.[237] The extremes thus are 2.3 and *c.* 10 kg, and there is no obvious way to solve the discrepancies. If anything, 2.3 kg seems very low when compared to Franz' own measurements of later types. It is difficult to imagine that very early bell cuirasses would have been so much better in this respect if it provided sufficient protection against penetration: there would not have been any incentive to invent new, more sophisticated types of armour which were heavier and therefore markedly inferior. Historical analogies suggest that, all other things being equal, the technology of a given type of weapon continues to improve and be refined over time. Decreasing the weight of armour is no exception to this rule; something that is corroborated by a general tendency to lighten the weight of armour throughout the Archaic and Classical periods.

Its immediate successor, both in terms of type and chronology, was the so-called muscled cuirass, attested in iconography from about 480.[238] Typologically speaking, it was much the same principle as the bell cuirass: two halves (breast and back) being joined by means of a set of hinges at the sides and shoulders, meant to be opened in one side only. Pausanias gives a description of a bronze cuirass – an antique in his day – preserved in a painting by Polygnotos in Delphi:

> κεῖται δὲ καὶ θώραξ ἐπὶ τῷ βωμῷ χαλκοῦς. κατὰ δὴ ἐμὲ σπάνιον τῶν θωράκων τὸ σχῆμα ἦν τούτων, τὸ δὲ ἀρχαῖον ἔφερον αὐτούς. δύο ἦν χαλκᾶ ποιήματα, τὸ μὲν στέρνῳ καὶ τοῖς ἀμφὶ τὴν γαστέρα ἁρμόζον, τὸ δὲ ὡς νώτου σκέπην εἶναι - γύαλα ἐκαλοῦντο· τὸ μὲν ἔμπροσθεν τὸ δὲ ὄπισθεν προσῆγον, ἔπειτα περόναι συνῆπτον πρὸς ἄλληλα. ἀσφάλειαν δὲ ἀποχρῶσαν ἐδόκει παρέχεσθαι καὶ ἀσπίδος χωρίς.[239]

It is noteworthy that this otherwise fairly detailed description – of a painting, no less – does not allow us to decide whether it is a bell cuirass or its successor.[240]

Nonetheless, the muscle cuirass represents a significant refinement over the earlier type and features very naturalistic, though perhaps somewhat heroically exaggerated, moldings of upper-body musculature, often including even such details as nipples and navel. For these reasons, the muscled cuirass was doubtlessly far more difficult to produce than the comparatively simple bell cuirass. At the same time it all but dispensed with the bell-shaped hem below and the 'collar' at the neck, emphasising the apparent drawbacks of the eponymous 'bell' rim. However,

---

237 Franz (2002) 345.
238 Jarva (1995) 32.
239 Paus. 10.26.5–6 ("On the altar lies a bronze cuirass. Today cuirasses of this form are rare, but they used to be worn in days of old. They were made of two bronze pieces, one fitting the chest and the parts about the belly, the other intended to protect the back. They were called *gyala*. One was put on in front, and the other behind; then they were fastened together by buckles. They were thought to afford sufficient protection even without a shield" [trans. Jones, modified]).
240 Jarva (1995) 32 believes it to be a muscle cuirass because of the descriptions specifying that the front plate is fitted to the breast and belly, perhaps not implausibly: however that may be, the operative principle in the cuirasses remains the same.

Fig. 10: A hoplite in combat wearing a 'muscled' bronze cuirass. The muscles of the torso are painstakingly reproduced, if somewhat exaggerated, and the lower edge of the cuirass 'dips' down toward the groin area for enhanced protection. Notice also the detailed representation of the inside of the shield, with porpax and antilabe, *on both this hoplite and the one in the background, as well as the shield apron in the background.* Red-figure kalyx krater by the painter of the Berlin hydria (c. 465). Cf. fig. 14.

in certain specimens, a downward curve at the bottom and in front is intended to offer crucial protection for the abdominal area without hampering movement of legs and hips.[241] The belly-guard is not matched at the back, which emphasises the need for protection against woundings in the lower body, while also revealing something about the needs and nature of hoplite fighting, which was predominantly frontal and head-on. Most woundings will have been frontal as a matter of course, and the logical response was to strengthen the armour at its most exposed areas. The downward-jutting belly-guard was an improvement over the bell rim, as well

---

241 These features can be clearly seen in the attempted ¾ view of a cuirass of this type on a red-figure lekythos by the Oinokles painter (Cleveland Museum [28.660]; Jarva [1995] 31). Jarva's description of the bell-guard is apt: "The downward curving part is not so broad, resembling an upside-down omega curve in appearance, as if a paragon to rendering of the belly area in classical statuary".

as an aesthetically pleasing solution to the problem of abdominal protection, that did not severely restrict movement.

2.3.2 Corslets

The other main type of armour is the composite corslet. Of this corslet type frustratingly little is actually known, since it was made almost exclusively of organic material, which has of course perished completely. The only extant specimen to reveal anything about its shape is in fact an iron cuirass discovered in the royal tombs at Vergina and believed to have belonged to Philip II.[242] Although made of solid iron, it is apparently a metal version of a shoulder piece corslet. The term λινοθώραξ seems to bear out that it was somehow manufactured from linen; and so does the fact that it is represented in vase paintings by a relatively rare bright white, otherwise reserved for representing female skin and the like.[243] This interpretation is further corroborated by Herodotos' explicit mentioning of linen corslets (of the Egyptian type) sent by the pharaoh Amasis to sanctuaries in Greece.[244] The connexion cannot be established with certainty, but scale-corslets and armour made from linen had been known in the Orient for a very long time.[245] The linen was probably glued or sewn together in several layers in order to lend the corslet some of the structural stability and protection against penetration that a bronze cuirass, irrespective of type, offered. Protection could possibly be further enhanced by preparing the corslet with salt and vinegar, a method found in Pliny and discussed by Isaac Casaubon and re-attempted in 1969 by Sylvia Törnkvist.[246] If so, the salt and vinegar will have added to the overall weight of the garment.

Occasionally, however, corslets are shown in black on vases, which might indicate another choice of material, though we should of course be cautious in assuming a different material on such slight evidence. This may be the σπολάς, a corslet in many ways similar to the *linothorax,* but made from leather.[247] It has been plausibly suggested that leather was a rather more accessible commodity than linen in Greece,[248] and furthermore ox-hide is an extremely tough and durable material, certainly giving just as much protection against penetration as a linen corslet. Xenophon's mention of Kleonymos, who was killed τοξευθεὶς διὰ τῆς ἀσπίδος καὶ τῆς σπολάδος εἰς τὰς πλευράς ("by an arrow which went into the side of his body

---

242 Connolly (1998²) 58–59.
243 λινοθώραξ: *Il.* 2.529–530, 2.828–833; Alc. fr. 140.10 Voigt; and cf. Paus. 1.21.7. For the toughness of linen, see Ael. *NA* 9.17; Pliny *NH* 19.2.11.
244 Hdt. 2.182, 3.47. Based on Hdt. 1.135, Anderson (1970) 23 has even suggested that the Greek linen corslet was modeled on the Egyptian type by Greek mercenaries serving the twenty-sixth dynasty in Egypt (late C7 – early C6).
245 Snodgrass (1964a) 84–85.
246 Törnkvist (1969) 82 admits that her experiments were only partially successful, "but with a knowledge of the right proportions perhaps there would have been more success". Pliny *NH* 8.192 says that the Gauls and Parthians treat wool with vinegar (*acetum*) and achieve similar effects.
247 Poll. *Onom.* 7.70.
248 Anderson (1970) 23.

Fig. 11: A hoplite taking leave of his family before going to war. He is wearing a linen corslet (linothorax). *The lacing system of the* epomides *is clearly seen, as are the overlapping layers of* pteryges *in the skirt of the corslet. Notice also the shield grip. Red-figure stamnos by the Achilles painter (C5m).*

through the shield and the jerkin") is surely not said disparagingly of the *spolas,* but rather meant to underline the incredible power of the Karduchian bows.[249] Eero Jarva suggests that if used economically, a single hide might suffice for two corslets, something that would certainly keep the costs of production down.[250] Meanwhile, a leather corslet will have been far less time-consuming to produce as the material was ready. Conversely, a metal cuirass or a quilted or glued linen corslet will have taken quite a long time to make.

The corslet was apparently laced tight in the left side,[251] a natural choice since this side would also normally be covered by the shield; but alternatively the corslet might be laced frontally, so that the two parts overlapped, much as in a modern double-breasted pea jacket – and doubling the frontal protection in a similar way.[252]

249 Xen. *An.* 4.1.18.
250 Jarva (1995) 37.
251 See, e.g., a skyphos by the Brygos painter (Museo Gregoriano Etrusco Vaticano [16583]) and an amphora by the Achilles painter (Museo Gregoriano Etrusco Vaticano [16571]).
252 Seen on a red-figure amphora in the Antikensammlung, Munich (J378). Cf. Anderson (1970) 22; Jarva (1995) 33 n. 199.

Two long shoulder flaps (ἐπώμιδες) protruded from the top edge of the rear side, of which they were an integral part. These were, as the name suggests, to be brought forward over the shoulders and fastened by lacing them on to the chest, thereby ensuring that the brunt of the corslet's weight was borne by the shoulders. This had the double advantage of not restricting movement and breathing overly (by squeezing the corslet too tightly around the upper body), while also affording excellent protection for the shoulders. *Epomides* were typically quite broad, but markedly narrower at the end, where they were to be laced to fastenings on the chest, giving the composite corslet a very characteristic look. The fact that the *epomides* are depicted as standing up straight in the air has led to the assumption that slats of bronze were sewn or glued between the linen layers,[253] but it is equally possible that the many layers of fabric were sufficiently rigid and tight to make the *epomides* stand straight, whether the layers were sewn or glued together – or that the vase images actually represent leather *spolades*.[254] Between the *epomides* the corslet has a sort of neck-guard standing straight up in the air, and protecting the back of the neck.

Often the corslets were reinforced with metal, such as bronze scales, discs or plates, which doubled effectively as decoration, restoring some of the splendid gleams of a well-polished cuirass to the more lacklustre (albeit brightly white) linen corslet; and in fact such metal reinforcements are the only remains of ancient corslets to survive.[255] They could be placed anywhere on the corslet, indeed all over it, but most often they were restricted to the chest and abdomen. The amount of metal on corslets was probably determined by such factors as increased weight and price.

Apart from the *epomides,* the main visual characteristic of the composite corslet were the πτέρυγες, a row of narrow, overlapping flaps, likely of metal or reinforced leather, hanging down from about waist level on the wearer and normally just covering the buttocks and genital area, thereby giving the corslet its characteristic skirt-like appearance. This was an attempt to remedy the lack of protection for the abdominal and groinal area which a metal cuirass entailed; a problem which had vexed armourers from the very beginning, and which had only been partially solved by the bell rim.[256] On the whole, the system of *pteryges* must be said to be more successful in retaining some protection while also allowing a maximum of flexibility when walking, sitting, kneeling or squatting. However, the level of protection afforded by the *pteryges* is still somewhat questionable: it is difficult to imagine that the curtain-like flaps would withstand much force, let alone a determined spear thrust. *Pteryges* might also be used with plate metal cuirasses, thus lending them some of the greater flexibility-cum-protection of the composite corslets. Several times, the *pteryges* seem to have been detachable, so that they were fastened around the waist like a belt.

---

253 Franz (2002) 187–189.
254 Snodgrass (1967) 90 believes the layers to have been quilted together, whereas Connolly (1998²) 58 has made a reconstruction of a linen corslet glueing the layers together.
255 Jarva (1995) 37–40.
256 Cf. Nierhaus (1938) 91–92 and n. 6.

Composite corslets are in evidence at least from the second quarter of C6 (and therefore introduced a good deal earlier than the muscled cuirass); but "since it can be identified on Athenian late Geometric and early Protoattic vases it is possible that it was used practically throughout the Archaic Period."[257] Their weight is difficult to assess, but Peter Connolly, who has tried to make a linen corslet of his own, reached a weight of 3.6 kg, though that is without any metal reinforcements.[258] The degree of protection is next to impossible to assess, though it is significant that metal reinforcements in exposed areas are so frequent in the iconography. It is certain, however, that the type offered decided advantages over metal cuirasses in terms of comfort and flexibility. The number of layers in a linen corslet of course cannot be known with any certainty, but 10–15 layers has been suggested; and this might result in a corslet not much inferior to one of leather or even bronze.[259]

2.3.3 Cuirasses and corslets in combat

Bronze cuirasses certainly offered very good protection against edged weapons, but also entailed a number of difficulties for the wearer. First of all, the necessary weight in a metal cuirass was a considerable encumbrance for the wearer, no matter how refined the methods of manufacturing. Furthermore, the simple fact that metal is a very nearly perfect conductor of heat meant that a polished bronze cuirass would not only become extremely hot on the surface, but would also conduct the heat directly onto the body of the person inside it.[260] It is possible, even likely, that the cuirasses were regularly padded on the inside with felt or leather, as is the case with helmets; but even though this may have helped to protect the wearer's skin from being outright burned, it will have done little to stifle the tremendous heat generated inside it: heat-stroke must have been a very real risk. The hoplite's arms will more often than not have been bare, and accidentally touching the surface of the cuirass – which was bound to happen constantly during combat – must have been quite painful.

What is more, the cuirass was ideally a quite tight-fitting construction, thus ensuring that ventilation or circulation of air was virtually impossible. The copious amounts of sweat generated inside it from the heat of the sun and the physical exertions of battle will have soaked the tunic presumably worn underneath in a matter of seconds. Due to the closed design, the accumulated moisture and sweat will have had no easy way out – apart from trickling down to the bottom and out, probably only via the arm holes and neck opening – and this will have worsened conditions inside the cuirass. Battles were usually fought in the prime of summer, in which temperatures can get very high and oppressive: data from the National Meteorological Service in Greece show that the *average* day temperature in the shade for the Peloponnese (measured at Corinth) in the period May – September is as much as 28.5° C. Similar data for Kalamata show 29° C. On any individual day, the tem-

---

257 Jarva (1995) 35.
258 Connolly (1998²) 58.
259 Franz (2002) 187–188.
260 Cf. Hanson (2000²) 79.

peratures may of course be far higher, and certainly are around noon: absolute maximum temperatures measured are 38.8° C and 42.6° C respectively.[261]

However, the heat conducting qualities of bronze naturally worked the other way as well: if worn in very cold weather, the cuirass would effectively siphon away body heat and be conducive to hypothermia, or at least severe freezing and general misery. This condition of course worsened if it rained and the tunic underneath was drenched with water.[262] We know of certain military activities carried out in winter or late autumn/early spring; and at least sieges, such as at Poteidaia, were naturally kept up for years on end, including the potentially very cold and inhospitable Mediterranean winter (though of course during siege operations, hoplites will probably only have worn armour when strictly necessary).[263]

It would seem that the impractical heat conducting capabilities of metal armour may have ensured that battles were, if possible, fought in the afternoon, when the air was somewhat cooler and the blaze of the sun less ferocious. The morning hours were of course also less oppressive; but from then on it tended to get hotter towards noon. Afternoon battles, on the other hand, also gave combatants a chance to take up formations, and dress their lines in time for the battle, so more than one purpose might be served by fighting later in the day.[264]

The textile (or leather) will have been much easier to suit to the individual wearer, and the softer material will have been less rigid and uncomfortable.[265] Furthermore, corslets will have caused less trouble by not conducting heat, and by being far more permeable to air and allowing moisture to be absorbed to a far greater degree. Also, transport and storage will arguably have been much less of a hassle with the lighter and more pliant material. Nonetheless, it will have been an encumbrance as well. The weight, although reduced in comparison with a metal cuirass, was still a factor to be reckoned with, and it certainly restricted movement of the upper body. Reaching across the chest with the shield to protect one's right side will have been awkward at best, since any kind of body armour adds considerably to the circumference of the torso and hinders movement of the arms. Xenophon's cautionary words in Sokrates' chat with the armourer (θωρακοποιός) Pistias, no doubt culled from personal experience, also reflect this:

> καλόν γε, ὦ Πιστία, τὸ εὕρημα τὸ τὰ μὲν δεόμενα σκέπης τοῦ ἀνθρώπου σκεπάζειν τὸν θώρακα, ταῖς δὲ χερσὶ μὴ κωλύειν χρῆσθαι ... ἧττον, ἔφη, τῷ βάρει πιέζουσιν οἱ ἁρμόττοντες τῶν ἀναρμόστων τὸν αὐτὸν σταθμὸν ἔχοντες. οἱ μὲν γὰρ ἀνάρμοστοι ἢ ὅλοι ἐκ τῶν ὤμων κρεμάμενοι ἢ καὶ ἄλλο τι τοῦ σώματος σφόδρα πιέζοντες δύσφοροι καὶ χαλεποὶ γίγνονται· οἱ δὲ ἁρμότ-

---

261 The data in question may be found at the HNMS website: http://www.hnms.gr/hnms/english/climatology/climatology_region_diagrams_html?dr_city=Velos_Korinthia; and http://www.hnms.gr/hnms/english/climatology/climatology_region_diagrams_html?dr_city=Kalamata
262 Bad weather: Thuc. 6.70.1; Xen. An. 4 passim, Hell. 4.5.3, 5.4.17–18; Dem. 50.23. See also Plut. Tim. 28.1–4 and cf. Diod. Sic. 16.77.4–80
263 See especially Thuc. 3.20–23, 4.103.1–2; Xen. Hell. 5.4.17–18; Pl. Symp. 219e – 220b (inv. no. 30).
264 Infra 207–211.
265 Connolly writes of his first experience with his new linen corslet: "It was difficult to put on because of its stiffness, but once one had got used to it, it was quite comfortable and easy to move about in" (Connolly [1998²] 58).

τοντες, διειλημμένοι τὸ βάρος τὸ μὲν ὑπὸ τῶν κλειδῶν καὶ ἐπωμίδων, τὸ δ' ὑπὸ τῶν ὤμων, τὸ δὲ ὑπὸ τοῦ στήθους, τὸ δὲ ὑπὸ τοῦ νώτου, τὸ δὲ ὑπὸ τῆς γαστρός, ὀλίγου δεῖν οὐ φορήματι, ἀλλὰ προσθήματι ἐοίκασιν.[266]

It is most likely that the θώρακες discussed here are composite corslets of either type, since Sokrates and Pistias agree that buyers of badly fitting but ποικίλοι and ἐπίχρυσοι armours purchase something spectacularly useless, something that is not flexible enough to allow movement of the upper body.[267] The passage demonstrates that an armour had to fit the wearer quite precisely in a number of places, in order to distribute the burden evenly on as many different points of the upper body as possible.[268] It further tells us that however much lighter a composite corslet was, it was still a considerable weight to carry around, a burden that could be made much heavier by an ill-fitting corslet. By extrapolation, this speaks volumes about bronze cuirasses worn by men they were not originally made for (or who had changed in their physical appearance): there was no way to adapt bronze armour once it was finished, but at the same time a fair number of them must certainly have been worn by men who had put on or lost weight, by men who had inherited them from their fathers as part of the family heirloom, or perhaps even by men who had captured them as the spoils of battle. Unless the armour was custom-made, then, we may assume that it was to some degree troubling, heavy and restricting, on no less an authority than Xenophon's.

2.3.4 Greaves

The greaves were included in the hoplite panoply from the earliest times, and apparently were a staple of defensive armour long before this: they certainly figure prominently in the *Iliad,* and, as Anthony Snodgrass has pointed out, the Mycenaeans clearly knew and used greaves or shin-guards.[269] The hoplite greaves were from the start hammered out of quite thin sheets of bronze, and made to fit the wearer's

---

266 Xen. *Mem.* 3.10.9–13 (" 'Upon my word,' he said, 'it's a splendid idea, Pistias, that the corselet should protect the parts of a man's body that need protection without preventing him from using his hands. ... A corselet that fits irks you less by its weight than an equally heavy one that doesn't fit. A badly-fitting corselet either hangs entirely from the shoulders or presses severely on some other part of the body, and that makes it clumsy and uncomfortable. A well-fitting one has its weight distributed between the collar and shoulder bones, the shoulders, chest, back and abdomen; so that it seems almost more like an accessory part than something to carry" [trans. Tredennick]). Cf. *Eq.* 12.1: πρῶτον μὲν τοίνυν φαμὲν χρῆναι πρὸς τὸ σῶμα τὸν θώρακα πεποιῆσθαι. τὸν μὲν γὰρ καλῶς ἁρμόττοντα φέρει τὸ σῶμα, τὸν δὲ ἄγαν χαλαρὸν οἱ ὦμοι μόνοι φέρουσιν. ὅ γε μὴν λίαν στενὸς δεσμός, οὐχ ὅπλον ἐστίν ("In the first place, then, you should, I think, have a breastplate made for your trunk. Now, a well-fitting breastplate can be supported by the whole trunk, but all the weight of one that is too loose falls on the shoulders, and one that is too tight is a straitjacket rather than a piece of armour" [trans. Waterfield]); and see also Ober (1991) 181.
267 Xen. *Mem.* 3.10.14–15.
268 The description of the snugly-fitting armour is somewhat reminiscent of modern hiking backpacks, which are highly refined pieces of equipment, adapted closely to the human physique and often even having different carriage structures for men and women.
269 Snodgrass (1964a) 86–88.

calves, running from about the knee to the instep, though naturally not covering the malleoli. Surprisingly, they were apparently not intended to be laced or otherwise fastened to the leg, but simply 'snapped' in shape around the calf muscles. This could be done because the metal was so thin; but it seems somewhat unpractical that the hoplite would have to bent them in and out of shape every time he needed them. Yet this is what must be inferred from the absence of proper holes for attaching straps or strings to the greaves.[270] Extant, early greaves are frequently supplied with rows of minute holes around the edges, which presumably made it possible to fasten padding, made from some sort of organic material, on the inside to prevent grating and chafing.[271] The holes gradually disappear from C7 onwards, however (and are completely gone by C5);[272] but rather than a complete abandonment, this is probably simply indicative of a new method of polstering, perhaps involving glue. The highly undesirable side effects of chafing and grating will have been practically impossible to avoid due to the flexing of the calf and foot muscles when walking or running,[273] making it all the more difficult to understand why some kind of attaching arrangement was not an integral part of the design.[274] We can only surmise that since it was apparently not perceived to be enough of an annoyance to develop a system with a better fit, hoplite combat must have been of a kind in which this did not make much of a difference.

Within the hoplite period, the greaves did not change significantly in appearance.[275] Such variation as there is is mainly due to a better understanding of the manufacturing process on the part of armourers, always refining and improving on the relatively simple principle. The ergonomic fit was gradually improved on the original design by imitating the indentations formed by the calf muscle, so that in its advanced form (reached already in C7[276]) the two flanges of the greave all but met on the back of the lower leg. The indentations in the metal made the greaves even more elastic, while not losing crucial solidity.[277] Owing to the high demands on ergonomics, left and right greaves of the evolved kind are easily distinguishable.[278]

---

270 It should be noted, however, that at least one extant pair of greaves have larger holes probably intended for laces, and laced greaves are also known from at least two artistical representations: Jarva (1995) 90 and n. 564, 99.
271 Greave padding will have been made from some kind of organic material, such as leather or felt. See also Arist. *Hist. an.* 548a 31 – b 4: Aristotle states that padding made of a certain kind of sea sponge dampens the sound inside helmets and greaves, but it is clear enough that this is merely careless writing, a sort of *zeugma:* the damping of sound naturally only applies to helmets, whereas the absorbing of blows is vitally important in both cases (see *supra* 60–61).
272 Jarva (1995) 99–100.
273 Cf. Franz (2002) 141.
274 The unpleasantness and unpracticality of the greaves is vividly evoked in Hanson (2000²) 75–76.
275 There is, however, a very thorough typology with five sub-groups in Jarva (1995) 85–100. For a depiction of greaves see, e.g., figs. 11 (71), 12 (87) and 14 (88).
87   Snodgrass (1964a) 88.
277 Jarva (1995) 96, 100.
278 See, e.g., Jarva (1995) 96 figs. 49, 50.

Over time, the greaves were also made thinner, and the upper end made to extend past and so include the knee-cap. The greatest variation is probably in length: early specimens tend to be shorter, about 30 cm, whereas the later versions were made to include and cover the kneecap, thus extending the range to 40 cm.[279] This necessarily altered the basic shape of the greave. Now the front reached higher than the rear, since the greave must stop just before the hollow of the knee, if the leg was to bend and function normally: this resulted in a graceful, slight S-curve from the rear to the front side. The part covering only the kneecap was thus put on display very conspicuously, and it lent itself naturally to decoration (embossed palmette patterns, gorgo heads etc.). Thickness of the metal in extant examples typically varies between 0.5 and 2 mm, and as a rule the trend is towards thinner, lighter and more elastic greaves; but, as Jarva's findings demonstrate, there are notable exceptions to the rule.[280] The greaves' protective abilities were aided in no small degree by their being rounded so that attacks tended to glance off, and this in turn squared nicely with the ergonomical design that favoured a snug fit on the wearer's (round) shin. The weight is difficult to assess, but Franz' estimate – between 1.2 and 2.2 kg for a pair, regardless of the period, does not seem unreasonable.[281]

While always useful against sharp-edged weapons, greaves were especially necessary for hoplites for precisely the same reasons that the shield was heavy and cumbersome. The greaves protected the lower leg, beginning at precisely the point where the shield 'left off', thus at least partly relieving the need for covering the legs and feet with the shield. This would mean exposing one's breast and head, which would otherwise be sufficiently covered under and inside the convex rim. The hoplite shield was perhaps even worse than other types of shield in this respect, because the double-grip system effectively prevented the bearer from merely parrying or blocking by lowering his shield. A hoplite shield held directly away from the body can only be lowered 10–15 cm; beyond this point it becomes necessary to rotate the lower arm outwards and downwards – clearly not conducive to good protection. The only two alternatives to this would be either to squat or to bend over, folding at the hip, and this was not advisable: apart from being too slow a process, it would result in the entire back being exposed to any kind of attack – a quick stab, or a slashing sword blow across the spine. Furthermore, the size, weight and shape of the shield made it unlikely that there would at all be time to react defensively and reach down to deflect an attack to the ankles or feet. However, the most important reason for the greaves was possibly that they gave protection against missiles of any kind: stones, whether from slings or thrown with the hand, arrows and javelins.[282]

At any rate, the ankle is particularly vulnerable, as the bone is directly beneath the skin, and the area is very sensitive: a wounding here was likely to be incapacitating or cause the hoplite to collapse in pain. For these reasons, any protection of the lower leg *independent* of the shield was very advisable.

---

279 Snodgrass (1964a) 88.
280 Jarva (1995) 141–142.
281 Franz (2002) 346.
282 Aesch. *Sept.* 675–676.

2.3.5 Additional body armour

Other items of body armour can be attested for C6f, though they do not seem to have been the norm. Accordingly, they will be treated only summarily, as their influence on hoplite fighting was arguably rather small and they were abandoned relatively quickly.

The armour for the upper body might be completed with guards for upper and lower arm (for the spear-arm only, since the left arm would be covered by the shield).[283] Some of the upper arm-guards have a large plate covering the shoulder; and they extend to the elbow. They might be fastened to the breastplate or fastened like a greave: Xenophon knew both types and favoured the latter.[284] Another type actually covered everything from the shoulder to the fingers, and was known as 'the arm' or 'the hand' (χεῖρ);[285] so this was preferred for the horseman's left arm, which did not need to be quite so flexible as the right arm (wielding the javelins). All the same, the χεῖρ apparently was a quite sophisticated piece of armour, as it was able to follow the arm's movements (ἐκτείνεται δὲ καὶ συγκάμπτεται).

Normally, however, the upper arm-guard was complemented by a separate lower arm-guard. This extended from the elbow joint to the wrist, where the ulna bones jut out – something that is clearly reflected in the design.[286] A lower arm-guard is a defensive supplement in that it enables the wearer to deflect attacks by warding them off directly with the protected weapon arm; but there are considerable advantages to enclosing the upper arm also.

Another addition to the armour was thigh-guards (called, it seems, παραμηρίδια[287]), which were relatively more common than the arm-guards: presumably they were viewed as a natural supplement to the ordinary greaves. Ordinarily, they would be formed with a rounded area near the knee, creating sufficient space for the greave so that they would not collide.[288] If anything, the thigh-guards indicate that the upper legs were perhaps not perfectly covered by the shield: numerous representations show the thigh-guard being worn on both legs, whereas at least the left leg should theoretically be sufficiently protected by the shield, at least with the sideways-on, behind the centre of the shield stance that van Wees argues.[289]

The greaves might apparently also be supplemented or replaced (as the case may be) by a pair of ankle-guards. Extant examples vary in length from (usually) 11–13 cm, to no less than 28–29 cm – though the longer variants may not be Greek

---

283 Jarva (1995) 72–76, with a catalogue and typology of Archaic upper arm-guards.
284 Xen. *Eq.* 12.7.
285 Xen. *Eq.* 12.5; cf. Poll. *Onom.* 1.135. Xenophon naturally had horsemen in mind, so the latter item was designed for the left arm, as a horseman was unable to carry a shield.
286 Cf. Jarva (1995) 77 fig. 34.
287 Xen. *An.* 1.8.6, *Cyr.* 6.4.1, 7.1.1–2 (though the latter example clearly refers to a part of the horse's armour); Arr. *Tact.* 4.1.
288 Jarva (1995) 82.
289 Jarva (1995) figs. 37, 83, 39, 40.

after all.²⁹⁰ Ankle-guards were rounded at the back, and were hammered out to fit the malleoli.

A foot-guard might be the final complement to the panoply. This was, however, apparently rare: a total of four foot guards have been found, and it is very likely that they were in fact meant to be fitted (laced?) on to some kind of footwear (sandal or boot), rather than worn directly.²⁹¹ They consist in a sheet of metal beaten out to fit a human foot, complete with separate toes (where even the nails are visible). One pair has a hinged toe section, which apparently enabled air circulation inside the foot-guard: otherwise it is difficult to see the purpose. Ankle- and foot-guards demonstrate the need for added *Panzerung* at the lower extremities; and it is quite plausible that spear-thrusts or missiles, which were already delivered at a sloping angle, might all too easily glance off the smooth surface of the greaves and continue directly into the ankle or foot.

All the additional extremity 'greaves' are tubular in shape, much like the actual greaves, and are also made to fit human anatomy. Like the greaves, they are punctuated along the edges in order to enable fastening of padding or lining to the inside, and most were possibly supposed to be laced on. Much like other items of armour, they are between approximately 0.5 mm and 2 mm thick, which ought to be sufficient protection in this area: thrusts delivered would have to travel quite long, and so probably were relatively weak.

2.3.6 Effectiveness of bronze armour

Experiments have been carried out by Richard Gabriel and Karen Metz to calculate the degree of protection afforded by bronze armour, using formulae adopted from the U.S. Army Ballistic Laboratory as the source material for the calculations.²⁹² The thickness of the bronze plates in question was 2 mm, and experiments showed that the energy required to penetrate such plates could not be produced with any kind of weapon known to and employed by ancient Greeks. Our scarce knowledge of composite corslets naturally makes it virtually impossible to judge the efficiency of this type of armour; but the evaluation of bronze armour yielded some interesting and rather surprising results. It could be shown that a mere 2 mm of bronze was in fact sufficient to protect the wearer from lethal injury. Greek armour could be thinner than 2 mm; but this is most often the case with such items as greaves or advanced helmets. These items need not be quite so thick, since their roundness tended to make weapons glance off. Shield facings were often extremely thin; but because of the solid wooden core beneath, they were scarcely more prone to penetration; indeed "the bronze shields of the ancient Greeks would have easily repulsed the gunfire of the Napoleonic rifle, and the severely angled helmet of the Assyrians would have made penetration by even a Civil War musket quite difficult."²⁹³

---

290 Jarva (1995) 102.
291 Jarva (1995) 105–106.
292 Gabriel & Metz (1991) xviii–xix.
293 Gabriel & Metz (1991) 56.

Another consideration would be that the weight of the spears and swords used in the experiments are seemingly somewhat below par: spear and sword are listed as weighing 0.68 kg and 0.82 kg respectively.[294] This will naturally have affected the overall performance of the weapons to a certain degree in that their impact will have been less, but the margins are so wide that the difference is insignificant; and Gabriel and Metz' experiments are therefore still of great value. The table runs thus:

Weapon	Speed	Impact area	Energy produced	Energy required
Spear (overhand)	16.76 m/s	0.2 cm²	96 Nm	186 Nm
Spear (underhand)	7.32 m/s	0.2 cm²	18 Nm	186 Nm
Sword (cut)	18.29 m/s	9.7 cm²	137 Nm	205 Nm
Javelin	17.68 m/s	0.2 cm²	91 Nm	134 Nm
Arrow	60.05 m/s	0.2 cm²	64 Nm	103 Nm[295]

It is interesting to note that both the thrusting spear and the sword, when used for cutting, are woefully insufficient to penetrate the armour; in fact only about half the required energy can be generated with these weapons. The javelin and arrow fare somewhat better, but are also far from attaining the desired objective. The weapons would naturally be quite adequate against unarmoured humans: it requires only 91 Nm of energy to break the breastbone or rib;[296] but bronze armour was plainly more than sufficient to protect the wearer, albeit at a cost. The especially weak energy that can be produced with an underhand spear thrust demonstrates that if it were to be effective, it would have to be directed at the opponent's unprotected groin.

Experiments have also been made by Franz, albeit under less scientifically strict conditions: using a German type roofing hammer and an axe (simulating spear and sword respectively) on a straight 1 mm bronze plate and a curved 0.5 mm plate (simulating a breastplate and greaves respectively), Franz himself has simply pounded the bronze plates as hard as possible at a suitable angle.[297] Franz is careful to point to the importance of Newton's first law ($W_{kin} = ½ \times m \times v^2$); but the equa-

---

294 Gabriel & Metz (1991) 59 (table 3.1).
295 Gabriel & Metz (1991) 51–63; esp. table 3.1 and 3.2, of which mine is a conflation. Gabriel and Metz' calculations operate with imperial standards. For the sake of clarity, I have taken the liberty of converting feet per second (fps.) to metres per second (m/s), and foot-pounds (ft-lb) to newton metres (Nm), the definition of a newton metre being "the torque exerted by a force of one newton acting at a distance of one metre from the axis of rotation". 1 ft-lb equals 1.335818 Nm.
296 Gabriel & Metz (1991) 60.
297 Franz (2002) 355–360. Jarva also discusses the effectiveness of Archaic armour, though without carrying out experiments himself (Jarva [1995] 139–144).

tion does not seem to play an actual role in carrying out the experiment: it does not appear as such in any calculations. Therefore, unfortunately, the amount of force or striking distance from the target cannot be accurately judged; but it is nonetheless revealing that even with the roofing hammer swung against the 0.5 mm thick 'greaves', only the slightest of perforations could be made (although it should be noted that a hammer-head cannot be compared to a spear-head: the volume of the bulky hammer-head is much greater than that of the slender spear-head). Despite the lack of accuracy in these experiments, they do point to the same conclusion as Gabriel and Metz' rigorously scientific findings: bronze armour, even when hammered to quite thin plates, afforded excellent protection against the damage that edged weapons can inflict on the human body.

It would seem that armour so effective in protecting the bearer against edged weapons must have been rather heavy in comparison with the still reasonable degree of protection against penetration that could have been achieved with considerably lighter armour. The fact that they were so heavy and afforded such a degree of protection gives an idea of the kind of fighting they were devised for. Unlike, e.g., medieval Japanese samurai armour, made of strips of metal laced together with leather, silk and even paper, which valued lightness and flexibility so high that swimming in armour was taught and used,[298] but offered considerably less in the way of resistance against penetration, Greek hoplites consciously piled on weight and gained massive protection at the cost of mobility and lightness. Furthermore, when we remember that the solid and heavy shield was always an indispensable part of the hoplite's equipment, it becomes clear just how much emphasis was laid on frontal protection: apparently, hoplites at least in the Archaic age and possibly up to C5l were virtually impregnable from the front, but at the same time were not very mobile.

All this points to a primarily defensive and still-standing type of fighting from the very outset – it is worth remembering that to the extent hoplite armour developed, it was in a steady evolution towards less and lighter armour, and doubtless increased mobility as well. Early hoplites, on the other hand, were truly heavily equipped by comparison. It would thus seem that the hoplite phalanx was a closed-order, close-range enterprise from the very outset.

2.4 OFFENSIVE WEAPONS

2.4.1 The spear

Judging from vase imagery, a hoplite's spear (δόρυ) was normally between 1.8 and 2.4 m long.[299] The wood of the shaft is often referred to as ash (μελία), though cornel (κράνεια), or more correctly cornelian cherry (*cornus mas*), was widely used

---

298 Ratti & Westbrook (1973) 184–208 (armour), 294–296 (swimming): several schools (*ryu*) throughout feudal Japan specialised in teaching this art.
299 Anderson (1991) 22–23; Franz (2002) 144.

and was certainly chosen for the Macedonian σάρισσα.[300] Quite a few sources testify to cornel as the preferred material for spear shafts, and certainly Xenophon singled it out above other types of wood as suitable for the horseman's *palta*.[301] Indeed, so commonly used was cornel that it could become altogether synonymous with 'spear' in poetry.[302] Cornel and ash share a number of characteristics in that they are quite dense and fairly elastic without being overly flexible; enough so, in fact, to make them both suitable for bows. They were in all likelihood relatively readily available in Greece and they enjoy a long season, which might also explain their popularity.[303] Nonetheless, the remains of wood inside a butt-spike found at Olympia were pinewood;[304] and if this is anything to go by, it is no wonder that shafts snapped time and again: as pinewood is relatively soft and pliable and prone to splitting, it probably would not have withstood much pressure before breaking.[305]

As for the spear-heads, things are somewhat clearer. Iron was the material of choice, and with good reason. Iron is much harder than bronze, and therefore preferable if the objective is to penetrate bronze armour; but strangely enough, bronze spear-heads are not unknown.[306] Moreover, the fact that iron is heavier than bronze was insignificant for so small an object as a spear-head, making it all the more reasonable to use iron. The spear-heads are usually beautifully leaf-shaped and have a characteristic pronounced central rib for structural strengthening, and the length varies between 20 and as much as 30 cm.[307] The diameter is in the vicinity

300 Theophrastus (*Hist. pl.* 3.12.2) states that the maximum length of a Macedonian *sarissa* was 'limited' to 12 πηχεῖς, since the cornel grows no higher than this. Xen. *Hell.* 3.4.14 surprisingly states that some Persian spears, made of cornel, were stronger than the Greek ones, which snapped at the first blow.
301 Xen. *Eq.* 12.12, cf. *Hell.* 3.4.14. Ranging from Herodotos to Strabo, the evidence is collected by Markle, III (1977) 324 n. 5.
302 E.g. in two dedicatory epigrams, *Anth. Pal.* 6.122, 6.123: ἔσταθι τᾷδε, κράνεια βροτοκτόνε, "stand there, man-slaying cornel".
303 *Pace* Anderson (1991) 23, who notes that neither wood is to be found in any significant quantity in southern Greece. The situation may have been different in antiquity; but the assumption that this is one reason why we so often hear about spear shafts breaking is interesting. Anderson lists a number of possible substitutes found in Grattius Faliscus and including yew, which has much the same characeristics as cornel and ash and is a bow wood of choice.
304 Kunze & Schleif (1938) 103.
305 Franz (2002) 144 n. 223 criticises Hanson ([2000²] 85): "Hanson scheint allerdings auch davon auszugehen, daß der Speer in der Mitte brach. Dies ist aber physikalisch nicht möglich" ("Hanson seems also to assume that the spear would snap in the midsection. However, this is physically impossible"). But Hanson in fact says nothing of the sort: no specific breaking point is indicated; if anything, he actually seems to imply – like Franz himself – that the spear-head broke clean off ("once the lance head was snapped off ..."; "once his lance head was lost"). Even if Hanson actually meant that the spear broke 'in the middle', it is impossible to imagine anyone believing that the break would occur at the exact geometrical midpoint, rather than between the sharp end and the place where the spear was held, i.e. from where the pressure was generated. This is in fact brought out by Hanson (1991) 72: "most spears probably broke between the lance-head and grip point".
306 Typically in C6 and C5: Snodgrass (1964a) 103, 133–134; Anderson (1991) 23–24.
307 Anderson (1991) 23; Snodgrass (1967) 38.

of 2.5 cm, as is evidenced by the specimens examined by Cahn (two 2.3 cm, one 2.5 cm and one 2.8 cm).[308] This, however, does not rule out the possibility that the shaft tapered at the socket.[309]

At the bottom end of the hoplite spear the butt-spike (σαυρωτήρ, στύραξ) was attached. This was an oblong triangular or pyramidal bronze spike which served a number of functions: it helped to balance the weight of the spear-head, so that the spear could be held further back and thereby have a greater reach, it enabled the spear to be conveniently stuck upright into the ground and it prevented the shaft from splintering.[310] Most importantly, however, it doubled as a makeshift point in case the shaft broke: the spear might be reversed and used with the *sauroter* in front instead. The spear-head was usually attached by means of a socket and strengthened by riveting, though the rivets might be further reinforced by lashing the socket tightly to the shaft with bronze thread wound through and around holes through shaft and socket.[311] The shaft was regularly wound with leather thongs in the middle, near the point of balance, to ensure a better grip.[312]

Tentatively putting the length of the shaft at 2.2 m, the weight of a spear-shaft may then be calculated as follows: $3.14 \times 1.61 \times 220$ ($\pi \times r^2 \times h$) = 1112.2 cm³. The density of *cornus mas* is 0.76 g/cm³, so the weight of a shaft of this size is $1112.2 \times 0.76 = 0.845$ kg (though a shaft of 2.4 m would bump up the weight to 1.213 kg).[313] The shaft, then, would typically weigh about 0.85–1.2 kg, largely irrespective of smaller variations in length (though a tapering shaft, being thicker overall, will certainly have added something to the overall weight). M.M. Markle, III gives the weight of an iron spear-head as 97 g; but surely a considerable amount should be added to this, as iron tends to corrode badly. A spear-head can scarcely have weighed less than 500 g,[314] and at least half that should be added for the often massive bronze *sauroter*. I would estimate both at possibly as much as one kg; certainly 750 g.[315] The weight of an average hoplite spear, then, would be approximately 1.6–1.95 kg (if the metal parts weigh 750 g) or 1.85–2.2 kg (if they weigh one kg).

---

308 Cahn (1989) 13, 15. Cf. also Hanson (2000²) 84: "about an inch [2.54 cm] in diameter".
309 Markle, III (1977) 324.
310 For this function, see *Il.* 10.151–154.
311 Anderson (1991) 23; Franz (2002) 143–144.
312 Visible on, e.g., an amphora in Brussels, Musées Royaux (R308).
313 The density of cornelian cherry is borrowed from Markle, III (1977) 324, whose own assessment of the weight of a hoplite spear is 1 kg. I have taken the liberty of converting all values to the metric system for the sake of clarity.
314 See, e.g., http://www.ancientbattlecrafts.com/weapons.htm, who sell steel replicas of ancient weapons: a steel spear-head typically weighs between 0.5 and 0.75 kg. The same goes for http://www.thesteelsource.com/html/mr9066.htm, where a set of a Greek spear-head and *sauroter* is offered. Both items weigh 1 lb. 4 oz., or 1.12 kg in all. Furthermore, Greek spear-heads were made of iron, not steel, and thus had a higher density. At http://www.manningimperial.com/item.php?item_id=244, a 'Greek' iron spear-head weighs only 150 g, but the matching bronze *sauroter* as much as *c.* 1.5 kg.
315 Markle, III's otherwise careful calculation leaves out the butt-spike altogether and accepts the present-day weight of merely 97 g for the spear-head, which is too low to be credible (Markle, III [1977] 324–325).

During C7 hoplites seem to have carried several spears into battle. This is clear from two things: the frequent representation of additional spears on vase paintings, and the relatively frequent grave finds of thrusting and throwing spears in pairs.[316] The famous Chigi olpe shows two opposing phalanxes armed with secondary spears (although these are very nearly erased from the vase and quite difficult to see in most reproductions), gripped with the left hand that also holds the *antilabe* of the shield. The physical presence of these 'superfluous' spears has been disputed, chiefly by Hilda Lorimer, who interpreted them as 'ghost' spears, a representation of additional spears held at the ready by the hoplites' servants, but not actually 'there'.[317] This explanation is not completely satisfactory, and the real reason is revealed by four spears stuck into the ground on the left, where two hoplites are arming themselves. Two of these spears are supplied with throwing-loops and thus are javelins, probably intended to be discharged before the collision. The easiest explanation by far is that they are quite simply just that: additional throwing spears, carried and discharged prior to the actual close combat. In a similar vein, Kallinos exhorts his fellows to ἀποθνήσκων ὕστατ' ἀκοντισάτω ("even as one is dying, let him make one final cast of his javelin") and strikingly captures the surroundings of battle as δοῦπον ἀκόντων ("the thud of javelins"), both in the same fragment.[318] Mimnermos may be expressing similar ideas (πίκρα ... βελέα ["bitter shots"]); but βέλος simply means 'missile', so arrows or slingshots may equally well be referred to here.[319] Tyrtaios also exhorts his fellow Spartans not to stand ἐκτὸς βελέων ("outside the range of missiles") but it is not at all clear who discharges these βελέα.[320] The references to throwing spears (ἀκόντια or ἀκοντίσματα) in Archaic poetry in fact cannot be linked to hoplites.

However, the logical inference is that javelins actually were used regularly during C7, but that this practice later ceased.[321] It is likely that hoplites found the damage inflicted by missiles insufficient; and certainly additional javelins will have meant extra weight to be carried, as well as the serious risk of breaking up the formation when pausing to throw the weapon. Furthermore, the fumbling necessarily entailed by transferring the javelin from the left hand (holding the shield) to the right, must have been awkward and impractical, and even more so if it was to be done while marching forward. The fact that both the *antilabe* and the javelin(s) had to be held in the left hand must also have accounted for trouble: dropped javelins, irrecoverable in the steady forward march of the phalanx, must have been a usual phenomenon.

---

316 The finds are listed in Snodgrass (1964a) 136–137.
317 Lorimer (1942) 83; *contra* Snodgrass (1964a) 138; Anderson (1991) 18–20 and cf. Franz (2002) 152 n. 252.
318 Callin. fr. 1.5, 14 West (trans. Gerber).
319 Mimn. fr. 14.8 West (trans. Gerber, modified): *pace* Franz (2002) 157.
320 Tyrt. fr. 11.28 West (trans. Gerber).
321 Salmon (1977) 90–92 suggests a transitional phase in C7m in which hoplites fought with javelins in addition to the 'usual' weapons. See also Snodgrass (1964) 62.

Additional javelins will typically have been shorter and slimmer.[322] They can, however, scarcely have weighed less than one kg each, so weight will certainly have been a natural consideration; especially if they were carried along with the shield in the left hand. Another consideration (to my knowledge not hitherto treated of) is the fact that a stage of throwing javelins would effectively have prevented a running charge (δρόμος). This squares nicely with Herodotos' insistent claim that the Athenians and Plataians at Marathon in 490 were the first Greeks to charge at a run.[323]

The longest realistic distance between the two phalanxes, if the javelins were to reach their targets, would be approximately 30 m,[324] and this would be too short a distance to create sufficient momentum for the *dromos* to be worth while. If, as seems likely, this distance requires a few steps of preliminary run, it would have been necessary to halt the entire formation and throw the spears at a command. An advantage of this procedure will have been an extremely dense volley of javelins, likely to inflict considerable casualties on the enemy immediately before the clash. Alternatively, a protracted phase of javelin-throwing actually preceded the closer combat, and there was no *dromos* as yet at this time. No matter how important a part of the fighting in C7, however, the use of javelins while bearing the large hoplite shield was probably awkward and difficult to carry out, and the damage inflicted may have been insufficient. At any rate, the overall combat value must have been negligible, since the procedure was eventually abandoned.

2.4.2 Swords and other secondary weapons

Hoplites normally had a secondary weapon in the sword (ξίφος). The hoplite sword was presumably a direct development of older Greek (and Middle Eastern) double-edged sword types with straight sides and usually with a noticeably narrower blade, dating back to at least Mycenaean times.[325] From the dawn of the hoplite era, the sword was typically made of iron, about 60 cm long and more or less leaf-shaped. The blade was characterised by a noticeable mid-rib. A cylindrical pommel was fixed at the end of the slightly almond-shaped hilt to create balance. The quite small cross-guard was an integral part of the sword itself, but the hilt might be fitted with wood or bone to ensure a better grip and, no doubt, for aesthetic reasons. The leaf shape, apart from being quite beautiful, is a way of ensuring that a maximum of power can be delivered near the end of the blade, thus concentrating the slashing force at the longest possible extension from the arm. The sword is often shown

---

322 See Snodgrass (1964a) 136–137.
323 Hdt. 6.112.1–3*quater* (inv. no. 19).
324 Franz (2002) 145–146. Franz has carried out experiments with javelins and throwing-loops himself and concludes that 25–35 m seems to be the maximum throwing distance. A similar result is reached by Harris, whose incisive study treats of, *inter alia,* the use of throwing-loops on Greek war javelins: Harris (1963), esp. 29–36. For comparison, Jan Zelezny's world record with a javelin (albeit a modern and light construction), achieved in 1996, is no less than 98.48 m.
325 Snodgrass (1964a) 98; Connolly (1998²) 63.

worn in a scabbard, which is draped over the right shoulder and across the body, so that it hung at the hoplite's left side. This was a practical measure, enabling the hoplite to draw the sword in the quickest manner possible, if he could somehow squeeze the scabbard under his left arm without letting go of the shield. The sword, like all pointed straight swords, was of course equally well-suited to thrusting, something that was probably more useful in close combat.

Another type of sword is the single-edged, lightly curved slashing sword known as κοπίς or μάχαιρα.[326] This weapon was presumably introduced at a later time and was a fearsome weapon, probably capable of slicing off a head or an arm entirely, or of splitting a shield. Its terrible cutting proficiency is demonstrated by its close kinship with a butcher's cleaver, as is evidenced on several vases; and quite possibly the *kopis* began its career as a tool being taken to war.[327] This might also explain the nickname μάχαιρα, 'knife'. The *kopis* varied in length between about 50 and possibly as much as 70 cm long, made of iron and curving slightly inwards, the blade being much broader near the tip than at the hilt. The hilt was in line with the back of the blade and apparently often curved back, so that the hilt more or less closed on itself. The blade was strengthened by a mid-rib along and below the dull back of the blade.

This forward-heavy single-edged sabre was a construction which was near ideal for slashing and chopping, but which was equally unsuited to stabbing or thrusting. The leverage would most likely be greatest in downward cuts, and this is borne out by Xenophon who recommends this type for cavalrymen: ἐφ' ὑψηλοῦ γὰρ ὄντι τῷ ἱππεῖ κοπίδος μᾶλλον ἡ πληγὴ ἢ ξίφους ἀρκέσει.[328] It is open to wonder how useful the *kopis* would have been in actual combat, as it is only truly effective when swung freely and with great leverage. This might be difficult to achieve in a tightly-packed phalanx, and the sensible suggestion has indeed been made that it was primarily useful for pursuit or 'mopping up', certainly both situations where great slashing blows could be used to terrifying effect.[329]

2.4.3 The spear in combat

The spear was always the mainstay of the hoplite's offensive weaponry. It is usually claimed that the sword was the hoplite's last resort if he lost his spear; and though this may be something of an exaggeration, it certainly contains a kernel of truth. Much like the sword in western European culture, the thrusting spear was *the* weapon for the Greeks, as can be appreciated from the fact that whereas most western European languages have expressions like 'taken by the sword', the Greek equivalent is αἰχμάλωτος. Aischylos neatly summed up the Persian wars as "the

---

326 Both terms are used synonymously in Eur. *Cyc.* 241–243.
327 Cf., e.g., National Museum, Athens (CC1175); a black-figured *oinochoe* previously in the Museum of Fine Arts Boston, MA (99.527); Friedrich-Alexander-Universität, Erlangen (486) See also Webster (1972) 248.
328 Xen. *Eq.* 12.11 ("because from the height of a horse's back the cut of a sabre will serve you better than the thrust of a sword" [trans. Waterfield]).
329 See Anderson (1991) 26–27.

2.4 Offensive weapons

*Fig. 12: A chaotic combat scene representing the sacking of Troy (*Iliou persis*). The two central figures are both armed with a* xiphos, *with clearly seen double-edged leaf-shaped blade, cross-guard and pommel. The dying man, already bleeding from wounds in his shoulder and thigh, is letting go of his* xiphos, *which is slipping from his limp fingers. Notice his scabbard, slung across the torso and shoulder with a baldric. The grip system inside his adversary's shield is extremely clearly displayed. Red-figure kylix by the Brygos painter (c. 480).*

Fig. 13: Apollo in the moment of killing Tityos with a kopis *or* machaira. *The god has brought the weapon back across his left shoulder for a forceful downward slashing blow, displaying the heavy, single-edged sabre-shaped blade with the pronounced tip and the clearly seen ridge line. This is probably very close to how the weapon was used in reality. Red-figure kylix by the Penthesileia painter (C5f)*

Fig. 14: An Iliou persis *scene, portraying confused mêlée scenes. Weapons and armour of both living and dead warriors are beautifully and accurately rendered, with great attention to detail. Notice especially the* linothorakes, *greaves, shield grips and the great* kopis *or* machaira *with which Neoptolemos is about to kill the old Priam in the central scene. Red-figure kalpis hydria by the Kleophrades painter (c. 480–475)*

spear versus the bow"; and Archilochos' hymn to his spear expresses in no uncertain terms that this weapon above anything was what earned him his livelihood as a mercenary.[330] This indicates that the spear was indeed the weapon that was primarily used in hoplite figthing. There are good reasons for this: with a 2.4 m long spear, hoplites had a very good reach indeed and stood a good chance of keeping all enemies at a considerable distance from themselves. Herodotos was greatly preoccupied with the tremendous advantage that the long shafts gave the Greeks against their Persian enemies: thus he claims that the desperate Persians at Plataiai resorted to breaking off the spear-heads with their bare hands because they were 'out-

---

330 Aesch. *Pers.* 239–240; Archil. fr. 2 West: ἐν δορὶ μέν μοι μᾶζα μεμαγμένη, ἐν δορὶ δ' οἶνος / Ἰσμαρικός· πίνω δ' ἐν δορὶ κεκλιμένος ("In my spear I have kneaded barley bread, in my spear Ismaric wine; and leaning on my spear I drink it").

reached' by them; and in a similar vein, Aristagoras calms king Kleomenes' fears about the Persians: ἥ τε μάχη αὐτῶν ἐστι τοιήδε, τόξα καὶ αἰχμὴ βραχέα ... οὕτω εὐπετέες χειρωθῆναί εἰσι.[331] However, spears apparently *were* rather prone to breaking, which is credible enough given their great length.[332] This resulted in a most unwelcome situation if we are to believe Euripides: ἀνὴρ ὁπλίτης δοῦλός ἐστι τῶν ὅπλων ... θραύσας τε λόγχην οὐκ ἔχει τῷ σώματι / θάνατον ἀμῦναι, μίαν ἔχων ἀλκὴν μόνον.[333] Even if Euripides is exaggerating the hoplite's actual helplessness without the spear here, it cannot be denied that hoplites in this situation clearly *felt* strongly deprived.

It has been ingeniously suggested by Hanson that hoplites might then reverse the grip on their spears and thus have an effective, if somewhat shortened, makeshift spear with which to continue fighting.[334] The suggestion has been criticised and the effectiveness of the shortened spear called into question; but it seems logical that a shortened spear would be very handy indeed in the mêlée. Less leverage could certainly be put on the shorter spike, but this would have been compensated for by the great increase in manageability. There is no good reason to suppose that this weapon will have been much inferior once close combat was a reality, and the penetration power of the *sauroter* should not be underestimated.[335] After all, it is highly likely that in the tumult of closely compressed combat, a shorter spear was of greater use, being more mobile and lighter; much like swords were deliberately kept relatively short. Also, it could be used to great effect for unexpected underhand jabs, possibly even delivered below the shield rim: this might certainly cause the kind of hideous wounds to the groin described so vividly by Tyrtaios.[336] It may also explain frequent thigh woundings.[337]

---

331 Hdt. 9.62.2 (inv. no. 29), 5.49.3–4 ("The Persian weapons are bows and short spears ... that will show you how easy they are to beat!" [trans. de Sélincourt]). This is often repeated in the description of the other contingents of the Persian army: 7.61.1, 7.64.1, 7.72.1, 7.77, 7.78 (short spears with long heads), 7.79.
332 Aesch. *Ag.* 60–67; Eur. *Phoen.* 1396–1399; Hdt. 7.224.1, 9.62.2; Xen. *Hell.* 3.4.14; Diod. Sic. 15.86.2, 17.100.6–7; Plut. *Alex.* 16.6, *Eum.* 7.3.
333 Eur. *HF* 190–194 ("A hoplite / is slave to his own arms. Suppose, to right or left, / the next man loses courage, he himself gets killed / through others' cowardice; if he breaks his spear-shaft, how / can he defend himself? He's lost his one resource" [trans. Vellacott]).
334 Hanson (2000²) 84–86, (1991) 72–74.
335 *Pace* Franz (2002) 143 n. 201. Testifying to the effectiveness of the *sauroter* is probably the fact that bronze breastplates with square holes in them have been found (Snodgrass [1967] 56, 80), and also the find of a skeleton from the Theban *polyandrion* at Chaironeia, which had suffered a penetrating wound; see Liston (forthcoming) "The weapon was nearly square in cross section, producing an opening approximately 6 × 8 mm. ... The size and cross section of the hole, and the surrounding flange suggest that this wound was produced by a spear butt-spike."
336 Tyrt. fr. 10.21–27 West; cf. *Il.* 13.567–569.
337 Xen. *Hell.* 7.4.23; cf. Lucian *Tox.* 55. A Corinthian krater from *c.* 600 shows two hoplites fighting over the body of a hoplite lying on the ground, dead or dying, red blood pouring out of a gaping wound in his thigh: Musée du Louvre, Paris (E 635). Thigh woundings were on the whole not uncommon. One of the Chaironeia skeletons examined by Maria Liston has received a sword cut to the anterior surface of the femur above the knee: Liston (forthcoming).

The *ancien militaire* Polybios shook his head in disbelief at the otherwise efficient Romans, who, unlike the Greeks, did not, at least initially, use butt-spikes:

πρὸς δὲ τούτοις ἄνευ σαυρωτήρων κατασκευάζοντες μιᾷ τῇ πρώτῃ διὰ τῆς ἐπιδορατίδος ἐχρῶντο πληγῇ, μετὰ δὲ ταῦτα κλασθέντων λοιπὸν ἦν ἄπρακτ' αὐτοῖς καὶ μάταια ... διόπερ ἀδοκίμου τῆς χρείας οὔσης, ταχέως μετέλαβον τὴν Ἑλληνικὴν κατασκευὴν τῶν ὅπλων, ἐν ᾗ τῶν μὲν δοράτων τὴν πρώτην εὐθέως τῆς ἐπιδορατίδος πληγὴν εὔστοχον ἅμα καὶ πρακτικὴν γίνεσθαι συμβαίνει, διὰ τὴν κατασκευὴν ἀτρεμοῦς καὶ στασίμου τοῦ δόρατος ὑπάρχοντος, ὁμοίως δὲ καὶ τὴν ἐκ μεταλήψεως τοῦ σαυρωτῆρος χρείαν μόνιμον καὶ βίαιον.[338]

It is revealing that Roman spears apparently often snapped after the *first* blow (although it must be admitted that Polybios explicitly complains of their relative structural weakness). An intact spear naturally had greater range, possibly allowing the hoplites of the first ranks to aim at enemies behind the first rank; but a shorter, already broken spear-shaft was likely much less prone to snapping, and thus for yet another reason was ideal for close combat: it had penetrating power combined with a greater reach than any type of sword, yet was considerably less unwieldy than a full-length spear.

It is a question in itself how spears were held. Vase paintings regularly represent hoplites holding their spears in an overhand grip, ready for a deadly downward jab toward the enemy's face or throat. Holding the spear this way seems natural and would have enabled the hoplite to deliver far more power to each thrust than if it were held underhand. Even though such underhand thrusts might prove effective, there are no actual references to them in the sources;[339] and I believe the most economical explanation is that spears were invariably wielded overhand, despite the fact that they were apparently carried underhand (or more likely over the shoulder) *before* the charge.[340] *How* the grip was altered remains a mystery, but it must have been carried out: the manoeuvre in later times even had its own name (ἐκ μεταλήψεως).[341] It would presumably have been impossible to put down the spear (or stick it in the ground) when the phalanx was on the move;[342] and letting go of the

---

338 Polyb. 6.25.6–9 ("Secondly, the butt-end was not fitted with a spike, so that they could only deliver one thrust, the first, with the point, and if the weapon then broke, it became quite useless ... Since this equipment proved so unsatisfactory in use, the Romans lost no time in changing over to the Greek type. The advantage of this was that in the case of the lance the horseman could deliver the first thrust with a sure and accurate aim, since the weapon was designed to remain steady and not quiver in the hand, and also it could be used to deliver a very hard blow indeed by reversing it and striking with the spike at the butt-end." [trans. Scott-Kilvert, modified]). Polybios also has a vivid example of the *sauroter* in active combat: Philopoimen killed Machanidas by first wounding him with the spear-head, then finishing him off with the *sauroter*: Polyb. 11.18.4, 16.33.2–3; cf. Plut. *Arist.* 14.5, *Phil.* 10.7–8; Paus. 8.50.2.
339 *Pace* Hanson (2000²) 84.
340 Anderson's suggestion (Anderson [1991] 31; and see also Hanson [2000²] 84), that spears were first leveled to a sloping position, then to a true underhand grip for the charge, and only in the last moments changed to an overhand grip, seems to me to be unnecessarily complicated and involving far too much risk of getting the weapons entangled or of bungling the complicated sequence of movements. The unsuitability of an underhand thrust because of its weakness is also brought out by Gabriel and Metz' experiments (*supra* 79–81).
341 Polyb. 6.25.9.
342 *Pace* Lazenby (1991) 93.

spear for a moment to catch it again with the grip reversed seems like a recipe for disaster: if the spear was lost, there was no easy way to pick it up again in the marching phalanx.[343] Expressions such as δόρατα καθίεσαν ("they lowered the spears") may merely mean that the spears were brought down from the normal carrying position (sloping over the right shoulder) to the overhand, charging position.[344] If this interpretation is accepted, it follows that spears were in fact never carried in an actual underhand position; but it does not remove the difficulty of how the reversal was achieved: either way, the normal carrying position meant holding the spear so that the thumb pointed towards the spear-head, whereas an overhand grip invariably means holding the spear with the thumb facing the spear-butt. The only remaining possibility seems to be that the spear was transferred briefly to the left hand (holding the *antilabe* of the shield) and then gripped again with the right hand, thumb now towards the *sauroter*. If we are to believe that hoplites' additional spears could be held in the left hand in C7, it is hardly a stretch of the imagination that such a relatively simple manoeuvre could be carried out easily.

Probably the easiest explanation is that the *metalepsis* was carried out prior to the charge, while the phalanx was standing still. In this way it could be carried out efficiently and safely, and without much risk of accidentally wounding fellow phalangites. This may be precisely the source of Plutarch's rendering of the terrifying sight that met the Persians at Plataiai: tens of thousands of Greek hoplites lowered their spears in something approaching unison, so that the first ranks of the phalanx suddenly bristled with spears, lending it all the appearance of a defiant beast.[345] This vivid image also nicely captures the sheer compact mass of the phalanx. It is just possible, then, that the hoplite switched grip by transferring it briefly to the left hand holding the *antilabe*; but the most economical solution to the problem is that the spears were simply brought down at once from the hoplites' right shoulders to an overhand charging and fighting position before the charge itself.

Whether his spear was intact or broken (or replaced by a hastily drawn sword), the hoplite's method of fighting would prove essentially the same at all times: cowering cautiously behind and beneath the rim of his shield, he would attempt to get a quick jab in against his immediate enemy, whether in the first or the second (or perhaps even the third) rank. Every time a hoplite made an attack, he was aware that he would lose his cover, since, as described above, any attack entailed at least a partial and temporary exposure of the hoplite as he twisted his upper right body sideways and forwards.

At the same time, it is difficult to see how accidental woundings of the hoplites in at least the second rank would not occur with butt-spikes moving back and forth as the hoplites in the first rank stabbed away at the enemy.[346] Hardly much in the way of technical prowess with the spear was needed in the phalanx at this point, as hoplites stabbed away more or less blindly at the equally closely packed ranks of the enemy: as Xenophon pithily observed, everybody had his chance as it was actu-

---

343 *Pace* Anderson (1991) 31; Lazenby (1991) 93.
344 Xen. *An.* 6.5.27; cf. 1.2.16–17; Polyaen. 3.9.8.
345 Plut. *Arist.* 18.2 (inv. no. 29).
346 Eur. *Supp.* 846–848; cf. Hanson (2000^2) 86–87.

ally rather difficult *not* to hit something or other in the enemy phalanx.³⁴⁷ Another reason for the quick jab may be the fact that a spear wielded with one hand only lacks penetration power compared to a spear wielded with both hands. Consequently, it is unlikely that sufficient power could be generated with a one-hand thrust to penetrate bronze armour (or shields); and the hoplite was therefore better advised to attempt to hit a vulnerable spot in his opponent's defence (such as the neck or the face) than to try in vain to force his spear through metal or wood which he knew would normally repel the spear easily. When and if such penetrations occurred, it was most likely in the collision, when the two phalanxes met: here, the running hoplite will possibly have gathered sufficient momentum to penetrate either shield or breastplate – if his spear withstood the tremendous pressure, that is. Perhaps Brasidas' anecdotal remark about his shield that "turned traitor on him" is to be understood in this light, although other references can be found to shields and/or breastplates being run through.³⁴⁸

2.4.4 The sword in combat

It is regularly assumed that the sword was an additional weapon; something the hoplite turned to only as a last resort if his spear was broken or lost.³⁴⁹ This belief is warranted to a certain degree in the sources, chief among which is the mention in Euripides' *Herakles* of the hoplite's helplessness if he lost his spear.³⁵⁰ When, in Xenophon's description of the ghastly aftermath of the battle of Chaironeia, emphasis is laid on the fact that swords *without their sheaths* (ἐγχειρίδια γυμνὰ κωλεῶν) were plainly visible on the ground, stuck in bodies and still gripped by dead hands, "does [it] not imply that you would normally have expected to see them still *in* their sheaths?"³⁵¹ Herodotos says that swords were drawn at Thermopylai only when the spears were finally broken: δόρατα μέν νυν τοῖσι πλέοισι αὐτῶν τηνικαῦτα ἤδη ἐτύγχανε κατεηγότα, οἱ δὲ τοῖσι ξίφεσι διεργάζοντο τοὺς Πέρσας.³⁵² This is rather reminiscent of Plutarch's description of the battle at the Krimisos in 339: here, the battle only deteriorated into sword-work *after* an initial bout of δορατισμός.³⁵³

We may assume that swords were drawn only when the spears were broken or lost; or possibly if the lines had become sufficiently entangled and jammed together to warrant the use of much shorter weapons. At any rate, this reinforces the notion that swords were normally employed only after an interval of spear-fighting. Furthermore, it would seem that when Archilochos says that an impending battle with

---

347 Xen. *Cyr.* 2.1.16–17, 2.3.9–11.
348 Tyrt. fr. 12.25–26, 19.17–20 West; Eur. *Heracl.* 685, 738; Xen. *An.* 4.1.18; cf. Plut. *Mor.* 219c (= 190b), *Alex.* 63.5; *Pel.* 2.3; Diod. Sic. 15.87.1. See also Blyth (1982) 17–21 and Salazar (2000) 231–235.
349 See, e.g., Snodgrass (1967) 58; Anderson (1991) 25; Lazenby (1991) 96–97; Hanson (2000²) 165.
350 *Supra* n. 333.
351 Xen. *Ages.* 2.14; Lazenby (1991) 96–97; inv. no. 3.
352 Hdt. 7.224.1 ("By this time most of their spears were broken, and they were killing the Persians with their swords" [trans. de Sélincourt]); cf. Eur. *Phoen.* 1382–1406.
353 Plut. *Tim.* 28.1–4; cf. Diod. Sic. 16.77.4–16.80.

Euboians will be ξιφέων δὲ πολύστονον ἔργον ("the woeful work of swords"), it may suggest something about the embittered and fierce nature of the fighting: normally, spears would be sufficient for the duration of a battle.354

The reasons for the preferment of spears over swords seem evident. Any type of sword had a much smaller range than even a broken spear, and it was far more difficult to get sufficient leverage with a sword. This is reflected by the many quips about the shortness of Spartan swords, which made for discomfort and nervousness, yet, to the Spartans' own mind, and probably to others' as well, proved their fighting bravery and prowess and their inimicable style of fighting extremely close to their enemies.355 The sword certainly had greater versatility than the spear, since it could also be used for slashing; but cuts are far less effective from a killing perspective. Simply put, cuts just do not penetrate deep enough; and they are far less effective with regard to piercing armour. This is because the force of the slash is spread and dissipated over a much greater area than if the full force of the thrust is centred squarely on the point of a spear or a sword. The late Roman military historian Vegetius is quite explicit on this point:

> *Caesa enim, quovis impetu veniat, non frequenter interficit, cum et armis vitalia defendantur et ossibus; at contra puncta duas uncias adacta mortalis est; necesse est enim ut vitalia penetret quicquid immergitur. Deinde, dum caesa infertur, brachium dextrum latusque nudatur; puncta autem tecto corpore infertur et adversarium sauciat antequam videat.*356

Though this is essentially an admonition to teach Roman soldiers to strike with the point rather than the edge, it contains *in nuce* the other principal reasons why the sword was considered inferior to the spear in Greece as well. Since the only real advantage of the sword lay in its cutting ability, and this ability was less needed than the capacity for powerful stabbing thrusts, it was not superior in any important respect. Furthermore, using the sword for slashing blows necessitates drawing the arm up and back, preferably above one's own head, thereby exposing the side and the armpit. The movement is easily noticed and can be quickly anticipated by the adversary who has time to take countermeasures and perhaps even get a quick thrust in himself to the unprotected right side. Furthermore, in a closely-packed phalanx line, the shield would have been difficult to shift to the left; but this would have been necessary in order to reach far enough with the sword: the right side simply needs to be shifted forward somewhat to create enough momentum in the blow.

---

354 Archil. fr. 3.3–5 West (trans. Gerber); cf. Donlan (1970) 138 n. 22; Lorimer (1942) 115. It should be added, however, that hoplites are not directly mentioned here (Anderson [1991] 16–17); but Archilochos' own status as a hoplite warrior otherwise seems to be safely enough established.
355 Reflected in Plutarch's *Spartan aphorisms:* Plut. *Mor.* 191e, 217e, 241f.
356 Veg. *Mil.* 1.12 ("A stroke with the edges, though made with ever so much force, seldom kills, as the vital parts of the body are defended both by the bones and armour. On the contrary, a stab, though it penetrates but two inches, is generally fatal. Besides in the attitude of striking, it is impossible to avoid exposing the right arm and side; but on the other hand, the body is covered while a thrust is given, and the adversary receives the point before he sees the sword" [trans. Clarke]).

The shield will likely have collided with other shields or the bodies of one's comrades and so made cutting blows even more difficult to carry out.

This perhaps also explains why axes and halberds or partisans never proved popular with the Greeks, despite the fact that such weapons had been used to great effect since the earliest history in other militarily advanced cultures, such as that of China:[357] such weapons need quite a lot of space to be swung if they are to function properly. Axes were certainly known as cutting *tools,* but apparently never found favour with warriors.[358] That it was recognised as a potentially powerful weapon may be seen from the fact that axes as weapons are mentioned a few times in the *Iliad*,[359] and that in iconography amazons are regularly represented brandishing them; but outside this mythical context, the only use of an axe as a weapon is the one thrown at Klearchos by an enraged mercenary who just happened to be chopping wood.[360]

It is presumably entirely deliberate on Plato's part that a 'sickle-spear' (δορυδρέπανον) is described so scornfully by the Athenian general Laches in Plato's eponymous dialogue.[361] During a hostile encounter at sea, the *hoplomachos* Stesileos suffered the disgrace of having to see his clever new weapon stuck in the rigging of the enemy ship which sailed safely past, the strange *dorydrepanon* dangling from its rigging. This appears to have been precisely such a 'prototype' of a weapon combining the cutting abilities of the sword with the range of the spear; perhaps something akin to the medieval Japanese *naginata*. The weapon may have functioned properly enough on other occasions – and indeed its failure in this context can be ascribed to no more than a mishap – but the ridicule is palpable, and we get the impression that such new-fangled weapons were not really taken seriously.[362]

In this light, it becomes somewhat difficult to explain the *kopis*. As mentioned above, it had no stabbing proficiency to speak of; but on the other hand the *kopis* was about as effective a slashing weapon as can be imagined; and it possibly actually had the ability to penetrate armour and sever limbs. It has even been suggested that the curved tip might reach over or beside the rim of the shield and be able to pierce helmets and armour, because most of the tremendous kinetic energy of a *kopis* blow would have been concentrated here.[363] The drawbacks outlined by Vegetius remained the same, however, so it must be assumed that the *kopis* was a secondary weapon just as much as the *xiphos* was, the main difference being that it

---

357 Yates (1999) 10, 18–19.
358 Anderson (1991) 24–25.
359 Hom. *Il.* 13.611–612, 15.711. It is telling that the ancient commentators suggest that the axes were for cutting up the Achaian ships: cf. Anderson (1991) 25.
360 Xen. *An.* 1.5.12 and cf. 4.4.16.
361 Pl. *Lach.* 183d.
362 Anderson (1991) 24–25 adduces Athenian vases on which such *dorydrepana* may be represented: they are indeed quite similar to the *naginata*, which was a particularly murderous weapon in the skilled hands of the Japanese samurai of the feudal period; see Ratti & Westbrook (1973) 244–253.
363 Franz (2002) 220–221.

was superbly effective as a slashing blade, rather than a stabbing one,[364] although it was used by other troop types as well, and especially horsemen – it is, after all, for them Xenophon recommends the *kopis* for its qualities in downward slashes.[365]

Possibly it was primarily intended for great slashing blows in the pursuit or otherwise on the retreat: in this situation, when the ranks were broken and there was room to swing it, the *kopis* would be both devastatingly effective and, owing to its terrifying appearance and effect, also a decided psychological asset that might win a pursued hoplite crucial seconds.

2.5 PHYSICAL LIMITATIONS – CONCLUSIONS

2.5.1 The weight of armour

At this point, an examination of the evidence concerning the weight of hoplite equipment and the physical measurements of people in ancient Greece will be of use. Hanson's students at California State University, attempting to recreate ancient Greek armour, have found it "difficult to keep the weight of their shield, greaves, sword, spear, breastplate, helmet, and tunic under seventy pounds [31.82 kg]." This figure is approached by Franz' estimates for the maximum hoplite equipment weight in C6, namely 29.1 kg (though his assessment for the minimum weight in the same period is as little as 18.6 kg).[366] Rüstow and Köchly judged that the weight would be approximately 26 kg, though Blyth sets the weight of a panoply at C5e as low as 12–13 kg.[367] J.F.C. Fuller estimated no less than 32.73 kg for the entire equipment; whereas the latest study, by Jarva, gives 40 kg and 12–13 kg respectively as the extremes for hoplites of the Archaic period (though with the latter indicated as likelier, given a supposed tendency to discard certain elements).[368]

We saw above that the weight of the full panoply was quite considerable: a bell cuirass might weigh anything between 2.3 and 10 kg,[369] while estimates for a linen or a composite corslet range between 3.5 kg and perhaps as much as 6 kg.[370] The presence of metal scales or other reinforcements on such armour will have meant a

---

364 Franz (2002) 222 claims that the *kopis* was not a secondary weapon in that it was so awkward and heavy: accordingly, it must be almost on a par with the spear. He next remarks: "Das Schwert läßt sich also nur insofern als Sekundärwaffe bezeichnen, weil es erst zum Einsatz kam, wenn der Speer nach längerem Gebrauch oder im Nahkampf auch einfach verloren ging." ("Accordingly, the sword can only termed a secondary weapon insofar as it was only employed when the spear snapped after long use or was lost in the mêlée.") It is difficult to see in what *other* ways the weapon could be conceived as 'secondary' than precisely this one.
365 Xen. *Eq.* 12.11.
366 Hanson (2000²) 56; Franz (2002) 348.
367 Rüstow & Köchly (1852) 11–20.
368 Fuller (1946) 37; Jarva (1995) 133–139 (esp. 138–139).
369 Connolly's assessment (for a lined cuirass) is right in the middle, *c.* 6 kg: Connolly (1998²) 58; Jarva assumes some 4 kg with a plate thickness of 1 mm: Jarva (1995) 135.
370 Connolly's reconstructed linen corslet weighed 3.6 kg (Connolly [1998²] 58), and Jarva (1995) 135–136, calculating the thickness of yarn, reaches the result of 4 kg. His attempts to treat linen

considerable increase; perhaps up to so much as a staggering 16 kg and one additional kg for the necessary leather backing.[371] The weight of a shield would typically be 7 or 8 kg (while a leather shield apron might easily add up to 1 kg to the weight of the shield); the helmet according to its type between 1 and *c.* 2 kg. Greaves would typically weigh between 1.2 and 2.2 kg; and for C6 there is also the possibility of added weight from arm-guards, thigh-guards, ankle-guards and foot-guards. The offensive weapons add something like 1.6–2.2 kg for the spear (with additional weight for possible javelins during C7), and the sword (with scabbard) something like 2–2.2 kg.[372] The minimum weight of a full panoply by these estimates, then, lies somewhere between 15 and 25–30 kg, neither estimate counting additional armour or additional spears.

As the great differences in these estimates amply demonstrate, the correct weight of a hoplite panoply at *any* given time is frustratingly difficult to judge, since the degree of decay and corrosion of extant items is subject to no small amount of speculation. However, modern reproductions such as those carried out by the CSU students and by Connolly indicate that most of Franz' assessments are, if anything, somewhat below par; and the corrosion factor, although impossible to gauge sufficiently precisely, should not be underestimated. These considerations suggest that 25–30 kg is probably a more realistic estimate than 15 kg: a 'guesstimate' would be something like 20 kg on average, depending of course on the period and the quality of weapons manufacture.

On the whole, hoplite equipment evolved remarkably little over a period of about 400 years. Certain items underwent minor changes, and almost all were – naturally enough – greatly improved upon; though not so much that they strayed from their initial design or were altered altogether. Almost none were discarded, although that fate befell the javelins that early hoplites apparently carried into battle, and certain elements of the very full early panoply, such as arm and ankle guards. Other, more central, items, such as the shield, remained essentially unchanged during this entire period, which proves the eminent aptitude of the original designs to their purpose and strongly suggests that there were more similarities than differences between hoplite fighting in 700 and 338.

It has been suggested that some percentage of hoplites had *always* been less than fully equipped. Jarva found that there were some 350 helmets, 280 shields and 225 greaves (i.e. *c.* 112 pairs) and only 33 cuirasses from the Archaic period dedicated at Olympia. He put forth the theory that this reflected the actual distribution of armour among the hoplites from whom they were captured.[373] This is in itself problematic: we have no way of knowing what was dedicated why; and we cannot

---

with vinegar and salt, as suggested by Törnkvist, resulted in an increase by as much as 2–3 kg.
371 Jarva (1995) 136, calculating this from 3500 metal scales (each measuring 2.5 × 4.2 cm and – probably – *c.* 0.5 mm thick), found in the Crimea and possibly belonging to the same corslet.
372 Cf. Rüstow & Köchly (1852) 19–20.
373 Jarva (1995) 110–111, 124–128. The sources cited amount to the expression that the men in the front and rear ranks should be the best; nothing is said of equipment. The suggestion has nonetheless been accepted by Storch (1998) 3–7 and van Wees (2004) 47–52.

be certain that everything was always dedicated. Sometimes only the 'best' items were selected for dedication, and quite possibly at other times the victors kept what they liked: after all, plunder was always a part of Greek warfare.[374] At any rate, the conclusion that "in the archaic period about one in three soldiers wore greaves and one in ten a metal cuirass" is not warranted.[375]

Be that as it may, this is unlikely to change our conception of the Archaic phalanx; for on Jarva's own logical suggestion the fullest armour was given to the men in the front ranks, decreasing as we move backwards.[376] Thus on any interpretation the hoplites in the front ranks would *still* be wearing the full panoply; those in direct contact with the enemy, who had to actually use their weapons, would be just as encumbered. That the rear ranks of the phalanx may have been equipped more lightly (or may have worn armour insufficiently glamorous to be dedicated, or too good to 'throw away') does not remove the objection that hoplites actually fighting would do so wearing heavy and restricting armour.[377]

Developments and improvements did mean that the equipment became somewhat lighter over the centuries, however; yet from late C5 it seems that some protective features were voluntarily discarded altogether in a continuous development, in some cases apparently resulting ultimately in complete rejection of all body armour: thus it appears in the iconography that hoplites from this period onwards regularly omitted body armour, and that the Corinthian helmet generally fell out of favour and was replaced with the open-faced and cap-like *pilos*.[378] It is difficult to interpret this in any other way than as a widespread dissatisfaction with traditional hoplite armour and the considerable burden and discomfort it entailed, despite the many improvements over the earliest specimens. The only possible interpretation, then, is that the earlier, 'canonical' versions of the hoplite armour, quite simply were not satisfactory in terms of weight and mobility. This may be linked to new demands in hoplite warfare – other strategic needs and an expanding theatre of military operations – but there is certainly no lack of sources from exactly this period describing hoplite battles fought in the 'canonical' manner: between hoplite

---

374 Jackson (1991) 229 and cf. 230: helmets may have been a cultural euphemism for head-hunting: "perhaps unconsciously helmets were felt to be the least impersonal part of panoply."
375 van Wees (2004) 50. Also, this fails to take into account armour made of organic material.
376 Jarva (1995) 125–128.
377 Hoplite equipment could be quite costly. An inscription in Salamis from C6l requires kleruchs to be able to present hoplite equipment worth 30 *drachmai*: *IG* I² 1.8–10 (= Meiggs-Lewis *GHI*² 14) and cf. Thuc. 8.97.1; Jarva (1995) 148–154; Morgan (2001) 22–24. It is entirely possible, however, that possession of *relatively* cheap equipment might still qualify the owner as hoplite: see especially van Wees (2001) 56–61 and Gabrielsen (2001) 212–217.
378 See especially Anderson (1971) 13–42. Van Wees (2004) 48–50 and n. 6 seems to maintain that hoplites of all periods wore little or no armour; but his two sources (Diod. Sic. 14.43.2–3 and Xen. *An.* 1.2.15–16, 3.3.20) testify to mercenaries about 399, past the Peloponnesian war and within the period discussed by Anderson. Moreover, Xen. *An.* 3.3.20 is explicit that the required cuirasses and corslets are for the newly formed *cavalry* corps, who need a different kind of protection; and I fail to see that the phrase σπολάδες καὶ θώρακες αὐτοῖς ἐπορίσθησαν ("they were provided with leather jerkins and breastplates" [trans. Warner]) warrants the interpretation "barely a few dozen [corslets] could be scraped together".

armies of roughly equal size, massed in closely-packed phalanxes, and in a recognisable, recurring pattern of phases.

It follows *a posteriori* from this that equipment prior to these developments was *not* easily manoeuvrable, and that it did, in point of fact, necessitate a static, defensive form of combat. It is unrealistic to assume that hoplites were at any point free to move about as they saw fit on the battlefield: not only the weight of armour and weapons, but also the incredible awkwardness of it, ensured that hoplite battles were not enacted as series of loosely connected duels. Hoplite arms and armour were designed specifically for close fighting in a massed formation, where every hoplite would profit from standing close to his neighbour, and where the unwieldiness of his shield and the severe limitations imposed upon his sight and hearing by his helmet would matter far less than strong frontal protection.

We have seen that the inclusion of hoplite armour in hoplite race and *pyrrhiche* are not sufficient to warrant the assumption that it was particularly 'light', and this is reinforced in no small degree by what we know can be done and what cannot be done with much lighter, modern police shields. The idea that hoplites were able to roam about freely, wielding their shields with perfect ease and singly taking on attackers from multiple angles, presupposes that the hoplite shield could be handled in such a way; but as we have seen, this is a fiction. The heavy shield would wear out the strongest arm in a matter of minutes, and this would be exacerbated by the considerable weight of the rest of the equipment which the hoplite had to carry as well. A sagging *hoplite* shield must be twisted to the left and sideways, because of the double-grip, leaving the bearer's front more or less exposed; a completely untenable position in an open battle order. A closed battle order, on the other hand, would mean that the shield rim could be safely hung on the hoplite's shoulder and be held still in this position more or less constantly, as he profited from the lateral cover afforded him by his right neighbour.

2.5.2 Physiology

Another factor worth considering is the physical characteristics of the men actually wearing the armour. When theories are put forward about what could and could not be done with hoplite armour, the tacit assumption must be that Greek males of antiquity are immediately comparable to moderns. These questions have great bearing on the understanding of the 'relative' weight of armour; yet the problem has seldom, if ever, been brought up by scholars. It is therefore well worth examining the available data supplied by skeletal remains from the Archaic and Classical period.

J.L. Angel, who has examined skeletal remains exhumed in Attika, has put the average height for the ancient Greek male at 162.2 cm (and the female at 153.35 cm), though this is based on rather few skeletons: 61 male and 43 female skeletons from Attika, as against a total of 225 dateable males and 132 females.[379] Similar

---

379 Angel (1945) 284–285 and n. 25. The male skeletons dating from the Classical period – a mere three – are on average 165.4 cm high: Angel (1945) 324. Foxhall & Forbes (1982) 47 warn

results accrue from Angel's 1944 analysis of all known Greek skeletal remains: here, the result is given as 162.19 cm for males, with a range between extremes of 148 and 175 cm. The result for females overall remains the same, 153.35 cm.[380] Angel, whose interest is primarily 'racial' analysis, lists crania from Attika, Boiotia, Corinthia and Macedonia; but unfortunately he does not indicate the distribution of more complete skeletons which may have formed the basis for the calculations.[381] Nevertheless it must be assumed that the average measurements represent the average geographically as well as chronologically.[382]

The comparatively small number notwithstanding, we would be well advised to keep in mind that "this sample may be biased in favour of higher socio-economic groups since it is the graves of the comparatively wealthy that are most likely to receive attention from archaeologists."[383] If this position is accepted, it follows that the average Greek male was probably less well nourished, and the skeletons examined by Angel may well belong in the absolute upper percentile.[384] According to Donlan and Thompson, however, the average height of Greek males in the Classical period was approximately 170 cm, with body weight between 65 and 67 kg.[385] Unfortunately, in their article no information is given about how these results were arrived at; so Angel's data must assume priority. The average modern European or American male, on the other hand, measures *c.* 179 cm.[386] Weight depends to a large extent on age (not to mention a host of other factors); but the average weight of modern males between the age of 20 and 60 is 80.97 kg.[387]

Men in antiquity were shorter and quite likely rather lighter, considering a diet largely consisting of cereals and vegetables.[388] This should be remembered when

against the statistic insignificance due to the relatively small sample material. Insufficient though it may be, however, there is no other way to estimate bodily proportions of ancients, so the material at hand will simply have to suffice.

380 Angel (1944) 334, table II a.
381 Angel (1944) 331, table I.
382 Garnsey (1999) 57–59 cites a number of analyses of ancient Roman skeletal material from, among others, Pompeii and Herculaneum. He circumspectly concludes that all that may be said for the average male height here is that it is "for the most part within the range 162–170 cm for men and 152–157 for women".
383 Foxhall & Forbes (1982) 47 n. 21.
384 Maria Liston points out that teeth from several skeletons excavated under the Stone Lion monument at Chaironeia bear signs of linear enamel hypoplasia, an indicator of systemic stress during childhood; the reasons include severe malnutrition or illness. "The presence of these lines in multiple individuals indicates that even for the future military élite, childhood could at times be stressful in ancient Thebes": Liston (forthcoming).
385 Donlan & Thompson (1976) 341 n. 4; the 170 cm apparently repeated in Stewart (1990) vol. I 75. Hanson posits some 1.67 m (Hanson [1991] 67–68 n. 14).
386 http://en.wikipedia.org/wiki/Human_height. The figure is the average of German, Dutch, British and American males.
387 http://www.halls.md/chart/men-weight-w.htm.
388 Foxhall & Forbes (1982) and Garnsey (1999) 17–21 suggest that 70–75% of the total consumption was cereals. A post-war survey carried out in Crete demonstrated that no less than 29% of the calorie intake was olive oil: Allbaugh, cited in Garnsey (1999) 19 n. 12. Even for heavyweight boxers it was rare and apparently deserving of mention to stray from the staple diet: Harris (1966) 88–89 relates the few instances; cf. Waterlow (1989) 6–9.

discussing the burden of hoplite armour: Greek men in antiquity were significantly smaller than is the case today, and consequently what seems comparatively light or small to us may have been quite a lot heavier and more cumbersome for smaller men to bear. Weapons will have been even more uncomfortable and unwieldy to men possibly as much as 15 cm shorter than the modern Western average; and the shield, some 90 cm in diameter, will have been even larger for such comparatively small men. This may help explain the presence of a seemingly larger shield, such as that mentioned by Tyrtaios, which covers μήρους τε κνήμας τε κάτω καὶ στέρνα καὶ ὤμους ("thighs, shins below, chest and shoulders"): "if the diameter ... was 3 ft (less than a metre), we can readily imagine only 2 ft 6 ins (76 cm) of unprotected flesh below and above the rim on a 5 ft 6 ins (1.67 m) crouching hoplite".[389]

Indeed, even the lightest possible of Greek panoplies will have been a disproportionately heavy burden to bear when compared to the physical norm of the average modern Western male. All this serves to underline the fact that hoplite weapons were not by any stretch of the imagination easily or comfortably manipulated or worn.

It should also be remembered that *polis* armies were normally comprised of citizens between the ages 18 and 60.[390] Due to this extremely long military service, there must have been quite many older men in the phalanxes, and perhaps even a majority: "after all, thirty of forty-two age classes liable to military service were composed of men over thirty years of age".[391] It should be taken into consideration, however, that these age groups will have been exponentially depleted of members because of the increasing mortality resulting from both natural causes and participating in more battles. References to older men fighting in the rank and file are strewn throughout the sources,[392] and, all things being equal, the weight of arms and armour must have been much harder to bear for men pushing sixty.

The sources to a large degree confirm this. Personal servants (ὑπασπισταί) normally carried the hoplites' weapons[393] and supplies[394] (apart from less appealing tasks such as picking up dead and wounded[395]); at least this was the case for higher-ranking and wealthier persons. Thucydides' description of the Athenian retreat from Syracuse in 413 is a key passage.[396] Even when hoplites did hold their own shields, they generally did not pick them up until the *last possible moment:* the command θέσθαι (τὰ) ὅπλα means to set down the shields (and spears), ready to be picked up

---

389 Tyrt. fr. 11.23–24 West; Hanson (1991) 68 n. 14. Cf. Lorimer (1947) 106.
390 Xen. *Hell.* 6.4.17; [Arist.] *Ath. pol.* 53.4, cf. *IG* II² 1926.
391 Hanson (2000²) 90.
392 Holoka (1997) 342; Hanson (2000²) 89–95.
393 Hdt. 7.229.1, 5.111; Xen. *An.* 4.2.20 (Xenophon's personal *hypaspistes* runs away with his shield, leaving him in a tight spot), *Hell.* 4.5.14, 4.8.39; Polyaen. 2.3.10; cf. Lazenby (1991) 89.
394 Hdt. 7.40.1; Thuc. 2.79.5, 4.101.2, 7.78.2; Xen. *Hell.* 3.4.22, *Cyr.* 5.3.40, 6.3.4.
395 Hdt. 6.80, 6.81, 9.80; Thuc. 6.102.2; Xen. *An.* 2.1.9, *Hell.* 4.5.14.
396 Thuc. 7.75.5; cf. 7.13.2. For other references to *hypaspistai*, see Hdt. 7.229.1, 9.10, 9.28–29; Thuc. 3.17.4, 4.16.1, 4.94.1; Xen. *Hell.* 6.4.9; Is. 5.11; Dem. 54.4; Antiphanes 16 Kassel-Austin; Theophr. *Char.* 25.4.

again quickly.[397] So well-known was this 'at ease' position that Chabrias, according to later historians, on one occasion could show contempt for the advancing enemy by letting his men stand brazenly like that as the enemy approached.[398] Even the Phokians guarding the Anopaia pass at Thermopylai – who, it must be supposed, should have been on maximum alert – only picked up their weapons *after* they saw the Persians approaching, surprising the enemy with the strange sight: ἀνά τε ἔδραμον οἱ Φωκέες καὶ ἐνέδυον τὰ ὅπλα, καὶ αὐτίκα οἱ βάρβαροι παρῆσαν. ὡς δὲ εἶδον ἄνδρας ἐνδυομένους ὅπλα, ἐν θώματι ἐγένοντο.[399] The verb ἐνέδυον ("put on", rather than, e.g., ἀνέλαβον "picked up") implies that they did not even put on their body armour until the last possible moment.[400] In Euripides' play *Herakleidai*, the weight of a "full set of armour" (ὅπλων παντευχίαν) seems to be a consideration. Iolaos' servant advises the old man to put it on in a hurry, since battle is near, but adds εἰ δὲ τευχέων φοβῇ βάρος, / νῦν μὲν πορεύου γυμνός, ἐν δὲ τάξεσιν / κόσμῳ πυκάζου τῷδ'· ἐγὼ δ' οἴσω τέως. Apparently relieved, Iolaos accepts (καλῶς ἔλεξας).[401] If anything, the problem of weight was reduced over the centuries, as more and more items were discarded: it is noteworthy that the complete hoplite equipment was heaviest and most cumbersome in the earliest stages. Interestingly, the Euripides passage shows that even by 430, at a comparatively late stage in the development, hoplite armour was still thought of as a serious burden.

---

397 Hdt. 1.62.3, 5.74.2, 9.52; Thuc. 2.2.4*bis*, 4.44.1, 4.68.3, 4.90.4, 4.91, 4.93.3, 5.74.2, 7.3.1, 7.83.5, 8.25.4, 8.93.1*bis*; Xen. *An.* 1.5.14, 1.5.17, 1.6.4, 1.10.16, 2.2.8, 2.2.21, 4.2.16, 4.3.17, 4.3.26, 5.2.8, 5.2.19, 5.4.11, 6.1.8, 6.5.3, 7.1.22*bis*, *Hell.* 2.4.5, 2.4.12, 3.1.23*bis*, 4.5.8, 5.2.40, 5.3.18, 5.4.8, 6.4.14, 7.3.9, 7.5.22; Diod. Sic. 11.5.4, 12.66.2, 14.105.2, 18.26.4, 18.61.1, 20.42.5, 20.88.8.

398 Diod. Sic. 15.32.5; Polyaen. 2.1.2. The Athenians later erected a statue of Chabrias in just this position in the Agora, the base of which has probably been found: Arist. *Rh.* 1411b 6–10; Anderson (1963) 411–413.

399 Hdt. 7.218.1–2 ("Leaping to their feet, the Phokians were in the act of arming themselves when the enemy were upon them. The Persians were surprised at the sight of troops preparing to fight" [trans. de Sélincourt]).

400 Cf. Thuc. 6.69.1; Xen. *Hell.* 4.8.37–39; Plut. *Pel.* 32.3.

401 Eur. *Heracl.* 720–726 ("However, if you dread the weight of it, / go unarmed for the present, and when you reach the ranks / put all this on there. Meanwhile I'll carry it" [trans. Vellacott]).

3. THE PHALANX

3.1 THE DEVELOPMENT OF THE PHALANX: MYTH OR REALITY?

3.1.1 The development of phalanx fighting: an overview

Phalanx fighting naturally did not appear out of nowhere: there has to be an origin. Due to the complete lack of sources from the dark ages, however, there has been much speculation about the appearance of this new aspect of Greek warfare. Since the hoplite phalanx was seemingly a radical innovation, it has been suggested that the transition occurred suddenly and was caused by the invention and swiftly following introduction of the weapons and equipment especially suited for phalanx fighting. This was formulated quite clearly by Lorimer, who claimed that

> the Carians, if they invented the πόρπαξ shield, must also have invented the cohesive tactics of which it was to be the instrument. ... Moreover, its spread over the Peloponnese must have been rapid, for when some new military device has been tried with success by one power, it is necessarily adopted by such other communities as may be called on to encounter it in use by the first.[402]

Lorimer thus claimed logical and chronological priority for the invention of the double-grip shield: in her analysis, the invention of the shield was the cause and the phalanx the effect, as it were. The time of transition would then be roughly the beginning of C7. This fits well with early iconographical representations of phalanxes; and there is therefore to this day widespread acceptance that this was the time the hoplite phalanx first emerged. In fact, this dating has become so entrenched that at one time an attempt was made at establishing a causal connexion between the appearance of the early hoplite phalanx and the rise of the Greek tyrants: these, it was argued, wrested political power from the nobility by courting the favour of the rising hoplite segment.[403] In Lorimer's view, accordingly, the phalanx was a logical consequence of hoplite weaponry, and hoplites were inseparably linked to the phalanx: without the phalanx, no hoplites either. However, there are very serious problems to this approach, the greatest of which is the chain of causality.

In opposition to this, Anthony Snodgrass put forth the theory that the weapons may have been developed over a longer time, rather than suddenly invented 'out of

---

[402] Lorimer (1947) 107–108; cf. (1950) 462: "Apart from the phalanx the hoplite was nothing; the phalanx therefore came together if it came at all." The connexion between hoplite weapons and Karians is tenuous at best, as hinted at in the quotation. Lorimer herself is sceptical, and Snodgrass (1964b) has run the notion of Karian origins decisively into the ground.

[403] Andrewes (1956), esp. 31–42; Forrest (1966) 112–114; Salmon (1977) 95; Snodgrass (1980) 111–113; Bryant (1990) 499 and cf. Schuller (1982²) 17. See also the particularly cautious approach in Cartledge (2001) 160–161.

the blue' by some military genius. In support of this, he argued that all the individual elements that make up the hoplite 'panoply' (with the exception of greaves) are represented in Greece for quite some time before 700, "though not probably earlier than *c.* 750",[404] even though the first iconographical representation of a warrior in full panoply can be seen on Protocorinthian vases only from *c.* 675, about the same time as the earliest acknowledged phalanx representations. Snodgrass argues that

> the chance statements of contemporary [i.e. C7] poets, and the evidence of the only relevant grave-group, both support the *prima facie* evidence of the artists: that the adoption of the 'hoplite panoply' was a long drawn out, piecemeal process, which at first did not entail any radical change in tactics.[405]

Especially crucial to Snodgrass' analysis of the sources is "the only relevant grave-group", namely the Argos grave finds (excavated after 1953 and thus not accessible to Lorimer). These grave finds, as we have seen, included weapons of a type that might be called 'proto-hoplitic', above all an early specimen of the 'bell' cuirass and an 'Illyrian' helmet.[406] The assumption therefore is that what was essentially hoplite weapons are attested before 700, thereby disproving the causality theory offered by Lorimer.

However, setting aside the argument from iconography for a moment, there is only the Argos armour finds and some Archaic poetry theoretically linking hoplite weapons to non-hoplite fighting, i.e. fighting without phalanxes. Furthermore, as we shall see, the various C7 poets – above all Tyrtaios, Kallinos and Mimnermos – cannot be cited to support the theory that something fundamentally different from phalanx fighting is meant. Quite the contrary: the evidence in Kallinos and Mimnermos may possibly be too tenuous to enable definitive statements on their behalf, but Tyrtaios certainly nicely captures the *polis* ethos and the realities of hoplite phalanx combat.

Of Snodgrass' three source types, it is thus only iconography which may reasonably be claimed to link hoplite weapons and non-hoplite fighting. Nonetheless, no one, as far as I am aware, has suggested the solution that the phalanx itself, and not merely hoplite weapons, might instead be pushed back to *c.* 750.[407] Yet it is my contention that Lorimer was essentially correct in assuming a close connexion betwen hoplite armament and close-order tactics, though not in her analysis of causality. Iconography aside, there is no good reason whatsoever for rejecting her solution: the alternative is forcibly separating weapons and tactics, only to push the introduction of the weapons some fifty years back. If hoplite weapons and phalanx formations are inseparably connected, the Argos finds must necessarily supply the *terminus ante quem* for both phenomena. Franz criticises Lorimer for vagueness of definition (*Begriffsunschärfe*) in her use of the term 'phalanx', because she employs it in a sense identical to the *Classical* phalanx.[408] However, hers is a logical

---

404 Snodgrass (1965) 110.
405 Snodgrass (1965) 110.
406 *Supra* 66–67.
407 Hanson (2000²) xxxi dates the "introduction of hoplites" to 700–650; cf. (1991) 63–68.
408 Franz (2002) 3–4.

way of thinking, since such is indeed the accepted and normal concept of 'phalanx': a discussion about any other theoretical battle formation (whether including hoplites or not) would do better to employ another term in order to avoid confusion and, indeed, *Begriffsunschärfe*.

Apart from the problems entailed in separating hoplite weapons and tactics, Snodgrass' approach is certainly theoretically and methodologically sound: the introduction of the 'piecemeal' theory, as it has come to be known, is a far more realistic way of addressing the problems of the early phalanx, even if Lorimer's 'sudden change' theory still has advocates.[409] A 'third way' – essentially a further extension of the 'piecemeal' theory – has recently been suggested by van Wees, namely that the evolution of phalanx tactics continued throughout the Archaic period and was not completed until perhaps as late as the Persian wars.[410]

Hanson's analysis of the usual causal process of inventing and adopting new military equipment seems essentially correct. The logical (and causally plausible) explanation is the reverse of that suggested by Lorimer: hoplite weapons were invented to suit the needs of massed fighting, already in existence.[411] Hanson's striking analogy with modern practice is worth citing *in extenso:*

> There seems always to be a symbiotic relationship between ancient Greek tactics and armament; usually the former determines the latter. ... Modern methods of procurement also reflect this sequence. The Pentagon usually publishes criteria for new weapons systems based on their own evolving particular tactical and strategic needs; the defense industry then develops technology that meets those military requirements. ... For example, fighter aircraft in the First World War were developed in response to pilots who desired a new technology superior to the aerial exchange of revolver and rifle fire: battle in the air antedated the appearance of true fighter aircraft. No one would suggest that air combat grew out of the discovery of novel aerially mounted automatic weapons.[412]

The keyword to this striking illustration is of course *evolution*. It is natural to assume that the heavy, cumbersome, awkward and uncomfortable hoplite weapons were developed as a response to very specific needs on the battlefield. Consequently, a careful analysis of the weapons themselves may be the best way to catch a glimpse of what fighting was truly like.

If we accept this analysis of the usual correlation between armament and tactics, it follows that phalanx tactics (or phalanx fighting) already existed when the hoplite shield and other items in the 'panoply' were introduced. On this interpretation, it is actually possible to combine the 'piecemeal' and the 'sudden change' theories, in a manner of speaking: provided that pre-existing phalanx tactics are ac-

---

409 The term 'sudden change' was coined by Cartledge (1977) 19–20, who professes to "subscribe" to the sudden change theory himself. The majority have followed Snodgrass: see, e.g., Salmon (1977) 90–92; Latacz (1977) 237–238; Donlan (1970) 137 n. 16; Hanson (1991) 63–64; Schwartz (2002).
410 van Wees (2000) 125, 134 ff., 155–156 f. His suggestion has been adopted by Krentz in a recent article (Krentz [2002] 25). A stance somewhere between this and the traditional view is taken by R.H. Storch (1998) 6–7. Storch argues in favour of a 'mixed' phalanx in which the rear ranks were not so completely equipped as the (wealthier) front ranks.
411 Hanson (1991) 68–69, 77 n. 28.
412 Hanson (1991) 64 and n. 5; cf. Latacz (1977) 237–238; Lorimer (1947) 107–108.

cepted, it is possible to argue that such technological breakthroughs as the double-grip shield may have been sudden inventions, and that they quickly found acceptance once introduced.[413] It seems certain, however, that the end result – a completely re-equipped hoplite phalanx – was not brought about immediately. Such a change would only occur after some time, as the necessary measures would naturally not be immediately put into effect: the invention would necessarily have taken time, possibly generations, to be put into effect.[414] Communications and exchange of ideas were comparatively slow in all respects, so the spread of the new technology will have taken some time. This effect will have been reinforced by the fact that the administration of armies (city military organisations) were not centralised, top-down organisations, and the initiative to adopt the new armament was therefore in all likelihood only private, that is to say: warriors adopted the armament individually and according as it suited them and their economy. The new weapons technology, as is the general rule in such cases, was most likely quite costly. To this should be added the process of learning the optimal use of the new weaponry.

Naturally, there will have been a powerful incitement in the very effectiveness of the new weapons and in their tremendous appropriateness for the kind of combat for which they were invented; but all the same, there must have been a certain phase of transition. It is impossible to determine just how long this phase was; but we cannot rule out the possibility that the change was in fact comparatively swift.

3.1.2 The Homeric poems

The dating of both composition and action of the Homeric poems is a hotly debated topic. Despite glaring linguistic and historical inconsistencies, it is a widely accepted belief that the society of 'Homer' as portrayed in the *Iliad* and the *Odyssey* was very real and that it should be placed in the iron age (more specifically in C8) and that the *Iliad* in its present state saw the light of day *c.* 750: the unitarian view of 'Homer', despite the findings of, *inter alios,* Parry and Lord,[415] still makes its presence felt.[416] This is in no small part because of Moses Finley's highly influential work *The World of Odysseus,* in which Finley argued in favour of a Homeric society and essentially paved the way for the re-emergence of a universally ac-

---

413 This is hinted at by Cartledge (1977) 20 and n. 71: "As an invention for use in pre-hoplite warfare the hoplite shield would not merely have been barely (if at all) superior to its single-handled predecessors but also in certain circumstances positively and dangerously inferior. ... [T]he Greeks *invented* the double-grip shield: why should it not have been invented with the phalanx in mind rather than the other way round? ... [W]hat the invention of the *porpax* and *antilabe* tells us is that concern for protection in the front was outweighing the need for manoeuvrability and for protection in the flank and rear – in other words, that a change in tactics in the direction of organised, hand-to-hand fighting was *already* in progress."
414 Latacz (1977) 238.
415 Lord (1960) 30–68; Parry (1971) 321.
416 For recent scholarship accepting the existence of a historical Homeric society, see Raaflaub (1997) 625–628 and also (1998); Osborne (2004), esp. 206–211. West in *OCD*³ s.v. "Homer" posits "some agreement" about dating the composition of the *Iliad* and the *Odyssey* to *c.* 750 and 725 respectively.

cepted unitarian view of Homer.[417] There may be general consensus that the poems are the legacy of a rich and well-established oral tradition; but nonetheless it seems that this view is seldom pressed to its logical conclusion. One cannot help but feel that chief among the reasons for this is a more or less conscious reluctance to acknowledge that poetry of such consummate beauty could rise from such a mishmash of ideas of (necessarily) lesser spirits, representing a brutal process of trial and error over the centuries, that literature at this level could in fact be the product of mere evolution.[418]

These powerful trends notwithstanding, Minna Skafte Jensen has convincingly re-analysed and interpreted the introduction of writing and Greek oral poetry in the Archaic period, and has argued plausibly that the date of final composition and commission to writing should in fact be placed as late as the reign of Peisistratos in Athens (c. 560–527).[419] It is entirely possible that the poems were actually composed about C8m – quite possibly by a poetic genius of the highest order – and that they essentially found their form in this period; but it was not until they were fixed in writing that the poems were 'frozen'. The distant and mythical 'dramatic date' is ostensibly as early as C13 (to the extent that it is meaningful to speak of an historical event at all), since the Mycenaean palaces in Greece show traces of having met with a violent end at this point. In other words, there is a possible interval of at least 450 years (and possibly, following Skafte Jensen, as much as 600 years) between the composition of the poems and the mythical past in which they take place.

Snodgrass' careful examination of both Homeric poems reveals many incongruencies in the social sphere: three drastically different and to a certain extent mutually exclusive marriage institutions (dowry, 'bride-price' and 'indirect dowry') confusingly seem to coexist in the poems; and the same applies to certain fundamental defining characteristics of two widely differing types of African and Eurasian society: both seem to be represented in the *Iliad* and the *Odyssey*.[420] At an even more fundamental level, there are insurmountable difficulties with phenomena such as metallurgy. Spears and swords in the poems are invariably made of bronze, a curious phenomenon if the action is indeed thought to take place after the bronze age. It is even more curious because Homer obviously knows and often enough refers to other household tools made of iron, which historically speaking came to be made of this material later than did weapons (and which was not known at all at the time of action).[421] It is also revealing that scholars disagree on the burial practices found in Homer: Finley found that "the *Iliad* and the *Odyssey* remain firmly anchored in the

---

417 Finley (1979²) 48–50 even argued for a historical Homeric society in C10–9.
418 See, e.g., Franz (2002) 24: "Homer hat in der Ilias und der Odyssee einen über mehrere hundert Jahre mündlich überlieferten Sagenstoff aufgegriffen, diesen in einer ganz besonders kunstvollen Art verarbeitet und schriftlich fixiert" ("In the *Iliad* and the *Odyssey*, Homer has treated subjects from myth, orally transmitted through centuries, shaped it into a piece of highly refined art and fixed it in writing").
419 Skafte-Jensen (1980), especially 96–103 and 128–171. *Contra* Morris (1986) 91–92, in my opinion unconvincingly.
420 Snodgrass (1974) 115–121.
421 Snodgrass (1974) 122.

earlier Dark Age on this point", whereas Kurtz and Boardman hold that the Homeric version "is wholly in keeping with Geometric and later Greek practice", two statements that cannot possibly be reconciled.[422] Snodgrass nicely sums up the ambiguities:

> Those who maintain that Homeric society is unitary and historical are bound to ask themselves the question, to what time and place that society belongs. The two answers which might seem, *prima facie,* to be the likeliest, can be shown to be improbable on other grounds: namely the historical period in which the story of the poems is ostensibly set, the later Mycenaean age, and the period in which the poems reached their final form and in which the historical Homer most probably lived, the eighth century B.C. ... A purely contemporary origin, though it may not be excluded by the ubiquitous and pervasive presence of formulae, affecting social life as much as other aspects, would surely be in utter conflict with the other evidence that we have for eighth-century society, from Hesiod and from archaeological sources. It is a surprise to encounter such primitive features as bride-price and polygamy in Homer at all; that they should have been taken, as normal features, from the Greek society of his own day is almost unthinkable.[423]

On a societal level, other problems make themselves felt. Homeric cities regularly feature both kingly palaces and temples;[424] yet while palaces were a regular feature of Mycenaean cities, temples were unknown. On the other hand, temples "were the most conspicuous type of monumental architecture from 700 B.C. onwards", while palaces all but disappeared after C12–11.[425]

Furthermore, both the *Iliad* and the *Odyssey* contain an absolute wealth of linguistic diversity, combining the most diverse dialects. Ionic, Lesbian, Aeolic and Arkado-Cypriotic can be found amongst each other, just as words and phrases of immense age coexist with expressions which were quite new at the time of composition. This complex mixture of different dialectal forms and different historical layers reflect the poems' origin in oral composition and transmission, and between them make up the so-called Homeric *Kunstsprache*.

The logical explanation to this seemingly hopeless jumble is the simple fact that the Homeric poems, for all their beauty and rich thematic contents, are essentially the products of a well-established oral tradition, with all the compository problems that this entails. The very fact that it is even possible to debate whether the 'Homeric society' is a bronze age or an iron age society not only provides an understanding of the turmoil of Homeric scholarship, but also exemplifies the very serious problems inherent in trying to date anything using the *Iliad* and the *Odyssey* as an historical source: the very large interval between the dramatic date and the date of composition (whether C8m or C6m) has resulted in a number of discrepancies in the poems.

It is therefore not unreasonable to assume that although (or perhaps precisely *because*) the poems deal with events from a distant and legendary past, details from the societies and cultures of the intervening periods have crept in by degrees; and

---

422 Quoted in Snodgrass (1974) 122–123.
423 Snodgrass (1974) 121. It has often been claimed that it is unrealistic to posit an historical Homeric society: see also Kirk (1962) 179–210.
424 Palaces: *Il.* 6.242–245; *Od.* 1.365–366, 3.386–389, 4.20–25, 7.78–83. Temples: *Il.* 1.37–42, 5.445–446, 6.297–300, 7.81–86; *Od.* 6.7–10.
425 Hansen (2000) 146b.

gradually and imperceptibly they have become inseparable parts of the poems. Rhapsodes of the intervening periods have interpreted and re-evaluated existing parts, jettisoning parts as they were no longer needed or did not make sense any more, and conversely adding phrases and verses which reflect the combat experience of their own time. These verses have then been absorbed into the 'storehouse' of the collective tradition.[426] Mogens Herman Hansen has aptly compared the poems to a coin-hoard: "the latest coin dates from a few years before it was buried whereas the oldest may have been struck centuries earlier."[427]

3.1.3 The *Iliad* and hoplite fighting

In 1977, Joachim Latacz published a seminal study carefully analysing and comparing the *Iliad* and the war poems of Tyrtaios and Kallinos.[428] He demonstrated convincingly that the language of the *Iliad* contains allusions to close combat between organised forces in closed formations, and he showed that though the epic narrative structure demands duels between protagonist heroes, there is in reality a narratologically subdued but crucial element of *Massenkampf* in between. On Latacz' interpretation, this is an attempt to reconcile the needs of the epic genre with the reality of the poet's contemporary (C8) warfare, so that the duels are in fact best viewed as combat between πρόμαχοι, i.e. the bravest warriors of the front rows of the opposing formations, exemplifying or crystallising elements of massed fighting.[429] They thus actually fight each other simultaneously, even though the 'duels' are represented in the narrative as following each other sequentially. The actual importance of the massed fighting may be gauged by the fact that there are no fewer than 23 instances of protracted and clearly described, decisive massed fighting, scattered throughout the *Iliad*.[430] This comes out especially clearly when we consider that the word *phalanx* in fact appears time and again in the *Iliad* (often used in the plural about one and the same formation): Latacz shows that the word *phalanx* means simply 'line' or 'file', quite on a par with the word στίξ; and these two terms are apparently used interchangeably in the *Iliad*.[431]

---

426 An excellent example is the formulaic verse αἵματος ἆσαι Ἄρηα ταλαύρινον πολεμιστήν ("glutted with his blood Ares, the warrior with tough shield of hide" [trans. Murray]), *Il.* 5.289, 20.78, 22.267; cf. 7.239, in which the word ταλαύρινος ("carrying a shield of oxhide"), used about the war god Ares, is evidently a remain from at least the bronze age: cf. Leumann (1950) 196–202. Ruijgh has argued that the formula βοῶπις πότνια Ἥρη ("the ox-eyed queenly Hera" [trans. Murray]) goes back all the way to Mycenaean times, while West suggests that "Mycenaean heroic poetry was cast in hexameters from at least the fourteenth century" (quoted from Edwards [1997] 267–269).
427 Hansen (2000) 146b.
428 Latacz (1977).
429 Raaflaub (1999) 133–134 shares this view and calls attention to the old age of massed fighting before the advent of the hoplite, and see de Jong (1987) 114–115.
430 For an overview and greatly detailed analysis, see Latacz (1977) 178–209, esp. 179–184.
431 See especially *Il.* 4.254, 4.280–282, 4.331–335, 4.422–426, 5.92–94, 6.83–85, 7.54–66,

The word φάλαγξ itself is very ancient and of Indo-European descent (from the stem *bhᵒl-ə₂-g-). In Greek, aside from the 'normal' use, it also carries the meaning 'cylindric piece of wood', a 'beam' or 'balk': in fact, the root is the same as that of the latter. Singor has attempted to connect the military sense of the word with the root meaning of 'wooden cylinder' by suggesting that this refers to spears or spearshafts, but admits that φάλαγξ is nowhere attested in this meaning.[432] Franz argues that since φάλαγξ is not found in the meaning 'roll' or 'balk' until Herodotos, whereas it appears repeatedly in a battle context in epic poetry, the original meaning must be military. This is an argument *e silentio*, however. Chantraine maintains, in my opinion correctly, that the situation is the reverse. The simpler sense of 'balk' must retain chronological priority over the other, better known one: "bien que l'acception militaire de φάλαγξ soit plus anciennement attestée, elle est sûrement métaphorique, donc secondaire."[433] Corroborating this is the fact that its cousins in the Germanic languages (German *Balken;* Danish *bjælke;* Old Icelandic *bjalki*) are directly related, via the form *belkan-; and so, of course, is the Latin word *palanga*. All of these carry the exact same meaning: that of 'beam' or 'balk', suggesting that this is indeed the root meaning.

The solution to this apparent problem is to interpret φάλαγξ as being transferred into a military context metonymically from the perception of the motion of a beam or roller; something that rolls forward with irresistible force, crushing or levelling all before it. On this interpretation, then, the etymology of φάλαγξ is an argument for the preexistence of closed-order tactics of a very old age.[434] Franz rejects this interpretation, again on the ground that it is "von einer klassischen Phalanx geprägt",[435] but this presupposes a qualitative difference between two such phalanx types, and that is scarcely sufficiently proved. The argument therefore is rather too close to *petitio principii* to be truly valid.

Yet the relevance of the Homeric poems in this connexion is forbiddingly difficult to assess. While Latacz has shown conclusively that fighting in massed formations does indeed go back to 'Homer', it is nigh impossible to judge how much overlap there is between 'massed formations' and actual hoplite fighting in the *Iliad*. The battle descriptions of the *Iliad* are highly confused, as is to be expected from a work of such staggering compositional and historic complexity. The images

---

12.415–426, 13.126–135, 13.145–148, 15.408–409; Latacz (1977) 26–44, 48–49 and Singor (1988) 12–15.
432 Singor (quoted in Franz [2002] 7).
433 Chantraine (1968) s.v. φάλαγξ.
434 Argued also by Lammert *RE* s.v. "Schlachtordnung", and see Latacz (1977) 237–238: "das Bestreben [ging] sicher dahin, die Phalangenformation, deren Effizienz man kannte, noch geschlossener und schlagkräftiger zu machen. Die Erfindung des Hoplon, des Hoplitenschildes, wird eine Folge dieses Bestrebens gewesen sein. Dieser Schild wurde nicht erfunden, um eine bestimmte Kampftaktik einzuführen, sondern um eine bestimmte Kampftaktik zu verbessern" ("surely, the effort was made with the intention of making the phalanx formation, whose efficiency was known, even more closed and forceful. The invention of the *hoplon*, the hoplite shield, was thus a result of this effort. This shield was not invented in order to introduce a certain battle tactic, but to improve a certain battle tactic").
435 Franz (2002) 7.

of armies in massed formations may be clear enough, but they do not seem to exert any influence over the actual action – noble kings and chieftains dueling in pairs, often after first riding to the front line in chariots, this odd transportation service being the only military role they play in the poem. In those few instances where heroes actually fight from their chariots, they are immobile: the poet seems unable to 'tear himself away' from the context of meaningful duels on foot between heroes.[436]

By any definition chariots should have been obsolete in a context of massed fighting, as they would not have had the necessary space to operate, and in truth the poems reveal utter confusion with regard to their presence: their main function is to conveniently transport the heroes to and from battle. This is in direct contradiction of what is known about the use of chariots from other war cultures of *c.* 1500–1000 (aptly termed an "eurasisches Streitwagenzeitalter" by Wiesner[437]), such as the Assyrians, Hurrians, Hittites, Syro-Levantine, Egyptians and even so geographically remote as the Chinese.[438]

There can scarcely be any doubt that the Mycenaeans employed the chariot in warfare: the Linear B tablets contain some 400 references to war chariots;[439] but it is equally clear that the *Iliad* is completely out of touch with the way massed chariots were normally used in battle as "an elevated, mobile firing platform for an archer standing beside the driver, and as a fast, flanking and pursuing arm".[440] Doubtlessly the presence of chariots in the *Iliad* represents remains of Mycenaean (or possibly even pre-Mycenaean) tradition, dimly understood at a later period;[441] and in an extremely interesting passage, the old Nestor harks back to the more coherent tactics of his own youth. The tactics in question is difficult to assess beyond the fact that it certainly intended to keep a rather tight formation, not allowing the charioteers to dismount at will in order to fight on foot.[442]

---

436 Heubeck (1966) 84–85 and cf. Lendon (2005) 23–24, 28–30. Crouwel (1992) 53–54 posits a transportation role for chariots ("conveyances for important warriors who actually fought on the ground").
437 Wiesner, quoted in Latacz (1977) 217 n. 114.
438 Yates (1999) 14.
439 Ventris & Chadwick (1973²) 361–375; Palmer (1963) 314–329; Heubeck (1966) 81–85; Greenhalgh (1973) 8–12.
440 Crouwel (1992) 54–55; Keegan (2000) 211–238; Heubeck (1966) 84–85; Latacz (1977) 216–217; *contra* van Wees (1994), esp. 9–12, (2004) 158–160. Van Wees claims that there are similarities between Homer's chariots and those of the Celts; but there are strong arguments against this: see Greenhalgh (1973) 14–17. Van Wees also adduces a passage from Xenophon's *Kyropaideia* on the allegedly similar Assyrian use of chariots (Xen. *Cyr.* 3.3.60); but Xenophon is completely out of touch with how the Assyrians actually used chariots: see a lifelike and gory account in the Assyrian annals of a late chariot battle from king Sennacherib's eighth Elamitic campaign in 689 (Luckenbill [1924] 43–47). For other interesting examples of the tactical use of the chariot arm as a shock weapon in various ancient cultures, see Lorimer (1950) 322–323.
441 Kirk (1962) 124–125; Snodgrass (1964a) 159–163. For a thorough examination of the use and technical specifications of chariots after Mycenaean times, see Crouwel (1992).
442 *Il.* 4.293–310; cf. Lorimer (1950) 324. Despite the ubiquitousness of the chariots in the *Iliad*, they are only used directly in the fighting in two brief passages: 11.150–153, 15.352–355.

Latacz admits that the tactical use of the chariots in the *Iliad* makes little sense, and instead argues that their other role in the poem, that of whisking wounded or fleeing heroes out of the battle, was more in keeping with actual warfare of C8 (which Latacz believes to be the time of composition); whereas van Wees claims that the mobility of the chariots is well suited to the perceived fluidity of all stages of fighting.[443] Latacz' assessment reveals his appreciation of the difficulty of moving chariots, drawn about on the battlefield by pairs of horses, which is why he relegates them to the second phase, the τροπή and pursuit; but this does not explain their virtual omnipresence in *all* combat phases.

According to Latacz, the chariots transport the heroes to the front and then retreat to a position behind the lines. Yet there are numerous examples of heroes killing first their immediate opponent and next the opponent's companion waiting with the chariot, demonstrating that the poet intended the chariots to be understood as used in direct fighting; and often, when heroes are wounded or threatened, they immediately jump in their chariot and hurry to safety.[444] This is difficult to reconcile with Latacz' interpretation, that heroes withdrawing (e.g., with minor injuries) went back – presumably through their own formation – to their chariots to be transported away. Also during the τροπή and pursuit, Latacz assumes that fleeing heroes escape on their chariots, but how do they manage this in the general rout, if their chariots are parked behind their own lines?[445] Nor is it easy to understand how chariots would manage to disengage from the fray of armies which, according to Latacz, are after all decidedly "infanteristisch", without adding considerably to the chaos on the battlefield. The same objections naturally apply to van Wees' interpretations, and to an even greater degree. Latacz himself, who is otherwise optimistic about the use of chariots in pursuit and flight, considers the use of chariots in actual combat problematic. It seems plausible enough that chariots may have been used for other purposes than warfare – such as transportation – in post-Mycenaean times, but that they should be employed *as such* in the turmoil of an infantry battle between phalanx formations still seems unlikely.

If we turn next to the weapons described in the *Iliad,* the resulting image is as unclear as that of the tactics. There is a superabundance of different types of helmet, for example: made of leather, made of animal skin, of bronze, trimmed with boar's tusks, with horns, with bosses, with horsehair crests.[446] The same applies to the shields: made of untanned leather, of bronze, round or as tall as a man and oblong, or the mysterious 'winged' λαισήϊον.[447] It may be objected that this simply

---

443 See note 440.
444 *Il.* 5.12–26, 5.43–48, 5.159–165, 5.275–310, 5.576–583, 5.608–609, 5.703–705, 6.16–19, 8.97–123, 11.86–112 (Agamemnon kills two pairs), 11.122–147, 11.264–274, 11.320–322, 11.328–335, 11.357–360, 11.396–400, 13.383–401, 15.445–458, 16.342–344, 16.652–658, 16.731–743, 20.484–489.
445 Latacz (1977) 218–219.
446 Helmets: *Il.* 3.316, 3.336, 3.369, 4.457–459, 5.4, 5.743, 6.9, 6.470–472, 6.494, 7.62, 10.257–259, 10.261–265, 10.335, 10.458, 11.41–42, 11.351–353, 12.160, 12.183–184, 12.384, 13.131–133, 13.188, 13.265, 13.341, 13.614–615, 13.714, 14.372, 15.480, 15.535–38, 16.70, 16.137–138, 16.214–216, 16.338, 16.413, 17.294–295, 18.610–612, 19.359, 22.314.
447 Shields: *Il.* 5.182, 5.452–453, 6.117–118, 8.192–193, 12.425–426, 13.611, 14.371, 14.402–

reflects a reality of non-conformity with no restrictions or specific demands on weapons, or that it reflects a period of transition and widespread diversity and experimentation of arms; but this is merely explaining away fundamental inherent difficulties in the Homeric poems.

Nevertheless, it is interesting that there is such a relative abundance of weapons and tactics which would not feel out of place in a hoplitic context. If we tentatively accept 750 as the date of composition for the *Iliad,* we also have the earliest possible date for observable traces of phalanx-like tactics; but, as we saw, it is likely that fighting between armies in close-order formations goes even further back. Moreover, there is also the distinct possibility that the date of composition (or at least of fixation in writing) is rather *c.* 550. This leaves an interval of *at least* 200 years during which the oral tradition, or rather its performers, was theoretically able to absorb elements from the new phenomenon of *hoplite* phalanxes.

Such elements, clearly describing hoplite phalanxes, are found here and there throughout the *Iliad.* Consider, for instance, the following passage:

οἳ γὰρ ἄριστοι
κρινθέντες Τρῶάς τε καὶ Ἕκτορα δῖον ἔμιμνον,
φράξαντες δόρυ δουρί, σάκος σάκεϊ προθελύμνῳ·
ἀσπὶς ἄρ' ἀσπίδ' ἔρειδε, κόρυς κόρυν, ἀνέρα δ' ἀνήρ·
ψαῦον δ' ἱππόκομοι κόρυθες λαμπροῖσι φάλοισι
νευόντων, ὡς πυκνοὶ ἐφέστασαν ἀλλήλοισιν,
ἔγχεα δὲ πτύσσοντο θρασειάων ἀπὸ χειρῶν
σείομεν'· οἱ δ' ἰθὺς φρόνεον, μέμασαν δὲ μάχεσθαι.[448]

The Achaians stand so close that their shields and bodies 'lean' on each other as they make a fence (φράξαντες) with spears and shields, marshalled (κρινθέντες) and waiting for the Trojans and Hektor; their shields touch, and even their helmet crests touch if they nod – apparently the formation is equally dense horizontally and vertically. This must be meant to evoke a formation with ranks at least as dense as those of a hoplite phalanx, and the term προθελύμνῳ ("with base advanced"[449])

406, 15.645–646, 20.267–272, 20.279–281, 21.581, 22.294. See also Snodgrass (1974) 123: "What need be said about Homeric fighting-equipment beyond the fact, today I hope accepted, that it is composite and shows internal inconsistency?" and Gray (1947) 113–116. On λαισήϊα, the meaning of which was disputed already in antiquity, see Lorimer (1950) 194–196.

448 *Il.* 13.128–133 ("for they who were the chosen bravest awaited the Trojans and noble Hektor, fencing spear with spear, and shield with serried shield; shield pressed on shield, helmet on helmet, and man on man; and the horsehair crests on the bright helmet ridges touched each other as the men moved their heads, in such close array stood they by one another, and spears in bold hands overlapped each other as they were brandished, and their minds swerved not, but they were eager to fight" [trans. Murray]). The verses are repeated more or less *verbatim* at 16.210–217, where the formation of the Myrmidons is likened to the well-fitted stone plinths of a solidly built house. Cf. also 8.60–63, 11.592–595 and 12.105–107, and see Fenik (1968) 123, who also adduces 15.615–616, 17.262–271 and possibly 17.364–365 as examples of what is "definitely a phalanx-line formation", and Bowden (1995) 52–54. Other passages interestingly seem to evoke the embittered in-fighting and *othismos* of Classic-age hoplite fighting: *Il.* 4.446–451, 8.60–65, 17.274–277. The entire passage 16.155–217 interestingly contains some of the latest linguistic phenomena in all of Homer: see Latacz (1977) 59 n. 31.

449 Cunliffe (1963²) s.v. προθέλυμνος.

suggests that the shields are held in the way hoplite shields were normally held, aslant and leaned on the shoulder. Moreover, phalanxes frequently appear in similes expressing their density and frightful potential destructive power, such as a terrifying storm front of massive thunder-heads threatening to break,[450] or a series of waves crashing on the shore.[451] Latacz also presents an interesting analysis of the stages of a hoplite battle of the Classical era – Syracuse 415 – which reveals surprisingly many points of similarity between a hoplite battle and the fighting described in the *Iliad*.[452]

These passages and others like them effectively demonstrate that elements of phalanx fighting were at some point included in the poem, but do not allow us to judge when; and there is still no certain way of knowing the extent to which hoplite warfare is treated in the *Iliad*. Despite Latacz' findings, van Wees has reached a different conclusion about the nature of the fighting portrayed in the *Iliad*, seeing instead a quite consistent and coherent image of warfare in C8.[453] He compares the fitful bursts of fighting, followed by slinking back, to the unorganised type of primitive tribal 'warfare' still found by anthropologists in Papua New Guinea.[454] In his opinion, the only way to accommodate both Latacz' findings and the other aspect of fluid combat between protagonists is to interpret them as "one and the same thing from different perspectives".[455] While van Wees is right to insist that the other type of fighting plays an important role in the poem, and that it does display similarities to primitive warfare, his explanation of the closed order passages is less successful. As seen by van Wees, these images are all of armies about to engage, whereas in the actual fighting which follows, the orderly armies seem to be more or less dispersed again, so that they apparently only rally for shorter periods of time. This is certainly the only way to understand it if we are to insist on a perfectly coherent and accurate rendering of the tactics employed; but it does seem to make too little of the frequent allusions to compactness and solidity. If van Wees' interpretation is accepted, these passages become almost irrelevant. It is difficult to imagine why an army would pack itself together for the attack only to disentangle completely moments later; and if this was in fact the inevitable outcome as seen by van Wees, one cannot help wondering why anyone bothered at all.

Rather, it seems there are basically two types of battle in the *Iliad:* one of fluid combat allowing warriors to roam freely across the battlefield and to slink back or attack as they see fit,[456] even moving about on the front line in chariots, from which the heroes dismount in order to fight duels in pairs or threes. But, as we have seen,

---

450 *Il*. 4.274–282: here, the phalanx of the two Aiases is "dense" (πυκιναί) and "bristling with shields and spears" (σάκεσίν τε καὶ ἔγχεσι πεφρικυῖαι); cf. 7.61–62 (τῶν δὲ στίχες εἴατο πυκναί, / ἀσπίσι καὶ κορύθεσσι καὶ ἔγχεσι πεφρικυῖαι).
451 *Il*. 13.795–801; cf. 4.422–429 and also 7.61–66. Latacz (1977) 63–67 argues that these formations are so dense that they are akin to the *synaspismos* known from later authors.
452 Latacz (1977) 226–229; inv. no. 36.
453 van Wees (2000) 139–154, (2004) 153–165, 170–183.
454 van Wees (1994) 1–9, (2004) 153–158; see also Gat (2006) 290–291 and n. 107.
455 van Wees (1994) 2–8.
456 See, e.g., van Wees (1994) *passim,* (2000) *passim* and Franz (2002) 23–109 and esp. 23–26.

there is also a battle type fought by massed formations, sometimes with apparently explicit hoplite references. The two types of combat are difficult to reconcile. Densely packed ranks of warriors, marching forward to fight or awaiting the enemy onslaught, are essentially different from warriors roaming freely across the battlefield and advancing, engaging or retreating as they see fit. If the ranks are indeed so densely packed, it would effectively hinder individual movement backwards and sideways: in fact, only the πρόμαχοι in the front ranks would be able to advance, and then only in one direction – forwards. Yet this is not the impression one gets from the apparently complete freedom of movement the heroes enjoy, even to the point of advancing or retreating time and again in their chariots. It may be argued that these represent different phases of the fighting; but if so, it is difficult to see how and when they are supposed to alternate. If such phases actually could succeed each other, it is open to wonder what would be achieved by loosening a close order formation; or how, indeed, it would be possible to re-group as frequently into a tight phalanx once again, as so often happens. In a typical passage, Hektor jumps from his chariot, apparently in the thick of the fighting, and runs across the front line of the hard-pressed Trojan army to rally them once more into an efficient battle line:

> Ὣς ἔφαθ᾽, Ἕκτωρ δ᾽ οὔ τι κασιγνήτῳ ἀπίθησεν.
> αὐτίκα δ᾽ ἐξ ὀχέων σὺν τεύχεσιν ἆλτο χαμᾶζε,
> πάλλων δ᾽ ὀξέα δοῦρα κατὰ στρατὸν ᾤχετο πάντῃ
> ὀτρύνων μαχέσασθαι, ἔγειρε δὲ φύλοπιν αἰνήν.
> οἳ δ᾽ ἐλελίχθησαν καὶ ἐναντίοι ἔσταν Ἀχαιῶν·
> Ἀργεῖοι δ᾽ ὑπεχώρησαν, λῆξαν δὲ φόνοιο,
> φὰν δέ τιν᾽ ἀθανάτων ἐξ οὐρανοῦ ἀστερόεντος
> Τρωσὶν ἀλεξήσοντα κατελθέμεν, ὣς ἐλέλιχθεν.[457]

Hektor is advised to try and stay the Trojan army, which is hard pressed and "being murdered" by the advancing Achaians. He accordingly takes courage and jumps from his chariot – which must obviously be present in the fray. If not, we must be supposed to imagine that this takes place *behind* the fighting lines, and that Helenos has sought out Hektor there, causing him to run around or through the Trojan lines before he can begin rallying the army. This does not appear likely. Next, Hektor runs along the Trojan lines, rallying the warriors, who suddenly turn and face the enemy, seemingly neatly coalescing into some sort of formation. At once the Achaians are dismayed by this performance: they yield and "cease from killing". To all appearances, their formation is lost just as easily and quickly as the Trojan formation is reestablished.

The two types of combat are completely different and not easily reconcilable, but this is hardly surprising. On the 'coin-hoard interpretation' the phalanxes reflect

---

457 *Il.* 6.102–109 ("So he spoke, and Hektor was in no way disobedient to his brother. Immediately he leapt in his armour from his chariot to the ground, and brandishing his two sharp spears went everywhere throughout the army, urging them to fight; and he roused the dread din of battle. So they rallied and stood facing the Achaians, and the Argives gave ground and ceased from slaying; and they thought that one of the immortals had come down from starry heaven to assist the Trojans, who had rallied so" [trans. Murray]); cf. 5.493–497.

a more recent period (*c.* 750–650), whereas the chariots may be a tradition from the bronze age. How much of this is 'historical' and how much is fantasies of later rhapsodes must remain unanswered.

Nonetheless, it seems that from a narrative point of view frequent descriptions of massed fighting are rather counterproductive, since the focus arguably is on duels between the major heroes. Moreover, phalanx fighting is a phenomenon far too complex to have been invented *e nihilo* for *artistic* purposes. It is therefore natural to assume that the inclusion of phalanx fighting in fact reflects near-contemporary social reality and military practice, or one that at least originated in a relatively recent past during the later stages of the poem's genesis.

There are thus too many unknown quantities in the equation for the result to be precise: the date of composition (or commission to writing) and the many internally irreconcilable cultural and linguistic layers in the poems all militate against the chances of ever finding out the truth behind the society and warfare described in the *Iliad* and the *Odyssey*. The most that can be said for the *Iliad* as a historical source, therefore, is that it can be shown conclusively that fighting in massed array is definitely represented, and that descriptions of hoplite fighting have been included. The practical impossibility of dating the *Iliad* cannot be sufficiently stressed in this connexion, nor the fact that it is a fallacy to claim any historical source value for it. To the extent that hoplites or phalanxes are concerned, however, we can see that a phalanx is in fact a closed-order formation, not easily reconcilable with dueling or fluid, open battles. The 'Homeric' society is basically an amalgam, and this extends to the warfare as well: all the elements that make up Homer's battles simply cannot be found together and at the same time.

3.1.4 The phalanx in Archaic poetry

As indicated above, van Wees sees the development of the phalanx as drawn out over an extremely long period of time, constantly evolving, perhaps until the time of the Persian wars. Van Wees' argument relies primarily on iconographic evidence which, he says, has "long played a curious part in the debate", in that the majority of C7 representations of combat have been dismissed as "a 'heroic' form of combat which bears no relation to contemporary warfare".[458] It may be so; but this deplorable situation may in fact rather be the consequence of insurmountable difficulties in correctly interpreting the notoriously shaky evidence found in obscure fighting scenes of early Greek vase-painting and, indeed, in separating historical fact from fiction.[459]

Van Wees turns to Archaic literature to find support for his theory, and especially Tyrtaios, whose Spartan war songs may be some of the earliest poetry apart

---

458 van Wees (2000) 125.
459 Cartledge (1996) 711–712: "… la convenzione artistica tendesse ad alterare la realtà o addirittura a inventarsela, rende particolarmente rischioso servirsi delle raffigurazioni dell'epoca a fini di ricostruzione storica" ("… the artistic convention tended to alter reality, or even to invent it anew, rendering it particularly risky to make use the iconography of the period for historical reconstructions"). Cf. Hurwit (2002) 14.

from Homer and Hesiod, possibly dating to the second Messenian war, but certainly belonging to C7.[460] Here van Wees finds further proof that the phalanx formation did not exist until well into the Classical period: rather, he argues, the combat found in Tyrtaios resembles that of the *Iliad*. He points out three significant themes in the elegies: hand to hand fighting, cohesion and the danger of fleeing.[461] These themes would all – as van Wees himself admits – be quite central to the hoplites of a Classical phalanx, and this may be exactly what has ensured the survival of the central fragments. This presumably means that other parts have not been handed down precisely because of their supposed awkwardness in a phalanx context; but this is an argument *e silentio*.[462] Indeed, it is open to wonder why Tyrtaios' war songs remained so popular with the Spartans (and in the Greek world in general) if they contained nothing that was of use in a hoplite context.[463]

The poetry of Tyrtaios is extremely exhortatory in kind, and his poetry resounds with the repeated demand: do not give way, stand together, die if need be.[464] Van Wees argues that Tyrtaios' constant exhortations to move close to the enemy and stay there, and not skulk back from the fighting,[465] are by implication indicative of warriors who *needed* this kind of pep talk because they had the freedom to roam about more or less freely on the battlefield (or at least to withdraw or slink back) as they saw fit.[466] It is an interesting interpretation and one that suggests itself to a certain degree, but it is not necessarily the only conclusion. What Tyrtaios wishes to induce in his listeners is certainly a general fear of giving way before the enemy, but the poet should not be construed as addressing the young men of Sparta *individually*: rather, the exhortation is directed at the army as a whole. The frequent imperatives in the text are to be understood as such: they are predominantly in the second person plural, and in those few cases where a third person singular imperative is used, it is, I believe, more likely a way of saying "let *every* man …".

Furthermore, as has been pointed out, even if phalanx fighting was not so open and fluid as suggested by van Wees and others, there was ample opportunity to display courage or cowardice in the throng of the phalanx by skulking behind or stepping up when the men in front were killed;[467] and this might well be what is hinted at in such verses as οἳ μὲν γὰρ τολμῶσι παρ' ἀλλήλοισι μένοντες / ἔς τ' αὐτοσχεδίην

---

460 Pl. *Leg.* 629a–b and Σ; Arist. *Pol.* 1306b 36 – 1307a 2; Diod. Sic. 8.27.1–2; Paus. 4.18.2–3; *Suda* s.v. Τυρταῖος.
461 van Wees (2000) 149.
462 This explanation is in fact dropped in favour of "the same change in tactics … in contemporary art" (van Wees [2000] 151).
463 Tyrtaios' songs were still drummed into Spartan youths centuries later: see Lycurg. *Leocr.* 106–107.
464 Bowie (1990) maintains that such poetry was primarily intended for a sympotic context; but some of the descriptions seem a little on the gory side for party entertainment.
465 See, e.g., Tyrt. fr. 11.11–13, 11.28, 12.15–17 West.
466 van Wees (2000) 149–150, (2004) 172–174, and see Snodgrass (1964a) 181–182. Luginbill (2002) 409 observes, "… the primary object of Tyrtaeus 12 is not to encourage warriors *in* battle, but to bring them into the ranks in the first place" (emphasis original).
467 Lazenby (1991) 93–94.

καὶ προμάχους ἰέναι, / παυρότεροι θνήσκουσι, σαοῦσι δὲ λαὸν ὀπίσσω.[468] Phalanxes might just as well fall apart because the rear ranks began to run away; and these would consequently need the same amount of encouragement as the front ranks. Everybody has an important role to play in a phalanx, and they are naturally addressed all at once.

Van Wees points to the similarity between Tyrtaios' poetry and the *Iliad*;[469] and certainly there are a great many points of contact between them at a linguistic level. This is not in itself so strange, however: as is the case with all the early Archaic poets, Tyrtaios is naturally deeply indebted to the Homeric poems; or, to put it more precisely: they are both indebted to the epic oral tradition. It is hardly surprising if Tyrtaios has taken all the words and phrases from epic poetry that he could to suit his purpose. Bruno Snell compiled a list of very similar exhortations from Tyrtaios and Homer and correctly concluded, "Solche ... Stellen zeigen zur Genüge, daß Tyrtaios Ansätze zu seinen eigenen Gedanken in älteren Kriegsgedichten finden konnte ...".[470] Again, the preponderance of Homeric phrases or whole verses demonstrates nothing but the enormous popularity and influence of the epic tradition. It also reveals an artist grappling with a new medium, employing bits and scraps from the abundant linguistic treasure of his great predecessors and trying to put these small bits of received wisdom to good use by adapting them. "Mere morsels from Homer's great table," as Aischylos poignantly said of his own tragedies:[471] this no doubt refers to the themes of early drama, but of course applies to an even greater degree to the language itself. Thus πρόμαχοι ("front row fighters") for instance, is quite simply a convenient word from an epic context which is used to refer to the front ranks; and an ἀσπὶς ὀμφαλόεσσα ("bossed shield") is similarly chosen for its convenient suitability to hexameter-based poetry.[472] Furthermore, traditionalism is presumably also at play here: due to the overwhelming influence of the epic poems, the epithet ὀμφαλόεσσα may simply have seemed authoritative and conventional (quite apart from metrically convenient) to the poet.[473]

Nevertheless, the poetry of Tyrtaios actually does differ from the *Iliad* in a number of respects, some of these very important, and all strongly reminiscent of the phalanx. Tyrtaios' poems abound in grim invitations to the Spartan youth not to φιλοψυχεῖν ("cling to life") but to actively *sacrifice their lives*,[474] as when he tells

---

468 Tyrt. fr. 11.11–13 ("Those who dare to stand fast at one another's side and to advance towards the front ranks in hand-to-hand conflict, they die in fewer numbers and they keep safe the troops behind them" [trans. Gerber]).
469 Van Wees (2000) 150–151.
470 Snell (1969) 50 ("Such ... passages sufficiently demonstrate that Tyrtaios could find a basis for his own ideas in older war poems"); cf. Jaeger (1961) 102: "The poems of Tyrtaeus ... are Homeric through and through."
471 Aesch. fr. 112 Radt *apud* Ath. 8.347e.
472 Tyrt. fr. 12.25 West; Gray (1947) 113–114. The word πρόμαχος crops up repeatedly in epigrams and epitaphs from the Classical period, and thus is hardly more than an epic-poetic phrase for 'someone fighting in the front lines': see, e.g., *CEG* I 10, 27 and 112 (ranging from C6 to C5s); Jaeger (1966) 135 and Anderson (1984) 152.
473 Cf. Cartledge (1977) 26; Luginbill (2002) 412–413.
474 Tyrt. fr. 10.1–14, 10.17–18, 10.27–30, 11.3–6 West.

them in no uncertain terms to "welcome the spirits of death like rays of the sun": μηδ' ἀνδρῶν πληθὺν δειμαίνετε, μηδὲ φοβεῖσθε, / ἰθὺς δ' ἐς προμάχους ἀσπίδ' ἀνὴρ ἐχέτω, / ἐχθρὴν μὲν ψυχὴν θέμενος, θανάτου δὲ μελαίνας / κῆρας <ὁμῶς> αὐγαῖς ἠελίοιο φίλας.[475] This theme is developed systematically and often in unpleasant graphic detail, and it mirrors the phalanx ethos perfectly: to keep one's place at all times, even at the cost of losing one's *own* life, so as not to jeopardise the entire army. In fact, it is most likely to Tyrtaios we owe the ghastly cliché about the glory of dying in battle for the fatherland, further immortalised by Horace, Tennyson and Wilfred Owen: τεθνάμεναι γὰρ καλὸν ἐνὶ προμάχοισι πεσόντα / ἄνδρ' ἀγαθὸν περὶ ᾗ πατρίδι μαρνάμενον.[476] Apparently, it found immediate favour with the Spartans (and possibly elsewhere);[477] but it certainly is an extraordinarily strong expression of the need for holding one's ground.

Exhortations of this kind would be out of place in the *Iliad,* and accordingly they are not to be found in Homer: there is nothing glorious about actually being killed in the *Iliad,* though of course noble warriors are required to show bravery on the battlefield. Tyrtaios, on the other hand, displays a rather macabre fascination with death on the battlefield, which is the highest praise a young man can win for himself and his kin. The ideal death is that offered on the battlefield – but in the proper, virtuous way, of course: facing the enemy.[478] Tyrtaios praises in high tones the body of a young man πολλὰ διὰ στέρνοιο καὶ ἀσπίδος ὀμφαλοέσσης / καὶ διὰ θώρηκος πρόσθεν ἐληλάμενος,[479] thereby winning glory for himself, his kin and his

---

475 Tyrt. fr. 11.3–6 West ("and do not fear the throngs of men or run in flight, but let a man hold his shield straight toward the front ranks, despising life and loving the black death-spirits no less than the rays of the sun" [trans. Gerber]).

476 Tyrt. fr. 10.1–2 West ("It is a fine thing for a brave man to die when he has fallen among the front ranks while fighting for his homeland" [trans. Gerber]). The entire poem, in a rather morbid fashion, is concerned with the inordinate glory gained by dying young while fighting for the fatherland. Cf. Hor. *Carm.* 3.2.13–16: *Dulce et decorum est pro patria mori. / Mors et fugacem persequitur virum, / nec parcit imbellis iuventae / poplitibus timidove tergo* and Wilfred Owen: "If in some smothering dreams, you too could pace / Behind the wagon that we flung him in, / And watch the white eyes writhing in his face, / His hanging face, like a devil's sick of sin; if you could hear, at every jolt, the blood / Come gargling from the froth-corrupted lungs, / Obscene as cancer, bitter as the cud / Of vile, incurable sores on innocent tongues, – / My friend, you would not tell with such high zest / To children ardent for some desperate glory, the old Lie: Dulce et decorum est / Pro patria mori."

477 Cf. Plut. *Cleom.* 2.3: Λεωνίδαν μὲν γὰρ τὸν παλαιὸν λέγουσιν ἐπερωτηθέντα, ποῖός τις αὐτῷ φαίνεται ποιητὴς γεγονέναι Τυρταῖος, εἰπεῖν· ἀγαθὸς νέων ψυχὰς κακκανῆν. ἐμπιπλάμενοι γὰρ ὑπὸ τῶν ποιημάτων ἐνθουσιασμοῦ παρὰ τὰς μάχας ἠφείδουν ἑαυτῶν ("For Leonidas in former times, we are told, when asked what manner of poet he thought Tyrtaios to be, replied: 'A good one to inflame the souls of young men.' And indeed they were filled with divine inspiration by his poems, and in battle were unsparing of their lives" [trans. Perrin, modified]); *GVI* 749, esp. 7–8 (an inscription from Thyrrheion in Akarnania): Τυρταίου δὲ Λάκαιναν ἐνὶ στέρνοισι φυλάσσων / ῥῆσιν τὰν ἀρετὰν εἵλετο πρόσθε βίου ("Guarding the words of Tyrtaios the Spartan in his chest, he chose virtue over life").

478 Tyrt. fr. 12.23–34 West. For a view of fr. 12 as a thoroughly romanticised "recruiting jingle", see Luginbill (2002) 412–414.

479 Tyrt. fr. 12.25–26 ("pierced many times through his breast and bossed shield and corselet from the front" [trans. Gerber]), cf. 10.27–30 West.

fatherland.⁴⁸⁰ This is of course put in stark contrast to bodies of men run through from behind (i.e. during flight), presenting an "attractive target"⁴⁸¹ or, worst of all, the ghastly sight of an old, grey-bearded man with a terrible, fatal wound to the groin – lying mortally wounded in front of the young men, precisely because he held his ground while they fled to safety farther behind.⁴⁸²

A brief comparison with the approximate contemporary, Kallinos, shows that Tyrtaios' new, grim hoplite ideals, however brutal, are not unique to Sparta. Kallinos does not hide the ugly facts about the possibility of death on the battlefield and in a similar manner extols all the good things that accrue from bravery in the face of danger; although he does not perhaps display the same morbid enthusiasm for death in itself: θάνατος δὲ τότ' ἔσσεται, ὁππότε κεν δὴ / Μοῖραι ἐπικλώσωσ'.⁴⁸³ Kallinos encourages his listeners by reminding them that although it is important not to think of death when fighting, the salient point is that luckily death may after all *not* be the result of combat. Nevertheless, he does represent death as an integral part of battle: οὐ γάρ κως θάνατόν γε φυγεῖν εἱμαρμένον ἐστίν / ἄνδρ' οὐδ' εἰ προγόνων ᾖ γένος ἀθανάτων.⁴⁸⁴ In other words, Kallinos also perceives death as a natural condition of combat.

Other features of Tyrtaios' poetry seem like a conscious and direct confrontation with the old nobleman's value system. The entire catalogue of a noble Homeric hero's virtues is dismissed as insufficient in a startling *priamel*:⁴⁸⁵ it is not that the virtues of strength, beauty, agility, wealth and the like are criticised as such – they are just not *enough* for the *polis,* which above all requires bravery and steadfastness of its citizens. This, needless to say, is the key requirement of the hoplite, and it brings about a "common good" (ξυνὸν ἐσθλόν).⁴⁸⁶

Conventional phalanx tactics also seem to be described in fragment 11: here, the famous description of the dense battle order given in the thirteenth song of the *Iliad* is mirrored to brilliant effect by a simple reversal. In a stunning piece of imagery, Tyrtaios borrows the famous lines from the *Iliad* describing the *synaspismos* of the Trojan heroes. He retains the description of pressure and proximity, but by a simple reversal emulates the verses to brilliant effect by adapting them to suit a context of bitter close combat or possibly even *othismos*:

---

480 Tyrt. fr. 12.20–21 West; cf. Simon. fr. 8 and 9 Campbell (= *Anth. Pal.* 7.253 and 7.251).
481 Tyrt. fr. 11.17 West (with Ahrens' excellent emendation of ἀργαλέον to ἁρπαλέον). Cf. Theocr. *Id.* 22.198–204; Snodgrass (1967) 56.
482 Tyrt. fr. 11.15–20, 10.21–30 West; cf. *Il.* 22.71–76.
483 Callin. fr. 1.8–9 West ("Death will occur only when the Fates have spun it out" [trans. Gerber]); cf., e.g., *Il.* 22.364–366.
484 Callin. fr. 1.12–13 West ("For it is in no way fated that a man escape death, not even if he has immortal ancestors in his lineage" [trans. Gerber]).
485 Tyrt. fr. 12.1–14 West. Cf. Eur. fr. 282.16–21 Nauck; and Xenoph. fr. 11 B 2.14–22 Diels-Kranz (*supra* 49 n. 157).
486 Tyrt. fr. 12.15 West; Luginbill (2002) 407–409 and cf. Jaeger (1961) 119–120.

> ἀλλά τις ἐγγὺς ἰὼν αὐτοσχεδὸν ἔγχεϊ μακρῷ
> ἢ ξίφει οὐτάζων δήϊον ἄνδρ' ἑλέτω,
> καὶ πόδα πὰρ ποδὶ θεὶς καὶ ἐπ' ἀσπίδος ἀσπίδ' ἐρείσας,
> ἐν δὲ λόφον τε λόφῳ καὶ κυνέην κυνέῃ
> καὶ στέρνον στέρνῳ πεπλημένος ἀνδρὶ μαχέσθω,
> ἢ ξίφεος κώπην ἢ δόρυ μακρὸν ἔχων.[487]

The proximity here is with the enemy, rather than with one's own comrades, and is the result of fighting at extremely close quarters. This is evident from the fact that Tyrtaios says that "breast shall be pressed against breast": no such touching could reasonably be said to occur within one's own lines. It is revealing that the two passages are so similar. The fact that the lines from the *Iliad* are echoed by Tyrtaios with only minor changes strongly suggests that the same type of fighting is described in both poems, and the same considerations apply here, to an even greater degree: the dense ranks and combat at extremely close quarter clearly betoken fighting in quite closed order.

Similarly with another instance: the hoplite is encouraged to stand so close that he can actually just "reach out" for the enemy (δηΐων ὀρέγοιτ' ἐγγύθεν ἱστάμενος),[488] presumably with his spear, as he stands by and encourages the man next to him in the phalanx (θαρσύνῃ δ' ἔπεσιν τὸν πλησίον ἄνδρα παρεστώς): in this way he will rout the bristling enemy phalanx (φάλαγγας τρηχείας).[489] The aggressively defensive way of fighting is also underlined by a passage stressing the need for a broad, firm stance:

> ἀλλά τις εὖ διαβὰς μενέτω ποσὶν ἀμφοτέροισι
> στηριχθεὶς ἐπὶ γῆς, χεῖλος ὀδοῦσι δακών,
> μηρούς τε κνήμας τε κάτω καὶ στέρνα καὶ ὤμους
> ἀσπίδος εὐρείης γαστρὶ καλυψάμενος.[490]

Incidentally, the position mentioned here is probably identical to the sideways-on stance discussed above,[491] and as I pointed out, the stance is eminently explicable from a defensive point of view in that it keeps the vulnerable parts of the body so far away from the enemy as possible while greatly decreasing the actual size of the target presented to the enemy. At the same time, however, 'digging in' like this is predominantly defensive and is not indicative of movement, let alone fluency, in the fighting. If one is to advance or retreat from this stance, it will be one step at a time, dragging either the hindmost or the foremost foot, as the case may be. But this

---

487 Tyrt. fr. 11.29–34 West ("but coming to close quarters let him strike the enemy, hitting him with long spear or sword; and also, with foot placed alongside foot and shield pressed against shield, let everyone draw near, crest to crest, helmet to helmet, and breast to breast, and fight against a man, seizing the hilt of his sword or his long spear" [trans. Gerber]).
488 Tyrt. fr. 12.12 West.
489 Tyrt. fr. 12.19–22 West. As mentioned above, Latacz sees no real difference between φάλαγγες and στίχες (Latacz [1977] 45–49).
490 Tyrt. fr. 11.21–34 ("Come, let everyone stand fast, with legs set well apart and both feet fixed firmly on the ground, biting his lip with his teeth, and covering thighs, shins below, chest and shoulders with the belly of his broad shield" [trans. Gerber]), cf. fr. 10.31–32, 12.15–19 West; Eur. *Heracl.* 836–837; Xen. *Hell.* 4.3.19.
491 *Supra* 34–35, 39–41.

does not seem to be of any concern to Tyrtaios, who sternly admonishes the young men to *dig in* their feet (εὖ διαβὰς ... ποσὶν ἀμφοτέροισι) and keep their ground at all costs (μενέτω).

Finally, the irresistible forward surge of an attacking phalanx is vividly captured in the astounding metaphor ἔσχεθε κῦμα μάχης[492] – which, interestingly, appears to be of Tyrtaios' own devising. Fragment 19, preserved on papyrus, is badly mauled, but the glimpses it affords us leave little doubt about the nature of fighting described by Tyrtaios. We hear of "making a fence with *hollow* shields" (κοίλης ἀσπίσι φραξάμενοι), which must refer to a phalanx line consisting of hoplite shields; and in a famous line, Tyrtaios promises that "all together we will crush (ἀλοιησέομεν, lit. "thresh") ... standing close to the spearmen. The din will be terrible (δεῖνος ... κτύπος) ... as both sides dash (?) well-rounded shields against shields" (ἀσπίδας εὐκύκλους ἀσπίσι τυπτ[ ).[493] This 'terrible din', described by ancient witnesses, of the terrifying crash of many shields being smashed together on impact, is likely to have occurred when two determined phalanxes met in a head-on collision, whereas a meeting between two open formations is unlikely to have entailed anything like a collision.[494]

In this fragment light-armed soldiers are also implied: here, they are throwing stones.[495] Light-armed troops appear quite frequently in Tyrtaios' poems, and they are often addressed alongside the hoplites. Van Wees believes that this indicates a fluid formation, where hoplites and light-armed (archers, slingers, javelin-throwers) mingle freely on the battlefield, engaging and disengaging at will, and usually opening their own chosen duels by throwing javelins at each other.[496] The key passage to this notion is fragment 11, where the light-armed warriors are addressed after the hoplites and are called upon to crouch (πτώσσοντες) under a shield, as they let fly with stones and javelins.[497] This van Wees takes to mean under a *hoplite's* shield;[498] but that is not necessary, since *aspis* is a very broad term for shield. The light-armed are told to crouch "here or there", or "on either side" (ἄλλοθεν ἄλλος) under *a* shield: why not their own? A hoplite shield, by van Wees' own reckoning, is in fact insufficient to cover more than its bearer,[499] and this would be the case to an even greater degree if the hoplite were supposed to fight successive 'duels': he would have to have completely free reign over his own shield. The Thracian peltast of later times carried his own light shield – the πέλτη, hence the name[500] – and there is no good reason why the light-armed of Tyrtaios' day should not carry their own (light) protection, which they could take with them as they scurried about on the

---

492 Tyrt. fr. 12.22 West.
493 Tyrt. fr. 19.7–15 West (trans. Gerber; emphasis mine).
494 Ar. *Ach.* 573 (where κυδοιμός is even used metonymically for 'war' or 'battle'); cf. Xen. *Cyr.* 7.1.35.
495 Tyrt. fr. 19.2, 19.19–20 West.
496 So van Wees (2000) 151–154, (2004) 169–170. This view is also held by Storch (1998) 5–6.
497 Tyrt. fr. 11.35–38.
498 Van Wees (2000) 151.
499 *Supra* 39–40 and cf. van Wees (2000) 129. Protection for light-armed warriors from hoplites' shields is also rejected by Lorimer (1947) 127.
500 Diod. Sic. 15.44.3.

battlefield and harassed the enemy. In this way, they will have had some protection for themselves, securing mobility and independence of action, while not interfering with the hoplites' use of their weapons.

Kallinos, of whose work only extremely disjointed fragments survive, develops some of the same themes: the appeal to the young, the notion that war may be terrible and that death, although not certain, is a natural consequence; but he appears at first glance to have different notions of fighting. No less than two times in the only somewhat complete fragment does he refer to javelins,[501] and the same applies to the only relevant extant fragment of Mimnermos.[502] There is also a fragment of Archilochos which mentions the 'thud of javelins' (ἀκόντων δοῦπον); but as the fragment contains a total of eight whole words (two of which are ἀκόντων δοῦπον), scattered over ten different lines, an interpretation of its meaning should not be pressed.[503] It does not necessarily follow, however, that the presence of javelins indicates a more open or fluid form of combat: as the Chigi vase shows, there are illustrations of hoplites marching into battle carrying javelins along with their thrusting spears,[504] and it is also clear from the same vase that perfectly orderly ranks are kept while the javelins are gripped in the left hand together with the *antilabe*. It can be imagined that these javelins were discharged before the collision, sometime during the march forward to meet the enemy. The presence of javelins in the hands of hoplites in the sources or iconography does nothing to prove or disprove an open-order phalanx: we can merely establish the fact that during C7, hoplites apparently carried javelins in addition to their thrusting spears. Their disappearance from later sources and iconography suggest that they were not very successful, or were too much of an impediment, and were therefore dropped. Nor is that all: there is actually no decisive link in either Tyrtaios, Mimnermos or Kallinos between javelins and hoplites. For all we know, the javelins in these poems may refer to the presence of light-armed troops of some kind participating in the fighting.

There is thus nothing to truly support the assumption that the fighting found in Tyrtaios was in any crucial respect more fluid and open than that of later times. On the contrary, quite a few things point to the phalanx fighting of Tyrtaios' world being exceedingly compact and pressed together; and if anything Tyrtaios urges his fellow Spartans to stand completely fast and to die sooner than yield as much as a step. In fact, his morbid enthusiasm for death and all the aspects of the apparent subsequent glory is a very strong indicator that the phalanx fighting was in practice as tightly locked and stubbornly defensive as could be imagined. In a loose-order formation, it would be illogical to encourage steadfastness to the almost absurd point that death was preferable to withdrawal: the common cause would be much better served if the better fighters were at liberty to pull out of sticky situations, if this meant that they could just catch their breath and rejoin the fray at a later time.

---

501 Callin. fr. 1.5, 1.14–15 West.
502 Mimn. fr. 14.5–8 West.
503 Archil. fr. 139 West. Despite van Wees' claim, this fragment also does not link hoplites to javelins: van Wees (2000) 147.
504 *Infra* 124–127 and fig. 16.

This notion is supported by the fact that early armour types, as we have seen, tended to be quite heavy and cumbersome compared to those a few centuries later. This is, of course, partly explained by the fact that such armour and weaponry was probably still relatively new and crude and thus subject to much experimentation on the part of the armourers. But this only makes it all the more probable that the fighting would naturally evolve in a way that was not incompatible with the given weapons. This is not to suggest that weapons dictated fighting style – quite the contrary[505] – but it is inconceivable that a certain type of fighting, which was ultimately incompatible with existing weapons and especially armour, would be able to be prevail.

The principles on display above all in the poetry of Tyrtaios are crucial to the cohesion and unity of a close-order phalanx. The notion that any withdrawal is *per se* an evil, likely to entail utter destruction and ruin, as well as eternal dishonour for the hapless individuals who show themselves to be cowards, is drummed into the audience again and again. While the concept of not withdrawing is self-evidently of fundamental value in any pitched-battle situation, it is even more important in close-order fighting such as that employed in a hoplite phalanx. A phalanx above all relied on cohesion: it was all-important that its lines were kept as intact as at all possible. If gaps opened between the hoplites and the sections were pried from each other, there was a very serious risk that the whole phalanx might crumble and fall apart.

On the other hand, if the lines could successfully be kept reasonably intact – that is, if parts of the line were not scattered or driven too far back to rally again – so that the front line still presented an unbroken, fairly straight line of shields to the enemy, there were good chances of rebutting any enemy attack. For this very reason, cohesion was of particular importance to closed-order formations like the phalanx, and this also explains Tyrtaios' vehement aversion against even the smallest concession to the enemy: it simply carried a risk of having the formation destroyed and the soldiers routed.

3.1.5 The phalanx in Archaic iconography

As mentioned above, the main support for van Wees' theory is to be found in iconography. He maintains that Archaic art in general displays an image of fighting that is at odds with the 'canonical' conception of the hoplite phalanx, and consequently this forms the main thrust of his argument. As I hope to show, however, such evidence is often ambiguous and in any case cannot attain priority over literary sources. A few examples will suffice to show that the iconography of C7 vase painting is open to differing interpretations.

The first example concerns a sherd from a Middle Corinthian krater.[506] In this early representation of phalanx fighting, it is just possible to see what is going on. Five hoplites are apparently fighting their way forward (the sherd is so badly bro-

---

505 On this subject, see *supra* 104–105.
506 van Wees (2000) 132–133 fig. 7.

ken that the two hoplites on the extreme left are visible only from the waist down and the knee down respectively). In their midst lies a sixth, fallen hoplite, his head with the helmet still on turned towards the viewer and one leg dangling limply over the other. Judging by the way he has fallen, it would appear that he belonged to the phalanx we can see. The hoplites are uniformly equipped with greaves, Corinthian helmet, corslet, spear and *aspis* except for one, who is wielding a so-called Boiotian shield. To the rear five hoplites, facing the same way, are visible.[507] They all assume the same posture: one knee strongly bent, the other significantly less bent. These hoplites' shields are all represented in profile. Van Wees concludes, "this unglamorous bunch must surely represent the rear ranks, standing back from the action and crouching behind their shields to present a smaller target to enemy missiles".[508] But there are two general objections to this interpretation.

(1) There is no reason to assume that the upper frieze on the sherd *must* represent different parts of one and the same scene. In between the two groups of warriors is a blot, which is probably where the handle was affixed: this would have served as an effective divider between different scenes. But even assuming that the two groups are connected, the scene may still be divided, perhaps displaying a chronological sequence to be 'read' in the direction of movement. In other words: the hoplites to the left may be the same as those to the right at different stages of the fighting, divided by a marker, much like in a modern comic strip.

(2) It is by no means certain that the second group of hoplites are standing at all. The posture with one leg strongly bent is well known in Archaic art under the name of *Knielauf* and is probably rather meant to illustrate running: on this interpretation, the rightmost hoplites are rushing to the aid of their comrades, already caught up in the battle to the left. This would explain the underhand grip used on the spears as well. That this is indeed the likeliest interpretation is suggested by a closer examination of the hoplites' stance or stride. It will be noticed that no hoplite's knee actually touches the ground (and only one is in fact even close to the ground). The same applies to the shields: none of the four visible shields touch the ground, though it must be supposed that a warrior, if crouching and resting, would immediately set down his knee and rest the shield on the ground. Furthermore, all the hoplites are represented with the same leg bent, the *left* leg. But this is a somewhat unnatural posture to assume if one is crouching with shield and spear: the natural posture would be exactly the opposite – resting on the right knee – which would produce much better balance and support for the shield, while keeping the right arm free to use the spear.[509]

The next example is the justly famous Chigi olpe, by the so-called Macmillan painter,[510] normally regarded as one of the finest extant representations of phalanx

---

507 van Wees sees six and five hoplites respectively (van Wees [2000] 132), but I can see no more than two times five.
508 van Wees (2000) 132.
509 The correct posture can be seen on a fragment of a funerary stele in the Ny Carlsberg Glyptotek in Copenhagen (I.N. 2787): here, two warriors (of which one is almost completely missing) can be seen crouching in exactly this position.
510 van Wees (2000) 134–139 fig. 9.

3.1 The development of the phalanx: myth or reality?   125

Fig. 15: Battle scene. Five hoplites are engaged in close fighting near the edge of the sherd, while a sixth lies dead on the ground beneath them. Behind them, five more hoplites are running towards the fighting, as can be seen from the Knielauf *representation. Sherd of Middle Corinthian column krater, attributed to the Cavalcade painter (c. 600–575).*

Fig. 16: Battle scene. Two opposing hoplite phalanxes engaged in close fighting. Notice the bronze cuirasses, shield blazons and insides of wooden shields. The 'Chigi' olpe by the Macmillan painter, Middle Protocorinthian (c. 650).

fighting.[511] Two groups of hoplites (four to the left, five to the right) are closing in on each other, spears poised for the downward thrust. Behind each group another group is apparently coming up from the rear, their spears still held in an underhand grip. Behind the left group a piper playing an *aulos* is calling the tune to the march, and the 'auxiliary troops' to the left have apparently only just completed donning their armour: a man is still arming, his spears stuck in the ground.

Although the frieze, van Wees agrees, *does* seem to represent phalanx warfare, "yet this picture is not what it seems".[512] The main objection is that the two groups of hoplites about to join battle carry second spears. To the left can be seen some weapons stuck into the ground, and some of the spears in fact appear to be javelins with an added throwing-loop. Therefore, van Wees argues, the picture cannot be taken to represent a 'traditional' phalanx battle: either the hoplites should have discharged their javelins long ago, or they are closing in on each other with the wrong weapons. But the artist probably wanted to represent both phalanxes and also their full panoply, and the argument that they are too close does not hold. Furthermore, as we have seen, hoplites and javelins were not necessarily mutually exclusive. Alternatively, the javelin might perhaps be retained as a reserve in case the thrusting spear was lost or broken: this would also explain the many twin spears despite the phalanxes' apparent proximity. At any rate, if the only obstacle to a phalanx identification here is the additional spear, it hardly seems insuperable.

A further problem is the seemingly superfluous legs and spears in the two opposing ranks.[513] Van Wees takes this to mean that the ranks are perhaps not so straight as they seem: some other hoplites must be 'hidden' behind those we can see – "whether in regular formations or bunched together in dense crowds, we cannot tell".[514] This technique, however, is the mark of an artist of genius struggling with the extremely complex task of expressing depth of vision without completely overloading the picture. His solution is completely successful: the additional legs (and spears) enhance the feeling of multitude by filling out less immediate areas of the picture above and below; as does one head too many in the second rank.[515] There is no good reason why we should surmise that the additional items represent anything but an artistically successful amplification of the regular formations we can see.

Yet another problem is the size of the fighting groups and their rear ranks. They are not completely equal; and besides, the second ranks seem to be running or at least trotting, and therefore they cannot be 'second ranks'.[516] Another obstacle is presented by the spears of the second ranks, held in underhand grip, suggesting a distance between the front ranks and their running comrades. It is still, however, possible to interpret the scene as a sequence of events: a hasty arming of one side, two armies rushing forth to give battle, and finally, at the centre of attention, the two armies seconds before the frightful clash. If this is so, the space between the fight-

---

511 See, e.g., Cartledge (1977) 19; Salmon (1977) 87; Anderson (1991) 18.
512 van Wees (2000) 136; cf. Krentz (1985b) 52.
513 van Wees (2000) 136.
514 van Wees (2000) 138.
515 Hurwit (2002) 15.
516 van Wees (2000) 138.

3.1 The development of the phalanx: myth or reality?   127

Fig. 17: Battle scene. Two opposing phalanxes engaged in close fighting. Aryballos by the Macmillan painter, Middle Protocorinthian (c. 650).

Fig. 18: Battle scene. Two opposing phalanxes in close fighting and different stages of rout and pursuit. Black-figure aryballos by the Macmillan painter, Middle Protocorinthian (c. 650).

ing hoplites and their running comrades may just as well be part of the concept, and in one case neatly enhanced by the piper. However, even if we choose to interpret the picture as one scene, there is no need to jump to conclusions. The easiest interpretation, to my mind, would be that two armies are about to join battle, their rear ranks rallying up to join them and therefore hastening somewhat more: perhaps the charge has come as a surprise to one side. It is equally possible that we see different stages of the battle, beautifully and almost seamlessly segueing into each other. However this may be, I cannot possibly see that this picture should represent disparate groups of hoplites, joining battle at will: the coherence, structure and unity of the picture are very clear indeed.

The two next examples are also attributed to the Macmillan painter.[517] These apparently earlier[518] works present a number of interesting features. The Berlin aryballos has a battle frieze in three sections; two apparently of the battle itself, the third of the rout of one side. As van Wees notes, there is a difference in the stride, which seems to be wider in the central section. Besides, the right-hand side generally has a wider stance than the left-hand side. Furthermore, the gaps between the hoplites are a little larger in the centre. Van Wees draws his conclusion: "it seems to me almost certain that the differences in spacing are by design rather than by accident, and indicate more and less open formations".[519] The difference may be there, but in reality it is hardly noticeable. Ultimately it may simply depend on the curious

---

517 van Wees (2000) 140–142, figs. 10–11.
518 Salmon (1977) 87–88, figs. 1–3; van Wees (2000) 137, 141, figs. 10–11.
519 van Wees (2000) 140.

difference in the representation of the shields: the frontal view of the right-hand side hardly allows for an even closer formation, if the viewer is still to see the shield blazons, a subject which apparently interested the Macmillan painter greatly. At any rate it scarcely warrants the notion that the left-hand side "are standing still, packed tightly together, while their opponents ... are moving towards them in rather more open formation".[520] The spacing is indeed greater in the centre; but this might as well indicate a later stage of the battle, in which the ranks have thinned because of the slaughter. The last stage would then quite naturally be the rout: the left-hand side are in full flight; two hoplites kneeling and one cowering, all three about to be run through by their adversaries.

The so-called Macmillan aryballos is next. A highly stylised piece of work, it is quite clear in its unflinching portrayal of the grim realities of hoplite warfare. The subject once again is the rout of an army on the left-hand side by its right-hand side adversaries, the absolutely last phase of a battle. The hoplites can be divided into three groups, and the battle scene here may again be viewed either as a sequence of events or as different aspects of one continuous scene, seen across the battlefield. In the leftmost group, the left-hand side is still putting up some resistance: two hoplites are still fighting, and one is dealing the death-blow to his opponent, who is on his knees. In the centre, things are going worse for the left-hand side: two kneeling hoplites are being finished off by their adversaries, while the third is fleeing, yet still fighting back. The last group is utter rout of the left-hand side, again with two kneeling hoplites being killed and no resistance being offered. The three sequences again segue imperceptibly into each other, the two scene transitions consisting in kneeling hoplites, who leave a blank space above them. Van Wees admits that this "is no doubt very much what a breakthrough in a Classical battle would have looked like, but it is no less close to what the *Iliad* describes ...". He accordingly concludes that the three vase paintings by the Macmillan painter are "superficially suggestive of the Classical phalanx, but in detail turn out to have closer analogies in Homeric battle scenes".[521] In my opinion, the three scenes represent hoplite warfare (and *eo ipso* phalanx warfare); and they do so very successfully indeed. The resemblance to phalanx formation is anything but superficial, and the artist seems to have been at great pains to express structure and regularity.

Yet another vase image, from the lid of a *lekanis* from Kyme, shows a battle between two phalanxes, much like the Chigi vase.[522] Two light-armed horsemen, slightly overlapping each other and in the act of throwing javelins (easily identifiable as such by the throwing-loop), lead the charge towards the left and towards the central scene. This is evidently the killing of Astyanax by Neoptolemos; so despite the presence of the horsemen, the scene most likely actually represents the sacking of Troy. Behind the cavalry follow three rows of seven hoplites, marching or running forward. They wear Corinthian helmets with high crests and bear shields with a wide array of different blazons, and greaves. They are uniformly armed with one

---

520 van Wees (2000) 140.
521 van Wees (2000) 142.
522 Beazley *ABV* 58.119. See Alföldi (1967) 14 and pl. 1–2; Gabrici (1912) 124–128 and pl. 5–6.

spear each held in front of them, spearheads pointing up. The three rows of hoplites are separated by single hoplites, apparently dismounting from their horses to join the phalanx, while mounted but unarmed squires hold their horses for them. The hoplites are depicted much like on the Chigi vase, in near uniform equipment, but with a variety of shield blazons; and their placing in rows of neatly overlapping hoplites in profile – so as to create a sense of depth and internal proximity – but with the shields turned directly towards the viewer is strongly reminiscent of the former artist's effort to portray the close ranks of a phalanx effectively.[523]

The differences between van Wees' interpretation and mine show that very different approaches are possible, and that an attempt towards interpretation is bedeviled by the ambiguity inherent in early iconography. There is no external criterion of control to ancient iconography; the iconographical 'code' may be altogether inaccessible to us. Both interpretations thus rest ultimately on 'modern' readings, and they are therefore, in principle, equally valid. It remains my belief, however, that my interpretation is much more easily reconcilable with the rest of the source material.

What is more, the reference to the *Iliad* is an argument that cuts both ways. Even if we grant that the vase paintings actually do represent scenes from the *Iliad*, this does not prove that the combat scenes of the *Iliad* are completely consistent and realistic. If the paintings really are meant to resemble the *Iliad*, this proves nothing but the popularity of the poem; and it certainly does not prove that the postulated *reality* is actually mirrored in the iconography. Furthermore, if a contemporary artist were to reproduce *any* literary, quasi-mythical scenes – from the *Iliad* or elsewhere – they would naturally appear in a more or less contemporary guise. It is therefore impossible to judge whether the postulated similarity between the battle narrative of the *Iliad* and the vases is imagined or real, incidental or deliberate. More importantly, however, it is impossible to establish a relation between reality on one side and iconography and literature on the other. This, in my view, renders the connexion reality – *Iliad* – iconography essentially null and void.[524]

Another reason why there are so few representations of phalanxes in action may be the difficulty of representing a phalanx correctly. This is something that is forbiddingly difficult to do with any degree of accuracy, and especially so given the lack of understanding of true depth and perspective. Even if the artist did in fact manage to present a clear side view of orderly ranks of hoplites 'behind' each other, this is perhaps not the most exciting or dramatic visual aspect of the hoplite phalanx. The clash of two opposing phalanxes may well have been an image too grand to reproduce in a way that did not discredit the original experience.

Not so, however, with single combat. Two single combatants can be portrayed in a number of ways that emphasise the toughness of the combat and which underline the physical, individual aspect of it. Thus, by focusing on smaller groups, the

---

523 See also Greenhalgh (1973) 111–112 and fig. 56.
524 Cartledge, with admirable clarity, sums up the difficulties: "the archaeological record is lacunose and insecure ... visual artists, despite their interest in the warrior as a subject, had no professional concern to represent his equipment or exertions with photographic fidelity, even when they possessed the requisite technical skill" (Cartledge [1977] 12).

artist can capture the attention of the viewer and centre it on something that better conveys the *personal* and individual experience of battle. By this synecdochical method a better understanding of the whole is obtained, and on a personal level, much more empathy is generated.[525] This is probably the best explanation why the Greeks produced so few accurate images of phalanxes, and even fewer reliable portrayals of phalanx *battles*.

The contention that there are few specific portrayals of recognisable phalanx fighting from the Archaic period can in fact be extrapolated to encompass the entire antiquity: with the Nereid monument as a possible exception, there are scarcely any representations of phalanxes from the Classical period either.[526] This is a significant stumbling-block to the idea that the reason there are no vase paintings from earlier periods is that there *was* no such thing as a phalanx in the 'classical' sense of the word. In this light it becomes rather difficult to explain why there should be such a dearth of accurate representations of hoplite phalanxes, whether in a combat context or not. The extreme technical difficulty of painting a massed formation in perspective, along with the loss of focus and personal interest this would entail, prevented battle arrays from ever becoming popular subjects for artistic expression. This is yet another reason, if any were needed, why analyses founded chiefly in iconography should be set forth with extreme caution.

3.1.6 The Amathus bowl

Nevertheless, it is not correct to assume that there are no very early representations of a hoplite phalanx in the classical sense. Apart from the Chigi olpe and countless other instances of hoplites fighting in pairs (not incompatible with certain phases of battle or the choice of focusing), there is in fact a very early representation of what seems quite clearly to be a hoplite phalanx. This is the so-called Amathus bowl, named after the Cypriot city in which it was found in 1875.[527] The bowl is made of

---

525 This has been the simple formula for war images all through history, right until Robert Capa's justly famous WW II photographs or the awful images of the Vietnam war: the power of such images as Capa's photograph of a Spanish Republican militiaman caught in the moment of his violent death in the Spanish civil war, or the horrifying snapshot of crying, burnt Vietnamese children running away from a napalm airstrike on their village, is such that they have become instantly recognisable and iconic, a part of our shared cultural legacy. Paradoxically, the magnitude of the suffering of war is best brought out by focusing the attention on single, crystal-lised, stand-out events of great suggestiveness. Hanson (1991) 65–66 n. 10 interestingly draws attention to the fact that the sources' dry accounts of hoplite battles are often interspersed with anecdotes of one or a few men.
526 The contention is not quite true. Andrew Stewart finds eight depictions of phalanxes in Archaic art (seven vase images and the frieze of the Siphnian treasury): Stewart (1997) 89–91, 247.
527 The bowl is in the British Museum (B.M. 123053): see Markoe (1985) 172–173, 248–249. Alternatively, the bowl can be seen at http://www.britishmuseum.org/explore/highlights.aspx, when the search string ANE 123053 is added. See also Myres (1933) 26 fig. 1, an excellent reproduction. Provenance: Markoe (1985) 173–174; Myres (1933) 25. I am deeply indebted to Nino Luraghi for calling my attention to the Amathus bowl. As he observes, it has gone completely unnoticed in discussions of early Greek infantry tactics – although it was reproduced already in Boardman (1980²) 50 fig. 19.

silver and somewhat fragmentary; almost half is missing. It is certainly of Oriental origin, possibly Assyrian or Phoenician,[528] though this cannot be said with certainty.[529] The Amathus bowl may be dated on stylistic grounds to the early part of the period of 710–675.[530]

The two inner bands are decorated with various Egyptian gods, sphinxes and also two figures in royal Assyrian garb, apparently picking leaves from a palmette.[531] A number of events are depicted on the outermost band, the most interesting for the present purpose being a great, walled Oriental city with three towers under siege: the enemy is closing in on the city from both sides, and to the left an orchard is being cut down with axes. The city itself is being attacked with scaling ladders from both sides, the attackers on the left holding their shields with spiked bosses upwards to protect them against missiles from above as they climb the scaling ladders.[532] On the right side, another man seems to be propping a ladder against the wall in preparation for the escalade. Immediately behind this figure follow four infantry warriors, and behind them again apparently four archers in long Assyrian overcoats, armed with short composite bows. The archers are followed again by two horsemen, one of them with elaborate headgear of Oriental or Egyptian appearance and holding a whip or perhaps a symbol of command.

The infantry warriors are represented as standing behind each other in a file, as is plain from the fact that their legs are crossing: the hindmost leg of the leftmost warrior is carefully engraved to be in front of the foremost leg of the man next to and 'behind' him, and so on.[533] The last warrior is sadly missing from the hip up due to a large hole in the side of the bowl (which has also removed all of the first archer except his legs and the Assyrian overcoat). The warriors are clearly equipped with weapons and armour instantly recognisable as that of Greek hoplites.[534] They are wearing close-fitting helmets of the Corinthian type (or perhaps, since it leaves the face free, more akin to the Illyrian type, though this could merely be a wish on the part of the artist to leave the face visible: the chin plates swing forward rather dramatically and give the helmets a strange resemblance to a modern crash helmet),[535] complete with horsehair crest across the crown, round shields with blazons, carried on the left arm, long thrusting spears with leaf-shaped spear-heads wielded overarm in the right hand, and finally greaves on their legs.[536] It is impossible to judge whether they are wearing cuirasses or corslets, since the shields cover

---

528 Markoe (1985) 173; cf. Luraghi (forthcoming) 15–16.
529 Myres (1933) 39.
530 Markoe (1985) 151, 154–156; Luraghi (2006) 36.
531 Markoe (1985) 172–173.
532 Incidentally, a shield of precisely this type was found together with the bowl and sundry weapons in a grave chamber: Myres (1933) 35 n. 26; Markoe (1985) 173–174.
533 Myres (1933) 29–30: "Similarly, in greater detail, each of the foot-soldiers occupies 10° between toe and heel, but the four together fill only 30° [of the bowl's circumference]."
534 See Myres (1933) 35–36; Markoe (1985) 173; Luraghi (2006) 37–38.
535 Snodgrass (1964a) 31 refers to this as a subvariant of the Corinthian helmet, the so-called Ionian helmet.
536 All this is radically different from the traditional Assyrian manner of portraying warriors: Markoe (1985) 52–53.

Fig. 19: Siege scene in a near-Oriental setting. A hoplite phalanx is storming a citadel, which counts among its defenders another contingent of hoplites. The Amathus bowl, silverwork; possibly Assyrian or Phoenician (c. 710–675).

their torsos completely; but underneath they are certainly wearing fringed *chiton*s, protruding under the shield edges. The blazons on the shield are seemingly all variants of ordinary 'geometric' patterns, common enough in the Archaic period: stars, whirls and so on.

These hoplites must be intended to create an impression of a phalanx: the perspective may be rather simple and the execution crude in comparison with later Greek art, but then the diameter of the bowl is only 18.8 cm, the height a mere 3.6 cm. In fact, the engraver has suggested depth in the hoplite file with considerable skill by means of the carefully executed criss-crossing legs – not unlike the technique employed by the painter of the Chigi vase. It would seem that so careful a 'pseudo-perspective' arrangement and so precise intervals and postures must be an attempt to capture the evenly intervalled ranks and files of a hoplite phalanx. This is borne out by the fact that none of the other warrior types – horsemen and archers – depicted on both sides of the phalanx are shown with 'overlapping' limbs, except perhaps for the significantly smaller warrior in Oriental equipment propping a ladder up against the citadel wall immediately next to the phalanx, or such minor details as the extreme tip of the hoof on the foremost horse, which is obscured behind the first archer: in fact the artist seems at pains to avoid getting the figures in all three bands mixed up by overlapping, except for the hoplites: even their raised right arms with the spears conspicuously fill out the small interval between the helmets. Inside the citadel under attack, various defenders look down from the parapets and towers, many of these equipped in the same manner as the hoplites outside (although only one has a crested helmet); but at least one appears to be an Assyrian archer.

The number of events taking place in the outermost band alone is staggering; and it is scarcely conceivable that an invading, besieging army would throw forward heavy battle infantry while attempting to scale the walls, even while yet others are busy laying waste to the trees and plants of the invaded country. Quite possibly the artist intended to suggest a sequence of events, reading from right to left: invasion, skirmishes between light-armed troops, advance of the invading forces' heavy infantry (possibly suggesting a battle between opposing forces of heavy infantry), and finally the assault on the citadel. The ravaging of the orchard could in all likelihood be carried out with impunity once the defenders had withdrawn behind their walls and the attackers had consolidated their positions.[537] This interpretation, then, is a parallel to that of the Chigi vase which I argued above: it is entirely possible that the iconography does in fact represent sequential portrayals of events.

This observation leads directly to another aspect of this fascinating artwork. It is noticeable that the three warriors who are doing the actual scaling are equipped quite differently from the hoplite phalanx: the two warriors on the left are not wearing any visible body armour, and their helmets (or possibly caps) may either be very snugly fitting metal skullcaps, or they are simply light protective headgear – caps made of sinew or leather, much like those apparently worn by Hannibal's Spanish mercenaries during the Punic wars.[538] It is also noteworthy that the two warriors scaling the ladder are not seemingly carrying any offensive weapons. This is of course absurd, but the engraver is careful to point out that their left hands are gripping their pointed shields, their right hand holding on to the ladder: possibly they carry smaller swords or daggers inside their shields.[539] The man with the ladder on the right may not be a member of the scaling party; at any rate he is not armed at all, save for a pointed (Assyrian) helmet with a neckguard, and possibly a cuirass or mail shirt of some type. Be that as it may, it is clear that the hoplites are not taking part in the escalade; and rightly so. As is apparent from the image here, hoplite weapons and equipment would be an all but impossible task to drag up the ladders: holding a shield of 7 or 8 kg raised high in the air with one hand while simultaneously clinging on to a rickety ladder with the other – and then somehow managing to also engage in active, fierce combat once the parapet is reached after an exhausting climb seems like both a superhuman effort and almost certain death.[540]

---

537 But see Hanson (1998²) 122–128.
538 See Connolly (1998²) 150–152.
539 Depictions of this practice exist, see *infra* 193–194. The spiked shield is a Cypriote type: Snodgrass (1964a) 31, 56.
540 This point is well made by Ober (1991) 180–183. It is in fact interesting that the Greeks of the Archaic and Classical periods never really developed effective siege tactics. The hesitation and perplexity of the Spartans facing the probably quite ordinary city walls at Plataiai is significant in this respect and reveals the weakness of Greek siegecraft at a relatively late time: see Thuc.

At the height of their imperial power, the Assyrians were very well versed in the art of storming cities; so it is maybe no surprise to find that these troops are doing the climbing; but at any rate the artist was careful to leave the Greek hoplites out of the escalade. Conversely, the towers and battlements are swarming with opposing, but identically equipped Greek forces inside: plainly, he was well aware of what could and what could not be done in hoplite armour. The hoplites inside, having the advantage of fighting defensively – in fact, doing precisely what their armour was designed for – and of being able to seek cover behind the battlements and parapets could be put to good use here, whereas letting the probably expensive foreign mercenaries scale the walls would be a costly exercise in futility.

These few observations have far-reaching consequences for our understanding of the early hoplite phalanx. Amathus may have been part of the Oriental Levant, and although it appears later as a Greek *polis*,[541] it was seemingly an independent kingdom earlier, as it features as such in the inscriptions of Esarhadon and Ashurbanipal.[542] The workmanship of the Amathus bowl is certainly Oriental, possibly Phoenician (the style is so-called Cypro-Phoenician), but it undoubtedly depicts Greek hoplites, arguably in a traditional phalanx, participating on both sides of a large-scale siege, and in two armies of Oriental origin. This very likely makes the Amathus bowl the earliest depiction of a hoplite phalanx at all.[543] Another important implication is that hoplites must have served abroad, quite possibly as mercenary soldiers for Oriental powers, at a very early date: 700 would be a conservative estimate for the age of the Amathus bowl, but already by this date we find largely uniformly equipped hoplites, apparently in large numbers, playing an active role in conflicts in the Levant. In fact, the spread of Greek hoplite mercenaries in the Orient was apparently such that already at this early period, two opposing powers might employ Greek hoplite mercenaries in significant numbers: "… the craftsman who made this bowl was familiar with the idea that Greek hoplites could be found fighting in Near Eastern armies, perhaps even on both sides of the same battle."[544] The notion that Greek mercenaries were a phenomenon not only of C4 but also a far earlier period – C7 – is often overlooked, but important nonetheless.[545]

2.75–78, 3.52.1–2 and compare 7.29.2–3: the fall of Mykalessos, a tiny hamlet by any standard, came about almost by chance and only because the walls and gates were decrepit and the city was inadequately defended; cf. Xen. *Hell.* 5.4.20–21 and 4.8.9–10: Peiraieus was unprotected by gates (ἀπύλωτος).

541 Pseudo-Skylax 103 calls Amathus a *polis* and adds αὐτόχθονές εἰσιν ("they are indigenous"); cf. Hdt. 5.104.1.
542 See Maier (2004) 1225.
543 Luraghi (2006) 37.
544 Luraghi (2006) 38; cf. Snodgrass (1964a) 31.
545 Luraghi (2006) 21–26. According to Herodotos (2.152), pharaoh Psammetichos (= Psamtik) I owed his accession to the throne to Ionian and Karian mercenaries; and see Alc. fr. 48 and 350 Voigt, describing Alkaios' brother Antimenidas' exploits as a mercenary, possibly fighting in the army of Nebuchadnezzar II and conquering Ashkelon in 601. Furthermore, several Greek mercenaries inscribed their names on a colossal statue of Rameses II at Abu Simbel in 591, apparently while in service of king Psamtik II (cf. Hdt. 2.161): Meiggs-Lewis *GHI*² 7a-g.

3.1 The development of the phalanx: myth or reality?

If that is so, it is only logical to assume that Greek hoplites and phalanxes, as they figure on the Amathus bowl, had been in existence for quite some time by *c.* 700. Their reputation will have required long time, perhaps as much as a few generations, to spread around the rim of the Mediterranean and to become established enough that several Oriental regents courted their favours. Also, artists from different cultures had taken notice of them and their unique appearance and tactics, incorporating them in their repertoire by this time, something that will also have required a certain period of time to do. These factors, along with the finds of the Argos grave, again points to *c.* 750 as the approximate date of origin of hoplite weapons and armour.

There is nothing substantial, then, in either iconography or *datable* Archaic literature to support the notion that hoplite fighting from *c.* 750 was in any respect more fluid or fought between more mobile warriors than those described in later sources; and there is nothing that necessarily and conclusively connects either of the two with what is portrayed in the *Iliad*. The contents of the *Iliad* itself are forbiddingly difficult (and ultimately probably impossible) to date with any reasonable degree of certainty due to its long-winded and extremely complex genesis.

As for the iconographical and literary evidence, there can be little doubt that there is in fact ample evidence of phalanx tactics. The presence of an additional spear for the hoplites and of light-armed troops is not in itself incompatible with phalanx tactics, and therefore not sufficient evidence to dismantle the prevalent notion of a hoplite phalanx, supported by other, more mobile troop types. If there are elements that point to open battle order in iconography and literature and the Homeric battles, there are even more similarities to hoplite phalanx fighting.

3.1.7 The phalanx during the Persian wars

The evidence for C7 on the whole points to an already well-established close-order type of fighting. Yet, as I have indicated above, van Wees not only accepts the *Iliad* as evidence for an open battle order in early hoplite fighting, finding similarities between this and Archaic source material for a different fighting style: the evolution of hoplite fighting, he asserts, continued until probably after the Persian wars. In Herodotos' battle descriptions, van Wees finds evidence that the hoplite phalanx, although by this time closer to the Classical form, was certainly still more loose and open: "At Marathon and Thermopylae, the Athenian and Spartan forces displayed a manoeuvrability unheard of in the battles of the Peloponnesian War."[546] He next adduces two passages from Herodotos.

The first is the celebrated pincer manoeuvre which doomed the Persian army at Marathon: the comparatively weak Athenian centre was broken by the Persian forces, whereas the Athenian wings routed their immediate adversaries, only to turn inwards and spring the trap.[547] If Herodotos' account is essentially correct, the Athenian battle array with a weakened centre is surely a sign of superior general-

---

546 van Wees (2000) 155, (2004) 180–181.
547 Hdt. 6.113 (inv. no. 19).

ship on the part of Miltiades and Kallimachos; but there is hardly anything done here that would require mobility above and beyond what could be performed by phalanxes of the Classical age. The movements are coordinated and do not as such require the individual hoplite to be extraordinarily mobile; and a quick glance at actual battle descriptions from the Peloponnesian war and later will suffice to demonstrate that manoeuvres just like this and more complicated were carried out time and again. Thus, at least equal mobility is shown by either whole phalanxes or separate contingents at, e.g., Delion in 424, where the Athenians surrounded a contingent of Thespians and spread out so much that they ended up attacking their own troops; at Amphipolis in 422, where Brasidas' troops forced the left Athenian flank to flee and then went on to engage their right; at Mantineia in 418, where king Agis ordered the leftmost contingents of his phalanx further to the left to avoid being outflanked and expected contingents from the right wing to fill up the gap; at Koroneia in 394, where king Agesilaos wheeled his phalanx about to re-engage the Thebans and collided with them head-on, and at Corinth in 392, where different contingents drove each other in various directions and returned to collide again; the Spartans, especially, waited while the returning Argives passed them by, then struck their flank.[548]

The second example concerns the battle of Thermopylai. Here the Spartan defenders of the pass are described as turning around and pretending to run away ever so often: every time, this convinces the Persians that victory is at hand, and accordingly they break ranks and pursue the retreating Spartans.[549] Now, the Greeks deliberately chose the pass of Thermopylai as their line of defence because the narrow pass was the only place that the enormous numerical advantage the Persians enjoyed would be negated.[550] This was a wise choice of strategic position, as later history, both ancient[551] and modern,[552] was to confirm.

---

548 Delion 424: Thuc. 4.96.1–9 (inv. no. 5.16); Amphipolis 422: Thuc. 5.10 (inv. no. 2.16), Mantineia 418: Thuc. 5.70–73 (inv. no. 17.16); Koroneia 394: Xen. *Hell.* 4.3.16–20; cf. *Ages.* 2.9–16 (inv. no. 8.16); Corinth 392: Xen. *Hell.* 4.4.10–12 (inv. no. 4.16).
549 Hdt. 7.211.3.
550 Hdt. 7.175. The width of the pass is given by Herodotos as ἡμίπλεθρον, Hdt. 7.176.2. A ἡμίπλεθρον is between approx. 14.78 and 16.65 m according to the definition of πλέθρον; so barely 17 m was the maximum width. If each hoplite took up 50 cm of space and the file interval furthermore was 1.83 m, there would be space enough for a file of no more than seven or eight men.
551 In 353, the Athenians prevented Philip II from crossing the pass, Dem. 19.84, cf. Aeschin. 2.132–133; Diod. Sic. 16.38.1–2; Just. *Epit.* 8.2.8. In 279 a coalition army under Athenian command fought to keep off the invading Gauls, Paus. 10.20.3–22.1, 10.22.8–13. Fighting also took place between the forces of Antiochos III and Marcus Acilius Glabrio in 191 (Livy 36.15.5–19.13; Plut. *Cat. Mai.* 13–14; App. *Syr.* 4.17–20) and between Philip V and the Aitolians in 207 (Livy 28.7.3), to name but a few.
552 During the German conquest of Greece in 1941, severely outnumbered and outgunned British and ANZAC forces successfully held the pass of Thermopylai and the higher passes from 22 April to 24–25 April against German mountain and armoured troops in a rearguard action, while the main force made good their escape from mainland Greece; see, e.g., Burn (1962) 409–410.

The overwhelming superiority of the Persian forces was a serious concern to the Greek coalition leaders, and the actual defenders never had any doubts about the inevitable result.553 An open order formation – particularly one allowing each hoplite 1.83 m – would in fact seem abject folly when facing an enemy so vastly superior in anything but terrain and possibly equipment. The Persians, by sheer pressure of mass in the narrow confines, would soon have overrun the scattered Greek forces. Indeed, if Herodotos can be believed, the Persian soldiers were actually driven forward under their masters' whips:

> τότε δὲ συμμίσγοντες ἔξω τῶν στεινῶν ἔπιπτον πλήθεϊ πολλοὶ τῶν βαρβάρων· ὄπισθε γὰρ οἱ ἡγεμόνες τῶν τελέων ἔχοντες μάστιγας ἐρράπιζον πάντα ἄνδρα, αἰεὶ ἐς τὸ πρόσω ἐποτρύνοντες. πολλοὶ μὲν δὴ ἐσέπιπτον αὐτῶν ἐς τὴν θάλασσαν καὶ διεφθείροντο, πολλῷ δ' ἔτι πλέονες κατεπατέοντο ζωοὶ ὑπ' ἀλλήλων· ἦν δὲ λόγος οὐδεὶς τοῦ ἀπολλυμένου.554

Hyperbolic as this description seems, it is nonetheless quite clear that Herodotos is at pains to create in the reader's imagination an image of the tremendous pressure of the swarming Persians: the throng was such that many Persians even trampled each other to death. The heavily outnumbered Greeks' only chance thus lay in fighting defensively by concentrating as many spear-points as possible within as little space as possible to keep the attackers at bay: ὡς δὲ καὶ οὗτοι [sc. οἱ Ἀθάνατοι] συνέμισγον τοῖσι Ἕλλησι, οὐδὲν πλέον ἐφέροντο τῆς στρατιῆς τῆς Μηδικῆς ἀλλὰ τὰ αὐτά, ἅτε ἐν στεινοπόρῳ τε χώρῳ μαχόμενοι καὶ δόρασι βραχυτέροισι χρεώμενοι ἤ περ οἱ Ἕλληνες καὶ οὐκ ἔχοντες πλήθεϊ χρήσασθαι.555 Herodotos is explicit that the Greek long spears and the throng in the pass rendered the numerical superiority of the Persians useless; and the only reasonable way to understand such fighting is to conceive the Greeks as packing themselves very closely together indeed. Their only hope of success lay in presenting as solid a bulwark of shields and spear-heads to the Persians as at all possible.

Therefore, it is a little surprising that the Spartans were able to perform such elaborate manoeuvres during a desperate and strictly defensive battle. Herodotos does say that the Spartans, on every occasion, μεταστρεφόμενοι δὲ κατέβαλλον πλήθεϊ ἀναριθμήτους τῶν Περσέων ("would wheel and face them and inflict in the new struggle innumerable casualties").556 How and Wells comment: "The pretended flight of the Spartans drew the Persians on, and thus made their losses heav-

---

553 Hdt. 7.223.2–4; cf. Diod. Sic. 11.4.3–5; Plut. *Mor.* 225a–d; but see Lazenby (1993) 146; Grant (1961) 25–27 and Hope Simpson (1972), esp. 9–11. A convenient overview of different scholarly approaches to Thermopylai can be found in Evans (1964) 231–235.
554 Hdt. 7.223.3 ("[N]ow they left the confined space and battle was joined on more open ground. Many of the invaders fell; behind them the company commanders plied their whips, driving the men remorselessly on. Many of them fell into the sea and were drowned, and still more were trampled to death by their friends. No one could count the number of the dead" [trans. de Sélincourt]); cf. 7.56.1 and 7.103.3–4.
555 Hdt. 7.211.2 ("But, once engaged, [the Immortals] were no more successful than the Medes had been; all went as before, the two armies fighting in a confined space, the Persians using shorter spears than the Greeks and having no advantage from their numbers" [trans. de Sélincourt]).
556 Hdt. 7.211.3 (trans de Sélincourt).

ier than if they had been merely kept at bay".[557] In this situation, the Greeks' primary concern must have been precisely to keep the Persians at bay if at all possible, but apparently performing such exploits of bravery and discipline was not impossible for the highly trained and professional Spartans. We may safely assume that the other Greeks present (Arkadians, Tegeans, Mantineians, Arkadian Orchomenians, Corinthians, Phleiasians, Mycenaeans, Thespians and Thebans[558]) fought in a more conventional manner during their relays.[559]

Furthermore, it must be assumed that these manoeuvres were carried out with perfect timing, since otherwise dense volleys from thousands of Persian bows[560] would quickly have decimated the scattered Spartans, when they turned to 'run away'. We must assume that the feigned retreat was quite short, also because the space was probably scant enough, possibly no more than 20–30 m, and consisted in the hoplites pivoting on their heels at a given signal, then shortly after spun round again at another signal. Probably the Spartans will have waited for a short lull, since it is difficult to imagine this carried out in the pressure from thousands of Persian soldiers. But if the Spartans were truly able to pull it off, the effect may well have been devastating.

Lazenby, probably acknowledging the inherent difficulties, offers another possible interpretation: Herodotos may have misunderstood a technique employed at least later by the Spartans, namely to let the front ranks of the phalanx (comprised of young men) charge in an effort to drive away light-armed adversaries.[561] This is an attractive solution, but does run counter to Herodotos' insistence that the Spartans did pretend to flee several times.[562] Although it should be borne in mind that one important purpose of Herodotos' account is the re-telling of larger-than-life heroics,[563] and the circumstances would make it difficult, the manoeuvre itself

---

557 How & Wells (1928²) vol. II 224.
558 Hdt. 7.202.
559 *Contra* Lazenby (1985) 88 who alleges that the tactics were "probably designed to nullify the effects of Persian archery by denying them a static target and by keeping the fighting fluid and at close-quarters". One wonders if the Spartans feigned their retreats constantly in order to deny the Persian archers a steady target, and if so, how long it would take for the Persians to grow wary of the rather simple stratagem.
560 The Persians were, above all, bowmen: see Aesch. *Pers.* 239–240 and the inscription on Dareios' tomb at Naqš-i-Rustam (quoted in Briant [1999] 112). The last Spartan survivors were undoubtedly finally picked off in a hailstorm of arrows (κατέχωσαν οἱ βάρβαροι βάλλοντες), Hdt. 7.225.2–3 and cf. Dienekes' coolly Laconic reply to the Trachian nervously assuring him that the Persian arrows would darken the sun (Hdt. 7.226).
561 Lazenby (1993) 138, comparing this passage to Xen. *Hell.* 4.5.15–16; see also 2.4.32 and 3.4.23.
562 See also Grant (1961) 16 for a sceptical view of this tactic.
563 Burn eloquently encapsulates Herodotos' approach: "Herodotos' story of the battle ... lies, in point of literary form, somewhere between sober history and the Chanson de Roland; nearer to history, admittedly, in that the principal facts are probably accurate; but 'fictionalised', not only in the accounts of the enemy's losses, but in the picture of that enemy, a cruder and more childish picture than is given elsewhere: Xerxes, the foolish tyrant, who waits four days for the enemy to run away, who has no plan except to hurl his troops against a fortified pass, until a countryman offers to show him a way round; who 'leaps from his throne in terror for his army',

would not be impossible to carry out for an army so highly disciplined as the Spartan. It is revealing in this respect that nothing of the sort is claimed for the other contingents.

However the case may be, van Wees' point that it is evidence of a phalanx somehow more fluid and mobile than those of the Peloponnesian war is easily refuted: there are actually later examples of such feats. Take, for instance, the battle of Haliartos in 395, when Lysander's Spartans were routed and hotly pursued by the victorious Thebans. When they began to ascend the slopes, the tables were quickly turned: the Spartan hoplites turned about and routed the Thebans by virtue of their higher position, actually throwing their spears like javelins.[564] The hoplites portrayed here are every bit as mobile as those in Herodotos' account of Thermopylai, and a lot more likely at that. The alleged evidence in Herodotos is thus insufficient to maintain the view that hoplites were more mobile at the time.

Moreover, there are the implications of Aristodemos who was denied the prize of valour despite fighting fiercely at Plataiai in 479, precisely because he recklessly left his station and threw himself into the thick of the fighting, thereby endangering the rest of the line: this indicates that the phalanx at this period was exactly as closed as later on.[565]

Van Wees' last allegation in this connexion, namely that every Spartan was actually surrounded by seven helots (περὶ ἄνδρα ἕκαστον ἑπτὰ τεταγμένοι)[566] in the battle formation at Plataiai, also faces certain problems.[567] Despite the fact that helots actually played a significant military role fighting as light-armed troops already at this time and did so on a regular basis later,[568] it is somewhat difficult to imagine that the Spartan army would run such risks of mutiny as this from the helots (who may have revolted as recently as 490 and definitely did so again in 464),[569] namely allowing every hoplite to be physically *surrounded* by no fewer than *seven* helots.[570] No matter how fearsome fighting men the Spartans were, this would have been inviting disaster: under no circumstances is one heavily armoured man a match for seven armed men.

and whose troops are no better than he deserves; slaves who have to be driven with whips to the slaughter; 'many bodies, but few men'. On the Greek side, the allies, except the Thespians, are fainthearted or worse. Only the Spartans shine." Burn (1962) 407.

564 Xen. *Hell.* 3.5.19–20, cf. Diod. Sic. 14.81.2; Plut. *Lys.* 28.6; inv. no. 7.
565 Hdt. 9.71.3–4 (inv. no. 29).
566 Repeated three times: Hdt. 9.10.1, 9.28.2, 9.29.1.
567 van Wees (2000) 155–156; inv. no. 29.
568 An interesting and apt parallel to the military employment of helots may be found in the use of recruited West Indian slaves in the British sugar colonies of the 18th century AD. Contemporary commentators were not blind to the fact that the reason slaves were willing to take up arms on behalf of their masters was that they were in all probability better fed and clothed in this capacity than otherwise, nor to the historical similarity with the helots: see Francis (1796) 43–45 and Woodfall (1796) vol. IV 54. I am much indebted to Stephen Hodkinson for calling these circumstances to my attention.
569 See Hunt (1998) 28–31, 38.
570 *Contra* Hunt (1998) 32–39, esp. 38–39. Hunt makes an excellent point that the helots' families back in Sparta functioned effectively as hostages; but there is no good military reason why the Spartans would have presented the helots with such a tempting opportunity in a battle context.

The reliance on helots as fighting men at Plataiai (and the inherent trust involved) should be weighed against other sources implying that the Spartans had good reasons to be wary of the helots.[571] According to a fragment of Kritias, they were in fact so suspicious that they removed the *porpages* of their shields when at home (οἴκοι) to prevent them from being used by the helots, while they themselves took their spears where they went and stood.[572] If conditions were like this during peacetime, control measures may well have been even more severe in the field, and it would at any rate be inadvisable to post seven potentially hostile hoplites around each Spartiate.

I suggest that the preposition περί must be understood in a *distributive* sense here: each Spartiate had seven helots *at his disposition*.[573] This is corroborated by an earlier mention of the 35,000 helots: the ephors ἐκπέμπουσι πεντακισχιλίους Σπαρτιητέων κατ' ἑπτὰ περὶ ἕκαστον τάξαντες τῶν εἱλώτων ("dispatched a force of 5000 Spartan troops, each man attended by seven helots").[574] Now, in this connexion an actual battle formation can scarcely be implied (despite the word τάξαντες), and I believe that the simplest solution is to regard the verb τάσσω as being employed in its other general meaning: to *appoint* or *assign*.[575]

Peter Hunt has recently argued that the helots did indeed fight alongside the Spartans. According to Hunt, the very specific ratio of seven to one can be neatly explained by the fact that this is just what is required for a first rank of Spartans with the helots making up the remaining seven ranks of the normal eight-deep phalanx.[576] A serious objection to this interpretation is Herodotos' insistence that the helots were equipped as and reckoned among the Greek *light-armed* troops:[577] this is neither reconcilable with van Wees' mixed phalanx (scarcely meant to be understood as one rank of hoplites and seven ranks of ψιλοί), nor with Hunt's apparent conception of an otherwise 'classical' phalanx – but with seven out of eight ranks unable to actually function as hoplites.[578]

---

571 See Lotze (1959) 34–35 for a discussion.
572 Kritias fr. 88 B 37 Diels-Kranz (= Lib. 25.63); cf. Ar. *Eq.* 843–859, 1369–1372; Xen. *Lac.* 12.4.
573 LSJ⁹ s.v. περί C.I.2.
574 Hdt. 9.10.1 (trans. de Sélincourt).
575 *Pace* Powell (1960²) s.v. τάσσω, who interprets these passages as meaning 'place' (albeit in the relatively rare non-military sense). The verb τάσσω is used 128 times in all in Herodotos; 75 of these in a military sense, and 33 times to mean 'appoint' or 'assign'.
576 Hunt (1998) 34–38; cf. (1997) 129–144.
577 Hdt. 9.29.1: ὁπλῖται μὲν οἱ πάντες συλλεγέντες ἐπὶ τὸν βάρβαρον ἦσαν τοσοῦτοι, ψιλῶν δὲ πλῆθος ἦν τόδε, τῆς μὲν Σπαρτιητικῆς τάξιος πεντακισχίλιοι καὶ τρισμύριοι ἄνδρες ὡς ἐόντων ἑπτὰ περὶ ἕκαστον ἄνδρα, καὶ τούτων πᾶς τις παρήρτητο ὡς ἐς πόλεμον ("This was the number of hoplites brought together against the barbarians. In addition to the hoplites there were light-armed auxiliaries: the 35,000 helots in attendance, as already mentioned, on the Spartans, all of them fighting men" [trans. de Sélincourt, modified]).
578 For a critical view of Hunt's analysis, see Figueira (2003) 219–220 and n. 48.

3.1.8 The Solonian census classes

A number of ancient sources attest to a new organisation of the Athenian citizens under Solon's archonship in 594/3. The Athenians were divided into four census groups: those citizens who produced at least 500 measures (μέτρα) of cereals, wine or olives were ranked as πεντακοσιομέδιμνοι; those who produced from 300 to 500 measures annually were ranked as ἱππεῖς; those who produced from 200 to 300 measures were ζευγῖται, and, finally, those producing less than 200 measures qualified as θῆτες.[579] The word *metra* was in this connexion apparently used to denote both dry and liquid measures: "*medimnoi* of 52 litres, by which corn was measured, and *metretai* of 39 litres, by which wine and olives were measured," the original standard probably being the dry grain measure, and presumably converted into a barley standard.[580]

From the outset the census qualifications were ostensibly employed chiefly to restrict eligibility to certain offices, but there has long been a tendency, based on linguistics and a number of seemingly cut-and-dried references in especially Thucydides,[581] to view them as the basis for assignment to particular arms in the Athenian army.[582] Specifically, *hippeis* are logically enough viewed as those possessing sufficient means to own and keep a horse, and thus to serve with the cavalry, while *zeugitai* are linked with the military meaning of ζεῦγος (or, more frequently, ζυγός or ζυγόν), i.e. a line of hoplites.[583] The *zeugitai* thus are the hoplites, 'yoked' together in the phalanx formation, with its connotations of interdependence and being locked closely together in battle. In this hierarchy, then, *pentakosiomedimnoi* are an 'artificial' superstructure, the only class of the four to be actually based on produce and included only for administrative and distributive purposes, whereas the *thetes* fall short of any qualification and only have the opportunity to serve as light-armed (peltasts or *psiloi*), or as rowers in the navy.[584]

The 'military' view was carried to its extreme by Geoffrey de Ste. Croix, who claimed that the Aristotelian *Constitution of the Athenians* had misunderstood the causality behind the four classes, and that there was in fact no fixed census: "I believe that a man had himself registered in the hoplite katalogos if he could afford to provide himself with arms and armour and was financially able to bear the burden of going on campaign when required ... I believe, then, that there was no fixed quantitative timēma possession of which made a man a hoplite."[585]

---

579 [Arist.] *Ath. pol.* 7.3–4.
580 de Ste. Croix (2004) 39, 42; Foxhall (1997) 129–131.
581 Thuc. 3.16.1, 6.43, 7.16.1, 8.24.2; Ar. fr. 248 Kassel-Austin; Antiphon fr. 61 Thalheim.
582 de Ste. Croix (2004) 13–14, 19–28; and see Whitehead (1981) 282 n. 8.
583 Thuc. 5.68.3; Polyb. 1.45.9, 3.81.2, 18.29.5.
584 Whitehead (1981); Rhodes (1981) 138; de Ste. Croix (2004) 13–28, 48–51.
585 de Ste. Croix (2004) 26, 23–24. In his opinion, the 2:3 ratio between these two classes is too insignificant to be taken as a basis for census qualification, and the *pentakosiomedimnoi* should be regarded as the only class to be defined on a basis of production and as a Solonian addition to a possibly pre-existing military division of three classes: de Ste. Croix (2004) 32–49, 51.

He thus reversed the traditional perception of the causality: a citizen was not eligible for cavalry or hoplite *because* he was a *hippeus* or *zeugites*. Rather, any citizen who served with the cavalry or the hoplite arm would automatically be classed with the *hippeis* or *zeugitai* respectively. Military stability would thus be ensured by the status inherent in serving with the military, as well as the social pressure to do so, which between them would be sufficient to induce any citizen to enrol.

Accepting this, however, clearly means rejecting the evidence of the *Constitution of the Athenians,* which states plainly that the defining criterion for *zeugites* and *hippeus* status was precisely the annual production of 200 or 300 *metra* respectively. The priority of this source has rightly been stressed by P.J. Rhodes (who had access to de Ste. Croix's articles in manuscript form). He comments, "we have no information which would justify us in rejecting A.P.'s figures as correct for Solon's definition of the classes."[586] The text of the *Constitution of the Athenians* must retain priority, and on the strength of this we cannot separate the original Solonian census from the *timemata* given there. Moreover, the use of the word ζευγῖται as referring to hoplites in a phalanx is attested only once, and in a late source,[587] while in some lexicographical notes it is said to denote a man able to keep a pair of oxen (a ζεῦγος), just as a *hippeus* is able to keep a horse.[588] When this distinction is upheld, we must accordingly accept the priority of the production-based census classes: *zeugitai* did in fact produce between 200 and 300 measures a year.

Now, the demographic results of accepting a correlation between *zeugitai* and a hoplite aspect of their status have been examined by van Wees. His analysis convincingly demonstrates that if a correlation between the hoplite status of the *zeugitai* and the production of a minimum of 200 measures is accepted, then there simply was not enough arable land in Attika for a hoplite army of the size indicated by Thucydides: 13,000 hoplites in the field army in 431 (not including the highest and lowest age groups, normally assigned to garrison duty, and possibly bringing the hoplite total up to between 18,000 and 25,000).[589] Van Wees' inescapable conclusion is that we cannot avoid assuming that a great deal of the hoplite force were in fact *thetes* by census class.[590] This is the only possible solution if we are to solve the problem of too little arable land for the hoplite population effectively.

---

586 Rhodes (1981) 145.
587 Plut. *Pel.* 23.4 (on Leuktra 371, inv. no. 16); de Ste. Croix (2004) 50 and cf. Rosivach (2002) 37.
588 Poll. *Onom.* 8.132. See also Hansen (1999²) 44, *contra:* Whitehead (1981) 285–286; Rhodes (1981) 138.
589 Thuc. 2.13.6–7; van Wees (2001) 47–54, esp. 48–51 and n. 47. The total population of male citizens at this time is variously assessed between 40,000 and 60,000: Hansen (1988) 14–28; van Wees (2001) 52 n. 50. Based on grain yield assessments van Wees estimates a size of between 8.7ha and 13ha for a plot sufficient to produce 200 measures of grain and the arable land of Attika at 96,000ha: accordingly, there was not enough land even for 13,000 hoplites. To this should of course be added the farms of those possessing even larger tracts of land (*hippeis* and *pentakosiomedimnoi*), and the land in the hands of those who fell below *zeugites* status. Even a tentative lowering of the census will not solve the problem entirely.
590 van Wees (2001) 51–56. A lowering of the census is also envisioned by de Ste. Croix (2004)

Since he thus effectively – and to my mind persuasively – disposes of the notion of a sign of equation between *zeugitai* and hoplites, it is all the more puzzling to note that van Wees agrees with the mainstream in contemporary scholarship in believing that the census classes were in fact rooted in military functions:

> Although we have no explicit evidence that the property classes already had a military dimension in early Greece, it is safe to infer that they did. The name *zeugitai*, 'yoked men', almost certainly refers to fighting in a rank, sometimes called 'a yoke' in Greek, which strongly suggests that from the moment it was created, whether as part of Solon's reforms or even earlier, this class was defined primarily by its duties in war.[591]

If we follow van Wees' argument that at least half the Athenian hoplites were *thetes*, the sign of equality between hoplites and *zeugitai* must necessarily disappear, and there is no longer any reasonable basis for interpreting *zeugites* as a military term. If, on the other hand, we follow the now generally accepted view – still upheld by van Wees – that *zeugitai* denotes 'yoked men', in a military context an obvious reference to closed phalanx fighting,[592] it follows that the closed phalanx must antedate by a certain margin the Solonian reform of 594/3.

3.1.9 The development of the phalanx: conclusions

The evidence thus for the most part militates against an interpretation of the early hoplite phalanx as evolving from a supposed fluid and more or less chaotic battle 'order'. Though there are certain differences from the Classical authors to the hoplite phalanxes described in early Archaic poetry – the many references to javelins and other missile weapons being one – the picture presented by the early evidence is on the whole consistent and does not differ in any crucial respect. Moreover, the iconography, such as it is, only serves to confirm the idea that artists struggled with the daunting task of accurately depicting two hoplite phalanxes engaging in combat.

The consequence of this is that trying to re-evaluate the age of the hoplite phalanx by analysing interpretations of iconography and a few passages in Archaic poetry and extrapolating it to cover the entire Archaic period from *c.* 700 to well after 480 is not a very fruitful method, nor is it strictly valid. The alleged inconsistencies are trifling; and the picture of hoplite fighting that emerges is on the whole remarkably compatible with the hitherto more or less accepted notion of phalanx fighting in the Classical period. On the strength of this, I submit that it would be a

---

51–56.
591 van Wees (2001) 46.
592 van Wees (2001) 46 n. 8 explains this crux away by stating that "the notion of men 'yoked' together on the battlefield need not be taken as evidence of a very close and rigid formation, but refers more generally to the solidarity and hard work of the soldiers, and perhaps also ... to an element of compulsion in military service." But the notion conveyed by the root word ζεῦγος is not one of 'hard work', but one of 'lashing securely together': see, e.g., Chantraine (1968) s.v. ζεύγνυμι: " 'atteler avec un joug' ... 'lier solidement, attacher ensemble' ". If a notion of 'hard work' were what was needed, Greek is more than capable of supplying an altogether less ambiguous term.

more viable and productive approach to recognise these consistent similarities and accordances for what they are and conclude that the hoplite phalanx and its fighting principles were in all crucial respects essentially 'right' more or less from the very beginning.

This squares with the observation that the invention of arms and armour generally happens in response to quite specific combat requirements, which is also true in the case of hoplite arms and armour. Hoplite weapons – above all the shield, closed helmet and bronze cuirass and greaves, in short the defensive weapons – may have been technical masterpieces in their own right, but they were also heavy, awkward and constrictive; they limited and hampered movement, hearing and vision, and they were incredibly uncomfortable under combat conditions, especially in Greece in the summer. It is unlikely that these weapons should have been introduced in a style of fighting where they would inevitably spell the bearer's doom, against all odds found favour here for the next three hundred years under circumstances that would surely have meant their rapid extinction – only to finally fit in perfectly with a new, closer type of fighting, which was, incidentally, diametrically opposed to the type which supposedly prevailed at the point of its introduction.

This is somewhat illogical. As Latacz' analyses demonstrate, *some* type of close-order fighting is clearly in evidence in the *Iliad,* so much in fact that it can be perceived even when we peer as through a glass darkly into the multi-layered and often self-contradictory confusion that is the world of Homer: the influence of fighting in *stiches* and *phalanges* over the centuries of the poem's composition was such that it makes itself felt, despite all the other jarring inconsistencies of the military sphere.

The invention of the concave hoplite shield, along with the other items of the canonical panoply, reveals certain specific needs based on a period of prior experience with close-range and close-order fighting. Quite possibly the eventual stimulus that produced the happy invention was the result of a contemporary tendency towards an even closer order; but it may equally well simply be just that – a happy invention. In any case, this invention relatively quickly proved its worth to the point of all but supplanting other types of armour and defensive weapons throughout the Greek world. Be that as it may, it should now be clear that the weapons were constructed sometime around C8m for this very specific type of fighting, and not the other way round.

We should also accept that phalanx fighting did not evolve in any fundamental respect over the next 400 centuries. It was, from the beginning, a close-order type of combat, and the hoplite equipment provided the very basic qualities required for this grueling fighting style: (1) massive frontal protection for thoroughly defensive, yet ultimately non-yielding brutal hand-to-hand spear and sword fighting; and (2) the ability to create a larger frame of protection, inasmuch as the very closed and compact nature of the phalanx made it physically possible for the hoplites to provide much-needed lateral cover for each other by means of the large shields. This was an extremely successful formula. On the battlefields of Greece, phalanxes composed of hoplites with the new equipment must have proved time and again from the very outset that they were a match for just about any known type of military land power known at the time. So astounding was their success, in fact, that their reputation spread around the rim of the Mediterranean and made it possible for Greek hoplites to go abroad and serve successfully as mercenaries, as is in evidence from different sources. These hoplites, naturally, brought with them the formula that had guaranteed their military superiority at home: the invincible constellation of arms, armour and tactics. This formula must have been successful enough abroad that it found favour with employers of mercenaries in a variety of military climates, and this again suggests that the hoplite phalanx was something unique in its day. Oriental employers cannot have experienced any shortage of warrior types who fought freely and in less organised, fluid formations: it should come as no surprise that what the Greek hoplite mercenaries offered was entirely different.

The evidence, as well as analogy with arms history, therefore points to one conclusion: the notion that the fighting method of the hoplite phalanx developed in any *significant* degree during the Archaic period may be dismissed. The consequences of this are that we should accept that hoplite phalanx fighting, inasmuch as it was a very good formula from the outset, never needed any major revisions. In the light of the typological consistency of the main offensive and defensive weapons and the way of fighting that they necessitate, it would seem that the concept of hoplite phalanx fighting did not alter in any significant respect at *any* point throughout its existence from C8m to C4m: there are simply too many points of contact between what can be assessed about the earliest hoplite phalanx and its relatively well-documented Classical parallel. The differences are so few and far between – the presence of javelins in C7e phalanxes in both iconography and literature being a good example – that they are more than made up for by the similarities.

The consequences of this observation are far-reaching. Rather than employing van Wees' method for C7 – i.e. interpreting especially the iconography and then extrapolating it forwards in time, probably all the way until after the Persian wars – we may in fact rely on the comparatively abundant sources for the hoplite phalanx from C5 and C4, thereby attaining a clearer view of *both* the Archaic and the Classical period by interpreting the comparative fullness of the 'late' literary sources chiefly from the latter – to extrapolate 'backwards' from the written sources, as it were. This has the added advantage of relying principally on sources we can be *certain* actually have phalanx combat as their object, rather than on the far vaguer iconography – vague partly because illustrations are often rather crudely executed, partly because we lack a certain point of reference and knowledge of the artist's intentions. This problem applies in a much smaller degree to the written sources, which, to all intents and purposes, transmit their information in a much more direct and immediate way.

To paraphrase: since the style of fighting must have matched the nature of the weapons and the range of their employment, and since these were developed for use in this style of fighting and did not develop significantly through the entire hoplite era, it is strictly speaking less relevant *when* the texts are composed: phalanx battle was a type of combat the basic elements of which it was difficult, if not downright impossible, to change noticeably. Our sources from the Classical period therefore form an excellent basis on which to formulate a generel hypothesis about the kind and nature of hoplite warfare, regardless of the specific time. It is certainly still possible to debate whether the transition to the Archaic-Classical phalanx happened by degrees, and if so, how quickly; but we can say with a very high degree of certainty that the dominating form of land warfare from *c.* 750 to *c.* 320 was *not* a result of the invention of hoplite arms and armour, but, on the contrary, the direct cause. Furthermore, we can say confidently that the type of combat, being dependent on the weaponry, did not alter significantly over these 400 years.

3.2 THE SHIELD IN PHALANX FIGHTING

3.2.1 Introduction

To truly understand the purpose of the hoplite's offensive and especially his defensive weapons, we must turn to the tactics and fighting principles employed in the phalanx. I have pointed out above that the defensive weapons – particularly the helmet and shield – were ill-suited to single combat and hand-to-hand combat in duel-like situations. On the other hand, the heavy equipment makes perfect sense in a phalanx context: it provided massive frontal protection (at the cost of freedom of movement and vision), and it enabled the bearer to withstand a considerable pressure from shields and bodies. Full freedom of vision was not necessary when the hoplite assumed his position in a close-packed formation, the only way was characteristically forwards, and when the hoplites, with few exceptions, were protected on all sides by other hoplites. The hoplite hardly ever needed to worry about being attacked in the back or side; for here his brothers-in-arms stood to protect him.

The shield was therefore especially important in phalanx warfare. As we have seen, it was very large indeed, roughly covering the bearer from chin to knee, and quite concave. Hanson has argued cogently that there is good reason to believe that the round shape was dictated precisely by the demand for concavity.[593] A rectangular shield, such as the later θυρεός or the *scutum* of the Roman legionary might theoretically be ideal, since its straight edges would make it possible for the shield edges to truly overlap. As Hanson notes,

> a wall of round shields left vital areas of the front lines unprotected; 'triangles' of exposure appeared between shields at both top and bottom. Indeed, the whole notion of a phalanx advancing with 'locked shields' is technically incorrect: it is impossible to create a continuous solid joint between curved edges, no matter how much shields overlap.[594]

But the gist is that the shield must of necessity be hollow all the way round, and this is infinitely more difficult to achieve with a quadrangular shape: here, awkward angles and corners cannot be avoided, and the resulting shape would render the shield thoroughly inexpedient. With the circular, concave hoplite shield the hoplite was afforded a maximum of protection all the way round, including the dangerous angles from above and below. The hollowness of the shield, furthermore, combined

---

593 Hanson (1991) 69–71.
594 Hanson (1991) 70.

with the angled shield edge, enabled the bearer to rest or hang the shield on his shoulder, thus relieving his arm temporarily. Since it was of paramount importance to hold out the shield at (half an) arm's length and not let it sink constantly, this was an absolute necessity, considering its weight and size. Besides, the hollowness allowed the bearer to push himself into the large shield, actually hiding inside it.[595]

There can be little doubt that the shield was absolutely central to hoplite warfare, not only in that the fighting style was built around it, but also to the extent that it played an extremely important part in the imagery of war altogether. This may also be reflected in Herodotos' puzzling story of the Athenian hoplite Sophanes, who allegedly had an iron anchor secured with a chain to his *thorax*. This anchor he would fix in the ground when he came close enough to the enemy ἵνα δή μιν οἱ πολέμιοι ἐσπίπτοντες ἐκ τῆς τάξιος μετακινῆσαι μὴ δυναίατο ("to prevent their attacks from shifting him"). Herodotos himself quite likely realised that this was perhaps a bit fantastic, for he immediately adds another version of the story – namely that the anchor in question was just a shield blazon, and that Sophanes merely whirled the shield round and round to create a simulation of a spinning anchor.[596]

How one can imagine spinning a shield weighing some seven or eight kg with nothing but the strap-like *porpax* at the centre is beyond comprehension, not to mention the reason why anybody would want to do so. Consequently, neither version of this anecdote seems to be necessarily true; however, the point of the anecdote remains much the same. Sophanes' strange invention, whether real or imagined, neatly sums up the ideal of phalanx fighting: to remain steadfast in the same place and not to flinch in the face of the enemy – or at least never to retreat, no matter what the situation. The anchor and chain express the almost superhuman ideal of the fighting in that this contraption would unerringly do what human nerves and muscles might not *necessarily* be expected to do: withstand the massive pressure and never yield as much as a step. The story of Sophanes may be unique in its almost surreal symbolism; but the principle of steadfastness, of not giving ground, is certainly ubiquitous in the literary sources. Chief among the phenomena that bring home the importance of the shield, of cohesion and interdependence, is that of *rhipsaspia,* throwing the shield down during combat.

3.2.2 *Rhipsaspia*

The centrality of the shield to phalanx fighting may – among other things – be gleaned from the quip ascribed to the Spartan king Demaratos;[597] but also from quite a few other sources. Archilochos' C7 poetry, for example, is steeped in stinging wit and confessions of outrageous behaviour. He lashes out against friend and foe alike, and particularly against a certain Lykambes, who had apparently betrothed his daughter Neobule to him and then gone back on his word. This is per-

---

595 *Supra* 41–42.
596 Hdt. 9.74.1–2 (trans. de Sélincourt).
597 *Infra* 164.

haps less surprising if we consider the Ionian iambic tradition, and there is every possible reason to suppose that he was serious when he wrote these famous lines:

> ἀσπίδι μὲν Σαίων τις ἀγάλλεται, ἣν παρὰ θάμνῳ,
>   ἔντος ἀμώμητον, κάλλιπον οὐκ ἐθέλων·
> αὐτὸν δ' ἐξεσάωσα. τί μοι μέλει ἀσπὶς ἐκείνη;
>   ἐρρέτω· ἐξαῦτις κτήσομαι οὐ κακίω.[598]

This, like much of his other poetry, was certainly intended to create a scandal as it went completely against the grain of everything hoplite fighting stood for. The frank admission of having thrown away the shield is tantamount to admitting that the poet had abandoned his part of the collective responsibility in holding the shield not only for himself but τῆς κοινῆς τάξεως ἕνεκα. The shocking part of it all is that he is deliberately mocking this ethos by stressing that the shield is essentially expendable, a mere tool, something which may easily be replaced. By refusing to see the shield as something more than the sum of its physical components he sends a very clear message, a message that certainly was not lost on his contemporaries.[599]

If, however, the contemporary martial values were similar to those of Homer, Archilochos could not count on making any special impression on his public: Homer's heroes take to their heels time and again in the *Iliad,* and although shying away from battle is not exactly preferred behaviour, retreating when the fighting gets too hard is not in itself something necessarily associated with disgrace. In fact, we often hear that mighty warriors such as Odysseus or Aias sling their shields around so that they cover their back with the useful strap intended for precisely this purpose – the τελαμών – which in itself implies that this exact situation was expected to occur. Hoplite double-grip shields, on the other hand, had no such thing as a *telamon,* so they could not be made to cover the hoplite's back in flight – apart from the difficulties of quickly getting the arm out of both *porpax* and *antilabe*. In a panicky situation, the immediate solution therefore was to discard the shield altogether, thus immediately getting rid of 7–8 kg dead weight, which could not be put to any good use anyway. Naturally, this would have exposed the hoplite's neighbours in the phalanx in no small degree: less so, of course, if they were also retreating, but the symbolism inherent in this was powerful enough that dropping the shield was syn-

---

598 Archil. fr. 5 West ("Some Saian exults in my shield which I left – a faultless weapon – beside a bush against my will. But I saved myself. What do I care about that shield? To hell with it! I'll get one that is just as good another time" [trans. Gerber]). The poem was famous enough in antiquity to warrant a nod by Horace, who dropped his shield and fled at Philippi: see Hor. *Carm.* 2.7.9–12: *Tecum et Philippos et celerem fugam / sensi relicta non bene parmula, / cum fracta virtus et minaces / turpe solum tetigere mento* ("With you beside me I experienced Philippi and its headlong rout, leaving my little shield behind without much credit" [trans. Rudd]).
599 *Contra* van Wees (2004) 172: "Judging by the tone of the poet, Archilochus ... running away from the enemy, though embarrassing, was not wholly unacceptable behaviour in the fluid, mobile battles of his time." This point of view, however, underestimates the shock value intended in the poem and overlooks its reception throughout antiquity. Lines such as τί μοι μέλει ἀσπὶς ἐκείνη; / ἐρρέτω· ἐξαῦτις κτήσομαι οὐ κακίω show a remarkable preoccupation with the shield, if such behaviour was really more or less par for the course.

onymous not only with *personal* cowardice, but carried overtones of treachery as well, since in theory the entire formation was put at risk.

The poem has been interpreted in a variety of ways.[600] Most recently, Thomas Schwertfeger has argued that other Archaic poets such as Anakreon and Alkaios sang about throwing away their shield as well, and that this in itself proves that it was in reality nothing to be ashamed of. The Anakreon fragment is *very* fragmented: the line ἀσπίδα ῥῖψ' ἐς ποταμοῦ καλλιρόου προχοάς ("throwing down his (my?) shield by the banks of the fair-flowing river") does not permit any interpretation of moral stance beyond the observation that its structure is apparently very similar to Archilochos, and so it may be dismissed in this context.[601] The Alkaios fragment runs as follows: Ἄλκαος σάος †ἄροι ἐνθάδ' οὐκυτὸν ἀληκτορίν† / ἐς Γλαυκώπιον ἶρον ὀνεκρέμασσαν Ἄττικοι ("Alkaios is safe, but the Athenians hung up his armour/shield that was his protection in the holy temple of grey-eyed Athena").[602] Alkaios famously reported the poem to his friend Melanippos, also a nobleman.[603] Schwertfeger maintains that this *ipso facto* reveals that *rhipsaspia* did not necessarily entail shame or loss of privilege: "Keine *zeitgenössische* Stimme der Kritik am Verhalten ... des Archilochos, Alkaios und Anakreon ist uns überliefert. ... Wir dürfen also folgern, daß dem Adel der archaischen Zeit die ῥιψασπία im Sinne unehrenhaften Handelns unbekannt war".[604]

This would seem correct at first sight. However, the Athenians, who had won this battle at Sigeion in the Troad (*c.* 600), apparently felt quite differently in that they made a spectacle of hanging his discarded weapons on display at the temple of Athena. They thereby publicised the event to good effect: certainly, this toothsome story was not lost on Herodotos.[605] In my view, Alkaios may well have sought to minimise the damage done by reporting the news himself before worse rumours could destroy his reputation altogether. Schwertfeger also touches on another important point: Alkaios was most certainly a member of the very highest circles in Mytilene, as proved by the fact that his two brothers at one point aided Pittakos in an abortive attempt to seize power there.[606] As such, it is reasonable to assume that in oligarchic Mytilene, a person of such elevated rank as Alkaios would not be overly affected politically and socially by any such events.[607] Another indication

---

600 Schwertfeger (1982) 253–254 gives an overview of the different modern approaches.
601 Anac. fr. 85 Gentili (trans. Campbell).
602 Alc. fr. 401 Ba Voigt *apud* Strabo 13.1.38 (trans. Campbell, modified).
603 Alc. fr. 401 Bb Voigt *apud* Hdt. 5.95.
604 Schwertfeger (1982) 264 ("No *contemporary* censure of the conduct of ... Archilochos, Alkaios and Anakreon has been preserved. We may accordingly conclude that ῥιψασπία in the sense of dishonourable behaviour was unknown to the nobility of the Archaic period"); cf. van Wees (2004) 172.
605 Hdt. 5.95.1–2.
606 Diog. Laert. 1.74; cf. Hansen, Spencer & Williams (2004) 1027.
607 Schwertfeger (1982) 280 touches on this: "Und Archilochos, und mit ihm Alkaios und Anakreon? Sie sind Dichter, die in der Zeit der Entstehung und Ausbildung der klassischen Polis leben. Und sie sind, was wichtiger ist, Aristokraten, denen das Polisethos fremd ist" ("And what of Archilochos, and Alkaios and Anakreon? They are poets living in the age when the *polis* emerged and took shape. Moreover, they are aristocrats, for whom the *polis* ethos is al-

that Alkaios' stance in this matter rather reflects the nobleman's haughty rejection of lowly and 'irrelevant' civic virtues accrues from a fragment of Archilochos, displaying the gritty reality of contemporary warfare far removed from the strange duels of the Homeric heroes:

> οὐ φιλέω μέγαν στρατηγὸν οὐδὲ διαπεπλιγμένον
> οὐδὲ βοστρύχοισι γαῦρον οὐδ' ὑπεξυρημένον,
> ἀλλά μοι σμικρός τις εἴη καὶ περὶ κνήμας ἰδεῖν
> ῥοικός, ἀσφαλέως βεβηκὼς ποσσί, καρδίης πλέως.[608]

Archilochos consciously rejects the nobleman's warrior ideal: a tall man with a dainty hairdo, a swaggering gait and carefully trimmed beard (?) simply does not cut it. As a seasoned hoplite mercenary, Archilochos knew exactly what was necessary in the phalanx. The general of his wishes actually has less to do with an Achilleus than with a Thersites;[609] but a commander who is small, knock-kneed and sure on his feet reflects a foot-soldier eminently equipped for the hardships of hoplite warfare, in which he participates himself. This is rather surprising, if the warfare of Archilochos and Alkaios' day respectively were in fact so similar to the chieftain duels portrayed in the *Iliad*.[610]

At any rate, Schwertfeger bases his conclusion on only two things: (1) the lack of any contemporary criticism, which is to all intents and purposes an argument *e silentio*, and (2) the fact that Alkaios apparently wrote a poem to his *hetairos* Melanippos. But as I argued, the cat was out of the bag no matter what: his weapons were on permanent display at Sigeion, and he could never hope to quell the talk about town in any way whatsoever. Better, then, to make a brazen display of charming recklessness and emulate the famous Archilochos.

Furthermore, while several of the heroes flee the battlefield in the *Iliad*, it is hardly an example for imitation, and the outrage caused by such behaviour as Archilochos' is tangible. Thus, Plutarch claims that Archilochos was expelled from Sparta when they learned that he was responsible for the verses stating that it was "better to throw away the shield than be killed."[611] The Athenian Kritias – one of the thirty tyrants and a pupil of Sokrates – shrilly criticised Archilochos for his

---

ien"). However, I cannot accept his apparent conclusion that their behaviour reflects the ideals of the community of their age.

608 Archil. fr. 114 West ("I have no liking for a general who is tall, walks with a swaggering gait, takes pride in his curls, and is partly shaven. Let mine be one who is short, has a bent look about the shins, stands firmly on his feet, and is full of courage" [trans. Gerber]); cf. Dio Chrys. *Or.* 33.17, where we find the variant ἀσφαλῶς βεβηκὼς καὶ ἐπὶ κνήμαισιν δασύς ("with shaggy shins").

609 *Il.* 2.217–219: φολκὸς ἔην, χωλὸς δ' ἕτερον πόδα· τὼ δέ οἱ ὤμω / κυρτώ, ἐπὶ στῆθος συνοχωκότε· αὐτὰρ ὕπερθεν / φοξὸς ἔην κεφαλήν, ψεδνὴ δ' ἐπενήνοθε λάχνη ("he was bandy-legged and lame in one foot, and his shoulders were rounded, hunching together over his chest, and above them his head was pointed, and a scant stubble grew on it" [trans. Murray]).

610 See also Rankin (1975) 197–198: Archilochos may be making the point that Homeric heroes by running away lived to fight again and that the contemporary hoplite ethos is exaggerated.

611 Plut. *Mor.* 239b (Ἀρχίλοχον τὸν ποιητὴν ἐν Λακεδαίμονι γενόμενον αὐτῆς ὥρας ἐδίωξαν, δι_ ὅτι ἐπέγνωσαν αὐτὸν πεποιηκότα ὡς κρεῖττόν ἐστιν ἀποβαλεῖν τὰ ὅπλα ἢ ἀποθανεῖν). *Contra* Schwertfeger (1982) 273, 275–276.

shocking accounts of his own scandalous life. He enumerates a host of less fortunate traits and concludes as follows:

> "πρὸς δὲ τούτοις" ἦ δ' ὅς "οὔτε ὅτι μοιχὸς ἦν ᾔδειμεν ἂν εἰ μὴ παρ' αὐτοῦ μαθόντες, οὔτε ὅτι λάγνος καὶ ὑβριστής, καὶ τὸ ἔτι τούτων αἴσχιον, ὅτι τὴν ἀσπίδα ἀπέβαλεν. οὐκ ἀγαθὸς ἄρα ἦν ὁ Ἀρχίλοχος μάρτυς ἑαυτῷ, τοιοῦτον κλέος ἀπολιπὼν καὶ τοιαύτην ἑαυτῷ φήμην."[612]

Schwertfeger sees this as a misunderstanding on the part of Kritias ("... denn es kann ja nicht ernstlich davon die Rede sein, daß Archilochos sich in seinen Gedichten selbst schmähen wollte");[613] but as I argued, Archilochos' concern was the shock value in the admission; not to 'smear himself' as such, but rather a conscious rejection of traditional *polis* values as insufficient or irrelevant to the poet's way of life.[614] The other points of Kritias' criticism include adultery, lewdness and just general ὕβρις; hardly examples of different moral values in any postulated past. Why should his ῥιψασπία be any different? Rather, I believe that Kritias easily recognised that Archilochos' casual rejection of traditional values was conscious, and this is precisely the source of his bitter but impotent railing.

The cases of Sparta and Athens in Classical times are enlightening. Spartans who had thrown down their shields in combat were punished along with individuals who had left their battle position or otherwise displayed cowardice (contemptuously referred to as 'tremblers', τρέσαντες). The punishment resulted in ἀτιμία, consisting in the loss of all civil rights, in practice combined with a social ostracisation, ensuring that the unhappy ἄτιμοι lived the rest of their lives as outcasts.[615] This makes it rather more understandable that Archilochos' verses should be considered so provocative.[616] The effects of *atimia* can be seen from Herodotos' account of the two sole Spartan survivors of Thermopylai. One of them, Aristodemos, had been excused from combat duty because he suffered from a disease which deprived him of his eyesight, but he was nonetheless punished with *atimia*. He suffered the disgrace and humiliation that no one would talk to him in Sparta; indeed, no one would kindle his fire. The other one, Pantites, unable to live with the shame and social 'outcastness', hanged himself in the end.[617] Xenophon, who knew the Spartan society intimately, coolly remarks ἐγὼ μὲν δὴ τοιαύτης τοῖς κακοῖς ἀτιμίας ἐπικειμένης οὐδὲν θαυμάζω τὸ προαιρεῖσθαι ἐκεῖ θάνατον ἀντὶ τοῦ οὕτως ἀτίμου τε καὶ ἐπονειδίστου βίου.[618]

---

612 Kritias fr. 88 B 44 Diels-Kranz (= Ael. *VH* 10.13) (" 'In addition', he remarks, 'we should not have known of his adultery if we had not learned of it from him, nor of his sexual appetite and arrogance, and – what is a great deal worse – that he threw away his shield. So Archilochos was not a favourable witness in his own case, given that he left behind him such renown and reputation' " [trans. Wilson]).
613 Schwertfeger (1982) 272–273 ("we cannot in seriousness accept that Archilochos wanted to smear himself in his poems").
614 See also Jackson (1991) 230.
615 Thuc. 5.34.2 and see *RE* s.v. Τρέσαντες.
616 Plut. *Mor.* 220a 2; Cartledge (1996) 710.
617 Hdt. 7.231–232.
618 Xen. *Lac.* 9.6 ("Small wonder, I think, that where such a load of dishonour is laid on the coward, death seems preferable to a life so dishonoured, so ignominious" [trans. Marchant]).

Such, or worse, was the punishment that befell men throwing away their shield in battle. Isokrates even listed the two very worst crimes of all: abandoning the king during battle, and, second only to that, throwing down the shield: ὑπὲρ ἐκείνων [sc. the Spartan kings] δ' οἱ μὴ τολμῶντες ἐν ταῖς μάχαις ἀποθνήσκειν ἀτιμότεροι γίγνονται τῶν τὰς τάξεις λειπόντων καὶ τὰς ἀσπίδας ἀποβαλλόντων.[619] A great number of Plutarch's *Spartan aphorisms* are concerned with the shield, cementing its centrality and value in both the reality and ideology of hoplite warfare.[620]

However, it should be mentioned that the circumstances of losing one's shield were important. Fighting at Pylos in 425, Brasidas was wounded and dropped unconscious to the deck of his ship. He thereby lost his shield, which slid off his arm and into the sea, only to be picked up as a trophy by the Athenians. Yet this in no way hampered his successful military career over the following few years. Clearly, the fact that he had lost it through no fault of his own, and while fighting bravely, outweighed the simple fact that he lost it.[621]

Similar laws applied in Athens, where throwing away the shield during battle was considered a very serious crime. Here, too, it was punishable by *atimia*. The same punishment was affixed for several other types of insubordination or cowardice (γραφὴ δειλίας) as well – there are at least seven types of it[622] – but is it noticeable that *rhipsaspia* is distinguished on several occasions by its own name as belonging to a separate category of cowardice.[623] Interestingly, the Athenians themselves were of the opinion that the military judiciary system dated back to the archonship of Solon, i.e. 594/3.[624] The official designation for the charge in Athenian jurisprudence is γραφὴ [τοῦ] ἀποβεβληκέναι τὴν ἀσπίδα, "charge of having thrown away [one's] shield".[625] Lysias offers a most interesting passage, illuminating the apparent wording of the law:

ἡδέως δ' ἄν σου πυθοίμην (περὶ τοῦτο γὰρ δεινὸς εἶ καὶ μεμελέτηκας καὶ ποιεῖν καὶ λέγειν)· εἴ τίς σε εἴποι ῥῖψαι τὴν ἀσπίδα (ἐν δὲ τῷ νόμῳ εἴρηται, "ἐάν τις φάσκῃ ἀποβεβληκέναι, ὑπόδικον εἶναι"), οὐκ ἂν ἐδικάζου αὐτῷ, ἀλλ' ἐξήρκει ἄν σοι ἐρριφέναι τὴν ἀσπίδα λέγοντι οὐδέν σοι μέλειν; οὐδὲ γὰρ τὸ αὐτό ἐστι ῥῖψαι καὶ ἀποβεβληκέναι.[626]

---

619 Isoc. 8.143 ("… those Spartans who are not ready to lay down their lives for their kings in battle are held in even greater dishonour than men who desert their posts and throw away their shields" [trans. Norlin, modified]); cf. also 5.80.
620 Plut. *Mor.* 220a, 234f – 235a, 241f, *infra* 164. Schwertfeger's claim that the word ῥιψασπία is not mentioned in – and therefore not known by – Tyrtaios is a glaring *argumentum e silentio:* Schwertfeger (1982) 273–274.
621 Thuc. 4.12.1 (inv. no. 31).
622 Hansen (1973) 81–83. Schwertfeger (1982) 264–265 is incorrect in following Lipsius and allowing only three.
623 Andoc. 1.74; Isoc. 8.143; Aeschin. 1.28–32; Poll. *Onom.* 8.40; *Lex. Seg.* 217.21–25 and cf. *Suda* s.v. ἀναυμαχίου; Hansen (1973) 81–85.
624 Aeschin. 3.175–176; cf. Gröschel (1989) 130–131. Ruschenbusch accepts the tenets of ῥιψασπία as genuinely Solonic; see Solon fr. 32b* and Ruschenbusch (1966) 59, 79.
625 Hansen (1973) 81–85.
626 Lys. 10.9 ("And I would be glad if you would tell me this, – since in this matter you are a past master, both in action and in speech: if a man said that you had flung away your shield (in the terms of the law it stands, – 'if anyone asserts that a man has thrown it away, he shall be liable to pay penalty'), would you not prosecute him? Would you be content, if someone said you

By total *atimia* in Athens "a number of privileges were lost in addition to those reserved for Athenian citizens."[627] More specifically, these rights included

> a) the right to move decrees (γράφειν), to speak in Assembly (λέγειν, δημηγορεῖν) and indeed to take part in the Assembly at all (ἐκκλησιάζειν); b) the right to serve as a juror (δικάζειν) to act as a prosecutor in both private and public suits (δικάζεσθαι, γράφεσθαι) and to give evidence (μαρτυρεῖν) c) the right to hold a magistracy (ἄρχειν) d) the right to enter the sanctuaries (εἰσιέναι εἰς τὰ ἱερά) and e) the right to enter the *Agora* (εἰσιέναι εἰς τὴν ἀγοράν).[628]

It is difficult to assess which crimes were punished with total *atimia* (παντάπασιν ἀτιμία) and which merited only partial *atimia* (ἀτιμία κατὰ πρόσταξιν);[629] but judging from speeches of Lysias, Demosthenes and Aischines, the punishment for another military offence, λιποταξίου, was full *atimia*.[630] Hoplites who had been found guilty of ἀποβεβληκέναι τὴν ἀσπίδα, then, in all likelihood lost all their legal and political rights and privileges and were to all practical purposes reduced to non-entities in the state.

It is therefore hard to overstate the importance attributed to the shield in hoplite warfare. The other known military offences – failing to report for duty, desertion, leaving one's battle station – can all easily be grouped into the category of behaviour that is detrimental to the common cause in war. It is not clear why the loss of the shield should be equally damning, unless the fact is accepted that it was the centrepiece of phalanx fighting altogether, precisely because it was a piece of equipment that did more than just protect its bearer: in this respect, it was in a very real sense the key to the success or failure of the entire army.

According to legend, when Epameinondas was carried mortally wounded from the field of Mantineia: πρῶτον μὲν γὰρ τὸν ὑπασπιστὴν προσκαλεσάμενος ἐπηρώτησεν, εἰ διασέσωκε τὴν ἀσπίδα. τοῦ δὲ φήσαντος καὶ θέντος αὐτὴν πρὸ τῆς ὁράσεως, πάλιν ἐπηρώτησε, πότεροι νενικήκασιν.[631] Epameinondas asked for his shield first, and only then about the outcome of the battle: such behaviour puts into perspective just how much the question of *rhipsaspia* engaged the minds of the ancients. Plato has quite a few things to say about it in his *Laws*, where regulations are to be set up for such despicable individuals,[632] who should be heavily fined and barred from further

---

> flung away your shield, to make nothing of it, because flinging away and throwing away are not the same thing?" [trans. Lamb, modified]); cf. 10.1.

627 Hansen (1976) 55; cf. Hansen (1973) 19.
628 Hansen (1976) 61–62; cf. Hansen (1973) 26–28. Hansen gives the sources for the loss of different privileges.
629 Cf. And. 1.75.
630 Lys. 14.9; Dem. 15.32; Aeschin. 3.176.
631 Diod. Sic. 15.87.6 ("First calling for his armour-bearer he asked him if he had saved his shield. On his replying yes and placing it before his eyes, he again asked, which side was victorious" [trans. Sherman, modified]; cf. Just. *Epit.* 6.8.11–13 (*veluti laborum gloriaeque socium osculatus est*); inv. no. 18. See Gröschel (1989) 129–130 n. 607: "Selbst wenn die Erzählung unhistorisch ist, ist bereits das Faktum aufschlußreich genug, daß Ephoros, wohl Quelle Diodors, diese Worte Epameinondas in den Mund legen konnte" ("Even if the anecdote is unhistorical, the very fact that Ephoros – probably Diodoros' source – could have Epameinondas say these words is illuminating").
632 Pl. *Leg.* 944c: ἐὰν καταλαμβανόμενός τις ὑπὸ πολεμίων καὶ ἔχων ὅπλα μὴ ἀναστρέφῃ καὶ

military service. The ῥίψασπις, presumably also much like in Sparta, is scornfully likened to a woman: indeed, the appropriate penalty would be to transform him into a woman ("like Kaineus"), since he has willingly adopted her prerogatives of safety and exemption from military service; but since this is impossible, the closest approximation will have to do.[633] Plato does, however, distinguish between the man who wilfully abandons his shield (ῥίψασπις, ἀφεὶς ἑκών) and the one who merely loses it in an accident (ἀποβολεύς) such as during a voyage at sea, bad weather or the like: the latter is, of course, to be exempt from the law.[634] The distinction was important, since, in Plato's words, it had happened in countless cases (μυρίοις συνέπεσεν) since the Trojan war that warriors saved their life, but lost their weapons in the process.

Turning next to old Attic comedy, we find that both Aristophanes' extant comedies and several fragments are littered with references to *rhipsaspia*. In particular, references in Aristophanes' comedies to a certain Kleonymos are ubiquitous. This unfortunate Athenian politician (of Kleon's retinue) was the butt of many a merciless joke and makes his appearance in almost all Aristophanes' comedies over 15 years, and all but a few are concerned with his having allegedly thrown away his shield at the battle of Delion in 424 (inv. no. 5). Almost any mention of the name Kleonymos in Aristophanes, if ever so passing, may safely be considered a joke or pun on his martial prowess or lack thereof.[635] Kleonymos was more than a personal pet peeve of Aristophanes, however: he was enough of a by-word for cowardice to be mocked also by Eupolis.[636]

All this clearly indicates that the shield was the hoplite's most important weapon by far; it was his distinguishing feature and hallmark. It was surrounded by a unique mythology which cannot easily be compared to any other hoplite weapon, and the proper treatment of the shield was important enough to be enshrined in law. Nonetheless, the laws and sanctions against *rhipsaspia* reflect a very real problem: the shield was hopelessly heavy and clumsy and spectacularly out of place outside the highly specialised fighting environment of the phalanx. No doubt hoplites on the

ἀμύνηται, ἀφῇ δὲ ἑκὼν ἢ ῥίψῃ, ζωὴν αἰσχρὰν ἀρνύμενος μετὰ κάκης μᾶλλον ἢ μετ' ἀνδρείας καλὸν καὶ εὐδαίμονα θάνατον, τοιαύτης μὲν ὅπλων ἀποβολῆς ἔστω δίκη ῥιφθέντων ("If a person having arms is overtaken by the enemy and does not turn round and defend himself, but lets them go, voluntarily or throws them away, choosing a base life and a swift escape rather than a courageous and noble and blessed death – in such a case of the throwing away of arms let justice be done" [trans. Jowett]). There is a striking resemblance between the phrase ζωὴν αἰσχρὰν μετὰ κάκης μᾶλλον ἢ μετ' ἀνδρείας καλὸν καὶ εὐδαίμονα θάνατον and Tyrt. fr. 11.3–6 and 10.1–2 West.

633 Pl. *Leg.* 944c – 945b.
634 Pl. *Leg.* 944a–c.
635 Ar. *Vesp.* 12–20, 821–825, *Pax* 444–449, 673–678, 1295–1300 (in which Kleonymos' son is actually portrayed as reciting Archilochos' infamous lines, quoted above, and asked "are you singing about your father?"), *Eq.* 1369–1372, *Nub.* 351–354, 670–680, *Av.* 289–290, 1473–1481. The comparatively few passages which mock Kleonymos for non-military matters, added here for the sake of completeness, are *Ach.* 88–90, 842–844, *Eq.* 953–958, *Nub.* 400, *Thesm.* 604–607.
636 Eup. fr. 100 Austin (= *P Oxy.* 1087): ῥιψάσπιδόν τε χεῖρα τὴν Κλεωνύμου.

run felt a natural urge to throw away this burden in order to be able to outrun their pursuers; and this is, on an overall military basis, extremely undesirable. This is the reason for the reprisals against the unhappy who had thrown away their shield on the battlefield, and who had perhaps survived for that very reason.

3.2.3 Comparison with a modern 'phalanx'

The interdependence of the hoplites making up the phalanx is also well brought out by the modern example. As I stated earlier, modern Danish police riot forces, in keeping with the trend through most of the 20th century, have used shield fighting extensively.[637] I was informed by chief inspector Olsen that the standard procedure was to form a line, making sure that shield edges touched if at all possible, in order to create as contiguous a line of shields as possible, thereby maximising the zone of protection for both the individual policeman and the line as a whole. This line was usually only one man deep, as the fighting was completely defensive and no collisions as such were envisioned in the tactics. The policemen would assume a halfway sideways-on stance, immediately next to each other. When they advanced, they would support the upper edge of the large Plexiglas shield with the free right hand, as they were usually pelted with all kinds of hand-thrown missiles, a toilet bowl being among the more unusual and famous. This is akin to the leaning on the left shoulder of the hoplite shield, as seen on numerous vases: it gave the upper edge of the shield something to lean on, thus preventing it from bouncing back against the bearer. The advance would be carried out by dragging the hindmost foot (in practice usually the right foot, although the shield allows for ambidexterity). The advance in this manner can only be done slowly because of the unsteady gait this position involves; but since surprise or momentum was not a factor in this type of combat, it was amply made up for by the increased balance which the broad stance afforded. On the other hand it was possible to lean the shield on the knee, precisely *because* the advance was carried out by dragging the hindmost foot, thus creating even more support for the unwieldy shield and ensuring that powerful barrages would not overturn the bearer or strike the shield from his hand.

During the advance, the line of policemen would try to make sure that the shield edges touched each other, even though this could be quite difficult. This is a sobering thought. If one line of comparatively well-drilled policemen could really be in danger of drifting apart and creating gaps in the 'wall' when traversing paved city streets (albeit littered with cobblestones and shattered glass), then a Classical Greek hoplite phalanx must have had far more problems of this kind, advancing eight ranks deep in an uneven tempo across the landscape of the Greek countryside, strewn with trees, bushes, rivulets, garden walls and generally uneven terrain, usually probably with nothing but perhaps a sung *paian* to aid the marching rhythm, each man fundamentally unable to see much more than his nearest neighbours on all sides. It is no wonder that phalanxes risked falling apart during the advance, even before battle was joined.

---

637 *Supra* 53–54.

During combat, the policemen would stay firmly put behind their shields, under no circumstances venturing out of the line. In their experience, this tended to expose those left out of the common shield shelter dangerously; and even with their highly protective combat gear, they were not sufficiently equipped to sustain single combat, as this would be. For the purpose of defensive, delaying combat in a tightly locked formation, on the other hand, it was exceptionally useful. Rioters would face a contiguous wall of police shields touching edge to edge, while they themselves were usually only loosely organised (if at all); and they would therefore constantly flounder against the immobile and therefore impenetrable wall of Plexiglas, exactly because the police had the advantage of holding a fortified defensive position, which cannot easily be overrun by scattered attacks singly or in small groups. The 'tortoise' in fact served to effectively 'draw fire' and keep the rioters busy while not suffering any damage to speak of itself, much like the outstretched left arm of an experienced boxer. The compactness and impenetrability of such a shield line can also be measured by the tantalising glimpse offered by Herodotos in his account of the battle of Plataiai in 479:

> λήματι μέν νυν καὶ ῥώμῃ οὐκ ἥσσονες ἦσαν οἱ Πέρσαι, ἄνοπλοι δὲ ἐόντες καὶ πρὸς ἀνεπιστήμονες ἦσαν καὶ οὐκ ὅμοιοι τοῖσι ἐναντίοισι σοφίην. προεξαΐσσοντες δὲ κατ' ἕνα καὶ δέκα καὶ πλεῦνές τε καὶ ἐλάσσονες συστρεφόμενοι ἐσέπιπτον ἐς τοὺς Σπαρτιήτας καὶ διεφθείροντο.[638]

It is clear that the lack of order and sufficient protection was the Persians' undoing. Faced with the contiguous wall of Greek shields, their insufficiently co-ordinated sorties ended predictably: unable to breach (or perhaps even to reach) the shield fence, they attacked in small groups of uneven size and were impaled on the Greek spears protruding over, under or between the shields, like waves crashing against a bulwark. The efficiency of a solidly consolidated, defensive infantry formation against attacks from lighter troop types or cavalry is a staple of the history of warfare: this is the very reason why pikemen were employed for so long in western Europe even in modern times;[639] and this is also why police forces have adopted and employed this principle until quite recently.

As I stated above, the shield was deemed far too large, heavy and awkward for solo fighting, and consequently it was applied singularly for such situations where it could be held still as part of a larger, defensive shield wall, in which they all together ensured the mutual (relative) safety. If it was necessary to make sorties, improvised lighter shields (normal shields sawed through horizontally to about half size) would have to be used, and usually the policemen singled out for such sortie tasks were not fully armoured to give them greater mobility and dexterity. What is true for the individual – that the shield is unwieldy and counterproductive outside the formation – is ostensibly also valid for the entire line: defensive formation fighting is what the shield is made for, and that is the limit of its use. The possible objec-

---

638 Hdt. 9.62.2 ("in courage and strength they were as good as their adversaries, but they were deficient in armour, untrained, and greatly inferior in skill. Sometimes singly, sometimes in groups of ten men – perhaps fewer, perhaps more – they fell upon the Spartan line and were cut down" [trans. de Sélincourt]); inv. no. 29.
639 Cf. *infra* 166–167 and n. 678.

tion that Danish police shields are rectangular is immaterial, since other countries' police forces operate with shields of other shapes, among them circular, as I have witnessed myself in Greece.[640] Clearly, then, it is possible to make a satisfactory shield line, with sufficient shelter for the members, with round shields.

3.3 DEPLOYMENT

3.3.1 Width of file

To understand how the phalanx worked in battle, it is essential to understand how it was drawn up and arrayed. We have no written sources informing us of the exact distance between the individual hoplites; at least not until the Roman period, when the so-called tacticians are represented by their tactical handbooks. Aside from the fact that they are much later than any hoplite phalanxes, their writings, as it has been aptly said, "do not fill one with gratitude or respect."[641] In general, all three of these later writers (Arrian, Ailian and Asklepiodotos) reckon with three standard intervals in respect of breadth, or three battle orders: very open (no separate name), closed (πύκνωσις) and tightly closed order (συνασπισμός):

τοῦτον δὴ τὸν τρόπον ἐξομοιωθέντων τῷ ὅλῳ τῶν μορίων ἑξῆς ἂν εἴη ῥητέον περὶ διαστημάτων κατά τε μῆκος καὶ βάθος· τριττὰ γὰρ ἐξηύρηται πρὸς τὰς τῶν πολεμίων χρείας, τό τε ἀραιότατον, καθ' ὃ ἀλλήλων ἀπέχουσι κατά τε μῆκος καὶ βάθος ἕκαστοι πήχεις τέσσαρας, καὶ τὸ πυκνότατον, καθ' ὃ συνησπικῶς ἕκαστος ἀπὸ τῶν ἄλλων πανταχόθεν διέστηκεν πηχυαῖον διάστημα, τό τε μέσον, ὃ καὶ πύκνωσιν ἐπονομάζουσιν, ᾧ διεστήκασι πανταχόθεν δύο πήχεις ἀπ' ἀλλήλων. γίνεται δὲ μεταβολὴ κατὰ τὰς χρείας ἔκ τινος τούτων εἴς τι τῶν λοιπῶν, καὶ ἤτοι κατὰ μῆκος μόνον, ὃ καὶ ζυγεῖν ἔφαμεν λέγεσθαι, ἢ κατὰ βάθος, τὸ καὶ στοιχεῖν, ἢ κατ' ἄμφω, ὅπερ ὀνομάζεται κατὰ παραστάτην καὶ ἐπιστάτην. δοκεῖ δὲ τὸ τετράπηχυ κατὰ φύσιν εἶναι, ὅθεν οὐδὲ κεῖται ἐπ' αὐτῷ ὄνομα· ἀναγκαῖον δὲ τὸ δίπηχυ καὶ ἔτι μᾶλλον τὸ πηχυαῖον. τούτων δὲ τὸ μὲν δίπηχυ κατὰ πύκνωσιν, ἔφην, ἐπωνόμασται, τὸ δὲ πηχυαῖον κατὰ συνασπισμόν. γίνεται δὲ ἡ μὲν πύκνωσις, ὅτ' ἂν ἡμεῖς τοῖς πολεμίοις τὴν φάλαγγα ἐπάγωμεν, ὁ δὲ συνασπισμός, ὅτ' ἂν οἱ πολέμιοι ἡμῖν ἐπάγωνται.[642]

---

640 See also, e.g., www.fotosearch.com/DGV464/766019/ for a photo of unspecified police riot control forces holding round shields.
641 Cawkwell (1989) 381.
642 Ascl. 4.1–3 ("Now that the parts of the army have been brought into due relation with the entire force, we may speak of the intervals in length and depth. The needs of warfare have brought forth three systems of intervals: the most open order, in which the men are spaced 180 cm apart, the most compact, in which with locked shields each man is 45 cm distant on all sides from his comrades, and the intermediate, also called a 'compact formation', in which they are distant 90 cm from one another on all sides. As occasion demands a change is made from one of these intervals to another, and this, either in length only (which, as we noted before, is called forming by rank), or in depth, i.e. forming by file, or in both rank and file (called 'by comrade-in-rank' and 'by rear-rank-man'). The interval of 180 cm seems to be the natural one and accordingly has no special name; the one of 90 cm and especially that of 45 cm are forced formations. I have stated that of these two spacings, the one of 90 cm is called 'compact spacing' and that of 45 cm 'with locked shields'. The former is used when we are marching the phalanx against the enemy, the latter when the enemy is marching against us" [trans. Oldfather & Pease, modified]). More or less similar accounts are found in Arr. *Tact.* 11.1–6 and Ael. *Tact.* 11.1–5.

The tacticians, then, use the πῆχυς (or cubit) as the basic unit for their formations. It is difficult to assign a precise value to a cubit, since it varied greatly geographically and chronologically; but for this purpose it will suffice that it corresponds roughly to 45 cm.[643] Accordingly, the three intervals are *c.* 45 cm, *c.* 90 cm and *c.* 180 cm.

The tacticians' value as a source for the hoplite phalanx, however, is questionable. They presumably wrote about the Macedonian phalanx in the first and second centuries AD, a period when the might of Macedon was long gone; and in fact at this late point tactics were regarded as a valid branch of philosophy, a palpable aspect of these works. The highly theoretical strain in the works can also be gleaned from the fact that, e.g., Arrian, although actually discussing only one type of 'phalanx', nonetheless distinguishes between different kinds of equipment, which are in reality not compatible or completely commeasurable.[644] The case has been made that their material can be traced back to Poseidonios;[645] but Poseidonios (himself a Stoic philosopher) lived *c.* 135 – *c.* 51 and thus was born more than thirty years after the Romans' great defeat of the Macedonian phalanx at Pydna in 168. It must be supposed that the Macedonian phalanx as such never truly reared its head again after this event, and certainly not after Greece passed under Roman dominion with Mummius' sacking of Corinth in 146. It is therefore practically impossible to judge exactly what kind of army the tacticians are dealing with, though there is a possibility that the subject matter is actually the "late Egyptian or Syrian phalanx".[646] Firstly, it is highly doubtful that the tacticians' body of work is intended to refer to anything but contemporary matters; and secondly, even if in fact they do point back to rules and precepts pertaining to the Macedonian phalanx, this brings us no further back than the late Hellenistic period.

On these grounds, it seems unlikely that, even if intended, the tacticians had any amount of authority on military matters which belonged in the remote past; at least some 200 years as far as the true Macedonian phalanx is concerned, and more than 400 years regarding the Classical hoplite phalanx. It may therefore well be true that the πῆχυς was the standard basic unit of whatever kind of Greek-style phalanx had survived to the first century AD; but this is a far cry from being certain that

---

See also Polyb. 12.19.7–8, 18.29.2: here, Polybios reckons with three feet between each man in normal fighting order: ὁ μὲν ἀνὴρ ἵσταται σὺν τοῖς ὅπλοις ἐν τρισὶ ποσὶ κατὰ τὰς ἐναγωνίους πυκνώσεις ... ("When the phalanx is closed up for action, each man with his arms occupies a space of three feet" [trans. Scott-Kilvert]); cf. 18.30.6.

643 Cf. Poznanski's note: "La traduction des unités de poids et mesures est toujours sujette à caution, car celles-ci variaient suivant l'endroit et l'époque. On a donc adopté le système attique soit 1 coudée (πῆχυς) = 1,5 pied (πούς) = 0,444 m" ("Translation of weights and measures is always subject to caution, for these varied according to place and period. I have therefore adopted the Attic system of 1 cubit = 1,5 foot = 0.444 m": Poznanski [1992] 10 n. 8).

644 Arr. *Tact.* 3.2: τὸ μὲν δὴ βαρύτατον <τὸ> ὁπλιτικὸν θώρακας ἔχει καὶ ἀσπίδας ἢ θυρεοὺς παραμήκεις καὶ μαχαίρας καὶ δόρατα, ὡς Ἕλληνες, ἢ σαρίσας, ὡς οἱ Μακεδόνες ("the heaviest arm, the hoplites, carries breastplates and large round or oblong shields, swords and spears, like Greek hoplites, or *sarissai*, like Macedonian hoplites").

645 Arr. *Tact.* 1.1; cf. Cawkwell (1989) 382–383.

646 Connolly (1998²) 76.

it is a reliable source for the Macedonian phalanx, let alone that of the Greek city-state in the Archaic and Classical period.

Nonetheless, it has been maintained that the fact that the 180 cm interval is referred to as κατὰ φύσιν by Asklepiodotos could be a recollection of a similar trait of the earlier hoplite phalanx, and, partly on the basis of this, assumed that such large spacings were also the norm in the Archaic and Classical period.[647] Even if such a transmission seems unlikely on the grounds mentioned above, the suggestion of the name may still be sound enough on its own. Peter Krentz argues:

> A hoplite needed to know that he would not be attacked from the side or rear while he was engaged with the man in front of him. How close did he need to be to his neighbor to feel reasonably protected? Within a spear's thrust, I should think. Consider the position from the point of view of the enemy hoplite: how far would hoplite A have to be from hoplite B for an enemy to enter the gap and attack A from the side? Far enough so that the enemy would not have to worry about a spear or sword in his back from B while his attention was directed toward A. The comfortable limit, therefore, would be about six feet per man ... .[648]

Certain problems face Krentz' solution. The almost two metres' distance to one's neighbour may seem sufficiently within reach of a spear; but it is doubtful whether this could be achieved in the press of battle. After all, when wielded with one hand, as it invariably was, it must have been quite difficult to change the direction of a hoplite spear; and the spear, gripped more or less in the middle, only had a practical range of approximately one m.[649] If it were to have greater reach than this, the hoplite would either have to over-extend his arm – and possibly lose his balance and cover – or run the risk of not having power enough behind the thrust at such a short distance.

Furthermore, there were very good chances that one's neighbour was engaged in desperate combat himself, and that at the critical moment he would not be at liberty to defend his comrade by bringing his spear to the opposite side – or perhaps that he did not even notice his comrade's predicament. A formation based on this principle might soon find itself in jeopardy, as hoplites could not feel safe knowing that their sides were inadequately covered. As we have seen, hoplite equipment, and particularly the shield, did not encourage soloist fighting: it is no wonder that hoplites strongly felt the need for lateral and dorsal protection, not only provided them by their own armour, but also by the bodies, weapons and shields of their comrades.

Another reason for an interval of three feet or 1.83 m is suggested by van Wees.[650] Van Wees' stance seems to be somewhat ambiguous: he does assert that there was a certain constriction of the Classical phalanx compared to its Archaic counterpart: "most classical Greek hoplite formations ... were ... very much tighter

---

647 See, e.g., Krentz (1985b) 54, assuming six feet (= 1.83 m). The statement is repeated (1994) 46–47; but here it reads "six m. per man". I suppose that this is a typographical error and that, as before, six feet is actually meant (cf. Goldsworthy [1997] 15 n. 53). Also Cawkwell (1978) 153 n. 9, (1989) 382–384.
648 Krentz (1985b) 53–54; cf. van Wees (2004) 185–186.
649 Franz (2002) 305–306 suggests a range of 1.4 m.
650 van Wees (2000) 128–130, (2004) 168–169.

than anything the Archaic age had seen"; yet the Classical phalanx, he maintains, "cannot have operated with intervals much less than six feet", explicitly referring the reader to the figure showing a six-foot interval in the *Archaic* phalanx.[651] On the whole, it seems that he prefers a six feet interval for the Classical period also; but it is difficult to see how one represents a tightening of the other.

At any rate, van Wees maintains that the interval need not be any smaller than 1.83 m because of the way hoplites held their shields: they twisted their torso half sideways-on with the left side facing forward, so that they were directly behind the centre of their shield. In this way, each man was covered by his own shield only, and consequently no one had any need for a closer formation. He thus acknowledges the problem unaddressed by Krentz – that lateral protection would in fact be a problem in a phalanx as traditionally conceived, i.e. with hoplites facing directly forward.

As we have seen, however, although this observation is partially correct, it cannot explain the difficulties of standing in this position *constantly,* since combative action must have required that the hoplite turned and twisted, according as he himself attacked or sought to fend off enemy attacks.[652] This will have had the consequence that the hoplites in the front line were frontally exposed about as often as they were covered by their own shields, if they were to attack and defend themselves effectively. Van Wees' solution still leaves a gap between each hoplite large enough that he could be reached, whether or not he had optimal use of his own shield.

This would also have severe consequences for the phalanx when it moved across the battlefield before the actual encounter. With 1.83 m between each hoplite there would be ample opportunity for the enemy to direct arrows, slingshots and javelins at the phalanx at an angle of *c.* 45°. These missiles would have a much greater chance of finding their targets with such large gaps between each man. If they also marched forward in the posture suggested by van Wees, their torsos will have been turned to the right and have made for a desirable target, one that could be covered neither by their own shield, nor by that of their neighbour, who was too far away. This opening would have presented rows of dangerous gaps for enemy archers and skirmishers to fire arrows and other missiles through at an angle, effectively isolating individual hoplites and depriving them of any chance to protect themselves properly. Worse, the hoplites' backs will have been equally unprotected in a similar way, but from the opposite side; and here there would be the additional disadvantage of not being able to see attacks directed at oneself.[653]

If the open battle order was to be maintained on this interpretation, then, the phalanx would have to march forward in the traditional, front-facing closed order and then open up just before (or just after) closing with the enemy. This is surely impracticable, though it has in fact been suggested by George Cawkwell. Cawkwell, although preferring an open battle order in the Classical (and possibly the

---

651 van Wees (2004) 187, 185, 195–196 and 169 fig. 17; cf. (2000) 130 fig. 5.
652 *Supra* 40–41.
653 Peltasts, when charging hoplites, actually sought to run in at an oblique angle, getting at their unprotected, right side: Xen. *Hell.* 4.5.16.

Archaic) period, argues sensibly that a close order would be the only practicable way of getting the phalanx across the battlefield intact: "it is open to wonder whether such a 'thinning' could occur when the advance was concluded and just before the battle was ἐν χερσίν, i.e. a matter of hand-to-hand conflict."[654]

This speculation is indeed open to wonder. One may wonder, for instance, precisely when the phalanx's forward movement was to be halted – or what would be the consequence of halting immediately before closing with the enemy, only to begin some quite intricate re-spacing between the columns – or how the signal was to be given or understood over the noise. All in all, Cawkwell's solution seems rather far-fetched: it would be suicidal to begin rearraying the formations immediately before the enemy and it would almost certainly land the phalanx in great confusion and a very real risk of panic spreading through the ranks. In all probability the phalanx would be at its weakest just as the enemy phalanx struck it. It is equally impossible to imagine that ranks opened up after closing: by this time, the hoplites would of course be engaged in close combat and would scarcely have the necessary general view or, for that matter, co-ordination to carry out so complicated a manoeuvre. It would require the columns to step any number of paces to the left in time, or else easily become entangled.[655]

The closing and re-opening of the phalanx thus was not practically feasible, but the open order also faces several problems, chief among which is the fact that hoplites were equipped in a manner thoroughly unsuitable for soloist fighting. Krentz' interpretation of the open order would not have made the intervals between hoplites safe zones, and neither would that of van Wees. The hoplite panoply did, however, offer tremendous frontal protection; and in this position the hoplite would profit from being able to 'hang' the rim of his shield on his shoulder. Each hoplite was incapable of ensuring crucial protection of his sides and back, and therefore had to rely on his comrades. The only sensible way of doing this was to stand close together, leaving as little space as possible for the enemy to enter or reach through. In this light, the heavy frontal protection and the awkwardness of the weapons makes perfect sense. The hoplite was ideally virtually encased in bronze, which afforded excellent protection; but this was necessitated by the severe restriction of mobility which the heavy armour entailed.

On this interpretation the best order would be the one that was as closed as possible, while ensuring a minimum of freedom of movement for each hoplite. The ideal file width would thus be one that approached the diameter of the hoplite shield; and if we rely on the archaeological sources for this, i.e. the extant shields, it would be between 80 and 100 cm; probably very close to 90 cm. This is Pritchett's attractive solution to the problem.[656]

---

654 Cawkwell (1989) 379–389; see also Goldsworthy (1997) 7–8. To be fair, Cawkwell himself does seem to recognise the problems inherent in this: "Now this sort of shuffling to right or to left is inconceivable in a hoplite battle ... in any case it could not be done once the fighting had begun": Cawkwell (1989) 382: but see 384.
655 For another criticism of Cawkwell's solution see Holladay (1982) 95.
656 Pritchett (1971) 148–154. Küsters (1939) 26 reached a similar conclusion (but surprisingly adduces Polybios as his source).

It may, however, be somewhat inaccurate to say that the interval *between the files* is determined by shield width: it is a far better description to say that the available space for each file was determined approximately by the width of a shield, and so each file commanded a space of 90–100 cm. Each file member would of course take up some of this space with his own body – perhaps some 45 cm across the shoulders, leaving another 45–55 cm to be evenly divided on the left and right side.[657] The space occupied by the hoplite's body might be considerably less than 45 cm, according as he turned or stood sideways as suggested by Snodgrass, van Wees and others: this would bring the occupied space down to perhaps a mere 25 cm. The point, not to be forgotten, however, is that the space to the right and left of our hoplite increased and decreased constantly as he himself and his immediate neighbours shifted and moved during the motions of combat: the 90 cm of space determined by shield width is a 'constant', as it were, whereas the actual file interval changed incessantly during battle.

This formation would make the shield edges touch and ensure that very little was left for the enemy to strike at, unless it be above or below the shields. It is of course correct that with circular shields there is no point in talking about a 'shield wall' since only the extreme ends of the shield edges wil have met ("it is impossible to create a continuous solid joint between curved edges, no matter how much shields overlap"),[658] but the reason for the shield's being circular is another entirely – namely that of having a concave shield and a protruding rim on which to hang the shield[659] – and bringing the shield edges near to each other was as good an approximation of a 'shield wall' as could be made with round shields. It does not change the fact that there must have been a feeling of greater security, increasing according as the shields were held closer together. On Pritchett's model, the width of the shield determined the breadth of the file (the result being between 90 and 100 cm) so that the shield edges touched (or almost touched), thereby creating a sort of fence or bulwark, which afforded the phalanx a maximum of frontal and lateral protection as they marched forward, or stood and fought. Xenophon's insistence that no expertise is needed in battle since it is all but impossible to avoid hitting *something* in the enemy phalanx is also revealing in this light.[660]

The pivotal source for intervals between the columns is a passage in Thucydides, nearly hackneyed with repeated quotation – something that, perhaps, proves its centrality. It is a remark made almost *en passant* by Thucydides, discussing an apparently normal tendency of hoplite phalanxes, in this case occurring just before the battle of Mantineia in 418:

τὰ στρατόπεδα ποιεῖ μὲν καὶ ἅπαντα τοῦτο· ἐπὶ τὰ δεξιὰ κέρατα αὐτῶν ἐν ταῖς ξυνόδοις μᾶλλον ἐξωθεῖται, καὶ περιίσχουσι κατὰ τὸ τῶν ἐναντίων εὐώνυμον ἀμφότεροι τῷ δεξιῷ, διὰ τὸ φοβουμένους προσστέλλειν τὰ γυμνὰ ἕκαστον ὡς μάλιστα τῇ τοῦ ἐν δεξιᾷ παρατεταγμένου ἀσπίδι καὶ νομίζειν τὴν πυκνότητα τῆς ξυγκλῄσεως εὐσκεπαστότατον εἶναι· καὶ ἡγεῖται μὲν

---

[657] *Pace* Pritchett (1971) 154.
[658] Hanson (1991) 70.
[659] Hanson (1991) 69–71.
[660] Xen. *Cyr.* 2.1.16; cf. 2.3.9–11.

τῆς αἰτίας ταύτης ὁ πρωτοστάτης τοῦ δεξιοῦ κέρως, προθυμούμενος ἐξαλλάσσειν αἰεὶ τῶν ἐναντίων τὴν ἑαυτοῦ γύμνωσιν, ἕπονται δὲ διὰ τὸν αὐτὸν φόβον καὶ οἱ ἄλλοι.[661]

On the strength of this source there is no getting around the fact that hoplites of the Classical age felt an intense desire to have their unprotected right side (τὰ γυμνά) covered and were uneasy if they felt deprived of this protection; so much so that they constantly tried to steal closer to their right-hand neighbour. In this case the 'overstretching' was so large that king Agis feared that his left flank might be outflanked and surrounded, and he had to resort to desperate and dangerous measures to extend his line and fill up the resulting gap.[662] This easing to the right by degrees was so common an occurrence that Thucydides might reasonably generalise that *all* armies were prone to it: in this instance, it even happened to the supposedly superior Spartan army, otherwise a paragon of discipline and self-control. There is no question about Thucydides' trustworthiness in this instance: as *strategos* of the year 424/3 he not only had thorough personal experience of warfare; he must have been regarded by his contemporaries also as a man of considerable military acumen and judgment.[663]

If the phalanx was drawn up in open order – assuming, with Krentz and van Wees, file intervals of almost two metres – then it is difficult to see just what it was that hoplites were afraid of, or what they sought to gain by constantly moving closer to their right-hand neighbour. If, on the other hand, a gap suddenly opened in an otherwise contiguous shield 'wall', it makes perfect sense that hoplites sensed that they were becoming exposed, and that they, more or less consciously, would try to close the gap themselves.

In Thucydides' description of the battle at Mantineia (inv. no. 17), the word used to describe the phenomenon is ξύγκλῃσις, a 'closing together'. It is natural to assume that the word means just that: a joining together of the files in order to close the unnerving gaps emerging between the hoplites entirely (ξυγ-). Van Wees claims that this is qualified by ὡς μάλιστα, however: "clearly what he meant by 'as near as possible' depends on how much room hoplites needed to wield their weapons effectively. ... The 'protection' ... was therefore not direct cover provided by a neighbour's shield, but the general protection of having a friend close by."[664]

Perhaps so. It still seems to me, however, that the phrasing προσστέλλειν τὰ γυμνὰ ἕκαστον ὡς μάλιστα **τῇ** τοῦ ἐν δεξιᾷ παρατεταγμένου **ἀσπίδι** is somewhat odd if there was a gap of two metres to the next hoplite, and all that was sought reached

---

661 Thuc. 5.71.1 ("It is true of all armies that, when they are moving into action, the right wing tends to get unduly extended and each side overlaps the enemy's left with its own right. This is because fear makes every man want to do his best to find protection for his unarmed side in the shield of the man next to him on the right, thinking that the more closely the shields are locked together, the safer he will be. The fault comes originally from the man on the extreme right of the front line, who is always trying to keep his own unarmed side away from the enemy, and his fear spreads to the others who follow his example" [trans. Warner]); inv. no. 17. Cf. Xen. *Hell.* 4.2.18–22.
662 Thuc. 5.71.2–72.2.
663 Thuc. 4.104.4.
664 van Wees (2004) 185–185.

was a vague 'general protection': on van Wees' and Krentz' interpretations, only the range of the neighbour's spear should matter, not his shield.[665] Why, then, is it mentioned at all?

In later literature we do come across the word συνασπισμός,[666] which, although infrequent in Classical authors, does occur;[667] and this word must imply physical contact between the shields: indeed, the usual English translation of the word is 'locking of shields'. The passage in Thucydides would indicate that the phenomenon, if not the word itself, was readily known at least also in the Classical period: that files were being kept close on purpose, so that the shields might give maximum cover for the line. The metaphorical use of συνασπίζοντες in Euripides' *Cyclops* to mean 'participate in something with someone' (of Silenos and the chorus of satyrs) indicates that the phrase was actually well known in C5.[668] An important passage in this connexion is found in Plutarch's *Spartan aphorisms:* Here the quite sensible question is asked why the unhappy lot of *atimia* befalls any Spartan who has thrown away his shield in combat, when no fixed disgrace clings to someone who has lost other parts of the armour; to which king Demaratos strikingly answers that a shield – as opposed to any other piece of armour – is worn *for the sake of the entire line* (τῆς κοινῆς τάξεως ἕνεκα).[669] This makes no sense unless we understand it literally: every hoplite in the line depended, to a certain degree, on the *shields* of others for his safety.

It is of course true that such a battle order to a certain extent will have deprived hoplites of the opportunity to 'brandish' their weapons properly – feinting, striking and dodging – but that was precisely the type of combat which was consciously avoided, in favour of a different mode of fighting altogether. Very much space simply was not necessary; and in this way, the considerable defensive qualities of the shield were best exploited. It allowed no by-way behind the heavy and awkward shield, which was too large and heavy to be quickly swung by the individual hoplite. Apart from the actual defensive qualities exploited in this way it ensured a tangible, reciprocal feeling of safety that was probably equally important for the

---

665 Krentz (1985b) 54 and (2007) 72–73; cf. van Wees (2004) 185–186.
666 See, e.g. Arr. *Anab.* 5.17.7, Plut. *Phil.* 9.2, *Flam.* 8.4; Diod. Sic. 16.3.2; Polyaen. 4.2.2 (with Lammert's emendation of συνησπισμένην for συνεσπασμένην); Polyb. 4.64.6–7, 12.21.3 and cf. 18.29–30. Since Greek effortlessly makes verbs of nouns and *vice versa,* I do not regard συνασπιδόω and συνασπισμός as essentially different words.
667 Xen. *Hell.* 3.5.11, 7.4.23.
668 Eur. *Cyc.* 39; cf. Helbig (1911) 8; Seaford (1984) 105.
669 Plut. *Mor.* 220a and cf. *Pel.* 1.5: οἱ δὲ τῶν Ἑλλήνων νομοθέται τὸν ῥίψασπιν κολάζουσιν, οὐ τὸν ξίφος οὐδὲ λόγχην προέμενον, διδάσκοντες ὅτι τοῦ μὴ παθεῖν κακῶς πρότερον ἢ τοῦ ποιῆσαι τοὺς πολεμίους ἑκάστῳ μέλειν προσήκει ("the Greek lawgivers punish him who casts away his shield, not him who throws down his sword or spear, thus teaching that his own defence from harm, rather than the infliction of harm upon the enemy, should be every man's first care ..." [trans. Perrin]). Similar instances can be found in other suitably Laconic phrases in Plutarch, such as the mother bidding her son to return either *with* the shield or *on* it (*Mor.* 241f 16). The latter was apparently quite famous in antiquity; cf. Σ Thuc. 2.39.1; Stob. *Flor.* 3.7.30 and see Hammond (1979–80). Cf. also *Mor.* 210f: Agesilaos emphasises the need for standing one's ground – to the point that lame warriors too are useful in the phalanx.

safety of the phalanx. Adrian Goldsworthy offers a final argument for a spacing of c. 90 cm (3 ft) between each hoplite, namely that such a formation would help keeping the formation reasonably intact through the advance and charge, if nothing else because it securely prevented the less confident hoplites from running away.[670]

Such a scarce amount of space may seem terribly insufficient, but in this connexion it is important to remember that the Macedonian armies possibly operated with even narrower file intervals: indeed, Philip II of Macedon was credited with tightening the formation by narrowing the file width, imitating the *synaspismos* of the heroes of the Trojan war.[671] With this particular choice of words, the only ἐν Τροίᾳ τῶν ἡρώων συνασπισμός Diodoros could reasonably be having in mind is the following passage in the *Iliad:*

> οἳ γὰρ ἄριστοι
> κρινθέντες Τρῶάς τε καὶ Ἕκτορα δῖον ἔμιμνον,
> φράξαντες δόρυ δουρί, σάκος σάκεϊ προθελύμνῳ·
> ἀσπὶς ἄρ' ἀσπίδ' ἔρειδε, κόρυς κόρυν, ἀνέρα δ' ἀνήρ·
> ψαῦον δ' ἱππόκομοι κόρυθες λαμπροῖσι φάλοισι
> νευόντων, ὡς πυκνοὶ ἐφέστασαν ἀλλήλοισιν.[672]

This justly celebrated passage is very clear that shields are actually touching (ἔρειδε), and the extreme proximity of the files is underlined by the fact that not only shields, but men and even helmets touch each other. Further, Diodoros' specific technical term *synaspismos* should be contrasted with the generic word πυκνότης (as opposed to the technical term πύκνωσις), probably meaning simply 'compression' or 'contraction'.[673] This indicates that it was rather the Macedonian *synaspismos* formation which was in fact the novelty devised by Philip. As we have seen, the *synaspismos* was the most constrictive formation in the tacticians with an interval of just one πηχύς, corresponding approximately to 45 cm. This would cause the rimless Macedonian shields, c. 60 cm across,[674] to overlap slightly (7.5 cm to

---

670 Goldsworthy (1997) 16 and n. 60: "There are eyewitness descriptions from the early eighteenth century of lines of infantry which had begun the battle 3–6 ranks deep becoming '40–80 men deep' or even '100 deep' as men drifted away from the firing line ... I have not come across similar descriptions from the battles of succeeding decades, when infantry formed in much closer formation. This may suggest that more closely packed ranks were better at preventing the less confident men from refusing combat [quoting Nosworthy]."
671 Diod. Sic. 16.3.2: ἐπενόησε δὲ καὶ τὴν τῆς φάλαγγος πυκνότητα καὶ κατασκευήν, μιμησάμενος τὸν ἐν Τροίᾳ τῶν ἡρώων συνασπισμόν, καὶ πρῶτος συνεστήσατο τὴν Μακεδονικὴν φάλαγγα ("Indeed he devised the compact order and the equipment of the phalanx, imitating the close order fighting with overlapping shields of the warriors at Troy" [trans. Sherman]). The same thought occurred to Polybios: Polyb. 18.29.5–7; cf. Curt. 3.2.13.
672 *Il.* 13.128–133 ("for they who were chosen bravest awaited the Trojans and noble Hektor, fencing spear with spear, and shield with serried shield; shield pressed on shield, helmet on helmet, and man on man; and the horsehair crests on the bright helmet ridges touched each other as the men moved their heads, in such close array stood they by one another" [trans. Murray]); cf. 16.210–217.
673 See Buckler (1985) 138: Diodoros uses the terms πυκνός, πυκνότης and πυκνοῦσθαι indiscriminately to mean 'density'.
674 Ascl. 5.1: τῶν δὲ φάλαγγος ἀσπίδων ἀρίστη ἡ Μακεδονικὴ χαλκῆ ὀκτωπάλαιστος, οὐ λίαν

each side); not unlikely for a formation of such extreme density as the Macedonian. Furthermore, this would result in an exceedingly compact mass of *sarissa*-heads protruding powerfully before the first line, obviously an asset with the Macedonian phalanx. On this interpretation, there was ample space for a tightening of the file intervals of the hoplite phalanx: in fact, the Macedonian *synaspismos* was half the file interval of a hoplite phalanx, not merely 10 cm as has been suggested.[675] Accordingly, hoplites of the Archaic and Classical age apparently had twice as much space between each man; and that should be sufficient for this particular type of combat.

An early 17th century AD infantry drill manual, *Kriegskunst zu Fuß*, by Oberst-Wachtmeister Johann Jacobi von Wallhausen of Danzig, furnishes a fascinating example of modern pike drill.[676] In the following, von Wallhausen reckons with one pace throughout as the basis of all determinations of width and depth in the pike 'phalanx' when operating with closed fighting orders:

> Das zweyte, in enger und geschlossener Ordnung stehen, ist dasjenige, so im Exercitio gewiesen, und in Schlachtordnung gegen Reuteren am bräuchlichsten, da denn auch diese zwey Stück gemercket werden: Erstlich mit geschlossener Schlachtordnung gegen Fußvolck streiten. Zum andern mit wol geschlossener Schlachtordnung gegen Reuterey streiten. Die erste gegen Fußvolck geschihet nach Gelegenheit etwas weiter und mit anderthalb Schritt in Reyen und Gliedern Distantien. Die zweyte gegen Reuterey *hart angeschlossen,* damit das im ein- und durchbrechen der Reuterey besserer Widerstand zu thun seye.[677]

Schneider admits that this may at first glance seem ample space, but next comments:

> … während wir heute 0,75 bis 0,80 m auf den Schritt rechnen, kann der damalige Schritt nur etwa 0,55 m betragen haben. Danach hatte der einzelne Pikenier in der gewöhnlichen Standfassung, 'mit zween Schritt von einander' 1,10 m Raum in der Front, davon nahm er selber 0,55 m ein, und 0,55 betrug der mannsbreite Abstand vom Nebenmanne. In der geschlossenen Stel-

---

κοίλη ("the best shields for the phalanx is the Macedonian bronze shield of eight palms' width"). ὀκτωπάλαιστος – eight palms – should probably be *c*. 60 cm: see Snodgrass (1967) 117–118; Connolly (1998²) 77–79; Anderson (1976) 3.
675 *Pace* Pritchett (1971) 154.
676 Found in Rudolf Schneider (1833). Curiously, Pritchett (1971) 153 n. 41 asserts that the subject is "armored Swiss pikemen": aside from the fact that federal Swiss pikemen more often than not were unarmoured, Schneider discusses a *German* pike drill manual, something that is unfortunately lacking for Swiss peasant armies, which flourished somewhat earlier. Another important characteristic of the Swiss *Igel* formation was the reliance on the deadly halberds, utterly incompatible with the use of a pike. The appearance and armament of Swiss mercenaries may be gleaned today by observing the Vatican's Swiss guard. It is puzzling to see this misunderstanding repeated by Poznanski (1992) 42.
677 von Wallhausen (quoted in Schneider [1893] 76): "The second item, standing in tightly closed formation (demonstrated in the section on drill) is particularly useful in a battle line against cavalry, and can be subdivided into two points: first, fighting in close order against infantry, and second, fighting in extremely close order against cavalry. The first formation, against infantry, may, if possible, be a little wider and with an interval in rank and file of one and a half paces; the second, against cavalry, is *firmly closed* in order to defend the better against the onslaught and breakthrough of a cavalry attack"). Emphasis original.

lung, wo die Mannsbreite Lücke durch Anschließen oder Duplieren ausgefüllt war, betrug der Frontraum des Pikeniers 0,55 m.[678]

The pikemen of von Wallhausen's day thus probably had more in common with the Macedonian phalanx than with the Classical phalanx. They were armed with pikes at least 4.26–4.42 m long, but more likely as long as 5.05–5.68 m,[679] and they carried no shields. In the narrowest of von Wallhausen's positions – "hart angeschlossen" – there is no excess space at all, only the 0.55 m that each pikeman took up in the formation: the pikemen's shoulders were actually touching. This was an attempt to concentrate a maximum weight of pike-heads in as little space as possible – a trait perhaps also recalled in the fact that the dreaded Swiss pikemen of the 16th century AD referred to their formation as 'hedgehog' (*Igel*), and of Diodoros' likening of Philip's reformed Macedonian phalanx to the Homeric *synaspismos* in which shields, helmets and spears actually touched.[680] Nevertheless it should suffice to demonstrate that we have a relatively exact modern equivalent to the ancient phalanx proving that it is indeed possible to fight with spears in extremely narrow confines. Hoplites did carry a shield, and it may be thought that a closed order would restrict its movement; but in a sufficiently tight formation this might in fact be a positive asset rather than a drawback. The shield was supposed to be held relatively still and instead be a part of a much larger bulwark, less mobile but efficient if they were kept together.

3.3.2 Phalanx depth

We are regularly informed about the phalanx depth in accounts of hoplite battles of the Classical age, and it therefore seems that it was considered of no small importance for the outcome of the battle. Almost invariably, the expression is ταχϑῆναι (*vel sim.*) ἐπὶ ... ἀσπίδων, with the relevant number of 'shields' (metonymically for 'men' or 'hoplites') inserted. The blank should almost always be replaced with the same number: the standard depth for hoplite phalanxes in the sources is eight ranks. It is interesting to note that the hoplite and his place in the phalanx – indeed entire ranks – may be summarised by the word *aspis*: the shield is what matters about the rank, and phalanxes are not composed of men or ranks but rather (ranks of) shields, emphasising the bulwark aspect of line upon line of shields.

---

678 Schneider (1893) 77 ("whereas we count 75–80 cm to the pace today, a pace then can hardly have been more than 55 cm. Accordingly, a pikeman in the normal position 'two paces from each other', had 110 cm in front of him, of which he himself occupied 55 cm, leaving an interval of 55 cm to the next rank. In closed order, where the gaps of one man's width were closed by moving up or doubling, a pikeman's space was 55 cm"). It is perhaps noteworthy that corps of pikemen were deemed effective enough to be regular components in the national armies of England (until 1691), France (until 1703), Sweden (until 1708) and Russia (where pikes were not abandoned until 1721): Schneider (1893) 70.

679 Based on Schneider's indications (13½–14 and 16–18 German *Füße* respectively). One German *Fuß* = 0.31 m: Schneider (1893) 71.

680 This similarity was also assumed by Machiavelli, who in his *Arte della guerra* (1520) likened the Swiss phalanx to that of the Macedonians: see esp. 2.18, 2.140–144, 3.21–37, 3.134–150, 3.165–171.

It is difficult to say precisely why eight should be the 'magic' number; but the relative wealth of sources leaves no possible doubt that this was indeed the *Urtiefe*.[681] Only on relatively rare occasions is this pattern altered. There are two instances of phalanxes arraying only four ranks deep; but this appears to be an exception. A cursory overview of the two sources reveal that one does not in fact describe a battle order as such but rather a military parade (Kyros' Greek mercenaries at Tyriaeion in Persia).[682] The other is found in a passage in Diodoros discussing the Spartan invasion of Attika in 408 and is highly suspect on several accounts,[683] although Rusch has argued reasonably, and by analogy, that a formation four deep may be reconcilable with what besieging armies normally did: after all, this was not a direct line-up for battle but rather an attempt to strike terror into the Athenians a mere kilometre away.[684]

Strangely, however, Xenophon himself specifically states that Kyros ordered the mercenaries to form up in the *normal* Greek fashion. The wording is important: ἐκέλευσε δὲ τοὺς Ἕλληνας ὡς νόμος αὐτοῖς εἰς μάχην οὕτω ταχϑῆναι καὶ στῆναι, συντάξαι δ' ἕκαστον τοὺς ἑαυτοῦ. ἐτάχϑησαν οὖν ἐπὶ τεττάρων.[685] The qualifying particles οὕτω and οὖν are particularly revealing: the Greeks were ordered to form up *just as* they usually did for battle and *therefore* formed up four ranks deep. At first glance, this seems to fly in the face of Pritchett's collected evidence, but Xenophon seems to have no qualms presenting this as Greek battle *nomos*. The reason for the shallow phalanx may have been a wish to present a very long front for Kyros to walk down; but with 11,000 hoplites[686] deployed eight ranks deep the front line would have been 1.23 km long. Deploying four deep would have resulted in a front line twice as long, fully 2.47 km long. Now Kyros did inspect the troops driving along the front in a chariot,[687] but more than a kilometre really should suffice for this purpose. One cannot avoid the impression that the Greeks actually did form up for battle as was their custom. Though the extant sources may differ, there is little to dispute the military expertise of Xenophon: hoplite phalanxes actually could draw up merely four deep, depending on the circumstances.

On occasion, we hear about exceptionally deep phalanxes. This could be merely a result of tangential circumstances, such as poorly chosen terrain: Thrasybulos' rebel army at the Peiraieus in 403 faced the forces of the thirty tyrants, reinforced by Spartan troops of the occupation army, on the Munychia hill. Because of the steep terrain and the fact that they were forced to deploy on a narrow road, the occupation army phalanx on this occasion was no less than fifty ranks deep, whereas

---

681 The sources are conveniently collected and tabulated in Pritchett (1971) 134–135.
682 Xen. *An.* 1.2.15.
683 Diod. Sic. 13.72.5–6; cf. Pritchett (1971) 134–135.
684 Rusch (2002) 292–294.
685 Xen. *An.* 1.2.15 ("He ordered the Greeks to fall in and stand in their normal battle order; each officer should see to the order of his own men. So they stood on parade in fours ..." [trans. Warner]).
686 Xen. *An.* 1.2.9.
687 Xen. *An.* 1.2.16.

Thrasybulos' men deployed 10 ranks deep.[688] In another case, again involving Spartans, Archidamos' relieving forces attacked a besieging Arkadian army outside the village of Kromnos in 365. Again, the only access to the area apparently was a road (ἁμαξιτός), and Archidamos' forces had to deploy in two columns. Predictably, they were beaten.[689] On the whole, unusual terrain called for unusual deployments: Kyros' mercenaries, on their retreat through the wasteland of the Persian empire, often found that innovative thinking was necessary.[690]

Despite the limited success these particular unfortunate deep phalanxes met with, other formations were intentionally kept unusually deep. The Boiotians experimented with such very deep formations and in particular the Thebans, who seem to have tried out 'irregular' phalanxes time and again during C5 and C4: 25 ranks deep at Delion in 424,[691] and again at Nemea in 395 with a phalanx described by Xenophon as "exceedingly deep" (βαθεῖαν παντελῶς): this despite the fact that the coalition army had apparently agreed to form up 16 ranks deep.[692] The phrase βαθεῖαν παντελῶς perhaps seems a little strong if the difference was no more than nine ranks; so accordingly the Boiotian phalanx on this occasion was probably a good deal more than 16 ranks, quite possibly even more than the 25 ranks seen at Delion. Again at Tegyra in 375, Pelopidas seems to have massed his men very deep (implied, since Plutarch speaks of 300 hoplites with a very narrow front).[693] The results obtained with these extraordinarily deep phalanxes must have been satisfactory enough that further experimentation was called for.

Accordingly, in some accounts of the famous battle of Leuktra in 371, the Thebans are reported to have massed no less than 50 ranks deep against the opposing Spartan army. Few battles of the Classical period have been analysed so often and debated so hotly as Leuktra, or, in the words of Pritchett: "ironically, there are more reconstructions of Leuktra than of any other Greek battle, and the end is not in sight."[694] This may be ascribed to the fact that the sources differ wildly and are rather difficult to reconcile.[695] One radical view is that of Hanson, who analyses the sources and maintains that the only one of any value is Xenophon's, largely due to the fact that he – as the only one – was a contemporary. On the basis of his terse, laconic and very straightforward account of the battle, Hanson rejects any possibility of a revolution in Greek phalanx tactics at Leuktra.[696] Hanson is correct in pointing out that the prevailing agreement to reject Xenophon's brief and apparently too

---

688 Xen. *Hell.* 2.4.10–12 (inv. no. 22).
689 Xen. *Hell.* 7.4.22–23 (inv. no. 10). Xenophon does not say how deep the phalanx was on this occasion, but it is clear that if Archidamos' men were forced to form up in two columns, the phalanx must have been many times deeper than normal.
690 Xen. *An.* 4.8.9–19, 5.2.3–13.
691 Thuc. 4.93.4; cf. Diod. Sic. 12.70 (inv. no. 5).
692 Xen. *Hell.* 4.2.13, 4.2.18 (inv. no. 25).
693 Plut. *Pel.* 17.1–10 (inv. no. 41).
694 Pritchett (1985a) 54 n. 159. For an overview of analyses of the battle, see Stylianou (1998) 398 (inv. no. 16).
695 Sources for the Theban formation at Leuktra (inv. no. 16) include Xen. *Hell.* 6.4.4–16; Diod. Sic. 15.53–56; Plut. *Pel.* 23.1–3; Polyaen. 2.3.2–3, 2.3.8; Din. 1.72–73; Frontin. *Str.* 4.2.6.
696 Hanson (1988).

straightforward account of a battle of almost mythical proportions can largely be ascribed to the perceived 'anti-Theban' strain in the *Hellenika*.[697] Recent work has shown, however, that despite allegations, there is not sufficient basis for assuming any anti-Theban tendency in Xenophon – although one is scarcely wide off the mark in implying that he is certainly preoccupied with keeping the image of Sparta untarnished.[698]

Be that as it may, Xenophon is quite explicit that the Thebans massed no less than 50 ranks deep, as opposed to the Spartan formation of 'only' 12 ranks.[699] This perhaps seems somewhat excessive, since it is difficult to imagine what use could be made of the majority of these ranks, and I believe the best solution to this number – which should not be disputed – has been furnished by P.J. Stylianou. His suggestion is that the Thebans (again) formed up 25 ranks deep, but that the Theban élite corps, the 'Sacred Band' (ὁ ἱερὸς λόχος) was stationed immediately behind the main Theban force on the left flank, the intention being to let the crack troops of the Sacred Band pin down the extreme right of the Spartan phalanx. Brilliant though it may be, such a use of reserves is certainly not without precedent: at Amphipolis in 422, Brasidas ordered his subordinate commander Klearidas to attack with the main force after he himself had led 150 picked men in a surprise attack, pinning down the Athenians outside.[700] A salient point of Stylianou's analysis is that it makes the two phalanxes fairly comparable in width: some 2,300 Spartans arranged 12 ranks deep would have had a front of 190 men, while the Thebans, 3,000 men strong, massed 25 ranks deep, giving them a front of 120 men (plus the 12 broad Sacred Band); in all 132 men wide. This addresses a crux in our understanding of the battle formation, as stated by J.K. Anderson: if the Thebans massed 50 ranks deep, it is dificult to understand the heavy Spartan casualties ("more killed than, on any probable reckoning of numbers, were opposed to the front of the Theban phalanx").[701] Even allowing 3000 Theban hoplites, they would have presented a front line of merely 60 men, if they were deployed 50 deep. Stylianou's solution makes the two phalanxes roughly comparable, even if it is still somewhat disconcerting that the Spartan phalanx should out-reach the Thebans by about 60 men. Nonetheless, this allowed the Theban phalanx to effectively engage almost the entire Spartan phalanx.[702] Stylianou's analysis has the added advantage of being relatively unaffected by the anti-Xenophontic trend in recent scholarship and so does not suffer from an excessively critical perception of Xenophon's account.[703]

The presence of the Sacred Band may quite possibly have caused the Theban phalanx to look as if it were actually fully 50 ranks deep – and it may have done so in particular to Xenophon's Spartan informants, who would probably have been

---

697 Hanson (1988) 191 n. 3.
698 Christensen (2001) 7–21 shows, to my mind conclusively, that there is no basis for claiming an anti-Theban bias in the *Hellenika*.
699 Xen. *Hell.* 6.4.12. No other source in fact mentions the 50 ranks on the Theban side.
700 Thuc. 5.8.4, 5.9.1–8; Anderson (1970) 179–180; inv. no. 2.
701 Anderson (1970) 215.
702 Stylianou (1998) 402–403.
703 Stylianou (1998) 398–400.

more or less directly opposite – when in fact the overall depth was 'only' the usual Boiotian 25 ranks. Whether or not one accepts Stylianou's suggestion, however, Xenophon is perfectly clear that the Theban sector at least *seemed* to be massed 50 ranks deep. That the Thebans did in fact concentrate their force on their own left flank – an unorthodox procedure – may be seen from Xenophon: the Thebans formed up 50 ranks, λογιζόμενοι ὡς εἰ νικήσειαν τὸ περὶ τὸν βασιλέα, τὸ ἄλλο πᾶν εὐχείρωτον ἔσοιτο,[704] and later in the battle, the Spartan left flank (οἱ τοῦ εὐωνύμου ὄντες τῶν Λακεδαιμονίων) became aware that their right flank (τὸ δέξιον), where the king himself was stationed, was being forced back by the immense pressure.[705] Therefore, if king Kleombrotos was on the Spartan right flank, the main Theban force must have been directly opposite it, i.e. on their own left flank.

The battle (and formation) thus is capable of being explained without having to resort to improbably dramatic tactical inventions; but already in antiquity an 'Epameinondas tradition' developed, suggesting implausibly advanced or complex formations and battle plans, such as the notion of the celebrated 'oblique phalanx' (λοξὴ φάλαγξ), something which may more likely have been extrapolated backwards from the later tacticians.[706] The fascinating concept of an oblique phalanx and just what it entailed has led to very different solutions, ranging from an oblique march across the battlefield to the somewhat far-fetched notion of the entire Theban phalanx arrayed in wedge formation (ἔμβολον).[707] It is more likely that the explanation for the origin of the 'λοξὴ φάλαγξ' is to be found in the uneven Boiotian phalanx. The other Boiotians probably did not mass quite so deep, so that the very deep Theban phalanx gave it an unbalanced appearance, concentrating tremendous weight and momentum on the left flank.

The upshot of all this is that if we accept that the Sacred Band was in fact used as a separate echelon to pin down the extreme Spartan right and that these 300 crack troops were in fact stationed immediately behind or in front of the main Theban force, the basic depth of the Theban phalanx at Leuktra was, as elsewhere in C5 and C4, 25 ranks. It is an open question what was satisfactorily achieved by massing in such extreme depth. After all, it made the front narrower, increasing the risk of outflanking; and the deeper the formation, the fewer men would actually be able to use their weapons.[708] These matters will be treated more fully later: suffice it to say that other factors weighing heavier than these considerations, however serious, must have determined the massing in depth. The *reason* for the Thebans' increased depth is thus a different matter altogether and is probably best left for later consideration.[709]

---

704 Xen. *Hell.* 6.4.12 ("They calculated that, if they proved superior in that part of the field where the king was, all the rest would be easy" [trans. Warner]).
705 Xen. *Hell.* 6.4.14.
706 Diod. Sic. 15.55.2; cf. Ascl. 10.1, 11.1; Onas. 21.8; Arr. *Tact.* 26.3. The many assignations of (internally differing) but ingenious and revolutionary tactics employed by Epameinondas are enumerated by Hanson (1988) 192–201; cf. Buckler (1985) 134–135.
707 Devine (1983) 205–210, soundly rebutted by Buckler (1985); see esp. 134–137.
708 Goldsworthy (1997) 2.
709 *Infra* 183–200.

3.3.3 The wings

The placement of different units was important in many respects and was consequently given very serious consideration. There was more or less universal agreement that the best and bravest men should be placed directly at the front, since this was the most dangerous place, immediately opposite the enemy and with absolutely nowhere to run once hostilities commenced. This may seem evident; but it was also a good idea to ensure that brave and experienced men were at the rear: they would calm the nervous and act as stabilising elements, hopefully preventing panic and dissolution of the phalanx from the rear. Xenophon actually actively recommends shutting the cowards up in between the braver men at the front and rear.[710] Here, they were protected from direct contact with the enemy's spear-heads while also effectively prevented from running away.

More surprisingly, perhaps, not only placement in relation to front and rear was important, but also which wing (κέρας) one was assigned to. Greek literature is permeated by the idea that the right wing was the more honourable of the two, and the reasons for this are plain. The right flank was more honourable simply because it was a significantly more dangerous post than almost any other. It was normally referred to as τὰ γυμνά, since the right side was insufficiently covered by the shield, as is clear from the Thucydides passage quoted above. The term for the shielded, left side was τὰ ὡπλισμένα.[711] There are many instances of successful flank attacks directed into the unshielded side of phalanxes, either by outflanking, stratagems or more or less haphazard encounters in the confusion of the battlefield.[712] Under no circumstances would an experienced general allow his troops to march past an enemy on the right side: at Kunaxa, Klearchos absolutely refused to move his Greek mercenary army further left to aid Kyros, since this would mean turning τὰ γυμνά towards the enemy.[713] Kleon at Amphipolis made the fatal blunder of allowing his troops to march past the city gates, right side facing the walls (τὰ γυμνὰ πρὸς τοὺς πολεμίους δούς). Brasidas was quick to arrange a two-pronged lightning attack against this extremely ill-advised manoeuvre, routing the Athenians in utter confusion and defeat.[714] The best example of this is probably furnished by the battle at Nemea in 394 in which the Spartans managed to outflank the Athenians with the extreme right of their phalanx and rout them. They next marched past them, at right angles to the front line. Athens' allies – Argives, Corinthians and Thebans – had

---

710 Xen. *Mem.* 3.1.8, cf. Arr. *Tact.* 12.11 and *Il.* 4.298–300.
711 Xen. *Lac.* 11.9 (cf. *supra* 162–164).
712 Thuc. 3.23.4; Xen. *Hell.* 4.4.11–12, 4.5.13–18, cf. *Cyr.* 5.4.45; Diod. Sic. 19.6.6; Arr. *Tact.* 37.5.
713 Xen. *An.* 1.8.13; cf. Plutarch *Artax.* 8.2–7. Plutarch blames Klearchos for Kyros' defeat and subsequent death in the light of Xenophon's description of Klearchos' 'insubordination'. Lendle (1966) 439–443 correctly acquits him of these charges, however. Kyros had no one to blame for the hopeless formation but himself (cf. *An.* 1.7.17–20), and no general in his right mind would allow 11,000 hoplites to march at right angles to the front line with the uncovered spear side facing the enemy (inv. no. 11).
714 Thuc. 5.10.2–12; cf. Anderson (1965); inv. no. 2.

broken through enemy lines on the other wing early on and were pursuing the Spartan allies directly opposite them:

> τοῖς δ' Ἀργείοις ἐπιτυγχάνουσιν οἱ Λακεδαιμόνιοι ἀναχωροῦσι, καὶ μέλλοντος τοῦ πρώτου πολεμάρχου ἐκ τοῦ ἐναντίου συμβάλλειν αὐτοῖς, λέγεται ἄρα τις ἀναβοῆσαι παρεῖναι τοὺς πρώτους. ὡς δὲ τοῦτ' ἐγένετο, παραθέοντας δὴ παίοντες εἰς τὰ γυμνὰ πολλοὺς ἀπέκτειναν αὐτῶν. ἐπελάβοντο δὲ καὶ Κορινθίων ἀναχωρούντων. ἔτι δ' ἐπέτυχον οἱ Λακεδαιμόνιοι καὶ τῶν Θηβαίων τισὶν ἀναχωροῦσιν ἐκ τῆς διώξεως, καὶ ἀπέκτειναν συχνοὺς αὐτῶν.[715]

These few examples are sufficient to demonstrate why the right side was rightly considered much more dangerous than the left. Of course, this also made it an extremely honourable place to be posted. If one could stay calm, not to mention alive, during a hoplite battle, even though posted at the extreme right and in the front rank, then certainly that was no small feat. As a consequence, the front-and-right place was almost always given to the best and most experienced men, and here, perhaps strangely, was also frequently the station reserved for the general.[716] If, as was often the case at least in the Classical period, the phalanx was a coalition army, consisting of contingents from several city states, it was *de rigueur* that leadership (ἡγεμονία) fell to those whose land was threatened or who had summoned the other allies to their aid in accordance with a treaty of alliance.[717] In such cases, ἡγεμονία more or less directly implied the right (and, indeed, duty) to command the right flank (τὸ δέξιον κέρας ἔχειν).[718]

This principle, again, is nicely illustrated by the battle of Leuktra: Epameinondas assumed – correctly – that the fiercely traditionalist Spartans would themselves occupy the rightmost sector of their own coalition phalanx, and furthermore that the king himself and his entire staff would be exactly there. He assembled a force at this point which would be not only equal to that of the Spartans, but completely overwhelming. Within a probably very short span of time, the Theban onslaught had all but wiped out the entire Spartan high command: king Kleombrotos fell mortally wounded, and the polemarch Deinon, Sphodrias, a member of the war council, and his son Kleonymos were all killed, certainly precipitating the swift collapse of Spartan morale and resistance.[719] The fact that his brilliant plan of concentrating a very great preponderance on the *left* flank worked so well merely goes to show just how deeply rooted the notion of responsibility, honour and tradition was under normal circumstances in the world of inter-city state warfare: the fact that these very

---

715 Xen. *Hell*. 4.2.20–22 ("But the Spartans did encounter the Argives on their way back from the pursuit. Here the story is that just when the first polemarch was going to attack them in front, someone shouted out, 'Let their first ranks go past!' This was done and then, as the Argives were running past, the Spartans attacked and struck down great numbers of them, since their blows were directed at their exposed right sides. They also attacked the Corinthians as they were returning and some of the Thebans too, many of whom they killed" [trans. Warner]); inv. no. 25.
716 Hanson (2000²) 107–116 gives the reasons for this and cites many relevant sources. I shall discuss this point more fully later.
717 Xen. *Hell*. 4.2.18, 7.1.14, 7.5.3; *IG* II² 112.34–35 (alliance treaty between Athens, the Arkadian federation, Achaia, Elis and Phleius).
718 Thuc. 5.47.7; cf. 5.67.2.
719 Xen. *Hell*. 6.4.13–14 (inv. no. 16).

things could be relied on to remain as always was Epameinondas' guarantee for success. According to Plutarch, the Thebans were so convinced of the superiority of this plan that they henceforth placed the command on the left flank in all their subsequent battles.[720]

The military dangers inherent in the right wing require no further explanation; but it has been speculated that there might also be reasons of a completely different nature why the right was considered propitious or honourable. Lévêque and Vidal-Naquet have assessed the religious and sociological implications for right and left wings of phalanxes, going so far in fact as to accuse Thucydides of "rationalisme abusif".[721] They argue that 'right' and 'left' in many cultures, including the Greek, are concepts commonly associated with fundamentally opposing qualities:

> Ainsi, chez Homère, la droite est toujours le côté de la force active et de la vie; la gauche le côté de la faiblesse passive et de la mort; de la droite émanent les influences vivifiantes et salutaires, tandis que de la gauche ne proviennent que des influences déprimantes et délétères.[722]

That this is essentially correct appears from the fact that 'left' is normally called εὐώνυμον, including in military contexts – as obvious a euphemism as one could ever hope for: even the mention of the word is best avoided. It seems perhaps a bit of a stretch that Pythagoreanism should be able to account for Epameinondas' battle plans; but there can be no denying that Greek popular morality was steeped in the notion that 'left' was simply bad or unfavourable.

Despite the ill-will that clung to the concept and name of left, the left flank was supposedly next in terms of honour: compared with the centre of a formation, the flanks will naturally always be more exposed to danger, and this applied also to hoplite phalanxes. The left flank was not a pleasant spot to be either, but it had the decided advantage of being covered by the shield. In the Spartan army, the Skiritai were regularly awarded the left flank, as a token of respect and honour.[723] If we can believe Herodotos, these matters were so important that allied states in coalition armies might quarrel about the right to obtain the most honourable post, the sectors being distributed according to a very complex and hierarchical system. Before Plataiai in 479 no one disputed the Spartans' right to hold the right flank (quite apart from the fact that Sparta had supreme command over the land forces), but the contingents from Athens and Tegea disagreed vehemently about the right to the left flank.[724] This account may well be exaggerated or even downright fictitious, but it

---

720 Plut. *Mor.* 282e.
721 Lévêque & Vidal-Naquet (1960) 299, also citing Thuc. 3.22.2 and Deonna's interesting comparison of this to many other cases of monosandalism as a symbolic ritual concerning chthonic deities. This makes rather better sense than Thucydides' own explanation that the left foot was bare to prevent slipping in the mud: if this was the case, why not avoid shoes altogether?
722 Lévêque & Vidal-Naquet (1960) 300 ("Thus, in Homer, the right side is always the side of active force and of life; the left is the side of passive feebleness and death. From the right side come salutary and life-giving influences, just as nothing but depressing and harmful influences come from the left"). Attention is also drawn to the fact that Aristotle (*Metaph.* 986a 15–21) lists right and left as fundamental opposites in the table of συστοιχίαι.
723 Thuc. 5.67.1.
724 Hdt. 9.26–28.1 (inv. no. 29); cf. 6.111.1, 9.28.2–6; Thuc. 5.71.1–2. Whenever Sparta partici-

serves to illustrate the importance of these principles. Even if these events are exaggerated, Herodotos must have known that he could count on his readers to recognise and appreciate the mechanism at work.[725]

3.3.4 Community ties in phalanx organisation

In phalanx fighting, mutual trust and a spirit of concord and cohesion was all-important, and it was therefore normal to organise the units according to pre-existing community ties, something that is to a great extent reflected in the battle orders. Within the phalanx of each individual city-state, posts were accordingly often distributed according to intricate principles of kinship or other relations, such as *phylai* or demes, though other structural or political relations could also be the determining factors.[726] It was probably the general who was ultimately responsible for assigning posts to the kinship groups, and perhaps even to individuals. In the case of Athens, after the Kleisthenic reforms in 501/0, men of the same tribe (φύλη) to all appearances were organised together,[727] although demes were also important on a smaller scale.[728] The ten στρατηγοί were in fact elected one from each *phyle*, termed τάξις in its military capacity.[729] It is scarcely a stretch of the imagination to suppose that the στρατηγός in question then in practice commanded his own tribesmen.[730]

According to an unlikely tradition, Kimon returned after 10 years of exile after an ostracism just in time to participate in the battle of Tanagra in 457: his fellow tribesmen were supposedly waiting for him, holding his place in the phalanx for him and keeping his weapons and equipment ready.[731] Another indication of the arrangement κατὰ φύλας is the fact that battle casualty lists are arranged with the

---

pated in coalition armies, there was apparently a universal consensus that the Spartans as a matter of course occupied the right flank: Hdt. 9.102.1–3; Thuc. 5.67.1; Xen. *Hell.* 4.2.16–23, 4.4.9, 5.2.40–41, 6.4.1–16; but see Diod. Sic. 15.85.2.

725 Pritchett (1974) 194–199 tabulates the sources' accounts of battles with indications of left and right wings, although both land and sea battles are included.

726 Plut. *Arist.* 5.3. An excellent overview of sources pertaining to the roles of *phylai* may be found in Jones' index, section L ("Representation of public units in the military organization"), Jones (1987) 394–395.

727 See, e.g., Hdt. 6.111.1; Lys. 13.79; Thuc. 6.98.4, 6.101.5, 8.92.4; [Arist.] *Ath. pol.* 42.1; Xen. *Hell.* 4.2.19; Plut. *Arist.* 5.5; Paus. 1.32.3. This principle can also be seen in Homer, *Il.* 2.362–363. Cf. Hanson (2000²) 121–125.

728 Lys. 16.14 (indicating a meeting of Mantitheos' demesmen before the ἔξοδος), 31.15; [Lys.] 20.23; Is. 2.42 (ἐστράτευμαι ἐν τῇ φύλῃ τῇ ἐκείνου καὶ ἐν τῷ δήμῳ ["I have served in his tribe and deme"]); cf. Theophr. *Char.* 25.3–6.

729 [Arist.] *Ath. pol.* 22.2. This practice endured until 441/0, when two *strategoi* from the same *phyle* were elected. Indeed, Aristotle points out that in his day the *strategoi* were elected from among the entire eligible population (see Jones [1987] 55).

730 For the military function of *taxeis/phylai*, see Thuc. 8.29.4; Lys. 13.79; Is. 2.42; Dem. 39.17; *IG* I² 1085 (446/5), *IG* II² 1155 (339/8).

731 Plut. *Cim.* 17.3–5 (inv. no. 40). The story is anecdotal, but nevertheless still reflects an idea that was not considered impossible.

names listed by *phyle*.⁷³² In a fascinating speech by Lysias, the defendant Mantitheos asserts that during a campaign he managed to be placed in the first line:

> μετὰ ταῦτα τοίνυν, ὦ βουλή, εἰς Κόρινθον ἐξόδου γενομένης καὶ πάντων προειδότων ὅτι δεήσει κινδυνεύειν, ἑτέρων ἀναδυομένων ἐγὼ διεπραξάμην ὥστε τῆς πρώτης [sc. τάξεως] τεταγμένος μάχεσθαι τοῖς πολεμίοις· καὶ μάλιστα τῆς ἡμετέρας φυλῆς δυστυχησάσης, καὶ πλείστων ἐναποθανόντων, ὕστερος ἀνεχώρησα τοῦ σεμνοῦ Στειριῶς τοῦ πᾶσιν ἀνθρώποις δειλίαν ὠνειδικότος.⁷³³

It is noteworthy that the entire *phyle* met with misfortune. The logical assumption would be that they were placed together in a sector where the fighting was particularly fierce.⁷³⁴ It is also significant that not only Mantitheos, but also all others are said to have exerted some measure of influence over their stations (or at least that they *tried;* ἀναδυομένων is presumably imperfect *de conatu*). Mantitheos certainly succeeded in obtaining his goal; but others at least tried as well (or so he and his witnesses claim). Possibly the ultimate responsibility of posting the *phylai* lay with the general in charge; other officers – perhaps taxiarchs⁷³⁵ – may have seen to the posting of individuals, even though it could be argued that the general himself might take a personal interest in the composition of the first few ranks, since so much depended on them.⁷³⁶

Other principles might also form the basis for battle orders. In Thebes was the famous 'Sacred Band' (ἱερὸς λόχος), an élite unit of 300 hoplites, which, by all accounts, was composed of 150 pairs of lovers.⁷³⁷ This was thought to achieve an optimal spirit of self-sacrifice and a maximum of mutual loyalty. The idea was that the strong ties between the couples and the mutual shame of appearing cowardly before their lover made them fight all the more fiercely.⁷³⁸ In Plutarch's *Pelopidas* we find a jocular remark on the strength of the ties within the Sacred Band:

> οὐ γὰρ ἔφη τακτικὸν εἶναι τὸν Ὁμήρου Νέστορα, κελεύοντα κατὰ φῦλα καὶ φρήτρας συλλοχίζεσθαι τοὺς Ἕλληνας, ὡς "φρήτρη φρήτρηφιν ἀρήγῃ, φῦλα δὲ φύλοις", δέον ἐραστὴν παρ' ἐρώμενον τάττειν. φυλέτας μὲν γὰρ φυλετῶν καὶ φρατόρων <φράτορας> οὐ πολὺν λόγον

---

732 Meiggs-Lewis *GHI*² 33 (= *IG* I² 929), 48 (= *IG* I² 943); Paus. 1.29.4 (deme), 1.32.3. Similar patterns may be found with other *poleis: IG* I² 931, Meiggs-Lewis *GHI*² 35 (= *IG* I² 932) and cf. Thuc. 5.59.5, 5.72.4 (Argos); Jones (1987) 97–103 (Corinth); *IG* IV² 1.28 (Epidauros); *IG* V 2.173, 174 (Tegea); Jones (1987) 94–97 (Megara). An inscription from Mantineia is believed by Pritchett (1969) 50–53 to be a casualty list from the battle of Mantineia in 418 (inv. no. 17); but see *contra* Solin (1974). See, in general, Pritchett (1985a) 139–145.

733 Lys. 16.15 ("After that, gentlemen, there was the expedition against Corinth; and everyone knew beforehand that it would be a dangerous affair. Some were trying to shirk their duty, but I contrived to have myself posted in the front rank for the battle. Our tribe had the worst fortune, and suffered the heaviest losses among our own men: I retired from the field later than that fine fellow of Steiria who has been accusing everybody of cowardice" [trans. Lamb, modified]). The battle is probably Nemea 394 (inv. no. 25), cf. Xen. *Hell.* 4.2.17, 4.2.21.

734 Cf. Plut. *Arist.* 5.3: Aristeides and Themistokles, each with their tribe, were stationed next to each other at Marathon (inv. no. 19).

735 Lys. 16.16.

736 Plato indicates that both scenarios were possible: Pl. *Ap.* 28d.

737 Élite units were relatively common in a number of *poleis:* see Pritchett (1974) 221–224 for the evidence.

738 Cf. Onas. 24.

ἔχειν ἐν τοῖς δεινοῖς, τὸ δ' ἐξ ἐρωτικῆς φιλίας συνηρμοσμένον στῖφος ἀδιάλυτον εἶναι καὶ ἄρρηκτον, ὅταν οἱ μὲν ἀγαπῶντες τοὺς ἐρωμένους, οἱ δ' αἰσχυνόμενοι τοὺς ἐρῶντας, ἐμμένωσι τοῖς δεινοῖς ὑπὲρ ἀλλήλων. καὶ τοῦτο θαυμαστὸν οὐκ ἔστιν, εἴγε δὴ καὶ μὴ παρόντας αἰδοῦνται μᾶλλον ἑτέρων παρόντων, ὡς ἐκεῖνος ὁ τοῦ πολεμίου κείμενον αὐτὸν ἐπισφάττειν μέλλοντος δεόμενος καὶ ἀντιβολῶν διὰ τοῦ στέρνου διεῖναι τὸ ξίφος, ὅπως ἔφη μή με νεκρὸν ὁ ἐρώμενος ὁρῶν κατὰ νώτου τετρωμένον αἰσχυνθῇ.[739]

This passage, however, demonstrates first and foremost that closeness and bonds of absolute solidarity were qualities considered supremely important in the moment of truth; and that such principles were important enough to have direct influence on the battle order. On the other hand it is likely that these principles came to exert this level of influence precisely because the order of the day was a method of fighting which relied on combining a maximum of impact force with a minimum requirement of tactical training and weapons prowess.

The sources indicate that Greek army structure was to a large extent based on hoplites' knowledge of each other, not only in terms of mobilisation and logistical organisation, but also battle order. This is perhaps puzzling inasmuch as such battle orders were bound to concentrate casualties of the units. Entire *phylai* (or their equivalents) might be all but wiped out, since the tribesmen were all gathered within a relatively small area. If disaster struck at this point, it could deal a devastating blow to the community in question, as seems to have happened to two Athenian *phylai* at the battle of Nemea, who were outflanked by the Spartan phalanx. They bore the brunt of the losses incurred, while the remaining four *phylai* got away almost completely unscathed.[740] A casualty list records the losses of the Erechtheid *phyle* for the year 460/59: no fewer than 185 were killed in action, including both generals.[741] The very fact that the names are inscribed on the same list testifies to the unusual numbers: normally the names of all *phylai* were inscribed on the same stele or a series of steles.[742] In a similar manner, the Theban Sacred Band élite unit of 300 were killed to a man in the battle of Chaironeia in 338.[743]

Poleis in general were relatively small city communities with comparatively small territories, and this also affected the number of adult, male citizens fit for military service. Even relatively slight losses might therefore be decidedly catastrophic to such small-scale communities, both in terms of man-power (militarily

---

739 Plut. *Pel.* 18 ("He said that Homer's Nestor was no tactician when he urged the Greeks to form in companies by clans and tribes, 'so that clan may aid clan, and tribe tribe,' since he should have stationed lover by beloved. For tribesmen and clansmen make little account of each other in danger, whereas a bond that is held together by the friendship between lovers is insoluble and unbreakable, since the lovers are ashamed to play the coward before their beloved, and *vice versa,* and both stand firm in combat to protect each other. And this is not strange, since men have more regard for their lovers even when absent than for others who are present, as was true of the man who, when his enemy was about to kill him where he lay, begged him to run his sword through his breast, 'in order,' as he said, 'that my beloved will not be ashamed to see my body run through from behind' " [trans. Perrin, modified]). Cf. Pl. *Symp.* 179a–b; Ath. 13.561e–f. The quote in the text is *Il.* 2.362–363.
740 Xen. *Hell.* 4.2.17, 4.2.21; cf. Lys. 16.15 (inv. no. 25).
741 *IG* I² 929 (= Meiggs-Lewis *GHI*² 33).
742 See Meiggs-Lewis *ad GHI*² 33 and Bradeen (1964) 21–29.
743 Plut. *Pel.* 18.5. See also Rahe (1981); inv. no. 3.

as well as civilly) and of many whole family structures (οἰκίαι) being deprived of their menfolk, often probably coinciding with the κύριοι. It is impossible to miss the horror expressed by Thucydides regarding the slaughter of the Ambrakiots at Idomene in 426, going out of his way to retell a comparatively lengthy conversation between a shocked Ambrakiot herald and one of his enemies.[744] The herald is so shocked by the disastrous losses that he goes away without having performed his errand. Thucydides concludes:

> πάθος γὰρ τοῦτο μιᾷ πόλει Ἑλληνίδι ἐν ἴσαις ἡμέραις μέγιστον δὴ τῶν κατὰ τὸν πόλεμον τόνδε ἐγένετο. καὶ ἀριθμὸν οὐκ ἔγραψα τῶν ἀποθανόντων, διότι ἄπιστον τὸ πλῆθος λέγεται ἀπολέσθαι ὡς πρὸς τὸ μέγεθος τῆς πόλεως. Ἀμπρακίαν μέντοι οἶδα ὅτι, εἰ ἐβουλήθησαν Ἀκαρνᾶνες καὶ Ἀμφίλοχοι Ἀθηναίοις καὶ Δημοσθένει πειθόμενοι ἐξελεῖν, αὐτοβοεὶ ἂν εἷλον.[745]

Ambrakian losses amounted to at least 1200 hoplites out of 3000;[746] but they were likely higher, since Thucydides is afraid that the true number will be received with disbelief.[747] This does not seem unlikely, given the fact that the Ambrakiot main force were taken by surprise in their sleep and that extensive traps, ambuscades and roadblocks had been prepared for those who escaped.[748] The structure and language of this unusual chapter[749] also leads to the natural conclusion that Ambrakia was a rather small *polis*, and that it was dealt a staggering blow on this occasion.[750] True,

---

744 Thuc. 3.111.3 – 3.113 (inv. no. 8).
745 Thuc. 3.113.6 ("In fact, this was, in all the war, certainly the greatest disaster that fell upon any single Greek *polis* in an equal number of days. I have not recorded the number of the killed, because the number said to have been destroyed is incredible, considering the size of the city. However, I do know that if the Akarnanians and Amphilochians had been willing to follow the advice of Demosthenes and the Athenians and to seize Ambrakia, they could have done so without striking a blow" [trans. Warner, modified]).
746 Thuc. 3.105.1, 3.111.3–4, 3.113.4. See also Gomme (1956a) 424–425. The 1000 killed in the second battle are a casual estimate made by the herald's interlocutor.
747 Thucydides may refuse to disclose anything specific, but he does say that ὀλίγοι ἀπὸ πολλῶν ἐσώθησαν. Hornblower (1991) 533 comments: "This closely echoes i.110.1 (the disaster in Egypt …) and looks forward to vii.87.6 (the disaster in Sicily). These verbal correspondences put the Akarnanian defeat in a very big league indeed".
748 Thuc. 3.112.3–8.
749 Gomme and Hornblower draw attention to the many points of similarity with contemporary tragedy: Gomme (1956a) 425 to Eur. *Supp.* 476–493; Hornblower (1991) 533 compares the rapid dialogue to tragedy, and more especially (and quite aptly) the recognition-like scene when the herald gradually realises the full extent of the disaster to Eur. *Bacch.* 1280–1300.
750 Assuming with Hansen (2006) 12 a maximum of 150 inhabitants per hectare inhabited space. Gehrke & Wirbelauer (2004) 354: Ambrakia was a 'type 4' *polis* with a χώρα of 200–500 km². *Poleis* of this size typically had an urban centre of *c*. 101 ha., see Hansen (2006) 10. Now, assuming that half the urban centre was inhabited and that one ha. of inhabited space could accommodate between 150 and 210 inhabitants, an urban centre would have accommodated a population between 7575 (50.5 × 150) and 10,605 (50.5 × 210) individuals. If we accept that the population was divided equally between an urban centre and the countryside, a conservative estimate for the population of Ambrakia may be 15,150–21,210. The hoplite segment typically made up 10% of the total population (see Hansen [2004] 114 n. 49), so it is not unrealistic to assume a hoplite army of up to 3000. However, there can be no doubt that Ambrakia suffered a disaster of the highest order.

this was certainly out of the usual, and under normal circumstances hoplite battles will not have taken such a heavy toll on the individual *polis*;[751] but the losses incurred here are at least 1200 killed (though possibly many more: this cannot be determined because of Thucydides' reticence).

This is by no means impossible for one side in a regular hoplite battle, and furthermore, the Ambrakiot herald's reaction demonstrates that heavy casualties were perceived as a very severe blow to the community. This rationale should also apply, *mutatis mutandis,* to communities or structural units *within* the *poleis*, such as *phylai, phatrai, hemiogdoa, triakades* and other types of subdivisions of the *polis* army, which ran a very real risk of being completely exterminated if they happened to be posted in an unfortunate sector.[752]

Accordingly, one would assume that it was in the *polis*' best interest to spread the *phylai* more evenly about the phalanx in order to ensure that the burden of casualties was somewhat 'fairly' distributed among the kinship groups. The fact that it was not should give pause for consideration. Evidently, other needs took priority over safeguarding against such veritable annihilations of entire communities. Since the purpose chiefly fulfilled by arranging hoplites according to kinship ties was, as we have seen, increased cohesion and spirit of unity, we must assume that the fulfilment of these needs were considered of paramount importance for the battle formation.

3.3.5 Place and function of the general

The function of the commanding officer is a phenomenon which reveals a great deal about the nature of hoplite fighting. As we saw, the right wing was the most honourable and dangerous post in the phalanx, and this was also often here the commanding officer (στρατηγός) stationed himself – or failing that, at least somewhere along the front line.[753] This is exemplified by the disastrous battle of Leuk-

---

751 Krentz has collected the relevant data from hoplite battles whose sources list the numbers of forces before and after the battle and the casualty rates afterwards and calculated the average losses on both sides. The ratio is a mere 3–6% (with an average of 5%) for the victors, as opposed to 10–20% for the defeated (14% on average) for the defeated: Krentz (1985a) 13–20, esp. 19–20 and cf. Hanson (1995) 308: the sum total of Athenian casualties in the century 470–370 perhaps does not exceed 24,000 all told.
752 Hanson (1999) 210–215 discusses the catastrophe that struck the small *polis* of Thespiai in three coalition battles (Thermopylai 479, Delion 424 [inv. no. 5] and Nemea 394 [inv. no. 25]), which between them all but wiped out the hoplite group of the adult male population and eventually brought about the obliteration of the city itself.
753 For concrete examples, see Marathon 490 (Hdt. 6.111; inv. no. 19); Poteidaia 432 (Thuc. 1.62.6; inv. no. 30); Olpai 426/5 (Thuc. 3.107.4; inv. no. 27); Peiraieus 403 (Xen. *Hell.* 2.4.30); Nemea 394 (Xen. *Hell.* 4.2.9, 4.2.19; inv. no. 25); Koroneia 394 (Xen. *Hell.* 4.3.16; inv. no. 9); Leuktra 371 (Xen. *Hell.* 6.4.13–14; Diod. Sic. 15.55.1; Plut. *Pel.* 23.1–3; inv. no. 16). Cf. Plut. *Mor.* 628d – 629b. These comprise six out of 41 battles in the Inventory (and one not included) or 17% of all the battles there. For battles where the general was demonstrably *not* posted on the right wing, see, e.g., Mantineia 418 (Thuc. 5.72.4; inv. no. 17); Syracuse 415 (Thuc. 6.67.1; inv. no. 36).

tra, where the concentrated Theban attack on the Spartan right wing wiped out their high command in probably a few minutes.[754]

During the entire hoplite era, the Greeks preferred a commanding officer who not only participated in the fighting but who was also placed right where the fighting was thickest. Two closely connected functions of hoplites are needed to understand this properly: a civic/political function (participating in the assembly and having access to office), and a military one where they partook as equals of the defence of the *polis* in a formation where everybody played an equally important part. Seen in this light, the hoplite phalanx was the logical extension of the egalitarian ideal of *polis* politics: militarily, as well as politically, all citizens were equals, and in this respect even the general was merely *primus inter pares*.[755]

Furthermore, it was surely motivating for the phalanx to see their general share the task on an equal footing. Xenophon, who all his life was strongly interested in the subject of leadership, made some useful observations which he willingly shared with his public, not least among which was the usefulness of setting a good example to the rank and file.[756] His experiences in Persia must have made a lasting impression, because his work above all addresses the question of how a leader should behave if he wants his underlings to perform optimally. The ideal general, participating in danger and hardship on an equal footing, is found repeatedly, Archilochos' wish for a warrior's general among them.[757] Aristophanes reveals precisely the same attitude when he ridicules Lamachos and other Athenian officers for their helmets "with three crests" and their dainty scarlet cloaks: they distance themselves too much from the common man and from the hardship they are expected to share.[758]

In later times, we typically find expressions of the direct opposite philosophy of leadership – namely that the commander-in-chief is so crucial in co-ordinating the tactical movements that he must be kept out of actual fighting at all costs;[759] but in Archaic and Classical times battles and battle strategy were far less complex: generally, there was no room for creative strategic thinking. Once the general had deployed his army in a phalanx which was reasonably calibrated against that of the

---

754 Xen. *Hell.* 6.4.13–14 (inv. no. 16).
755 Hanson (1995) 257–262.
756 Xen. *An.* 3.4.46–49, 4.4.11–12, 6.5.14 and esp. 3.1.37. Cf. *Ages.* 5.3, *Cyr.* 1.4.18, *Oec.* 21.4–7.
757 Archil. fr. 114 West, and see *supra* 149–150. Cf. Diod. Sic. 12.70.3, 15.39.1, 87.1; Plut. *Mor.* 639f.
758 Ar. *Ach.* 1071–1234, *Pax* 1172–1178. Van Wees' contention that generals might fight on horseback in the Classical period (van Wees [2004] 187 n. 17) is backed only by a reference to Xenophon's horse at the battle of Kunaxa (Xen. *An.* 1.8.15; inv. no. 11); but Xenophon is careful to point out later that at that time οὔτε στρατηγὸς οὔτε λοχαγὸς οὔτε στρατιώτης ὢν συνηκολούθει ("he accompanied the expedition neither as a general nor a captain nor an ordinary soldier"), and at any rate he was not chosen as general until several weeks after Kunaxa: Xen. *An.* 3.1.4, 3.1.47. Finally, Xenophon points out the awful results when he tried to capture a hill, leading on horseback: one hoplite pointed out that it was "unfair" (οὐκ ἐξ ἴσου), effectively compelling him to continue on foot in his heavy cavalry armour: Xen. *An.* 3.4.46–49.
759 Onas. 23, 33.1.

enemy in terms of width and depth, and which took maximum advantage of the ground, there was no pressing need for a separate battle co-ordination.

The best way a general could exert himself was therefore rather by being among his men and setting them a good example in order to boost morale and fighting spirit.[760] Once the two phalanxes marched against each other, there was no time or space for sophisticated strategic manoeuvres; and owing to the 'pre-programmed' nature of hoplite fighting, they were not strictly necessary. From this point the survival or death of the general mattered little as far as command was concerned; but perhaps did so on another level. The front ranks (or at least parts thereof) could probably see, or at least catch glimpses of, their leader, and his performance was likely a determining factor for the feats of arms at the 'sharp end' of the phalanx. It was not necessarily a military disaster if he fell; on the contrary his death might rouse the front ranks to fight all the more fiercely. This process is analogous to the one making them follow the general's good example in the fight: if he could fight with such contempt for death that he was actually killed, they, as rank and file, could not in decency do less.[761] Generals realised this and took their responsibility seriously. The Spartan Anaxibios led his troops into an ambush; and when he realised the seriousness of the situation, he demanded to be left alone with a few troops, trying to make good an escape for his men. He fell fighting, along with 12 Spartan harmosts.[762]

Just *how* much a general always assumed his more than equal part of danger and hardship can be appreciated by considering the many instances of generals being killed in battle: Anchimolios at Phaleron in 512; Kallimachos *and* Stesileos at Marathon in 490; king Leonidas at Thermopylai in 480; Sophanes at Daton in 465/4; Dikaiogenes at Eleusis in 459/8; Anaxikrates in Kilikia in 451/0; Tolmides at Koroneia in 447; Kallias at Poteidaia in 432; Melesandros in Lykia in 430; all three Athenian generals – Xenophon, Hestiodoros and Phanomachos – at Spartolos in 429; Asopios at Nerikos in 428; Lysikles at Sandion in 428; Charoiades on Sicily in 427; Prokles at Aigition in 426; Eurylochos *and* Makarios at Olpai in 426/5; Epitadas (besides his next-in-command, Hippagretas, severely wounded and left for dead in a pile of bodies) on Sphakteria in 425; Lykophron at Solygeia in 425; Hippokrates at Delion in 424; the generals of both sides – Brasidas and Kleon – at Amphipolis in 422; Xenares at Herakleia in 420/19; Laches *and* Nikostratos at Mantineia in 418; Diomilos at Syracuse in 414; Lamachos at Syracuse in 414; Chalkideus at Panormos in 412; Mindaros at Kyzikos in 410; Labotas at Herakleia in 408/7; Hippokrates at Chalkedon in 408/7; three ἄρχοντες (including two of the 'thirty tyrants', Kritias and Hippomachos) at Munychia in 403 as well as both Spar-

---

760 G.M. Paul speaks of an 'exemplary' (as opposed to a 'directorial') role (Paul [1987] 307–308 and n. 4).
761 There are references in literature to hoplites who – entirely in the style of Homer – fight furiously to keep the dead body of their fallen general: the Spartans at Thermopylai recaptured the body of king Leonidas (Hdt. 7.225); but even so late as 371, Spartan hoplites managed to salvage the body of king Kleombrotos at Leuktra (inv. no. 16): Xen. *Hell*. 6.4.13; cf. Diod. Sic. 15.55.5 – 15.56.2.
762 Xen. *Hell*. 4.8.35–39.

tan polemarchs, Chairon and Thibrachos; Lysander at Haliartos in 395; Polycharmos (a hipparch) in Thessaly in 394; Agesilaos at Koroneia in 394 (only "severely wounded all over by all kinds of weapons"); Pasimachos at Corinth in 392; Gorgopas on Aigina in 388; Teleutias at Olynthos in 381; Phoibidas at Thespiai in 378; Alypetos in Boiotia in 377; Gorgoleon *and* Theopompos at Tegyra in 375; Kleombrotos (and his entire staff!) at Leuktra in 371; Stratolas at Olympia in 364; Pelopidas at Kynoskephalai in 364; Epameinondas (and possibly also Daiphantos and Iolaidas) at Mantineia in 362; Mnaseas near Naryx (?) in 352/1; Kleinios in Egypt in 352/1; Leosthenes before Lamia in 323.[763] The high mortality rate among generals in Archaic and Classical times speaks volumes.[764]

The general's falling did not in itself ensure defeat: of the above, Kallimachos and Stesileos, Kallias, Brasidas, Pelopidas and Epameinondas fell in the moment of victory: "again, perhaps a suggestion that his presence in the front ranks, rather than his safety from injury, was important if his men were to fight well: so much for the idea that the battlefield general's survival was always vital for military success."[765] Demosthenes ironised the fact that generals of his day preferred dishon-

---

763 Anchimolios: Hdt. 5.63; Kallimachos and Stesileos: Hdt. 6.114 (inv. no. 19); Leonidas: Hdt. 7.224; Sophanes: Hdt. 9.75; Dikaiogenes: Is. 5.42; Anaxikrates: Diod. Sic. 12.3.4.; Tolmides: Diod. Sic. 12.6.2; Kallias: Thuc. 1.63.3 (inv. no. 30); Melesandros: Thuc. 2.69.2; Xenophon, Hestiodoros and Phanomachos: Thuc. 2.79.7 (inv. no. 33); Asopios: Thuc. 3.7.4; Lysikles: Thuc. 3.19.2; Charoiades: Thuc. 3.90.2; Prokles: Thuc. 3.98.4 (inv. no. 1); Eurylochos and Makarios: Thuc. 3.109.1 (inv. no. 27); Epitadas and Hippagretas: Thuc. 4.38.1 (inv. no. 34); Lykophron: Thuc. 4.44.2 (inv. no. 32); Hippokrates: Thuc. 4.101.2 (inv. no. 5); Brasidas and Kleon: Thuc. 5.10.8–9 (inv. no. 2); Xenares: Thuc. 5.51.2; Laches and Nikostratos: Thuc. 5.74.3, cf. Diod. Sic. 12.79.1 (inv. no. 17); Diomilos: Thuc. 6.97.4 (inv. no. 37); Lamachos: Thuc. 6.101.6, cf. Plut. *Nic.* 18.3; Chalkideus: Thuc. 8.24.1; Mindaros: Xen. *Hell.* 1.1.18 (inv. no. 13); Labotas: Xen. *Hell.* 1.2.18; Hippokrates: Xen. *Hell.* 1.3.5–6; three ἄρχοντες and both polemarchs: Xen. *Hell.* 2.4.19, 2.4.33, cf. *IG* II² 11678 (inv. no. 22); Lysander: Xen. *Hell.* 3.5.19, cf. Plut. *Lys.* 28.5 (inv. no. 7); Polycharmos: Xen. *Hell.* 4.3.8; Agesilaos: Xen. *Hell.* 4.3.20, cf. *Ages.* 2.13; Plut. *Ages.* 19.1, 36.2 (inv. no. 9); Pasimachos: Xen. *Hell.* 4.4.10 (inv. no. 4); Gorgopas: Xen. *Hell.* 5.1.11–12; Teleutias: Xen. *Hell.* 5.3.6 (inv. no. 28); Phoibidas: Xen. *Hell.* 5.4.41, 5.4.44–45; Diod. Sic. 15.33.6; Alypetos: Xen. *Hell.* 5.4.52; Gorgoleon and Theopompos: Plut. *Pel.* 17.3 (inv. no. 41); Kleombrotos: Xen. *Hell.* 6.4.13 (inv. no. 16); Stratolas: Xen. *Hell.* 7.4.31; Pelopidas: Diod. Sic. 15.80.5, cf. Plut. *Pel.* 32.7 (inv. no. 12); Epameinondas: Diod. Sic. 15.87.1; Daiphantos and Iolaidas: Plut. *Mor.* 194c (inv. no. 18); Mnaseas (a Phokian, wrongly termed Boiotian by Pritchett [1994] 132): Diod. Sic. 16.38.6–7; Kleinios: Diod. Sic. 16.48.5; Leosthenes: Diod. Sic. 18.13.3–5; Just. *Epit.* 13.5.12. Cf. Meiggs-Lewis *GHI*² 33.5–6, 33.63, 48.4: both generals, Ph[ryni]chos and Hippodamas, of the *phyle* Erechtheis in Athens fell in battle in 460/59; and also *IG* I³ 1162.4: the general Epiteles fell fighting in the Chersonese in 447 (erroneously given by Pritchett [1994] 131 as 'Epitales' and the year as 439 at *IG* I³ 439.4; cf. Hamel [1998] 205, Develin [1989] 82 and Meiggs-Lewis *GHI*² 48.4). Hamel (1998) 204–209 lists all 38 war fatalities of Athenian generals between 501 and 322. Generals executed afterwards or who fell in naval battles have not been included. Hanson (2000²) 113 lists some of these and curiously includes Klearchos, who (as is evident from Xenophon) was captured and executed long after Kunaxa (Xen. *An.* 2.6.1), and so does Franz (2002) 292 n. 341 ("ohne Anspruch auf Vollständigkeit"). Pritchett (1994) 127–133, on the other hand, has most of these.
764 Wheeler (1991) 146–151; cf. Pritchett (1994) 111, 130–141.
765 Hanson (2000²) 114. Hanson cites a long list of interesting modern parallels, in which soldiers'

ourable penalties in court to risking their lives on the battlefield: κακούργου μὲν γάρ ἐστι κριθέντ' ἀποθανεῖν, στρατηγοῦ δὲ μαχόμενον τοῖς πολεμίοις.[766] The gnomic style underlines the truism apparently inherent in the statement. Expectations were so high that when Lysikles had the face to be among the Athenian survivors after the battle of Chaironeia in 338, Lykurgos preferred a charge against him. Diodoros has a part of the indictment:

ἐστρατήγεις, ὦ Λύσικλες, καὶ χιλίων μὲν πολιτῶν τετελευτηκότων, δισχιλίων δ' αἰχμαλώτων γεγονότων, τροπαίου δὲ κατὰ τῆς πόλεως ἐστηκότος, τῆς δ' Ἑλλάδος ἁπάσης δουλευούσης, καὶ τούτων ἁπάντων γεγενημένων σοῦ ἡγουμένου καὶ στρατηγοῦντος τολμᾷς ζῆν καὶ τὸ τοῦ ἡλίου φῶς ὁρᾶν καὶ εἰς τὴν ἀγορὰν ἐμβάλλειν, ὑπόμνημα γεγονὼς αἰσχύνης καὶ ὀνείδους τῇ πατρίδι.

Lysikles was condemned and executed.[767]

3.4 OTHISMOS

3.4.1 A vexed question

Among the questions of how a hoplite battle actually took place one of the most hotly debated is that of ὠθισμός. This verbal noun, derived, of course, from ὠθέω, means 'pushing' or 'shoving'. It or its cognates appear frequently in the sources,[768] and the question what precisely is meant by this part of the fighting has occupied scholars for most of the last century. The prevalent notion, most cogently put forward by Hanson,[769] is that *othismos* designates a common effort, ostensibly a common push or shove (often referred to as a mass-shove) of the entire phalanx (or parts thereof) towards and into the enemy in order to drive them back, disrupt their lines and thus break their ranks entirely. If this could be achieved, the outcome of battle was as good as decided.[770]

This is where the rear ranks enter the picture. Xenophon recommended that the bravest men be posted in the front line *as well as* in the rear: καὶ γὰρ ἐν τῷ πολέμῳ

---

morale (and thus their performance) has been improved greatly by officer corps leading their men into battle, rather than directing them from a safe distance, 108–110, 115–116.
766 Dem. 4.47 ("criminals are condemned to execution, generals should die on the field of honour" [trans. Vince, modified]).
767 Diod. Sic. 16.88.1–2 ("You were general, Lysikles. A thousand citizens have perished and two thousand were taken captive. A trophy stands over your city's defeat, and all of Greece is enslaved. All of this happened under your leadership and command, and yet you dare to live and to look on the sun and even to intrude into the market, a living monument of our country's shame and disgrace" [trans. Bradford Welles]); inv. no. 3. On the role of the general, see also Wheeler (1991), who concludes that the active role of generals was due, in large part, to a more or less conscious cultural influence from the palpable heroism of the Homeric poems.
768 See *infra* 185 n. 779 for a survey of passages.
769 Hanson (2000²) 28–29, 156–158, 171–184.
770 See, e.g., Hanson (2000²) 28–29, 156–158, 169–178, (1991) 69 n. 18; Delbrück (1900) 26–27; Holladay (1982) 94–97; Luginbill (1994) 51–61; Lazenby (1991) 97–100; Anderson (1984) 152, (1991) 15–16; Pritchett (1985a) 65–73, 91–92.

τοὺς ἀρίστους δεῖ πρώτους τάττειν καὶ τελευταίους, ἐν μέσῳ δὲ τοὺς χειρίστους, ἵνα ὑπὸ μὲν τῶν ἄγωνται, ὑπὸ δὲ τῶν ὠθῶνται.[771] The cowards in the middle were not only to be enticed forwards by example, they were also to be bodily pushed forwards. This stage of battle accordingly required the entire phalanx to perform an enormous physical effort, not only the two or three front ranks. The hoplites in the front line stemmed their left shoulder against their shield and thrust it against the shields or bodies of the enemy with all their might; and the ranks behind them in turn stemmed their shields against the back and right side of the man in front in a ¾ angle, as it were.

In this way, a tremendous pressure could be generated and conveyed through the entire phalanx from the rearmost rank, its force increasing on the way. It is easy to imagine that such shoving might undulate back and forth for some time before it yielded a result on either side; and consequently hoplites had to step over other hoplites from both sides lying on the ground, dead, dying, wounded or merely fallen to the ground. They most likely also trampled some of them to death: once a man had fallen to the ground inside the phalanx, it was exceedingly difficult to get up again (although Xenophon relates that Kleonymos fell down and got up again no fewer than three times before he was finally killed at Leuktra).[772] In this situation, the hindmost hoplites could be of service with their spears, which they were forced to hold vertically in the throng: stepping over a fallen enemy, they could despatch him by means of a powerful, descending jab with the *sauroter* of their intact spear.[773] Bronze cuirasses with square holes have been found (though these may be nail holes from suspension as trophies or votive gifts).[774] Also, the erect spears might serve to deflect such arrows and other missiles as might be falling on the phalanx: at any rate, this was the case with the Macedonian phalanx.[775]

In many ways, then, *othismos* was the logical extension of the tightly packed fighting in the phalanx. The objective was to break the enemy ranks at all costs, thereby forcing them to expose their vulnerable sides. This would make them unable both to cover themselves and their comrades sufficiently, and also to generate a counter-pressure to relieve the pressure they were subjected to. Since breaking the enemy ranks was the main tactical objective of the battle, this was a way of effectively achieving that objective, rather than standing still and trying to decimate them with spears and swords, a task that was not only both risky and unpleasant, but would also take long time.

This is the 'canonical' interpretation of *othismos;* but it has been challenged on the assumption, among other things, that *othismos* is used about a 'driving back' in

---

771 Xen. *Mem.* 3.1.8 ("In war too you should post the best troops in front and in the rear, and the worst in between, so that they may be led by the one lot and pushed forward by the other" [trans. Tredennick]), cf. *Cyr.* 6.3.25; Plut. *Pel.* 19.3; Arr. *Tact.* 12.11 and Lazenby (1991) 97; Hanson (2000²) 172; *contra*: Goldsworthy (1997) 12.
772 Xen. *Hell.* 5.4.33 (inv. no. 16).
773 Hanson (1991) 67–74.
774 *Supra* 89 n. 335.
775 Polyb. 18.30.3; Hanson (2000²) 171–172.

a purely metaphorical sense, entirely without the physical aspect.[776] An argument for this case is the fact that *othismos* or *otheo* is sometimes used in a context of naval battles, where it can hardly mean a physical shoving as such.[777] While this shows that *otheo* and its compounds can *also* be used figuratively, it does not prove that they are not used in an entirely concrete sense also. If not, what are we to make of Herodotos' remark that the Athenians at Plataiai *pushed through* (διωσάμενοι) the Persians' wall of wicker shields, which had until that point kept them at a safe distance?[778] Surely the shields were not metaphorically 'put to flight'. The examples of *othismos* meaning bodily push are too many and too unambiguous to be safely ignored or explained away: of the 41 battles in the Inventory, 12 contain *explicit* references to *othismos,* making up for 29.27% of the battle narratives: Plataiai 479, Mykale 479, Pylos 425, Solygeia 425, Delion 424, Mantineia 418, Syracuse 415, Miletos 412, Koroneia 394, Corinth 392, Leuktra 371 and Kynoskephalai 364. Three more battles, Thermopylai 480, Peiraieus 403/2 and Olympia 364, which are not listed in the Inventory, also describe the phenomenon.[779] Remaining battles of course cannot be judged accurately *e silentio*.

Another indication is furnished by the later Arrian, who asserts that cavalry is nothing like the infantry, precisely because the horses cannot press or shove with their bodies:

ἐπεὶ οὐδ' ἐκεῖνο χρὴ ἀγνοεῖν, ὅτι οἱ ἐς βάθος ἐπιτεταγμένοι ἱππεῖς οὐ τὴν ἴσην ὠφέλειαν παρέχουσιν, ἥνπερ τὸ ἐπὶ τῶν πεζῶν βάθος· οὔτε γὰρ ἐπωθοῦσι τοὺς πρὸ σφῶν, διὰ τὸ μὴ δύνασθαι ἐπερείδειν ἵππον ἵππῳ, καθάπερ ἐκεῖ κατὰ τοὺς ὤμους καὶ τὰς πλευρὰς αἱ ἐνερείσεις γίγνονται τῶν πεζῶν, οὔτε συνεχεῖς γιγνόμενοι τοῖς πρὸ σφῶν τεταγμένοις ἕν τι βάρος τοῦ παντὸς πλήθους ἀποτελοῦσιν, ἀλλ' εἰ ξυνερείδοιεν καὶ πυκνοῖντο, ἐκταράσσουσι μᾶλλον τοὺς ἵππους.[780]

This accords well enough with Polybios' excellent description of the phenomenon: αὐτῷ γε μὴν τῷ τοῦ σώματος βάρει κατὰ τὴν ἐπαγωγὴν πιεζοῦντες οὗτοι τοὺς προηγου-

---

776 Fraser (1942) 15.
777 Krentz (1994) 48.
778 Hdt. 9.102.2–3 (inv. no. 29).
779 Thermopylai: Hdt. 7.225.1; Plataiai 479 (inv. no. 29): Hdt. 9.26.1, 9.62.2; Mykale 479 (inv. no. 23): Hdt. 9.102.3; Pylos 425 (inv. no. 31): Thuc. 4.11.3; Solygeia 425 (inv. no. 32): Thuc. 4.43.3; Delion 424 (inv. no. 5): Thuc. 4.96.2, 4.96.4; Mantineia 418 (inv. no. 17): Thuc. 5.72.3; Syracuse 415 (inv. no. 36): Thuc. 6.70.2; Miletos 412 (inv. no. 21): Thuc. 8.25.4; Peiraieus 403/2: Xen. *Hell.* 2.4.34; Koroneia 394 (inv. no. 9): Xen. *Hell.* 4.3.19, *Ages.* 2.12 and cf. Plut. *Ages.* 18.2–4; Corinth 392 (inv. no. 4): Xen. *Hell.* 4.4.11; Leuktra 371 (inv. no. 16): Xen. *Hell.* 6.4.14; Olympia 364: Xen. *Hell.* 7.4.31; Kynoskephalai 364 (inv. no. 12): Plut. *Pel.* 32.4. See also Arr. *Tact.* 12.3, 12.10–11, cf. 16.13–14; Ael. *Tact.* 14.6; Paus. 4.8.2 and cf. Ar. *Vesp.* 1081–1085. On this background, it is somewhat puzzling to see Fraser (1942) 15–16 maintain that there are "but three literary references" to the mass-shove.
780 Arr. *Tact.* 16.13–14 ("For we must not overlook the fact that massed cavalry cannot aid each other in a way equal to that of massed infantry. For they cannot push those in front (the horses cannot lean on each other), as infantry troops can lean on each other with their shoulders and sides, nor can they coalesce with the ranks in front of them and create one impetus from the entire force – if they tried to lean on one another and tighten the formation, it would merely result in confusion of the horses"). Luginbill (1994) 52–53 and n. 5 adduces this passage and draws attention to Mauricius, who makes the exact same point about 'the ancients'.

μένους βιαίαν μὲν ποιοῦσι τὴν ἔφοδον, ἀδύνατον δὲ τοῖς πρωτοστάταις τὴν εἰς τοὖπισ-
θεν μεταβολήν.⁷⁸¹

It has been objected that the relevance of such Hellenistic sources for the physical pressure is negligible, because they deal with a different style of fighting – that of the Macedonian army – and that therefore they have no bearing on the Archaic and Classical phalanx.⁷⁸² The implication must be that the chronological difference somehow makes the *physical* implications for the Archaic and Classical phalanx irrelevant. Such arguments overlook the simple fact that the Macedonians somehow *did* manage to include the bodily push in the fighting, and that it is therefore physically possible. Oliver Goldsworthy comes closest to realising this, but shies away from the logical conclusion: "This comment [Polybios' remark about the bodily weight] concerns expressly ranks 6–16, not 2–5, whose pikes projected in front of the formation …".⁷⁸³ I fail to see the relevance of this objection. The remark certainly does concern the rearmost ranks, but Polybios clearly states that they *physically press on the leading five ranks with their bodies* (αὐτῷ … τῷ τοῦ σώματος βάρει … πιεζοῦντες … τοὺς προηγουμένους). In the preceding paragraph Polybios says that only the pikes of the first five ranks project in front of the phalanx; yet he adds ἐκ δὲ τούτου ῥᾴδιον ὑπὸ τὴν ὄψιν λαβεῖν τὴν τῆς ὅλης φάλαγγος ἔφοδον καὶ προβολήν, ποίαν τιν' εἰκὸς εἶναι καὶ τίνα δύναμιν ἔχειν, ἐφ' ἑκκαίδεκα τὸ βάθος οὖσαν.⁷⁸⁴ Why is it so easy to imagine the "tremendous force and impetus" (ἔφοδον καὶ προβολήν) of the additional 11 ranks, if they are merely there to ensure nobody runs away?

It simply will not do to claim with Goldsworthy that the rear ranks only prevented the file leaders from fleeing: besides, this accords poorly with his own observation that such file leaders were "chosen from the most reliable, and received higher pay".⁷⁸⁵ Physical forward shoving was a reality in the Hellenistic armies, and no matter how different the fighting styles, this conclusively shows that fighting while being pushed is apparently physically possible after all. What is more, the 'collective push' by way of shoulder-and-shield as a mode of attack of an entire army is certainly well known also in Latin literature.⁷⁸⁶ Thus, the *a priori* argument

---

781 Polyb. 18.30.4 ("These rear ranks by the sheer pressure of their bodily weight against those in front of them greatly increase its momentum and make it impossible for the foremost ranks to face about" [trans. Scott-Kilvert, modified]); cf. Ascl. 5.2; Arr. *Tact.* 12.10; Ael. *Tact.* 14.6 who give the same information almost *verbatim*.
782 Krentz (1985b) 51–52, (1994) 47; Franz (2002) 300.
783 Goldsworthy (1997) 12–13.
784 Polyb. 18.29–30.2 ("From these facts we can easily picture the nature and the tremendous power of a charge by the whole phalanx, when it advances sixteen deep with levelled pikes" [trans. Scott-Kilvert]).
785 Goldsworthy (1997) 12 (citing Ascl. 3.2–5; but see also Arr. *Tact.* 12.1–2).
786 Livy 30.34.3–4: *Igitur primo impetu extemplo movere loco hostium aciem Romani. Ala deinde et umbonibus pulsantes in summotos gradu inlato aliquantum spatii velut nullo resistente incessere, urgentibus et novissimis primos ut semel motam aciem sensere, quod ipsum vim magnam ad pellendum hostem addebat* ("Consequently by the first attack the Romans at once dislodged the enemy's line. Then beating them back with their shoulders and the bosses of their shields, being now in close contact with men forced from their position, they made considera-

– *othismos* is physically implausible or impossible – carries no weight, and can be disregarded.

Finally, the reality of the physical, bodily thrust is clearly brought out by a passage in Xenophon's make-believe 'novel', the *Kyropaideia:*

ἐπλεονέκτουν μέντοι οἱ Αἰγύπτιοι καὶ πλήθει καὶ τοῖς ὅπλοις. τά τε γὰρ δόρατα ἰσχυρὰ καὶ μακρὰ ἔτι καὶ νῦν ἔχουσιν, αἵ τε ἀσπίδες πολὺ μᾶλλον τῶν θωράκων καὶ τῶν γέρρων καὶ στεγάζουσι τὰ σώματα καὶ πρὸς τὸ ὠθεῖσθαι συνεργάζονται πρὸς τοῖς ὤμοις οὖσαι. συγκλείσαντες οὖν τὰς ἀσπίδας ἐχώρουν καὶ ἐώθουν. οἱ δὲ Πέρσαι οὐκ ἐδύναντο ἀντέχειν, ἅτε ἐν ἄκραις ταῖς χερσὶ τὰ γέρρα ἔχοντες, ἀλλ' ἐπὶ πόδα ἀνεχάζοντο παίοντες καὶ παιόμενοι, ἕως ὑπὸ ταῖς μηχαναῖς ἐγένοντο.[787]

The entire *Kyropaideia* passage has a suspiciously hoplite-like ring to it, as might indeed be expected from Xenophon's 'mirror of princes', teaching Greek readers practical and applicable lessons in the favourite theme of leadership – both civic and military. Most likely Xenophon is in fact advocating the hoplite tactics of "long and powerful spears" and "shields [that] cover their bodies much more effectively than breastplates and wicker shields". Quite apart from this, however, the passage actually effectively refutes the notion that *otheo* is a mere metaphor: the pushing involved here is very concrete indeed. As Luginbill points out, "the reason for the longstanding assumption about the primary importance of *othismos* in hoplite battles should be clear: *it flows from a natural reading* of the best available contemporary witnesses to this sort of combat."[788]

3.4.2 The case against the mass-shove

The chaotic and ungracious element that was an integral part of hoplite battles has certainly not had universal appeal, despite frequent admissions that "the notion is superficially an attractive one".[789] In 1978, Cawkwell suggested that the idea of a mass-shove was incompatible with normal use of arms (and therefore impossible),

---

    ble progress, as no one offered any resistance, while as soon as they saw that the enemy's line had given way, even the rear line pressed upon the first, a circumstance which of itself gave them great force in repulsing the enemy" [trans. Moore]). See also Tac. *Hist.* 2.42: *In aggere viae, conlato gradu corporibus et umbonibus niti, omisso pilorum iactu, gladiis et securibus galeas loricasque perrumpere; noscentes inter se, ceteris conspicui, in eventum totius belli certabant* ("On the raised road they struggled at close quarters, pressing with the weight of their bodies behind their shields; they threw no spears, but crashed swords and axes through helmets and breastplates. They could recognize one another, they could be seen by all the rest, and they were fighting to decide the issue of the whole war" [trans. Moore]).

787 Xen. *Cyr.* 7.1.33–34 ("The Egyptians, however, had the advantage both in numbers and in weapons; for the spears that they use even today are long and powerful, and their shields cover their bodies much more effectively than breastplates and small shields, and as they rest against the shoulder, they are a help in shoving. So, locking their shields together, they advanced and shoved. And because the Persians had to hold out their little shields clutched in their hands, they were unable to hold the line, but were forced back, foot by foot, exchanging blows, until they reached protection from the moving towers" [trans. Miller, modified]).

788 Luginbill (1994) 52 (emphasis mine).

789 E.g. Goldsworthy (1997) 3.

thus renewing an idea originally put forth by Fraser in 1942.[790] Hoplite battles, Cawkwell suggested, rather played out like a series of duels: one hoplite taking on one adversary, hand-to-hand or one against two, two against three or other such combinations of very small groups fighting each other, but with no group action to speak of. This stance, often dubbed a 'heresy' by its proponents to emphasise the difficulty of dispelling the prevalent notion, has nevertheless had increased influence in the debate. It has been accepted and defended by Krentz, Goldsworthy and, most recently, van Wees,[791] whereas the 'traditionalist' conception of the *othismos* has been argued especially by Anderson, Hanson, Luginbill and Schwartz.[792] Most of the opponents, like Krentz, have denied the existence of any kind of mass-shoving altogether;[793] whereas Cawkwell adopted a position in between, agreeing that hoplites might shove – later in the battle, and "whenever a shove was called for".[794]

The criticism against the idea of *othismos* – that a massed bodily push is unfeasible or unlikely – may generally be divided into two main categories. (1) The tremendous physical pressure generated from behind by seven or more ranks pressing each other on would simply be too great to bear. (2) Even if it were tolerable, it would defy the purpose of allowing the hoplites to fight, since the immense pressure would badly impede use of weapons.

(1) Let us examine the first objection. "By virtue of this rear-to-front pressure, the battleline of either side becomes compressed to a thickness of not above 15 or 20 feet, and the front-liners are subjected to a degree of squeezing that is distressing to contemplate."[795] The pressure would be enormous, it is alleged, and either cause hoplites to suffocate or at least prevent them from fighting effectively. Franz makes the interesting observation that it is incorrect, from a physical point of view, to speak of the pressure as generated by bodily *weight*:

> Der Massendruck wurde nicht durch das Gewicht der Krieger erzeugt, sondern durch ihre Muskulatur, insbesondere die der Beine. Das Gewicht – oder genauer die Masse – der Phalangiten spielte dem gegenüber nur eine untergeordnete Rolle. Es wirkte sich vor allem dann aus, wenn kurze, stoßförmige Impulse von einem Krieger zum anderen weitergegeben wurden.[796]

---

790 Cawkwell (1978) 150–157; Fraser (1942) 15–16.
791 See, e.g., Krentz (1985b), (1994), (2007) 74–75; Cawkwell (1978) 150–157, (1989); van Wees (2000) 131–132, (2004) 152, 180–181 and esp. 188–191; Goldsworthy (1997).
792 Anderson (1984); Hanson (2000²) 68–69, 152–159, 171–184; Luginbill (1994); Schwartz (2002) 44–49, (2004) 153–157.
793 Krentz (1985b) 51–55, (1994) 47, (2007) 74–75.
794 Cawkwell (1978) 152–153; (1989) 376–378. Cawkwell's assumption that *othismos* took place in the later stages of battle seems chiefly to be based on the fact that the *results* of shoving (one side being turned and driven back) often come near the end of the historians' battle narratives. However, this does not warrant the inference that *othismos* was not part of the fighting.
795 Fraser (1942) 15; cf. van Wees (2000) 131.
796 Franz (2002) 304 ("The mass pressure was not achieved by the weight of the warriors, but by their muscles, especially their leg muscles. The weight – or rather, the mass – of the hoplites played a relatively minor role. It came into play chiefly when brief, thrusting impulses were transmitted from one warrior to another").

This corrects a common enough oversight in the *othismos* debate. The pushing itself is often rejected on this ground, but Franz is correct in asserting that the pressure generated, while great and inhibiting, is not generated by *weight* and would not necessarily be impossible to resist. Finally, hoplites would in many cases be wearing armour; and such armour, especially if made of bronze, would help alleviating the pressure somewhat.

Another relevant observation by Franz is that pressure does not increase linearly with file depth, and that crowd pressure, however awful, is never a constant. Anyone who has ever found himself inside a large, tightly-packed crowd of people moving in a certain direction can testify to this:

> Wenn die Hinterleute feststellen, daß das Drücken im Augenblick keine Vorteile bringt, beenden sie es. Das Resultat ist dann eine Art Rückstoß. Die Druckverringerung breitet sich aufgrund der Trägheit einer Menschenmasse langsam nach vorne aus. Die entlasteten Vorderleute werden möglicherweise nun ihrerseits Druck nach hinten ausüben, um sich für den Augenblick etwas Luft zu verschaffen. Solche Phasen nachlassenden Drucks sind insbesondere bei längeren Massenveranstaltungen nötig, damit sich die Teilnehmer erholen können.[797]

Most people have certainly experienced this. It is often very unpleasant (not least owing to the complete loss of control), but every so often, the pressure is relieved, only to be renewed yet again a short while later. This non-linear, undulating pattern cannot be gauged regularly, but is most certainly a real part of crowd movements. The pressure in a phalanx could likely at times be very oppressive, but in essence it was usually like any other crowd moving in a particular direction: rippling movements, always in one direction, and often quite powerful, but increasing or decreasing in force as it was perceived to gain headway.

Yet sources testifying to substantial pressure are plentiful. At Solygeia, the Athenians and Karystians "just barely" pushed back the Corinthians (ἐδέξαντό τε τοὺς Κορινθίους καὶ ἐώσαντο μόλις),[798] just as in Xenophon's fictional battle at Thymbrara the Persians were pushed back "step by step" (ἐπὶ πόδα).[799] Generating the necessary pressure and forward force was a tough and slow process; a sort of "'tug-of-war' in reverse".[800] According to legend, Epameinondas cried ἓν βῆμα χαρίσασθέ μοι καὶ τὴν νίκην ἕξομεν ("Give me just one step more, and victory is

---

[797] Franz (2002) 302–303 ("When people behind sense that the pushing does not bring about any immediate advantage, they stop pushing. The result is a kind of reverse thrust. The pressure drop spreads slowly forwards, due to the inertia of a crowd of people. The relieved front rank men may even press backwards in order to procure some air temporarily. Such phases of decreasing pressure are necessary, especially in long mass actions, to allow the participants to recover"). Conversely, crowd pressure where no progress can be made – such as at Corinth in 392 (inv. no. 4) – may often result in accidents: this can be compared to not entirely dissimilar conditions at modern rock concerts where spectators often suffocate, faint or even die from asphyxiation due to the pressure: nine people were killed and 25 injured in this way at Denmark's Roskilde festival in 2000. A list of these as well as many other recent deaths during concerts from such causes as 'crowd crush' or 'ingress' may be found at http://www.crowdsafe.com/thewall.html#denmark.
[798] Thuc. 4.43.3 (inv. no. 32).
[799] Xen. *Cyr.* 7.1.34.
[800] Luginbill (1997) 56.

ours!") at Leuktra, thereby inspiring his men to break the deadlock and drive back the Spartans by throwing in their last reserves.[801] This cry may be found only in the "child-minded chatterbox" Polyainos,[802] but at least it clearly shows that such a situation – one last step being all that separated the phalanx from victory – was not unthinkable in Greek tradition. Similarly, in Xenophon's contemporary description of Leuktra he is quite clear that the Spartans were pushed back *by the mass* of the Thebans (ὑπὸ τοῦ ὄχλου ὠθούμενοι ἀνεχώρουν),[803] just as the description of violently smashed shields on the deserted battlefield at Koroneia suggests far more than mere individual shield-nudging.[804] The Spartans, attacking in column along a road at Kromnos in 365, were unable to hold back the mass (πλήθει) of the Arkadians, who were advancing ἀθρόοι συνασπιδοῦντες.[805] The initial failure of Pausanias' forces at Peiraieus in 403, driven back by an eight-deep Athenian phalanx (suggesting that the Spartans were *not* massed eight deep) was remedied by a redeployment, making the Spartan phalanx παντελῶς βαθείαν: now, the Athenians were pushed back (ἐξεώσθησαν) right into the swamp at Halai, losing 150 men in the process. If weapons prowess was the only factor here, one wonders why the Spartans fared so differently before and after making their phalanx "exceedingly deep".[806]

This again points to *othismos,* much like the fact that Theban successes on the battlefield were later sometimes ascribed to their physical fitness from wrestling training: this, above all, indicates that they were strong and trained in long, locked bouts of digging in with the feet and stemming themselves against an adversary.[807] Much in the same vein, Euripides describes the mark left on the shield rim where the hoplite leans against it in combat in an immediately recognisable posture of extreme exertion; and the fact that his head is bent downwards, his face resting on the rim, shows that he had no need to hold up his head.[808]

Othismos was certainly exceedingly hard, but the objection that it is impossible to shove for several hours is unwarranted.[809] While it would most certainly be impossible to push for hours on end,[810] the evidence for such long battles is tenuous at best, and Cawkwell's assertion that descriptions of 'long' battles means hours' worth of fighting is unfounded.[811] Pushing, along with fighting with spear and sword, can have been over in considerably less time than that; and to this should be added the fact that the pressure under normal circumstances probably was not con-

---

801 Polyaen. 2.3.2, cf. 3.9.27, 4.3.8 (inv. no. 16).
802 Fraser (1942) 16.
803 Xen. *Hell.* 6.4.14 (inv. no. 16).
804 Xen. *Ages.* 2.14 (inv. no. 9).
805 Xen. *Hell.* 7.4.23 (inv. no. 10).
806 Xen. *Hell.* 2.4.32–34: the phrasing παντελῶς βαθείαν is the exact same (with the word order reversed) as that used of the Theban phalanx at Nemea (inv. no. 25): Xen. *Hell.* 4.2.13, 4.2.18.
807 Plut. *Mor.* 639f – 640a, 788a, 233e; cf. *Pel.* 7.3; Xen. *Hell.* 3.4.16; Diod. Sic. 12.70.3, 15.39.1, 15.87.1; Pritchett (1985a) 64–65; Hanson (1995) 273–274.
808 Eur. *Tro.* 1196–1199.
809 Fraser (1942) 16; Cawkwell (1989) 376; Krentz (1994) 47; van Wees (2000) 132.
810 *Pace* Luginbill (1994) 55–56.
811 Cawkwell (1989) 376: see *infra* 202–203.

stant, but fluctuating, allowing the pushed and pushing hoplites a little respite at short intervals.

Another indication that a massive forward surge was crucial for victory, both physically and mentally, is the fact that phalanxes apparently always charged (or countercharged), no matter how superior their ground. In most types of battle there is a considerable advantage in consolidating one's position ('digging in') and simply awaiting the enemy charge[812] – and more especially so for a formation so defensively oriented as a hoplite phalanx – but this tactical advantage was willingly abandoned in favour of momentum. At Delion, the Athenians even charged δρόμῳ against the Thebans who could advance downhill.[813] The opposite happened at Peiraieus in 403, where Thrasybulos' rebel army, deployed on the very steep hill of Munychia, enjoyed such superiority of terrain that the Spartans on top of this had to attack in column; but even here the Athenians abandoned their favourable position to meet the enemy – Thrasybulos' warnings to the contrary notwithstanding.[814]

(2) The other objection to a mass-shove is the contention that the pressure would prevent hoplites from using their weapons effectively. The hoplites in the front rank, thrusting their shields against those of the enemy and being themselves pushed in the back, would have had no chance to wield their spears. In Cawkwell's words, this is "wildest folly", since "the front ranks would have been better able to use their teeth than their weapons when a broad shield was jammed against their back with the weight of seven men".[815] This is certainly at least *partly* correct: if mass-shove could and did occur, spears will have been of *relatively* little value in that particular phase.

Van Wees further denies the possibility of a bodily push altogether on the basis of his observation of shields borne leaned against the shoulder and held at a slant: "in very close combat the lower edges of opponents' shields were liable to touch … 'Pushing' would thus consist of shoving the protruding lower part of one's shield against the corresponding part of the enemy's shield …".[816] But this is an oversimplification. The slanting position, leaned against the shoulder, may be the shield's initial position; but the vase images do not show any shields actually being pressed together in this way, and with good reason. The smooth and convex surface of the enemy's shield would make a shield thrust glance off and entailed the risk of throwing the attacker off balance because of the weight of a shield of some 7 or 8 kg; and besides, as we have seen, two shield bottoms pushed against each other will result rather in the two shields being pressed together in an upright position.[817]

---

812 A later example: the English defenders had but one chance against the cavalry of William the Conqueror at Hastings in 1066, namely to retain their position on Senlac Hill, in which they succeeded. It was not until king Harold's infantry broke ranks to pursue retreating Normans that William's forces were able to eliminate English resistance; see Oman (1924²) vol. I 152–158.
813 Thuc. 4.96.1; inv. no. 5.
814 Xen. *Hell.* 2.4.11–19 (inv. no. 22). Other instances: Xen. *Hell.* 3.4.23, 3.5.18–19, 4.3.17; Onas. 29.1; Polyaen. 3.9.26; but see Diod. Sic. 15.32.4. Hanson (2000²) 136–140.
815 Cawkwell (1978) 150–153; cf. Krentz (1985b) 58–59, (1994) 47; Goldsworthy (1997) 18.
816 van Wees (2000) 131 with a citation of figs. 3 and 4: *supra* 44–45 n. 136.
817 *Supra* 45.

Besides, the pressure on the shield was mainly directed into the centre at the *porpax,* which is why the shield, along with the upper arm bending at the elbow, would be slowly but surely pressed flat against the body. Finally, the rear ranks had no need to hold their shields obliquely. Collective *othismos* or not, the men in these ranks quite simply were not in the thick of the fighting and therefore had not the same urgent need to protect themselves against a frontal attack. Thus, they were not in any way compelled to hold out their shields in an oblique position; and indeed just 'hanging' it on the shoulder would save the bearer a lot of energy until he joined the mêlée.

Van Wees in fact cites the *Kyropaideia* passage quoted above to illustrate his point; but it really does not prove it: Xenophon merely says that the Egyptians' shields contribute to the shoving by resting against the shoulder (πρὸς τὸ ὠθεῖσθαι συνεργάζονται πρὸς τοῖς ὤμοις οὖσαι).[818] Rather, this is reminiscent of an entirely different way of 'pushing'. A far better way of throwing one's enemy off balance would surely be to throw oneself into the suitably hollow shield – shoulder first – with as much force as possible: the shoulder or arm in itself did not lend any noticeable leverage to the shoving; the bodily pressure delivered *through* the shoulder would do so. The shield was not rested in front of the shoulder but *on* it, to lessen the burden. This would make it impossible to 'push' with the shoulder and arm alone – and much more so if the shield was actually held obliquely. Consequently, Xenophon's meaning in fact probably is that the *shoulder* is thrust against the shield, rather than the reverse, even if it means that the shield must at this particular point be held in a vertical position. At any rate, a sloped shield in itself is insufficient to create momentum enough to drive a man backwards, whether held by a single or a double grip.

Again it is useful to take into consideration the design of hoplite weapons. From a structural view-point, the hoplite's shield was uniquely suited to this highly specialised task. It was rigid (although, as we have seen, somewhat pliable under pressure[819]) and dramatically concave; and these qualities enabled the bearer to stem his left shoulder against its centre, directing the force forwards. As he did so, the hoplite achieved three subordinate goals: he could rest his arm (since the rim was hung on the shoulder), gain a maximum of forward thrust, and obtain a maximum of protection from the hollow shield, since almost all of his upper body could be tucked into the hollow of the bowl-shaped shield. *Othismos* was thus quite likely one of the purposes for which the hoplite shield was designed and made.[820] As we noted above, the hollowness is the essential feature, determining the other, less obvious characteristics, such as its weight and circularity.[821] This fits extremely well

---

818 van Wees (2000) 131; Xen. *Cyr.* 7.1.33–34. Pritchett (1985a) 66 remarks that "if the Greek battle was merely a matter of a huge scrimmage, it seems as if the Greeks would have adopted a shield like that of the Egyptians, which Xenophon says was more suited for that purpose." This misses the point that the 'Egyptian' shields described in all probability *are* meant to represent Greek hoplite shields: see *supra* 187.
819 See *supra* 29 and n. 73.
820 Hanson (2000²) 68–69.
821 Cf. Hanson (1991) 68–69, 76–77, *supra* 146 and n. 593.

with the fact that the shield was the only item of the hoplite's equipment that did not change *at all* for more than 400 years.[822] The shield design was correct from the beginning.

Another consideration may help explain how pressure did not necessarily interfere with weapons use. Arrian has a description of the actual stance quoted above that accords well with what we have seen earlier. His exact phrasing makes it clear that in the Macedonian phalanx the pressure (αἱ ἐνερείσεις) is directed into the "shoulders and sides" (κατὰ τοὺς ὤμους καὶ τὰς πλεύρας) of the man in front.[823] During *othismos*, we must assume, hoplites thus stood more or less at a ¾ angle, receiving the thrust in their sides as much as in their back. This facilitated breathing in the press, and it also to a certain degree enabled the hoplite to use his weapons. We should be clear, however, that fighting of this type inevitably meant losing even more freedom of movement: the point is that *othismos,* if accepted, was a completely different tool in the fighting; one that relied on the shields for making headway instead. Spears were useful for other purposes; that they cannot have been used during *othismos* – or only in a rather limited way – surely does not prove that collective *othismos* did not occur.

Franz suggests that although a reach of 0.4 m (which was the maximum that could be achieved with the hoplite thrusting spear gripped at the middle) was too much within the compressed ranks during *othismos,* the second rank might easily reach the enemy with their spears – and the first rank was still at liberty to strike not the man opposite him, but rather *his* neighbour, or the man behind him.[824] The additional objection that a collision would result in the second and third ranks being impaled on the *sauroteres* of those in front[825] can fairly easily be laid to rest: if spears were held at a sufficiently steep angle, pointing downwards over the bearer's shoulder, they would be poised for a deadly stab against the face or throat of the immediate adversary, and the butt would point into the air. Upon impact, the spear would be jammed into a vertical position.

Otherwise, such extremely close infighting was certainly a situation in which the sword would be of great use (although this of course demanded that the spear be discarded or had already broken): it is interesting that the Spartans were famous for their very short sword, usually called ἐγχειρίδιον – the word suggests that it was actually closer to a dagger than to an actual sword.[826] It crops up frequently in Plutarch: it is twice given as the reason (or the effect) of the Spartans' extremely close fighting style;[827] and when king Agis III overhears a jibe to the effect that mountebanks easily swallow such swords when performing, he caustically remarks καὶ

---

822 Snodgrass (1964a) 64, 68.
823 Arr. *Tact.* 16.13–14; cf. 3.5 and *supra* 185 and n. 780.
824 Franz (2002) 305–306. The photomontage of opposing rows of hoplites, showing a likely distance between ranks, is most instructive (Franz [2002] 304 fig. 6.6]).
825 Hanson (2000²) 162–165; Lazenby (1991) 96–99; Goldsworthy (1997) 17.
826 Sekunda (1986) 27 has a picture of what is most likely a Cretan model of a Spartan short sword: it is 32.3 cm long and thus "perhaps slightly larger than life-size".
827 Plut. *Mor.* 217e, 232e.

μὴν μάλιστα ... ἡμεῖς ἐφικνούμεθα τοῖς ἐγχειριδίοις τῶν πολεμίων.[828] A young Spartan complaining about the shortness of the sword is told by his mother to add a step forward to it.[829] Such ἐγχειρίδια must have been designed for extremely close combat, and a number of Classical reliefs show them in the hands of Spartans, sometimes gripped in the left hand along with the *antilabe*. This may be in order to enable a quick draw if the spear should break,[830] but it might equally well be held easily in this manner, ready for immediate use during the press of shields.

3.4.3 Another purpose of the rear ranks?

A central point in the *othismos* debate is the problem of the rear ranks. As we have seen, the standard file depth was eight ranks, sometimes more and seldom less. Yet if we accept the sceptics' analysis of *othismos,* we are given to understand that only the first one or two ranks actually saw action. If this is so, why would there be any need for rear ranks at all? The undeniable presence of rear ranks has brought about a number of different explanations. These rear ranks could not participate actively in the fighting, short of assisting in the shove. Since the shove has been denied by *othismos* sceptics, they have therefore been interpreted as nothing more than so many reserve troops, intended to fill up any gap that might be left when hoplites in the front rank fell, thereby helping to simply "intimidate the enemy and boost the morale of their comrades by their very presence".[831] Ready to step up to fill the gaps of hoplites killed in the front line, they thus provided mainly *psychological* support for their comrades in the line of fire.

Goldsworthy feels that the question of morale has been severely underplayed in the debate over hoplite battles.[832] His approach is fresh, drawing on a number of analogies from modern warfare; and this leads him to believe that hoplite battles were primarily of a psychological nature – a battle of nerves, as it were, with the opposing phalanxes drifting ever closer to the edge of panic as the battle progressed. As a consequence of this, he rejects outright the idea that the rear ranks act as reserves (since the sources are not specific on this point), arguing that their sole function was to bolster the morale of the front line fighters by their very presence.[833] While it is certain that morale and the pressure of nerves played vital roles in phalanx battles, we must suppose that at a minimum the rear ranks – despite the reticence of the sources – were expected to fill up the gaps and thus keep the front line

---

828 Plut. *Lyc.* 19.2 ("And yet we reach our enemies with these daggers" [trans. Perrin, modified]), cf. *Mor.* 191e.
829 Plut. *Mor.* 241f.
830 Sekunda (1986) 24.
831 van Wees (2000) 132; cf. Fraser (1942) 16; Krentz (1985b) 60, (1994) 45–46; Goldsworthy (1997) 15, 23.
832 While acknowledging Hanson's contributions in this particular area, Goldsworthy does disagree with his conclusions (Goldsworthy [1997] 5 n. 21). After reading the works of both scholars, however, I would say that Hanson takes the question of morale into consideration no less than Goldsworthy, drawing also on several modern works on this subject: see Hanson (2000²) 234 for a list of works consulted.
833 Goldsworthy (1997) 23, cf. Krentz (1994) 45; van Wees (2000) 132, (2004) 189, 191.

ἀρραγή, if at all possible.⁸³⁴ It is scarcely conceivable – not to mention a terribly uneconomical use of troops – that seven-eighths of a battle formation should be intended to act as nothing more than a road block.

One cannot help but feeling that morale, as well as fighting, would be more effectively boosted by comrades actually taking a part in the grisly work, rather than merely standing idly around. Van Wees quotes two passages of the *Kyropaideia,* both to the effect that the rear ranks do not in any way contribute to the fighting.⁸³⁵ The notion of an army arrayed two ranks deep resisting the charge of another a hundred deep is of course hardly tenable, with or without pushing; but even more interestingly, it is actually directly gainsaid by Xenophon himself in the *Anabasis,* a work that portrays actual military practice in historical events (presumably verifiable to his readers).⁸³⁶ Here, a shallow phalanx in Xenophon's own opinion simply risks being overrun by the numerically superior enemy; and it is more than likely that Xenophon was at great pains to express at least *verifiable* occurrences accurately to the prospective readers of his military memoirs. Depth of phalanx apparently *did* matter after all; and therefore the hindmost ranks must necessarily play a role. Why not that of lending *active* support to the front ranks?

3.4.4 Maintaining cohesion

Goldsworthy offers yet another good reason why Greek phalanxes massed in depth at all. As phalanxes advanced – usually probably in poor marching order, if at all – there was a pronounced inherent tendency towards disruption of the formation. Aristotle has an interesting political analogy fetched from the realm of hoplite warfare: ὥσπερ γὰρ ἐν τοῖς πολέμοις αἱ διαβάσεις τῶν ὀχετῶν, καὶ τῶν πάνυ σμικρῶν, διασπῶσι τὰς φάλαγγας, οὕτως ἔοικε πᾶσα διαφορὰ ποιεῖν διάστασιν.⁸³⁷ He almost seems to recall Thucydides, who describes the opening stages of the battle of Delion, in which the wings of the Theban and Athenian phalanxes were prevented from ever clashing by the presence of ῥύακες.⁸³⁸ The vulnerable phalanx, then, was better protected against disruption if it formed up to at least a certain depth: "A

---

834 Arr. *Tact.* 12.4.
835 Xen. *Cyr.* 6.3.21–23, 6.4.17; van Wees (2000) 131–132.
836 Xen. *An.* 4.8.11. Xenophon sometimes lapses into the realm of the fantastic to bring home a point; cf. the pep talk delivered to the desperate hoplites in Persia, belittling the very real threat of the strong enemy cavalry: εἰ δέ τις ὑμῶν ἀθυμεῖ ὅτι ἡμῖν μὲν οὐκ εἰσὶν ἱππεῖς, τοῖς δὲ πολεμίοις πολλοὶ πάρεισιν, ἐνθυμήθητε ὅτι οἱ μύριοι ἱππεῖς οὐδὲν ἄλλο ἢ μύριοί εἰσιν ἄνθρωποι· ὑπὸ μὲν γὰρ ἵππου ἐν μάχῃ οὐδεὶς πώποτε οὔτε δηχθεὶς οὔτε λακτισθεὶς ἀπέθανεν (*An.* 3.2.18–19: "If any of you feel disheartened because of the fact that we have no cavalry, while the enemy have great numbers of them, you must remember that ten thousand cavalry only amount to ten thousand men. No one has ever died in battle through being bitten or kicked by a horse" [trans. Warner]).
837 Arist. *Pol.* 1303b 12–14 ("We know how in warfare the crossing of watercourses, even of quite small ones, tends to cause phalanxes to split up. So it seems that every distinction leads to division" [trans. Saunders, modified]); cf. Xen. *Oec.* 8.4–7; Polyb. 18.31.5.
838 Thuc. 4.96.2 (inv. no. 5).

deeper, and therefore narrower, phalanx encountered fewer obstacles and could as a result move faster and further, whilst retaining its order."[839]

This is no doubt a correct observation, but Goldsworthy seems overly pessimistic about the effect of such obstacles on the overall formation. After all, if the marching tempo was not set too high, each individual in any given rank should be able to keep up by aligning his shield edges with those of his neighbours. This would probably have resulted in *tolerably* straight, if not perfect, lines. It is entirely possible that phalanxes sometimes veered dangerously close to disintegrating during the final charge (as at Kunaxa);[840] but on the whole, the historians seem to be careful to inform us of those occasions when one side closed with the enemy in such a state of disarray that it brought about defeat.[841] In particular, the Argives seem to have been prone to 'running' too early, thereby tearing holes between themselves and the other contingents,[842] whereas the Spartans famously forwent the 'running' final charge in order to preserve their battle order intact, advancing as they did to the sound of *auloi* so that they might march in time.[843] On the other hand, cohesion and closed ranks might save a phalanx in a dangerous situation.[844]

This purpose was generally served by means of the marching *paian*, however.[845] Normally armies struck up a *paian* during the advance, apparently typically at a distance of some three or four stades (*c.* 550 – 725 m). Besides its religious, apotropaic function[846] the *paian* had a very distinct rhythm and thus set the beat for the marching phalanx: according to Thucydides, the Spartans set great store by it.[847] The *paian* thus fulfilled a useful as well as a ritual function. In all probability every *polis'* marching *paian* was always the same and as such surrounded by a certain ritualised mysticism. It therefore served as a moral support for the *polis* pha-

---

839 Goldsworthy (1997) 6–14, esp. 7–8. This also highlights the need for finding suitable (cultivated) land for hoplite battles: see Cary (1949) 40 and Hdt. 7.9β.1.
840 Xen. *An.* 1.8.17–18 (inv. no. 11).
841 See, e.g., Thuc. 3.108.3, 4.126.5, 5.9.3, 6.97.4, 7.53.2, 8.10.4, 8.25.3; Xen. *Hell.* 5.1.12; *Hell. Oxy.* 4.1 and cf. Diod. Sic. 15.85.6.
842 Thuc. 8.25.3 (Miletos 413); Xen. *Hell.* 4.3.17 (Koroneia 394 [inv. no. 9]).
843 Thuc. 5.70: Λακεδαιμόνιοι δὲ βραδέως [sc. χωροῦντες] καὶ ὑπὸ αὐλητῶν πολλῶν ὁμοῦ ἐγκαθεστώτων, οὐ τοῦ θείου χάριν, ἀλλ' ἵνα ὁμαλῶς μετὰ ῥυθμοῦ βαίνοντες προσέλθοιεν καὶ μὴ διασπασθείη αὐτοῖς ἡ τάξις, ὅπερ φιλεῖ τὰ μεγάλα στρατόπεδα ἐν ταῖς προσόδοις ποιεῖν ("The Spartans came on slowly and to the music of many flute-players in their ranks. This custom of theirs has nothing to do with religion; it is designed to make them keep in step and move forward steadily without breaking their ranks, as large armies often do when they are just about to join battle" [trans. Warner]). On the subject of the Spartan *auloi*, see also Plut. *Lyc.* 22.2–3; Polyaen. 1.10 (who claims that Leuktra in 371 [inv. no. 16] was the first pitched battle lost by the Spartans and that this was because they were *not* accompanied by *aulos*-players); Quint. *Inst.* 1.10.14.
844 Thuc. 1.63.1, 3.108.3.
845 Xen. *An.* 1.8.17 (Kunaxa 401 [inv. no. 11]); cf. Diod. Sic. 14.23.1, 5.34.5. See also Pritchett (1971) 105–106 for a table of *paian* singing in literature.
846 Σ Eur. *Phoen.* 1102; Ath. 15.701d–e; Pritchett (1971) 106.
847 Thuc. 5.69.2–5.70. Much later, Plutarch mentions the awe-inspiring spectacle of the Spartan king leading the *paian* as he leads the army out to battle: Plut. *Lyc.* 21–22, cf. *Mor.* 238b.

lanx; and young inexperienced hoplites probably found support and comfort in the familiar *paian* of their *polis* and its proud traditions.

This also explains *paian* singing at apparently inconvenient times: on several occasions, we hear of armies giving away their position by singing in situations where they could have surprised the enemy by keeping silent, such as at Nemea in 394.[848] Another puzzling incident of this is the botched Athenian night raid on the Syracusan outposts in 413: adding to the confusion was the fact that the Athenian allies (from Argos and Korkyra), being Dorians, sang a Doric *paian* that was virtually indistinguishable from that of the Syracusans.[849] It would seem apparent that the attackers would profit from remaining silent; but other considerations – the *paian*'s function as a watchword, a means of frightening the enemy and keeping the formation intact – assumed priority. In this capacity the *paian* might be sung repeatedly during fighting;[850] but one of its most important functions no doubt was to set the marching rhythm.

Furthermore, although the 'run' (δρόμος) at the end of the advance probably should be interpreted as precisely that, the evidence for 'running', such as it was, indicates that it was generally reserved for the very last dash towards the enemy.[851] Despite the understandable desire for maximum impetus, it is unlikely that hoplites would run several hundred metres in earnest. It simply is not very practical to run while wearing some 20–25 kg worth of wood, leather and bronze in Greek summer heat of perhaps above 35 °C. Herodotos famously claimed that the Athenian hoplites at Marathon ran "no less than eight stades" (= 1400 m); but it is significant that he also insists that they were the first Greeks he knew of to ever attack at a 'run'.[852] Donlan and Thompson have shown conclusively that this must in fact be a myth: 200 m is the *absolute maximum* running distance (for well-trained young men carrying a load of only 6.8 kg) if they are to have any strength left for combat. Running 565 yards (= 516.63 m) with objects simulating hoplite equipment carried in isometrically correct postures, as Donlan and Thompson's test subjects did, exhausted them completely and left them incapable of fighting once they reached the 'enemy'. To this should be added the much higher stress factor in a true battle: real running was not an option.[853]

If they ever truly ran, phalanxes probably only did so for the last 50 m or so, due to the tremendous nervous pressure, but also in part to practical considerations: the need for a maximum of penetration power at the collision and for the first spear

---

848 Xen. *Hell.* 4.2.19 (inv. no. 25).
849 Thuc. 7.44.6.
850 As Pritchett correctly observes, the frequentative optative ὁπότε παιανίσειαν signifies that the *paian* was sung more than once on this occasion (Pritchett [1971] 107–108 n. 15).
851 How (1919) 40–42, examining the evidence for the use of βάδην and δρόμῳ, found that the former must mean 'quick march'; but this is warranted only in Xen. *Hell.* 3.4.23. The most that may be said with certainty is that βάδην always refers to some type of pace within the category of 'walking'.
852 Repeated no less than four times in Hdt. 6.112.1–3 (inv. no. 19).
853 Hdt. 6.112.1–3. Donlan & Thompson (1976) 341; (1979) 419–420; cf. Delbrück (1900) 54–55; Hignett (1963) 62. See also Hanson (2000²) 56.

thrust.⁸⁵⁴ At any rate, the distance was reduced to half by the fact that the enemy phalanx ran the *other* remaining half of the μεταίχμιον, so that a short run-up of perhaps 20–25 m would create sufficient forward momentum. It was at this time that the marching rhythm became unnecessary and the hoplites altered their singing of the *paian* to the war cry (usually referred to as ἐλελεῦ, originally an invocation of the war god Enyalios, or of Ares).⁸⁵⁵

Apart from Goldsworthy's suggestion of the need for keeping the formation intact during the advance, *othismos* sceptics have trouble explaining the very *need* for cohesion, often emphasised in the sources. As we saw, the historians generally point out when insufficiently orderly ranks resulted in defeat; and Aristotle even adduces this simple fact in explanation of why the first hoplite ranks were allegedly inferior to cavalry: ἄνευ μὲν γὰρ συντάξεως ἄχρηστον τὸ ὁπλιτικόν ("for without order the hoplite arm is worthless").⁸⁵⁶ The γάρ should give pause for thought: seemingly self-evident to Aristotle (and, presumably, his public), this very basic principle is rather difficult to understand if the battle order was open and the fighting style in any way fluid. As has been observed, it is a strange thing indeed that so much store was set by an orderly formation if the lines were broken before or on impact anyway.⁸⁵⁷ On the other hand, if a collective pushing formed part of the fighting, relatively orderly lines would be of the utmost importance.

3.4.5 *Othismos?*

The advocates of the 'heretic' view, rejecting the traditionalist approach, must nevertheless reconcile their theory with the relative frequency of *othismos* references in the sources. As such, they do not dispute the fact that *something* called *othismos* occurred; but while it is at times interpreted as 'pushing with the shield', it is seen as an element of individual weapons handling, rather than tactics: "Rather, the 'shoving of shields' involved individual men in the front rank striking opponents with their shields, seeking to unbalance them or knock them over, so that they could be more easily killed with spear or sword."⁸⁵⁸ As we saw above, van Wees' belief that the shield was *always* held aslant leads him to conclude that *othismos*, such as it was, consisted in single bouts of thrusting only with the shield rim; but this, as was explained, is not physically feasible.⁸⁵⁹

Cawkwell hesitatingly admitted that hoplites might sometimes shove in the later stages of the battle, whereas Luginbill argues that employed from the outset, *othismos* might break the opposing ranks very early on (since "no loose formation

---

854 At Koroneia in 394 (inv. no. 9) the Thebans began 'running' c. 180 m from the Spartan lines, and at a distance of some 90 m the Spartans followed suit: Xen. *Hell.* 4.3.17.
855 Hes. *Theog.* 686; Hdt. 8.37.3; Ar. *Av.* 364; Xen. *An.* 1.8.18, 1.2.17, 6.5.26, *Hell.* 4.3.17, *Ages.* 2.10. Pindar uses the word metonymically about war, Pind. *Nem.* 3.60, *Isthm.* 7.10.
856 Arist. *Pol.* 1297b 20–21.
857 Holladay (1982) 96; cf. Luginbill (1994) 58–60.
858 Goldsworthy (1997) 19–20; cf. Krentz (1985b) 56, (1994) 48–49.
859 van Wees (2000) 131.

could hope to resist a mass-shove") and thus win the battle.⁸⁶⁰ Holladay has, more or less *en passant*, pointed out a problem with this, sensibly noting that the transition between such distinct phases would be difficult to execute in practice ("there was no referee with a whistle").⁸⁶¹ While it is indeed difficult to visualise hoplite battles as determined *only* by such pushing, it is of course equally difficult to imagine *othismos* as a set, clearly defined and limited phase of battle, whether early or late.

In analogy with observations of crowd behaviour, it should perhaps rather be construed as a phenomenon occurring in intervals, caused by the intense nervous urge to force a way forward, as happened at the 'Tearless Battle' at Melea in 368, where Spartan file leaders actually had trouble keeping back the rear ranks (ἀνείργειν τοὺς στρατιώτας **ὠθουμένους** εἰς τὸ πρόσθεν).⁸⁶² By pressing the shield into the side and shoulder of the men in front, the nervous rear rank hoplites could jam them even closer against the enemy, possibly even propelling them *into* the enemy ranks, where they would have a better chance of killing right and left. Such *othismos* may have occurred in short bursts, and at random intervals, as the rear ranks felt that they might help their comrades by applying pressure.⁸⁶³ And not all seven ranks needed to participate in the shoving simultaneously: it is certainly possible to imagine that in some parts perhaps the front five ranks shoved, in other parts perhaps only two or three, in yet others perhaps all seven, according as the battle raged back and forth.

On this interpretation, *othismos* was not a separate phase from the use of weapons or δορατισμός: they supplemented each other, both being vital instruments in bringing about victory. The objective was to break the enemy ranks, and this could be done by whittling down the enemy hoplites by weapons use, or it could be done by violently shoving against them in an attempt to drive back parts of the first rank, so that a path forward opened. Frequently, I suspect, they occurred more or less simultaneously or at least alternated, according as hoplites in the front or rear ranks felt that progress was possible. This is the image we get from Thucydides' account of Delion, which was determined καρτερᾷ μάχῃ καὶ ὠθισμῷ ἀσπίδων ("the fighting was stubborn, with pushing of shields")⁸⁶⁴ and from Xenophon's chilling account of Koroneia: ἀντιμέτωπος συνέρραξε [sc. ὁ Ἀγησίλαος] τοῖς Θηβαίοις· καὶ συμβαλόντες τὰς ἀσπίδας ἐωθοῦντο, ἐμάχοντο, ἀπέκτεινον, ἀπέθνῃσκον.⁸⁶⁵ Also, Thucydides insists that the entire battle of Solygeia was "hard-fought and hand-to-hand" (καρτερὰ καὶ ἐν χερσὶ πᾶσα), only to add in the next sentence, καὶ τὸ μὲν δεξιὸν

---

860 Cawkwell (1978) 152, cf. Cartledge (1996) 713; Luginbill (1994) 54–55.
861 Holladay (1982) 95. However, Holladay ([1982) 96–97) also opts for *othismos* from the outset.
862 Xen. *Hell.* 7.1.31–32 (emphasis mine).
863 Goldsworthy (1997) 21.
864 Thuc. 4.96.2; inv. no. 5 (*pace* Cawkwell [1989] 378–379 who maintains, in spite of this passage, that fighting and *othismos* are not coextensive).
865 Xen. *Hell.* 4.3.19 ("he crashed into the Thebans front to front. So smashing their shields together they pushed, fought, killed and were killed" [trans. Warner, modified]); cf. *Ages.* 2.12 (inv. no 9).

κέρας τῶν Ἀθηναίων καὶ Καρυστίων ... ἐδέξαντό τε τοὺς Κορινθίους καὶ **ἐώσαντο μόλις**.[866] Tyrtaios' description of this same scenario is strikingly similar.[867] When the Eleans and Arkadians (with their Argive allies) clashed in Olympia itself during the Olympic games in 364, the victorious Eleans advanced on their enemies, "fighting as bravely as ever and pushing the enemy back towards the altar" (ἐμάχοντο μὲν οὐδὲν ἧττον καὶ ἐώθουν πρὸς τὸν βωμόν).[868]

Hoplite battles were essentially chaotic in nature, and once the mêlée was joined there could be no turning back. The time for giving orders and executing tactical manoeuvres was past at this point, and so the outcome of the fighting – advance and retreat – essentially lay with the entire phalanx with no one to direct the movements of the enormous 'organism'. *Othismos,* then, may well have been a more or less conscious group effort, but not necessarily directed by a commanding officer or indeed by any principle other than the instinctive urge to advance: in other words, *othismos* was probably every bit as chaotic as the hand-to-hand mêlée, executed in bursts of varying length with varying degrees of success, and with different numbes of file members participating in the various sectors of the battlefield. Consequently, it is not a question of only either *othismos* or 'normal' fighting: rather, hoplite battles comprised both. One does not in any way exclude the other, provided that they are regarded as perfectly compossible elements in the constant and chaotic flux of battle.

---

866 Thuc. 4.43.3 ("The right wing of the Athenians with the Karystians ... stood up to the Corinthians and, though with difficulty, *pushed them back*" [trans. Warner; emphasis mine]); inv. no. 32.
867 Tyrt. fr. 11.29–34 West; see *supra* pp. 119–120 and n. 487. *Contra* Wheeler (1991) 130 and n. 48.
868 Xen. *Hell.* 7.4.31.

4. DURATION OF HOPLITE BATTLES

4.1 THE PROBLEM OF TEMPORAL DESIGNATIONS

4.1.1 Introduction

Time is a crucial factor for the understanding of any military operation, and perhaps even more so for a pitched battle. A particularly important aspect is the actual *length* of the battle, since the time elapsed may reveal a good deal about the nature or type of the actual combat, such as, e.g., the intensity of the fighting, and quite possibly also the maximum of effective fighting time that could be obtained with a hoplite phalanx. It is a reasonable assumption that in an 'average' battle, which was decided by sheer fighting (i.e. not by ambush, effective stratagems or superior generalship), the losing side kept fighting as long as at all possible, i.e. until they were completely exhausted and broke ranks.

Today, battles may last for several days or even weeks; but in antiquity, they seem to have been decided within a day's worth of fighting at the most. The sources never once attest to a protracted battle that lasts for more than one day, and normally apparently a lot less than that. Nonetheless, the exact duration of ancient battles is next to impossible to assess. The historians generally tend to describe the length of battles extremely cursorily and in highly conventional phrases, if at all. They rarely provide any real information about the *normal* temporary limitations of hoplite battle, although such limitations must surely be implied: it is unlikely that an ancient author could expect his readers to understand the vague phrasing of Classical historiography unless some tacit underlying consensus already existed – in many cases naturally derived from personal experience – as to what was understood to be the 'normal' duration of a hoplite battle.

Another possibility is of course that the question simply was of no interest to writer or reader, and that the point was accordingly omitted. This speculation may safely be put to rest, however, when we consider the large amount of battle narratives specifically describing the fighting as 'long' (or, more rarely, 'short'). In other words, it may be inferred that there was in all likelihood a shared conception of the 'normal' duration of a hoplite battle, and that the historians probably only commented on the length of battles insofar as they departed from the usual in any way. Thus it may be maintained that if we can determine the normal duration with any certainty, we can fairly safely assume that battles which are not specifically commented upon in regard to length must have lasted roughly the 'usual' amount of time. The problem of determining this amount, however, seems almost insurmountable.

4.1.2 The phraseology of battle duration

Let us look at the language normally used to describe the temporal extension of battle. By far the most common phrasing to denote length of battle is any of a number of variations on the theme of πολὺν χρόνον or ἐπὶ πολύ, or another extremely general phrase vaguely signifying 'a large amount of time'.[869] This, needless to say, is frustrating for a modern reader: phrases such as 'for a long time' in themselves prove nothing about the temporal extension of battle in the absence of a fixed standard. The ancient audience would in many cases have had first-hand experience of hoplite battles and therefore have had a shared conception of what was understood by a long or a short battle. There probably existed a rather fixed standard which the historian could trust his audience to know and share, and consequently there is seldom any *truly* definite temporal designation. On the other hand, it is entirely impossible for modern readers to ascertain just what is meant by 'long' or 'short'; and similarly, it is all too easy to fall into the trap of judging the term by modern standards.

In 1899 J.A.R. Munro, discussing the battle of Marathon as described by Herodotos 6.113, made the following astute observation: "As to the duration denoted by πολλός opinions differ. Probably most men would find an hour's hand to hand tussle ample, and the small Athenian loss points to a short estimate. The time might be measured in minutes and still be long under the conditions."[870] Munro's lucid point can hardly be stressed enough. Strange though it may seem, it is theoretically possible that hoplite battles – i.e., two phalanxes engaging each other in close combat – usually lasted no more than half an hour to 45 minutes, and that consequently a battle of an hour was what the Greeks would call hard-fought and extremely long. We should therefore be very careful when estimating what constituted a 'long' battle.

The opposite view – that the same standards of what is long and short apply then and now – is nevertheless advocated by Cawkwell, who finds fault with the 'orthodoxy' in the established conception of hoplite battles: "the first point to establish is that battles were not short in duration." This point decided, Cawkwell goes on to admit that the Greek historians generally "contented themselves with the phrase 'for a long time', which some might be inclined to dismiss. But we know what the Greeks knew as a long battle." Cawkwell next adduces examples from the most extreme ends of Greek literature, ranging from Homer to Diodoros (and thus from fiction to history), to show that a 'long battle' might last an entire day; and on the strength of this concludes that "there is no reason to think of battles as quickly

---

869 πολὺν χρόνον or ἐπὶ πολύ(ν χρόνον): Hdt. 6.113.1, 9.32.2 (Plataiai 479 [inv. no. 29]); Thuc. 3.97.3 (Aigition 426 [inv. no. 1]); Thuc. 4.35.4 (Sphakteria 425: χρόνον μὲν πολὺν καὶ τῆς ἡμέρας τὸ πλεῖστον [inv. no. 34]); Thuc. 4.44.1 (Solygeia 425 [inv. no. 32]); Thuc. 6.70.1 (Syracuse 415 [inv. no. 36]). Most of the evidence – though far from all of it – is collected and conveniently tabulated by Pritchett (1985a) 47–50.
870 Munro (1899) 196, also quoted in part by Pritchett (1985a) 49. The phrase in question is χρόνος ἐγίνετο πολλός (inv. no. 19). See also Burn (1962) 251.

over".[871] In those precious few instances where an all-day battle is mentioned, however, *the two types of stock phrases are never actually once coupled*. Thus, to carry the point to its logical extreme, it is in fact not even possible to say that such a battle actually was what the Greeks knew as a 'long battle', let alone that we 'know' it as such.

The same misguided certainty applies to N.G.L. Hammond's analysis of the battle of Marathon, in which he complains that both Munro and Delbrück with their rival hypotheses of a few minutes and several recognisable phases respectively "are both in conflict with the evidence of Herodotus".[872] One wonders how they manage to be so; certainly Herodotos himself offers no definite designation apart from (once again) χρόνος ἐγίνετο πολλός; but Hammond offers no further explanation. Presumably he believes his own hypothesis to be more in line with what Herodotos meant than that of either Munro or Delbrück: "I imagine the Greeks defeated the Persian wings and centre within an hour or so" (to which he adds, rather oddly: "the analogy is with a bout in fencing rather than with trench warfare").[873] This is an exact parallel to Cawkwell's position: Hammond berates his colleagues for being subjective and hypothetical and claims to be more in accordance with the evidence himself, yet he is unable to justify his verdict beyond *imagining* it to be so.

Therefore, when Cawkwell asserts that Thucydides "clearly was not thinking of conflicts of a few minutes' duration", it may well be true – it probably *is* true – but the very point is that this can be no more than guesswork.[874] We quite simply cannot tell from such evidence what Thucydides and his contemporary readers were 'clearly thinking'. It may seem fastidious to labour this point, but it serves to show how easy and tempting it is to make a foregone conclusion about such seemingly trivial and self-evident matters. The fact remains, however, that it is hazardous to ignore Munro's prudent words of warning.[875] Equally importantly, neither the *Iliad* nor Diodoros are reliable sources for hoplite battles. Diodoros' credibility in these matters, despite his access to a wide array of contemporary sources, is highly questionable at the best of times, and a quick glance at Pritchett's useful table of Diodoros' accounts of battle lengths will suffice to show that his descriptions of battles are almost entirely made up of the most tired clichés and stock phrases. Variants of the ἐπὶ πολύ type are by far the most common, but it is especially noteworthy that, as far as Diodoros is concerned, very few battles in the history of Greece appear to have been fought to their conclusion quickly.[876]

---

871 Cawkwell (1989) 376.
872 Hammond (1973) 171 (inv. no. 19).
873 Hammond (1973) 196 n. 1. Hammond presumably means that a 'long' battle accords better with the evidence of Herodotos, which makes his comparison with Plutarch's description of the battle of Pydna (cf. Plut. *Aem.* 22.1) altogether stranger, since Plutarch is quite explicit that this battle was fought to its conclusion uncommonly quickly: *infra* 204.
874 Cawkwell (1989) 376.
875 Cf. also Stylianou (1998) 552–553.
876 On the problem of the influence of Diodoros' style on the historical contents, see *infra* 217–218.

There is, however, one accurate specification of battle length. It is found in Plutarch's account of the battle of Pydna (168): καὶ κρίσιν μὲν ὀξυτάτην μέγιστος <ὁ> ἀγὼν οὗτος ἔσχεν· ἐνάτης γὰρ ὥρας ἀρξάμενοι μάχεσθαι, πρὸ δεκάτης ἐνίκησαν· τῷ δὲ λειπομένῳ τῆς ἡμέρας χρησάμενοι πρὸς τὴν δίωξιν, καὶ μέχρι σταδίων ἑκατὸν καὶ εἴκοσι διώξαντες, ἑσπέρας ἤδη βαθείας ἀπετράποντο.[877] Plutarch expressly states that this battle, fought to its conclusion in less than an hour, was decided "extremely quickly" for a battle on so grand a scale, which demonstrates that the norm was battles a great deal longer than an hour. Unfortunately this rather late battle, fought between the Roman legions of Aemilius Paullus and king Perseus' Macedonian phalanx, cannot reasonably be compared to a hoplite battle of the Archaic or Classical period centuries earlier. The battle tactics and fighting methods of the Macedonian phalanx, although superficially comparable with those of hoplite-style fighting, are ultimately of a different nature; if not so much in the actual formation and handling of weapons, then at least certainly in the integration and complete coordination of the several arms, such as cavalry and light-armed troops, and in the fact that Hellenistic and later armies operated with reserves, something that has a tendency to prolong fighting drastically.[878] The Romans certainly fought in a different manner entirely, and cannot be cited to indicate normality in a 'hoplite-style' battle. In fact, the far greater flexibility of the Roman legion probably was precisely that which secured the victory on this occasion as well as in other encounters with the Macedonian phalanxes. The same applies to Vegetius' famous statement – otherwise convenient – that pitched battles were usually decided "within two or three hours of fighting".[879] There is no way of ascertaining whether the battle conditions of Vegetius' day – the late Roman empire – were in any way comparable to those of hoplite battles almost 800 years earlier, but it does not seem likely.[880]

4.1.3 Hoplite battles described as long

The relevant source material for temporal designation is limited to battles expressly stated in the sources to be *hoplite* battles, i.e. battles between two phalanxes, irrespective of the presence of other troop types on the battlefield. If the examination is to have any value at all, we must therefore be able to ascertain that the conflict described was in fact between hoplites fighting in phalanx formation: accordingly, the word ὁπλίτης (descriptive of both sides and thus precluding, e.g., Marathon in 490),

---

877 Plut. *Aem.* 22.1 ("And this greatest of all struggles was extremely quickly decided; for the Romans began fighting at three o'clock in the afternoon, and were victorious within an hour; the rest of the day they spent in the pursuit, which they kept up for as many as 20 km, so that it was already late in the evening when they returned" [trans. Perrin, modified]). It is interesting that in this case the battle actually began very late in the day and that, much like at Delion (inv. no. 5), the pursuers only returned "in the late evening".

878 Goldsworthy (1997) 4 n. 15.

879 Veg. *Mil.* 3.9.2: *Conflictus publicus duarum aut trium horarum certamine definitur, post quem partis eius quae superata fuerit spes omnes intercidunt* ("A battle is commonly decided in two or three hours, after which no further hopes are left for the defeated army" [trans. Clarke, modified]).

880 *Contra* Krentz (1994) 47.

or another word used regularly in a context of hoplite fighting (such as ἀσπίς, ὠθισμός or φάλαγξ), must be found in the account of the battle. Furthermore, and most importantly, some sort of temporal designation must necessarily be assigned to the battles. A list produced by these criteria yields the following nine results, chronologically arranged: Plataiai 479, Poteidaia 432, Olpai 426/5, Pylos 425, Solygeia 425, Delion 424, Laodokeion 423, Syracuse 415, Koroneia 394. The list could in theory be considerably longer; but for reasons discussed below, Diodoros' assessments of battles lasting πολὺν χρόνον or ἐπὶ πολύ are excluded.[881]

It is also important to establish exactly how the phrase describing the battle duration is employed. It is by no means certain that, e.g., πολὺν χρόνον in fact describes the actual hand-to-hand fighting and only that: the phrasing may be intended to express the duration of the *entire* battle, including (but not necessarily limited to) marching up, deployment and possible waiting time before the encounter. Such phases as pursuit of the beaten enemy, identification, exchange and retrieval of the corpses following immediately after it may also be meant.

It is therefore necessary to determine whether they are intended to encompass the entire battle with all of its phases, or actually refer directly to the combat phase and to that *only*. For this purpose, it is vital that the sources be – as far as possible – contemporary. Simultaneousness in itself is of course not sufficient to guarantee veracity, but contemporary sources such as Thucydides and Xenophon are far more likely to include and be based on eyewitness accounts, and thus are to be preferred *a priori*. As stated above, Diodoros presents a special problem in this connexion and will be treated separately below.

(1) Plataiai 479 (inv. no. 29). This is, of course, not a hoplite battle as such; but the fighting was divided: the Spartans closed with the Persians, while the Athenians were fighting the medising Boiotians, who were, naturally, hoplites. They fought it out χρόνον ἐπὶ συχνόν, until the Thebans turned and fled back towards Thebes.[882] Earlier we are informed that the Athenians were pinned down by an attack (ἐπιτίθενται) from the medising Greeks;[883] so it would follow naturally that this is a description of the fighting itself.
(2) Poteidaia 432 (inv. no. 30). After joining battle, victory came about fast (διὰ τάχους)[884] for the Athenians, who routed the coalition forces of Poteidaia and Corinth opposite them: again, it seems that the fighting proper is understood.
(3) Olpai 426/5 (inv. no. 27). In this instance, it is far more difficult to assess the length. It is interesting, however, that Thucydides' statement that the battle lasted until evening (καὶ ἡ μὲν μάχη ἐτελεύτα ἐς ὀψέ) follows immediately after a description of the pursuit, which appears to have been intense (χαλεπῶς διεσῴζοντο ἐς τὰς Ὄλπας, καὶ πολλοὶ ἀπέθανον αὐτῶν, ἀτάκτως καὶ οὐδενὶ κόσμῳ προσπίπτοντες).[885] In this case, it appears that the indication of time is

---

881 *Infra* 217–218.
882 Hdt. 9.67.
883 Hdt. 9.61.1.
884 Thuc. 1.63.2.
885 Thuc. 3.108.3 ("It was only with great difficulty that they managed to get back to Olpai. Many

understood to include the pursuit, and that this was precisely the reason why the battle ended so late. What is more, Thucydides does not actually say when the fighting began.

(4) Pylos 425 (inv. no. 31). The Spartan attack on the Athenian stronghold met with fierce resistance. Thucydides claims that the attacks went on "that day and part of the next" (ταύτην μὲν οὖν τὴν ἡμέραν καὶ τῆς ὑστεραίας μέρος τι),[886] presumably interrupted at least by nightfall. It seems that only the fighting is meant. An important point here, however, is that the Spartans were repeatedly trying to force a landing in the harbour and not succeeding. This is clearly demonstrated by the case of Brasidas, who was driven back and actually collapsed on the gangway (παρεξειρεσία) from his wounds as he attempted to get ashore.[887] This unusual fighting – in relays, as it were – must consequently have given the combatants on both sides rather more time to catch their breath than would a 'standard' hoplite battle.

(5) Solygeia 425 (inv. no. 32). Thucydides' wording seems to indicate that this battle was particularly fierce (καὶ ἦν ἡ μάχη καρτερὰ καὶ ἐν χερσὶ πᾶσα),[888] and the meticulous description of the reversals of fortune in the changing battle phases seems also to point in that direction. The phrase χρόνον μὲν οὖν πολὺν ἀντεῖχον οὐκ ἐνδιδόντες ἀλλήλοις[889] is unambiguous: here, only pitched battle can be the point of reference.

(6) Delion 424 (inv. no. 5). The Theban *strategos* Pagondas, we are told, quickly marched out his army, since it was already late in the day (ἤδη γὰρ καὶ τῆς ἡμέρας ὀψὲ ἦν).[890] After battle was joined, the Athenian forces were routed and pursued all the way to Delion or to the sea. The pursuit was interrupted only by nightfall, which helped obscure the fugitives (νυκτὸς δὲ ἐπιλαβούσης τὸ ἔργον ῥᾷον τὸ πλῆθος τῶν φευγόντων διεσώθη).[891] Here, then, the designation of time must include both actual fighting and pursuit.

(7) Laodokeion 423 (inv. no. 14). Thucydides' description of this battle is very short indeed: we are merely informed that, after heavy losses on both sides, the battle was terminated by nightfall (ἀφελομένης νυκτὸς τὸ ἔργον).[892] It is impossible to say whether only actual fighting is meant to be included in this description. Furthermore, we are not told when the fighting began.

(8) Syracuse 415 (inv. no. 36). After the Athenian and Syracusan forces collided, they both held their ground for a long time (γενομένης δ' ἐν χερσί τῆς μάχης ἐπὶ

---

of them were killed, since in trying to break through they kept no order and showed no discipline" [trans. Warner]).
886 Thuc. 4.13.1.
887 Thuc. 4.12.1.
888 Thuc. 4.43.2 ("It was hard hand-to-hand fighting throughout" [trans. Warner]).
889 Thuc. 4.44.1 ("So for a long time both sides stood firm and yielded no ground" [trans. Warner]).
890 Thuc. 4.93.1.
891 Thuc. 4.96.8.
892 Thuc. 4.134.2.

πολὺ ἀντεῖχον ἀλλήλοις).[893] The careful phrasing and the sequence of events points naturally to ἐπὶ πολύ covering only the fighting.
(9) Koroneia 394 (inv. no. 9). Xenophon's account of this grisly battle includes the point of termination: the Spartans withdrew for the night after having prepared their supper – καὶ γὰρ ἦν ἤδη ὀψέ.[894] This is even more vague than usual. It is theoretically possible that it was late only after supper; but Xenophon's ἤδη seems to indicate that both the preparations and the actual meal took place ὀψέ.[895] We are neither informed of the commencement of fighting, nor whether only the fighting is to be understood – preparatory phases might be included.

There are thus in fact only five instances out of nine – 1, 2, 4, 5 and 8 – in which we can be more or less certain that the relevant phrase actually refers to the hand-to-hand fighting. This in itself suggests that the conventional phraseology surrounding accounts of battle cannot simply be counted on to deal only with actual fighting, unless the source *explicitly* states that this is the case. This important aspect of battle descriptions will be dealt with in greater detail below.

4.1.4 Ambiguous points of reference

Another, somewhat less frequent type of description seemingly operates with more fixed boundaries. Occasionally we are informed that a battle commenced late in the day or that it lasted εἰς ὀψέ, and sometimes both; or even that the fighting was interrupted only by the darkness.[896] But again the vagueness of the designations bedevils the analysis: if the account lacks a fixed starting point, then 'until darkness' is not very helpful, and the casual phrase may signify anything from half an hour to eight hours. The equally ambiguous 'late' or 'early' in the day do nothing to make matters any clearer; again, the interpretation of such designations turn on a vaguely defined absolute meaning, perhaps real and accessible enough to contemporary Greeks (though most likely not), but certainly beyond our grasp. We have no way of knowing when battles normally began; but if, as seems likely enough, they did in fact commence late in the afternoon, it is hardly surprising that they could drag on until nightfall.

That fighting should cease when darkness fell is equally understandable. Even today, night combat is difficult under the best of conditions, and this naturally applied to antiquity to an even greater extent. In the absence of any kind of visual aid

---

893 Thuc. 6.70.1.
894 Xen. *Hell.* 4.3.20 ("It was late by now"), cf. *Ages.* 2.15.
895 Warner's translation successfully brings out the meaning of the Greek: "It was already late in the day, and so they had their dinner and rested for the night" (Warner [1979²] 206). This is confirmed by the description in Xen. *Ages.* 2.15.
896 Mykale 479 ("about sunset"): Hdt. 9.101.2 (inv. no. 23); Tanagra 457 (apparently two different situations): Diod. Sic. 11.80.2, 11.80.6 (inv. no. 40); Olpai 425/6: Thuc. 3.108.3 (inv. no. 27); Delion 424: Thuc. 4.93.1–4.96.8, cf. Diod. Sic. 12.70.4 (inv. no. 5); Laodokeion 423: Thuc. 4.134.2 (inv. no. 14); Kunaxa 401: Xen. *An.* 1.10.16 (inv. no. 11); Koroneia 394: Xen. *Hell.* 4.3.20 (inv. no. 9); Nemea 394: Diod. Sic. 14.83.2 (inv. no. 25).

or even sufficient lighting the night was very dark indeed, and night operations were generally deemed risky and as such best avoided. Thus Thucydides relates that the Athenian attempt to reconquer the Epipolai plateau at Syracuse after the advent of Demosthenes' relief force in 413 was in fact the only one of that magnitude in the entire war.[897] Understandably so, since even in this battle – planned well in advance by the Athenians as a surprise attack – indescribable confusion reigned: the attackers were unfamiliar with the terrain of the plateau, they were confused by the chaotic intermingling of passwords and Doric *paianes* from both sides, and most importantly they were unable to visually distinguish friend from foe in the dark. What is more: this particular night was in fact as good as at all possible, since there was a clear sky and full moon.[898] One can only guess at the conditions at new moon, or on a clouded night.

Conversely, such conditions might be deliberately exploited for small-scale operations which required soldiers to remain unseen, as when some 220 Plataians escaped from the besieged Plataiai in the winter of 428/7.[899] In such rare cases, impenetrable darkness, rain and stormy weather may have been a positive asset, but on the whole, large-scale military operations could not be carried out with any efficiency in the dark. Consequently, battle at night was generally avoided at all times, and fighting naturally came to a standstill when the oncoming twilight made it impossible to see what was before one. Thus when the sources mention that the fighting was interrupted by the sunset, this may well be taken to mean simply that the battle just was not decided yet when darkness fell; and this, in turn, may imply nothing more than that the two sides engaged each other in actual battle very late in the afternoon.

4.2 CONTRIBUTING FACTORS

4.2.1 Necessary phases

Despite appearances, then, the sources on the whole do not permit the conclusion that battles were long simply because the source states that it was long: there is no external criterion of contol. The sources' vagueness militate against the chances of judging how long 'long time' actually was. So what was the likelihood that a hoplite battle, as Cawkwell believes, might actually last from sunrise to sunset? In the summer months in Greece, the length of day is approximately 14 hours and 40 minutes (at the latitude of Athens);[900] so if we are to take such phrases as διημέρευσαν ἐν τῇ μάχῃ literally, as Cawkwell wants us to, it would mean that hoplites were engaged in actual combat for more than 14 hours on end. One wonders if anyone could be left alive on either side after such a tremendously long fight, and if so, whether he would be able to stand on his legs. Hoplite battles were generally hard

---

897 Thuc. 7.44.1 (inv. no. 38).
898 Thuc. 7.43.2–44.8
899 Thuc. 3.20.1–24.2.
900 Kubitschek (1928) 182.

fought, and the manner of fighting left little or no time for the combatants to catch their breath, and no occasion to slink back from the fray, so naturally this must mean full-scale, furious non-stop fighting.

The flaws in this assumption are immediately apparent. Even if we concede that battles might last all day, this would leave no time for preparation beforehand or for all the necessary operations – for both the victors and the defeated – of the aftermath, for which daylight must have been necessary. When, for example, would the two armies march what was probably usually several kilometres out to the battlefield? When did they deploy their forces and take up positions? And again, when the fighting was over, when did they pursue the beaten enemy, attend to their wounded, strip the enemy dead, pick up their own dead and erect a trophy?[901] Usually, we are in no position to decide whether the historians in their blanket statements of the 'μάχη ἐπὶ πολὺν χρόνον' variety did not in fact include all the necessary phases of any armed conflict, including the above-mentioned.[902] If, as seems at least possible, these stages can somehow be included in the equation, the time left for the actual fighting is greatly reduced.

As it happens, Xenophon offers a striking example of these time-consuming initial phases before battle in his *Anabasis,* which, strange to say, seems to have escaped the attention of everybody in the debate over battle length.[903] At the battle of Kunaxa in 401, Xenophon was present himself without participating actively, and thus offers an ideal eyewitness account.[904] Kyros' army, in loose marching formation, was warned about the advancing enemy army ἀμφὶ ἀγορὰν πληθοῦσαν ("about the time when the market-place is full"), which is usually taken to mean something like forenoon; roughly between 10 and 11 or 12 a.m.[905] This caught Kyros by surprise, as the troops had not yet had their ἄριστον (a meal somewhere between our breakfast and lunch, apparently eaten just at late forenoon).[906] Following this nasty surprise, Kyros' army immediately went about taking up their battle stations. Noon came and went, but Artaxerxes' army still was not within sight. Then, by afternoon,[907] while Kyros' enormous army was still trying desperately to get in line for battle, Artaxerxes' forces were first glimpsed, then clearly seen across the plain.[908]

---

901 The significance of possession of the dead should not be underestimated: "in cases of ambiguous outcomes, the technical verdict of defeat or victory frequently hinged on which side had control of the war dead" (Vaughn [1991] 47).

902 Cf. Franz (2002) 234, 320: already Herodotos may have included such phases as *Aufstellung* and *Totenbergung* in his all-day battles.

903 Xen. *An.* 1.8, 1.10.1–17.

904 Xen. *An.* 3.1.2, cf. 1.8.15 (inv. no. 11).

905 See, e.g., Pind. *Pyth.* 4.85; Thuc. 8.92.2; Pl. *Gorg.* 469d and esp. Hdt. 4.181.3 and 3.104.2: the term ἀγορῆς διαλύσιος at the latter seems to indicate that the market was emptied again before the midday heat; cf. How & Wells (1928²) vol. I 252 (ad Hdt. 2.173.1); Lendle (1966) 433, (1995) 62.

906 Xen. *An.* 1.10.19.

907 The word is δείλη, which may indicate anything between early afternoon and late evening; cf. LSJ⁹ s.v. Δείλη is ambivalent even within Xenophon, cf. Lendle (1966) 445 n. 52.

908 Xen. *An.* 1.8.8.

From this point on, it becomes more difficult to establish the lapse of time; but in Xenophon's description, it seems as if quite a long time still went by as the royal army closed in ever so slowly (ἐν ἴσῳ καὶ βραδέως): presumably the range of vision was also exceptionally good on the level plain and in the dazzling sunlight.[909] At any rate, there was still time to note the different companies and arms in the advancing enemy army, time for Kyros to argue with Klearchos about the battle formation, time for Kyros to inspect his own and the enemy's forces, riding back and forth between them, time for the watchword to be passed back and forth two or more times, time for Xenophon to ask Kyros about last-moment orders, and time for the ritual sacrifices to be carried out and the result to be declared to the troops.[910] More importantly, at this late point – Lendle assumes four or five hours after Pategyas' sighting of the king's army – Greek mercenary hoplites were *still* pouring up from the rear to take their position in the phalanx.[911] The strange battle lasted very long indeed; and it was not until about sunset that the undefeated Greek mercenary 'Kyreans' swept the last shreds of resistance from a small hillock (σχεδὸν δ' ὅτε ταῦτα ἦν καὶ ἥλιος ἐδύετο).[912]

In this exceptional case, the battle by all lights actually *did* drag on most of the day, while the Greeks pursued their enemies some five km, then had to return the same distance again to drive plundering Persians from their camp; and finally to beat back the Great King, rallying his forces yet again. No matter how we look at it, these many kilometres' worth of marching (even if pursuing or charging) must have taken several hours for the heavy, slow hoplites; indeed, Lendle's estimate of 9 km/h for the initial five-kilometre charge and pursuit seems somewhat optimistic.[913] At any rate, it is sufficiently clear that the arraying of the Greek phalanx alone lasted several hours, from about, say, 10.30 a.m. to δείλη – at the very least 1 p.m., but more likely 3 p.m. The drawing up of at least the 10,600 Greek mercenary hoplites thus lasted two and a half hours at an absolute minimum, but quite possibly as much as four or five hours. The 'fighting' on this occasion, such as it was, can have lasted, at a high estimate, six hours, from 2 p.m. until nightfall about 8 p.m. The

---

909 Xen. *An.* 1.8.12. Lendle (1966) 445 n. 52 believes that about three hours may have gone after noon: "Man wird nach allem kaum fehlgehen, wenn man das Auftauchen der Staubwolke etwa auf 14 Uhr festsetzt. Bis der langsam marschierende ... Gegner bis auf 600 Meter heranwar ..., darf man wohl mindestens eine weitere Stunde veranschlagen" ("It is hardly wrong to set the appearance of the dust cloud to *c.* two p.m. Before the slowly marching ... enemy came to a distance of 600 m ... we may assume that at least another hour lapsed"). Cf. Anderson (1974) 103–105.
910 Xen. *An.* 1.8.8–17.
911 Xen. *An.* 1.8.14; Lendle (1966) 433 n. 20, 434, 445.
912 Xen. *An.* 1.10.15.
913 Xen. *An.* 1.10.4; Kromayer (1903) 236 n. 4; Lendle (1966) 444. On the five km pursuit, Lendle comments: "Bei Berücksichtigung der schweren Rüstung der Hopliten wird man das Tempo nicht allzu schnell ansetzen dürfen, aber mehr als eine Dreiviertelstunde war schwerlich vergangen ..." ("When we take into account the heavy hoplite armour, the pace cannot be put too high, but more than three quarters of an hour had scarcely passed"). Delbrück (1900) 58–59 gives the standard double time (*Laufschritt*) of the Prussian army as 165–175 m/m, (some 10 km/h); but soldiers in 1900 doubtlessly carried lighter equipment than fully armed hoplites of the Classical period.

objection that this was an impossibly large force and therefore out of the usual can also be laid to rest, for large battles, especially between coalition armies, often enough involved similarly large numbers, normally probably also arraying themselves by contingents.[914] It thus seems reasonable enough to conclude that the drawing up alone might last for several hours; and Xenophon's vivid description also illustrates that other necessary phases of preparation, such as meals, pre-battle sacrifice and the general's harangue to his army, normally had to be fitted in somehow: all this will naturally have taken a lot of time.

Some battles were decided abnormally quickly; and some even before the actual fighting began: thus, for instance, the so-called Tearless Battle between Spartans and Athenians on one side and Arkadians and Argives on the other at Melea in 368.[915] The Arkadians did not await the Spartan onslaught, and almost all ran for their lives before the phalanxes met. The battle derived its name from the fact that not a single Spartan – according to Xenophon – fell in the battle. Such exceptionally low casualties seem to warrant the extreme brevity of the 'battle'. A similar case can be made for Mantineia in 418, where king Agis and his 300 élite *hippeis* charged and scared off the enemy before them well in advance: ἔτρεψαν οὐδὲ ἐς χεῖρας τοὺς πολλοὺς ὑπομείναντας, ἀλλ' ὡς ἐπῆσαν οἱ Λακεδαιμόνιοι εὐθὺς ἐνδόντας καὶ ἔστιν οὓς καὶ καταπατηθέντας τοῦ μὴ φθῆναι τὴν ἐγκατάληψιν.[916]

In battles like these – certainly quickly decided by any standard – the fighting was over before it even began, at least, as at Mantineia, in certain sectors. The enormous nervous pressure was too much to bear for one side, and a collective panic will have spread through the ranks and dissolved the phalanx, perhaps from behind. The thought of crashing into the enemy's wall of wood, bronze and iron spearheads

---

914 Cf., e.g., Marathon 490 (inv. no. 19): rather more than 10–11,000 vs. many more Persians (Just. *Epit.* 2.9.9; Nep. *Milt.* 5.1; Paus. 4.25.5, 10.20.2, cf. Xen. *An.* 3.2.12); Plataiai 479 (inv. no. 29): 38,700 hoplites and 69,500 light-armed vs. at least as many Persians (Hdt. 9.10.1, 9.28.2–32.2, cf. Diod. Sic. 11.30.1); Tanagra 457 (inv. no. 40): 14,000 vs. ? (Thuc. 1.107.5; Diod. Sic. 11.80.1); Delion 424 (inv. no. 5): 7000 hoplites on both sides, several thousand ψιλοί (Thuc. 4.94.1–2); Mantineia 418 (inv. no. 17): at least 8000 on each side (Thuc. 5.68.1–3, cf. 5.64.3, 5.68.1; Diod. Sic. 12.28.4, 12.79.1); Nemea 394 (inv. no. 25): 13,500 vs. 24,000 (Xen. *Hell.* 4.2.16–17, cf. Diod. Sic. 14.82.10, 14.83.1); Koroneia 394 (inv. no. 9): possibly as many as 20,000 on each side (Pritchett [1969] 73–74); Leuktra 371 (inv. no. 16): 11,000 vs. 6000 (Diod. Sic. 15.52.2; Plut. *Pel.* 20.1, cf. Xen. *Hell.* 6.1.1, 6.4.15); Kynoskephalai 364 (inv. no. 12): 7000 vs. 14,000 (Diod. Sic. 15.80.2; Plut. *Pel.* 32.1); Mantineia 362 (inv. no. 18): 20,000 vs. 30,000 (Diod. Sic. 15.84.2–4); Chaironeia 338 (inv. no. 3): 15,000 infantry and 2,000 horse (Dem. 18.237; Plut. *Dem.* 17.3).

915 Xen. *Hell.* 7.1.31–32; cf. also 6.2.20 (on Korkyra): this *seems* to have been very short indeed, although the only 'evidence' for this is the sentence structure, which seems to indicate inordinately short duration: (ὁ δ' ἐπεὶ παρετάξατο, αὐτὸς μὲν τοὺς κατὰ τὰς πύλας τῶν πολεμίων τρεψάμενος ἐπεδίωκεν ["After he had formed his men in line Mnasippos himself defeated the enemy troops in front of the gates and pressed on after them in pursuit" (trans. Warner)]); Diod. Sic. 15.72.3; Plut. *Ages.* 33.3.

916 Thuc. 5.72.4 ("[they] did not even stand up to the first shock, but gave way immediately when the Spartans charged, some being actually trampled underfoot in their anxiety to get away before the enemy reached them" [trans. Warner]); inv. no. 17. See also Xen. *Hell.* 4.3.17; Eur. *Bacch.* 303–305.

simply became unbearable; and the result was that, as in these cases, one phalanx simply turned and ran. Such encounters were naturally extremely short, and it is highly likely that they played an important part in forming the collective conception of what was a long or a short battle: if a battle that did not even truly take place ranked as a short battle, then a battle lasting, e.g., an hour may easily have been thought rather long.

Another interesting indication of relatively short battle duration is furnished by the twin battles of Euripos and Chalkis that Athens fought in 506 against a Boiotian coalition and Chalkis respectively.[917] According to Herodotos, the Athenians first defeated the Thebans near the river Euripos and then, on the same day, crossed into Euboia, engaging the Chalkidians and defeating them as well. The historicity of these battles seems to be confirmed by an epitaph:

> Δίρφυος ἐδμήθημεν ὑπὸ πτυχί, σῆμα δ᾽ ἐφ᾽ ἡμῖν
> ἐγγύθεν Εὐρίπου δημοσίαι κέχυται·
> οὐκ ἀδίκως, ἐρατὴν γὰρ ἀπωλέσαμεν νεότητα
> τρηχεῖαν πολέμου δεξάμενοι νεφέλην.[918]

The inscription is arguably earlier than the 470s, when the Athenians first began repatriating their fallen soldiers for public burial. There seems to be some consensus that the epigram was placed on a *polyandrion* or mass grave for the Athenian fallen in these two battles, apparently corroborating their historicity by virtue of the mention of a war grave at the Euripos.[919] Consequently, the battles apparently actually did take place; but the notion that they should have been fought on the same day may of course be part of later Athenian propaganda, or simply a result of a good story getting better with every retelling. The important notion, however, is that Herodotos evidently saw no reason why it should be impossible for one army to fight two battles in one day. This would of course involve doubling all the necessary phases of battle – deployment, sacrifices, harangues, erection of trophies, mopping-up operations – but in this case also the embarkation, crossing and debarkation on Euboia. All this would have left the barest minimum of time for actual hand-to-hand fighting, divided between two separate battles. This would be an impressive feat by any standards, but not, after all, felt to be impossible – and accordingly, Heorodotos must have counted on his public to accept it as well. Certainly Plutarch, for all his vitriol against Herodotos, has nothing to say against this. This, then, reinforces the notion that battles in general were of rather short duration.

---

917 Hdt. 5.77.
918 *GVI* I (= *Anth. Pal.* 16.26) ("Under the cleft of Dirphys, we were subdued, and upon us was piled a grave mound near the Euripos at public expense. This was not without due cause, for we lost our lovely youths when we welcomed war's rugged cloud" [trans. Anderson]). The epigram is ascribed to Simonides (fr. 87 Diehl), although the last two lines of the *Anth. Pal.* 16.26 have been thought spurious since Schneidewin. Clairmont (1983) vol. I 88–89 discusses the validity of the text and persuasively dismisses the alternate view, that the epitaph might be set over the Euboians.
919 For an overview, see Anderson (2003) 151 n. 15.

4.2.2 Different sectors faring differently

Related to this is the fact that in a small number of cases, separate sections of the same phalanx actually fared differently, so that one section might actually be winning while other sections were being defeated. This is all in keeping with the common Greek practice of arraying the best contingents on the right, where danger and glory was greatest. Time and again, however, this had the peculiar result that the crack troops on the two opposing right wings might actually both defeat their immediate opponents and commence pursuit, while the rest of their respective phalanxes were still engaged in desperate fighting or, as was often the case, well on the way to being defeated. This certainly seems to have been the case in the battles of Poteidaia 432,[920] Mantineia 418,[921] Nemea 394,[922] Koroneia 394[923] and Corinth 392.[924] In these battles, we may therefore be fairly certain that actual fighting went on from the moment the battle became ἐν χερσί and until the *pararrexis*. Furthermore, it also continued across most of the battlefield while the victorious sector pursued their defeated opponents, and all the time it took them to return to their original positions.

The case of Nemea will suffice to demonstrate this point. The coalition forces managed to catch the Spartan forces by surprise, since there was much dense undergrowth (καὶ γὰρ ἦν λάσιον τὸ χωρίον).[925] This may explain why all the Spartans' allies (save the Pellenians) failed to withstand the charge and were immediately routed by the Argives, Corinthians and Thebans posted opposite them. The Spartans defeated the Athenians opposite them, joined ranks again and went after the Argives, just as these were returning from pursuing the Spartans' allies. The Spartans next attacked the returning Corinthians and then the Thebans. This carefully arranged sequence seems to be meant to indicate that the Argives, Corinthians and Thebans returned at slightly different times; and probably also shows that the Spartans marched at right angles across the battlefield, so that they caught their adversaries in the uncovered side, one by one. Be that as it may, we can be almost certain that fighting went on right through all this – this seems to indicate that certain battles lasted quite a lot longer than was usual, since normally fighting stopped when one side collapsed decisively in the *pararrexis*.

Interestingly, however, Nemea is also the only possible case where we can be *certain* that one sector of the phalanx kept on fighting while the rest of it had been defeated. Such occasions may have been very rare indeed; and the disastrous result of this particular incident may be judged from Xenophon's coolly casual observation that the hapless Pellenians, pitted against the equally unfortunate Thespians, were virtually annihilated "on the very spot" (Πελληνεῖς δὲ κατὰ Θεσπιέας γενόμενοι

---

920 Thuc. 1.62.6–63.1 (inv. no. 30).
921 Thuc. 5.72.3–73.3 (inv. no. 17).
922 Xen. *Hell.* 4.2.20–22 (inv. no. 25).
923 Xen. *Hell.* 4.3.17–19; cf. *Ages.* 2.12; Plut. *Ages.* 18 (inv. no. 9).
924 Xen. *Hell.* 4.4.9–11 (inv. no. 4).
925 Xen. *Hell.* 4.2.19 (inv. no. 25).

ἐμάχοντό τε καὶ ἐν χώρᾳ ἔπιπτον ἑκατέρων).[926] Normally, phalanxes probably crumbled entirely as soon as one part of it gave way or was pushed back sufficiently. This again indicates that the duration of the actual fighting was, if anything, rather short.

4.2.3 Pursuit

The central part of the battle itself, the actual fighting, can at most have taken up part of the day, and as we have seen, it is possible that the initial phases took up comparatively long time, perhaps even most of the day. Another important phase, which might take up an inordinately long time, was the pursuit, immediately following the collapse and rout of one phalanx. The Spartans may have been unique in declining to keep up the pursuit for very long; certainly we hear often enough of hot pursuits being kept up for a long time.[927] However, such pursuits would have to be relegated to the light-armed and the cavalry if they were to be effective, since it is unlikely that hoplites in heavy armour should be able to overtake men who probably had discarded their shields and whatever else they could and were running for their lives – or at least not for a very great distance. Nevertheless, we often hear about long pursuits, and it is probable that at least the first stages of pursuit were handled by the hoplites, following immediately behind the receding enemy.[928]

The military value of a long pursuit may in any case be questioned, but the sources nonetheless very often describe just that. The explanation is probably relatively simple: such an opportunity to vent one's pent-up frustration and anger at the helpless, panic-stricken enemy after the deadly struggle was something that the average Greek warrior rejoiced in.[929] It is no wonder that a vigorous pursuit would be kept up for a long time: now was the chance to really do harm to the helpless enemy, to attempt to decimate the enemy forces (which in turn might help to prevent similar battles for a long time to come), and to make sure that the enemy's fighting spirit was thoroughly broken, so that a rallying of their lines was completely out of the question. Probably the opportunity for plunder at this stage should be reckoned with as well.

These things may help explain the high frequency of long pursuits in the battle descriptions. It is not infrequent that we hear of pursuits actually lasting until nightfall.[930] In Plato's *Symposium,* there is a matter-of-fact description of how hoplites

---

926 Xen. *Hell.* 4.2.20.
927 Short Spartan pursuits: Thuc. 5.73.4 (but cf. Diod. Sic. 12.79.4 and 12.79.6–7). Long pursuits: Hdt. 6.113.2–115; Thuc. 1.62.6, 2.79.6, 3.98.4, 3.112.6–8, 4.96.8 (cf. Pl. *Symp.* 221a–c), 5.10.9–10; Xen. *Hell.* 2.4.6 (pursuit over 6–7 stades); 5.3.5–6, 7.5.25, *An.* 1.10.16, 3.3.8–19; Plut. *Lys.* 28.11, *Pel.* 32.7 (cf. Diod. Sic. 15.80.6). *Contra* Hanson (1995) 268–269.
928 Pursuit *expressly* by hoplites: Xen. *Hell.* 1.2.2–3, 1.2.16, 2.4.6, *An.* 3.3.8, 3.4.3–4, 5.4.24.
929 Xenophon says that the Spartans regarded it as a positive godsend when they were able to massacre the hapless Argives in droves, pressed up against the long walls of Corinth in 392 (inv. no. 4): Xen. *Hell.* 4.4.11–12; cf. Thuc. 7.84.3–85.1.
930 At Delion 424 (inv. no. 5), the Athenians were saved by the dark, which precluded further pursuit, Thuc. 4.96.8 (cf. 4.93.1); Diod. Sic. 12.70.4. At Epipolai outside Syracuse in 413 (inv. no. 38), Syracusan cavalry the next day rounded up and killed Athenian hoplites who had lost their

behaved after the τροπή in what was probably an all too well-known situation. Alkibiades relates how Sokrates remained composed and calm and thus made good his escape, because he made it clear that he would fight back: σχεδὸν γάρ τι τῶν οὕτω διακειμένων ἐν τῷ πολέμῳ οὐδὲ ἅπτονται, ἀλλὰ τοὺς προτροπάδην φεύγοντας διώκουσιν.[931] This well illustrates a certain, rather gruesome, appetite for destruction, and in all probability it made scenes of relentless pursuit over long distances and for a long time a normal occurrence. Even if Thucydides insists that the Spartans kept their pursuits short,[932] this may in fact serve to show that they were an exception to the rule, and that the unpleasant reality was that a defeated hoplite faced pursuit and harassment over several kilometres.[933]

If, as seems likely, battle normally commenced late in the day, there is nothing surprising in the fact that the very last phase of battle, the pursuit, might last until darkness intervened and made further slaughter impossible. It is thus possible that two phalanxes would take the field against each other in the early morning or forenoon (including the march to the battlefield), spend several hours in preparation, including arraying and dressing of the lines, battle harangue and plain and simple waiting as they eyed the enemy.[934] Next, battle would ensue, and when the outcome was decided, one side would turn and run, hereby triggering a pursuit, which might last for hours. It seems at least likely, then, that fighting, no matter how important, was probably just one stage out of several in a hoplite battle. It may in fact even have been the shortest one.

4.2.4 Physical limitations

As Cawkwell ironically enough implies himself, battle in general, and hoplite battle especially so, takes an enormous physical toll on the combatant; and this is the case to an even higher degree if we accept – as Cawkwell does not – that *othismos* was an integral part of the hoplite battle, or even the central part:

> If [battles] lasted for a couple of hours, that is an improbably long time to be pushing, and, one may add, for a man to be fighting. Fifteen three-minute rounds with rests exhaust the fittest

---

way during the night attack, Thuc. 4.44.8. At Abydos in 409, Alkibiades' troops pursued those of Pharnabazos "until night took over" (μέχρι σκότος ἀφείλετο), Xen. *Hell*. 1.2.16. At Kunaxa (inv. no. 11), the battle for the Greeks consisted mostly in pursuit of the defeated Persians all over the plain, Xen. *An*. 1.8.8–29, 1.10.11–15. At Naryx in 395 (inv. no. 24), the victorious Boiotians pursued and slaughtered the beaten Phokians until nightfall, Diod. Sic. 14.82.9.

931 Pl. *Symp*. 221c ("for this is the sort of man who is never touched in war; those only are pursued who are running away headlong" [trans. Jowett]).

932 Polyaen. 1.16.3. claims that Lykurgos urged the Spartans not to kill fleeing enemies in the hope that this would *encourage* Sparta's enemies to run away: such considerations may have resulted in official policy.

933 Xen. *Hell*. 2.4.6: Thrasyboulos' men surprised the enemy at Phyle in an attack at dawn, pursuing their enemy for "six or seven stadia", i.e. between 1,071 and 1,250 m. Whether this unusual specificity can be ascribed to an extreme length of pursuit, or merely reflects Xenophon's wish to be accurate, is a matter for speculation.

934 This seems to be implied by the fact that hoplites often put down their shields (θέσθαι τὰ ὅπλα) and leaned them against their knees: see *supra* 100–101 and n. 397.

pugilists in the world, and those who have had experience of scrummaging and forward play generally in 'the Rugby game of football' must be sceptical about how long such intense efforts could have been maintained without the rests provided by 'line-outs', 'half-time', and 'injury-time'.[935]

By Cawkwell's own admission, then, it is inconceivable that even the "fittest pugilists" should have the stamina to keep fighting for several hours, let alone all day. It should not be overlooked, however, that a hoplite battle by any reckoning required a lot more physically than can be expected from any boxer. The weight and awkwardness of the weapons and armour will surely have taken their toll on the hoplite and sapped his strength more quickly than even the fiercest fist-fight.

The interesting experiments carried out by Donlan and Thompson are sufficient to demonstrate that running or fighting in hoplite armour required an enormous energy expenditure on the hoplite's part: their results show that young men in peak form, when asked to 'charge' at a run, carrying loads less than 7 kg and in terrain simulating the conditions at Marathon, were able to run only very short distances.[936] The experiments were carried out with the intention of disproving the myth of the Athenian hoplites' charge at a run over eight stades at Marathon, and, to my mind, they have done so successfully; but in the debate over hoplite fighting, it has been generally overlooked that the results are equally crucial for appreciating the physical limitations for hoplites' battle capacity in general.

Furthermore, it must be stressed that the enormous strain of fighting with spear and shield – including the tremendous exertions of *othismos* – under the fierce Greek summer sun would have depleted the hoplites' energy reserves rapidly. Indeed, Hanson's students – again, fit, young men, which naturally not all hoplites on either side will have been – who carried out a mock battle with replicas of hoplite armour and weapons in the Californian summer, were utterly exhausted within half an hour.[937]

It should be stressed, however, that the two situations are not immediately comparable. Modern students – however physically fit – in mock combat will have rather less inner motivation to find extra reserves of energy to keep fighting than actual hoplite soldiers, whose very existence, possibly along with that of their families and fatherland, was at stake. Moreover, it must be remembered that the bulk of the hoplites in any given army will have been farmers, used to hard physical all-day, agricultural work in the field, perhaps lending them a kind of stamina and physical toughness all but unknown in the modern world.

Nevertheless, given that we are unlikely to ever be able to compare such situations in a meaningful way, this comparison may well be the closest we could ever

---

935 Cawkwell (1989) 376.
936 Donlan & Thompson (1976) 341; cf. (1979); *contra* Hammond (1973) 194 n. 4 (inv. no. 19).
937 Hanson (2000²) 56: "My own students at California State University, Fresno, who have created metal and wood replicas of ancient Greek and Roman armor and weapons, find it difficult to keep the weight of their shield, greaves, sword, spear, breastplate, helmet, and tunic under seventy pounds. After about thirty minutes of dueling in mock battles under the sun of the San Joaquin Valley they are utterly exhausted." See also Goldsworthy (1997) 21; Dawson (1996) 49–51.

hope to come an assessment of the physical hardships involved; and accordingly we are faced with the choice of not comparing at all, or comparing two different situations with the necessary caveats in mind. At the very least the modern recreations serve as an indicator that fighting a hoplite battle in the summer heat *was* gruelling physical work. It is not difficult to imagine how a battle situation could affect the average hoplite's capacity for combat adversely: even if we concede that the charge was probably very short, the strain brought about by weapons and armour, as well as the tremendous *nervous* strain produced by having to fight for one's life under the grimmest of conditions, make it abundantly clear that battles would have to be brief and effective, as well as fought in an economical manner in regard to energy expenditure.

4.3 UNUSUALLY LONG BATTLES

4.3.1 All-day battles: a problem of sources

Despite the many objections and the essential ambiguity in the phraseology of duration, there are a few explicit references to fighting going on all day. In his account of the battle of Oinophyta between Athens and Thebes in 457, for example, Diodoros states that the battle raged "the entire day" (διημέρευσαν ἐν τῇ μάχῃ), and he makes a similar claim for a battle at Pelousion in Egypt in 350/49.[938] Apologists of *a priori* long battles have claimed these particular battles as their own. Cawkwell certainly takes both of these at face value and includes them in his list of battles the Greeks allegedly "knew as long"; but it is always risky to put much faith in Diodoros, who had quite specific goals with his history. In Underhill's neat phrasing: "Diodorus, it is but too apparent, cares for history, not so much for its own sake, but as an *opus oratorium,* in which he can display his own surprising talents."[939]

Consequently, more often than not Diodoros' battle accounts are notable for their formulaic language, maybe derived from rhetorical conventions about battle accounts.[940] This is especially evident when his accounts are compared to other, more contemporary sources. Even the most unorthodox battles are very often described in the same standardised manner: the battle is almost always hard-fought (καρτερά), and for a long time the outcome is uncertain (ἰσόρροπος): then one side, by their ἀρετή and/or general steadfastness and military prowess, secure the victory and put their enemies to flight.[941] Indeed, the cookie-cutter nature of Diodoros' battle accounts may be guessed from the sheer amount of repetitions: the phrase πολὺν χρόνον appears in a battle context 31 times, whereas battles are described as καρτερά

---

938 Diod. Sic. 11.83.1 (inv. no. 26); cf. 16.46.9 (Pelousion 350/49: διημερεύσαντες).
939 Underhill (1906) xxxiv; cf. Sacks (1990) 93.
940 Cf. Diod. Sic. 20.1.3, 20.2.1–2.
941 Instances of this tendency may be seen, e.g., at Diod. Sic. 12.74.1–2 (the battle of Amphipolis 422 [inv. no. 2]: for another typically rhetorical description of the general's death, cf. 15.80.4–5 and Hornblower [1996] 448–449); 13.51.4, 16.86.2. There are, however, a few exceptions to the rule, such as 14.84.1 (μικρὸν ἀντισχόντες χρόνον ["resisting for a short time"]).

μάχη no fewer than 42 times.⁹⁴² Diodoros' tendency towards repetitiveness in both structure and phraseology has been examined by Jane Hornblower, who has the following assessment of his style and method:

> Diodorus made up his own clichés and applied them everywhere mechanically, even when they were actually inappropriate to the context ... Formulaic language was more suited to epic than to history; and in his efforts to silhouette the general moral truth behind a historical situation, Diodorus took much of the colour out of the narrative he drew upon. His main purpose being instructive, it did not matter if battles were standardized or individuals stereotyped: rather, this method facilitated his didactic aims. He has a style of his own, but it is a style characterized by monotony and repetition, the colourless style of bureaucratic prose.⁹⁴³

Of course, the formulaic repetitiousness of Diodoros' battle accounts does not in itself rule out entirely the possibility that he may be correct, but it is significant that the stereotyped battles are scattered throughout the entire *Bibliotheke* and thus cannot possibly be borrowings from Diodoros' sources, such as Ephoros or the Oxyrhynchus historian, both far more trustworthy in such matters.⁹⁴⁴

Another case of an entire day's worth of fighting can be found in Herodotos, normally a far superior source for military engagements to the notoriously untrustworthy Diodoros. Herodotos is very explicit that in the battle of Himera in 480, the Carthaginians – according to themselves, no less – fought the Greeks ἐξ ἠοῦς ἀρξάμενοι μέχρι δειλίης ὀψίης. Clearly, however, Herodotos was aware that his audience was unlikely to buy such a tall tale (and perhaps he even doubted it himself), for he immediately goes on to add ἐπὶ τοσοῦτο γὰρ λέγεται ἑλκύσαι τὴν σύστασιν by way of a rather apologetic afterthought. Apparently Herodotos himself had doubts that the battle could possibly have lasted this long; which, in turn, indicates that battles in general must have been rather shorter than this.

Herodotos' interesting claim that this particular λόγος is told by the Carthaginians themselves, for one thing, presents a serious problem. It is highly unlikely that Herodotos spoke or read Phoenician, and perhaps even that he had been there, de-

---

942 καρτερὰ μάχη: Diod. Sic. 3.54.7, 4.16.2, 4.66.4, 11.7.1, 11.30.2, 11.32.2, 11.74.3, 12.6.2, **12.80.8**, 12.82.6, **13.51.4**, 13.59.8, 13.60.7, 13.64.1, 13.66.2, 13.72.7, **14.33.2**, 14.80.3, (**14.82.9**: ἰσχυρά), 14.90.4, 15.13.3, (**15.34.2**: ἰσχυρά), 15.37.1, 15.62.2, 15.78.3, 15.80.5, (**15.86.4**: ἰσχυρά) **16.4.5–6**, (16.31.3: ἰσχυρά), 16.39.5, 16.48.5, 16.79.6, **16.86.2**, 17.103.3, 18.14.3, 18.70.6, **19.76.2**, 19.83.4, **19.89.2**, 20.87.3, (**20.88.8**: ἰσχυρά), 20.89.2.

πολὺν χρόνον: Diod. Sic. 2.18.3, 11.7.2, 11.79.3, **12.80.8**, 13.17.1, 13.46.2, **13.51.4**, 13.56.6, 13.67.5, 13.79.4, 13.80.6, 13.87.1, 13.99.5, 13.110.3, 14.12.7, **14.33.2**, **14.82.9**, **15.34.2**, **15.86.4**, **16.4.5–6**, **16.86.2**, 17.11.5, 17.63.2, 18.15.3, 18.34.4, **19.76.2**, 19.84.1, **19.89.2**, 20.38.5, **20.88.8**, 31.19.5. References in bold are instances in which battles are described with both phrases. Borderline cases are, e.g., 9.16, 11.9.4, 11.83.4, 12.46.2, 13.62.3, 13.78.2, 15.41.5, 19.108.3, 22.10.3, 37.2.13. Pritchett's table (*supra* 202 n. 869), however useful, is not exhaustive.

943 Hornblower (1981) 272; cf. Barber (1935) 142–144; Vial (1977) xx–xxii; Stylianou (1998) 15–16.

944 A certain amount of similarity does exist between Diodoros 11.60.1–2 and Ephoros (*FGrH* 70.191); and also, following Hornblower ([1981] 27–32), parallels between Diodoros and other sources such as Agatharchides and Polybios. The wholesale application of irrelevant terms in military matters is also noticed by Sinclair (1966) 255.

spite a few citations of Carthaginians here and elsewhere.⁹⁴⁵ Moreover, the story, such as it stands, is rather romanticised, and the detail of Hamilkar's self-sacrifice smacks rather of sensationalism, possibly influenced by what little was known about Carthage's practice of human sacrifice.⁹⁴⁶ The battle of Himera was clearly shrouded in the mists of myth already by Herodotos' time. There are altogether too many strange and convenient coincidences in these conflicts: the battle of Himera famously took place on the same day as the battle of Salamis,⁹⁴⁷ Hamilkar had 300,000 troops at his disposal, precisely the same number as Mardonios had at Plataiai;⁹⁴⁸ and Gelon's 200 ships exactly matched the Athenian naval forces at Salamis.⁹⁴⁹ Furthermore, as Walter Ameling correctly points out, there are quite remarkable similarities between the numbers and varieties of subject peoples led to war against the Greek world by Xerxes and Hamilkar respectively.⁹⁵⁰

Diodoros' account of these same events, although to a certain extent equally rhetorically exaggerated, is altogether different: Hamilkar is simply killed in his camp while sacrificing, by Syracusan horsemen posing as Carthaginian allies, at the very outset of the fighting. The battle (needless to say) drags on, and no fewer than 150,000 Carthaginians are killed, but nothing at least is said about battle from dawn till dusk.⁹⁵¹ In both Diodoros and Herodotos, everything about this battle seems to be larger than life – the suicidal heroism of Hamilkar, the unlikely coincidence of exactly simultaneous victory for Greece and Sicily, and the strongly exaggerated numbers at the disposal of the barbarians.⁹⁵²

945 Hdt. 4.43.1, 4.195.1, 4.196.1–2. Herodotos has in all probability heard stories cited by Greeks – for example in Sicily – as having Carthaginian provenance (cf. 7.165–166); *contra* Ameling (1993) 51–64, who takes all of Herodotos' account at face value, and does so in an attempt to postulate a certain, unspecified Semitic type of royal dignity for Hamilkar. The problem of Herodotos' direct access to Carthaginian stories and the improbability of a subsequent Hamilkar-cult in Carthage are not adequately addressed. Herodotos' trustworthiness in maintaining this particular source has been called into serious question by Fehling (1971) 12–15, esp. 14; 92–94. Herodotos' story of Hamilkar's self-sacrifice is offered in explanation of the fact that his body was never found despite Gelon's best efforts, Hdt. 7.166.

946 Cf. Hdt. 1.86; Just. *Epit*. 18.6.11: *Cruenta sacrorum religione et scelere pro remedio usi sunt; quippe homines ut victimas immolabant et inpuberes quae aetas etiam hostium misericordiam provocat aris admovebant ...* ("they adopted a cruel and abominable religious ceremony as a remedy for it; for they slaughtered human beings as sacrificial victims, and brought children (whose age excites pity even in enemies) to the altars ..." [trans. Watson, modified]). On human sacrifice to Ba'al Hammon, see Fantar (1993) vol. II 277–278; Huß (1990) 374–383 with an excellent overview of the wildly exaggerated accounts of human sacrifice by Greek and Roman authors.

947 Hdt. 7.166, cf. 9.3.2; Plut. *Arist*. 19.7.

948 Hdt. 7.165, cf. 9.32.2 (inv. no. 29); Diod. Sic. 11.30.1.

949 Hdt. 7.144.1, cf. 7.158.4: Gelon promises the Greek coalition 2000 ships, 20,000 hoplites and four different types of light-armed soldiers, all in neat groups of 2000 each. A double πόλις μυρίανδρος is mentioned by Plato as the ideal size of an army, Pl. *Criti*. 112d–e.

950 Ameling (1993) 23–26.

951 Diod. Sic. 11.21.4–22; cf. Polyaen. 1.27.2. Diodoros' account is probably based on another tradition, quite likely a Sicilian one. Polyainos' version also contains some rather silly elements, but nothing about all-day battle.

952 This in turn is directly connected to Herodotos' explicit thematic manifesto at the very begin-

In consequence, we can be almost certain that the temporal extent of the fighting has been telescoped as well. There is nothing at all to support Herodotos' claim that fighting really went on from sunrise to sunset. In the light of the mythologising that surrounds other aspects of this battle narrative, then, it is no doubt the better course to err on the side of caution. In summary, due to demonstrable weaknesses and exaggerations in these battle narratives, neither Herodotos nor Diodoros can be trusted completely with regard to the duration of battles. Many aspects have been maximised or distorted out of proportion, and grandiloquence often takes precedence over probability, particularly in the case of Diodoros. Diodoros and Herodotos' descriptions are embellished and exaggerated for the sake of emphasis and to give the reader an impression of the mighty deeds done. It is better to dismiss these accounts as evidence of extremely long battles.

4.3.2 All-day fighting: Sphakteria

There is, however, another account of seemingly impossibly long fighting. Thucydides, who otherwise seems to maintain his supreme cool when recounting and analysing battles, has a single instance of all-day fighting. This is the skirmish on Sphakteria in 425 (inv. no. 34) between Demosthenes' light-armed Athenian troops and the Spartan hoplites trapped on the island. Thucydides' account of the fighting is rather long, and quite clear and detailed: the influence of eye-witness accounts is plain. Fighting here went on καὶ χρόνον μὲν πολὺν καὶ τῆς ἡμέρας τὸ πλεῖστον ("for a long time, even for most of the day") and indeed right from dawn, since Demosthenes and Kleon deliberately sought to take the Spartan sentinel posts near the shore by surprise in the early morning.[953] But the point here, as Thucydides is at pains to point out, is exactly that this was *not* normal fighting. The Spartan hoplites, true to their training, wished to close at once with the Athenian hoplites in the conventional manner of hoplite battle, but this was exactly what Kleon and Demosthenes sought to avoid. The traditionalist Spartans here had their first taste of fighting against light-armed troops in terrain that did not favour the hoplite phalanx. The Spartans were constantly harassed by the much more agile light-armed Athenian troops and exhausted themselves trying in vain to pursue them, only to be showered with missiles once again when falling back to their own ranks.[954] Once they realised this, they closed their ranks and retreated to a stronghold in order to avoid being encircled and attacked in the flank, and were now able to defend themselves somewhat better until the Athenians were shown a path leading behind the Spartan position and could attack them from the rear as well.[955]

In this light, it is considerably less surprising that the 'battle' should go on for an exceptionally long time. The Spartans, although probably very exhausted in the first phases, were never really engaged in true hoplite fighting, precisely because

---

ning of the *Histories:* his intention of recording "great and marvellous deeds" (ἔργα μεγάλα τε καὶ θωμαστά) performed by both Greeks and non-Greeks: Hdt. *Prooem.*
953 Battle narrative: Thuc. 4.30–38; points of reference for the duration: 4.35.4, cf. 4.32.2.
954 Thuc. 4.33–35.
955 Thuc. 4.35.1–4.

the Athenians refused them this opportunity. Although constantly harassed and with dwindling numbers, the Spartans could do little more than close their ranks (ξυγκλῄσαντες) and ward off any light-armed who ventured too close. Their situation, although frustrating, was hardly comparable to the extreme physical duress of hoplite *othismos* or hand-to-hand fighting in the phalanx. This enabled them to stave off defeat for "most of the day". Even so, Thucydides' explicit battle narrative leaves little doubt that the unusual length of fighting combined with heat and the dense cloud of dust and ashes from the recently fired forest put the combatants of both sides – but especially the Spartan hoplites – under tremendous strain:

> γενομένης δὲ τῆς βοῆς ἅμα τῇ ἐπιδρομῇ ἔκπληξίς τε ἐνέπεσεν ἀνθρώποις ἀήθεσι τοιαύτης μάχης καὶ ὁ κονιορτὸς τῆς ὕλης νεωστὶ κεκαυμένης ἐχώρει πολὺς ἄνω, ἄπορόν τε ἦν ἰδεῖν τὸ πρὸ αὑτοῦ ὑπὸ τῶν τοξευμάτων καὶ λίθων ἀπὸ πολλῶν ἀνθρώπων μετὰ τοῦ κονιορτοῦ ἅμα φερομένων. τό τε ἔργον ἐνταῦθα χαλεπὸν τοῖς Λακεδαιμονίοις καθίστατο· ... καὶ χρόνον μὲν πολὺν καὶ τῆς ἡμέρας τὸ πλεῖστον ταλαιπωρούμενοι ἀμφότεροι ὑπό τε τῆς μάχης καὶ δίψης καὶ ἡλίου ἀντεῖχον, πειρώμενοι οἱ μὲν ἐξελάσασθαι ἐκ τοῦ μετεώρου, οἱ δὲ μὴ ἐνδοῦναι.[956]

The unusually richly detailed narrative here indicates that Thucydides was trying very hard to convey to his audience the unusual nature and uncommon duration, and, most of all, the excessive stress it put the soldiers under. If it was not so much the usual physical pressure from phalanx fighting which exhausted the Spartan hoplites, then, it seems that the duress and suffering Thucydides is trying to express resulted precisely from the extraordinary length of the battle, which in turn aggravated the corollary sufferings of combat: those of heat and dehydration.

Thucydides' carefully structured narrative thus points to the conclusion that both these factors were unusual; and since this is the only 'good' instance of a very long battle with a definite designation in Thucydides,[957] we may infer that fighting actually lasting 'most of the day' was exceedingly rare. This *a fortiori* indicates that more normal (i.e. 'pure' hoplite) battles of this length, without certain 'extenuating' circumstances such as these, were even more unheard of.

Everything thus seems to point to the logical conclusion that hoplite battles, or, to be more precise hoplite battle *fighting,* were in reality very short in duration. The frequent references to long battles are at best ambiguous, as designations such as

---

956 Thuc. 4.34.2–35.4 ("... shouting as they charged down upon him in a mass and letting fly with stones and arrows and javelins and every weapon that came to hand. The Spartans were not used to this kind of fighting, and they were thrown into consternation by the shouting which accompanied the attacks; great clouds of dust rose from the ashes where the wood had been recently burned, and what with the arrows and stones loosed from so many hands and flying through the dust-cloud, it became impossible to see in front of one. Things now began to go hard with the Spartans ... For a long time, indeed for most of the day, both sides held out, tired as they were with the fighting and the sun, the Athenians trying to dislodge the enemy from the high ground and the Spartans struggling to maintain their position" [trans. Warner]).

957 Franz (2002) 320 makes an interesting observation: the only battle in which Thucydides himself participated is described as rather short: "Ist es nun Zufall, daß Thukydides ausgerechnet für das Jahr, in dem er selbst Stratege war und für das er wohl die besten Informationen besaß, eine Schlacht von kurzer Dauer schildert (Thuc. 4.54.2)?" ("Is it mere coincidence that Thucydides describes a short battle in the exact same year he himself held a command, a year for which he accordingly probably had the best access to information?)"

long and short are impossible to define in the absence of a definite criterion. Battles that are said to last 'all day', however, can mostly be shown to be either under heavy influence from mythologising and rhetoric, or to be the result of highly unusual circumstances, such as the 'irregular' troop types faced by hoplites at Sphakteria. Alternatively, they reflect the many phases surrounding the battle itself and which may normally be understood to be included in the narrative, such as the arraying of the various contingents, as at Kunaxa. The apparent solutions offered by Plutarch's 'very short' battle at Pydna, or Vegetius' remarks about two or three hours of fighting, are tempting, but best left out of consideration, as they depict altogether different battle realities in a much later day.

What *can* be demonstrated, on the other hand, is that large hoplite armies (and to a certain degree surely smaller armies too) took a very long time to deploy and array properly, so that the best men were at the front and rear, and so that the demands of tribal associations were met to the hoplites' and the general's satisfaction. Furthermore, pursuit was probably kept up for a very long time in order to inflict a maximum of damage with a minimum of risk for oneself. Battles lasting 'until nightfall' should be viewed in the light of relatively many accounts of *pursuit* being interrupted only by nightfall. I therefore suggest that the fighting itself was usually confined to a rather small amount of time, and that other, necessary phases of battle took up much of the day. This, in turn, would naturally result in the conception of battles as lasting most or all of the day. Nor is this necessarily wrong. *Battles* might indeed last all day. The problem arises only when scholars in modern times attempt to determine the duration of hand-to-hand *fighting* from these vague and ambiguous statements.

4.4 EXCURSUS: THE POSSIBLE INFLUENCE OF LITERARY CONVENTIONS

As has been stated above, at least Herodotos and Diodoros are to a certain extent prone to rhetorical and mythologising embellishments and exaggerations. This may suggest the pre-existence of conventions within the sphere of rhetoric, and, possibly, literature in a wider sense. Such conventions, like any type of cultural norm, can be very powerful, and anyone within their confines is necessarily affected by them, more or less consciously. If, for example, there was a general consensus that certain fixed elements 'suited' a battle narrative, there would be a strong incentive for the author to include these traits in the work. Much the same parameter is operative on a linguistic level in the dialectal peculiarities of Greek literature: choruses were best written in Doric, history in Ionic and epic in the wholly fictitious *Kunstsprache* of this genre. It is consequently of some interest to examine the traditions of battle narratives more closely; particularly within the genres of rhetoric and epic (and to a certain extent historiography), as these will have been stylistic models for the nascent prose genre.

If we turn first to epic, we need look no further than Homer to notice certain similarities. The action in both the *Iliad* and the *Odyssey* is of course epic in its scope, and consequently there are a great deal of exaggerations that serve to raise the action out of the realm of the mundane and into a truly heroic sphere. Quite apart from the frequent instances of gods and *daimones* interfering in human affairs, we are also repeatedly reminded that these were different times, when the lives of men were lived on a far greater scale than is the case in the poet's own day. A typical instance is the recurrent motif of Aineias or another hero lifting an enormous boulder off the ground, ὃ οὐ δύο γ' ἄνδρε φέροιεν, οἷοι νῦν βροτοί εἰσ', or of Hektor's five-metre (ἑνδεκάπηχυ) spear.[958] These exaggerations serve a narrative purpose in that they enlarge the scale of the action and emphasise the distant past in which it takes place. It is therefore to be expected that similar mechanisms are operative in other aspects of the poems, and especially so with regard to the fighting at the core of the poem. An examination of the text of the *Iliad* reveals numerous central passages with instances of fighting which lasts a very long time.[959]

(1) The duel between Hektor and Aias is only interrupted by the heralds of both armies who proclaim that it is ἀγαθὸν καὶ νυκτὶ πιθέσθαι.[960] We are not informed, however, when the duel actually began, so this particular instance may be of lesser importance.
(2) The Trojans and Achaians both arm themselves at sunrise[961] and a protracted battle takes place in which the Trojans have the upper hand. They are disappointed when the fighting is interrupted by sunset, whereas the Achaians are 'saved' by it.[962]
(3) Within the larger narrative framework of the ferocious fighting around the Achaian ships, Nestor tells of a raid in his youth in which he fought all day from morning to evening.[963]
(4) The centrepiece of the action in the *Iliad*, the fighting for the Achaian ships – almost six whole songs, 2936 verses – in fact comprises *one day's worth of fighting*.[964]
(5) A whole day's fighting (πανημέριοι) until the beacons are lit is represented in a simile about siege operations.[965]
(6) The fighting is expressly said to last all day from sunrise to sunset.[966]
(7) Food and drink are necessary if the warriors are to have the stamina to fight all day long.[967]

---

958 *Il.* 5.302–304 ("one that not two men could carry, such as mortals now are" [trans. Murray]). Cf. 1.271–272, 12.381–383, 12.443–449, 20.285–287. Spear: *Il.* 6.318–319, 8.493–494.
959 The present examination, then, does not address the problem of descriptions of phalanx fighting in the *Iliad*, as suggested by Latacz (1977), but merely regards the cultural implications of stock phrases more or less invariably describing fighting as long-lasting.
960 *Il.* 7.273–282.
961 *Il.* 8.1, 8.53–74.
962 *Il.* 8.485–502.
963 *Il.* 11.735–758.
964 *Il.* 11.1 – 16.777–780.
965 *Il.* 18.207–213.
966 *Il.* 11.1 – 18.239–245.
967 *Il.* 19.164–170.

(8) Apollo has been allowed by Zeus to aid the Trojans in the fighting until evening, showing that for the poet, fighting until evening was a possibility.[968]

Similar instances of extremely long fights are found in Hesiod's *Theogony*, when the gods, and especially Zeus – not to be outdone by mere mortal men – fight the titans for 10 years on end.[969] Then, but not before, the fighting subsides; but until then ἀλλήλοις ἐπέχοντες / ἐμμενέως ἐμάχοντο διὰ κρατερὰς ὑσμίνας.[970]

There are thus quite a few instances in the cream of the epic tradition in which fighting is said to go on unabated for extremely long periods. Given the exceptionally central position of particularly the Homeric poems in all of antiquity, it is hardly far-fetched to claim that this will have had repercussions throughout all subsequent literature. The field of epic influence on historiography is still in need of serious study, but an important work – "Homer und die Geschichtsschreibung" – was furnished in 1972 by Hermann Strasburger, who skillfully pointed out numerous points of similarity in such areas as basic rules and subject treatment between epic and historiography.[971] Several traits which are central to (Ionic) epic poetry, such as ἀλήθεια or ἀτρέκεια, aetiology, a basic objectivity, are all readily assimilated into and employed in historiography; and even the very idea of the heroes' superhuman powers are treated from an essentially historical point of view in that they are described as belonging to a distant past, the 'heroic' age.[972]

Furthermore, the subject matter at the very heart of the *Iliad* is warfare and fighting, death and destruction; and these are the exact subjects that form the core of historiography: Herodotos' *Histories* are thematically arranged to form a crescendo towards book 8 and 9 in which the Persian wars are described; and the theme of warfare as the starting point and backbone of research in contemporary history is arguably crucial to an appreciation of Thucydides' historical work. By this time, the subject matter of warfare – at least from a thematical view-point – may fairly be said to have become canonical in ancient historiography.

The cultural and especially the literary importance of epic in general and of the Homeric poems in particular thus cannot be ignored, and as Strasburger's analysis demonstrates, it has had a profound effect on the notion of historiography from the earliest times. It is therefore a quite reasonable assumption that many basic narrative elements in epic would 'spill over' into historiography as already canonical literary traits, and among them naturally the notion of combat on a 'heroic' scale – including impossibly long battles and fighting all day. In many instances, this may have felt natural and suitable for the subject matter at hand, just as at least until Herodotos, Ionic was the suitable linguistic garb for writing history. Furthermore, the presence of formulaic verses in epic poetry may have played a part in generating a tradition for applying stock phrases to descriptions of standard situations even

---

968 *Il.* 21.229–232.
969 Hes. *Theog.* 629–640, 646–648.
970 Hes. *Theog.* 711–712 ("But until then, they kept at one another and fought continually in cruel war" [trans. Evelyn-White]).
971 Strasburger (1972).
972 Strasburger (1972) 21, 23.

in early Greek prose.⁹⁷³ It is perhaps significant that Herodotos' battles are uniformly long and hard-fought.

Finally, the word μάχομαι itself (and consequently μάχη) may in fact have been felt to convey a sense of length or duration. A quick search in the TLG word index thus reveals that there are a mere 299 instances in the entire Greek literature of μάχομαι used in the aorist aspect, whereas the imperfect or durative aspect of the same verb is represented by no fewer than 5406 instances. In other words, the imperfect of the verb outweighs the aorist fully 18 times in the sources. This in itself proves nothing; but it is a significant piece of circumstantial evidence that to the Greek mind, fighting was perceived as essentially long-lasting to the degree that it embedded itself in the very language. Fighting simply took time.

The upshot of all this is that the routine description of battles lasting 'all day' or similar exaggerations may be heavily influenced by very rigid traditions inherited from epic. This trait may in fact have become a literary convention to the degree that historians more or less consciously felt obliged to accompany any description of battles with diatribes on the length and the grimness of the fighting: at any rate, this should certainly be taken into consideration when assessing the ancient historians' battle accounts. Nor is it even certain that the ancient historians themselves and their public were even *aware* of these tendencies: they may have merely subconsciously felt something lacking from a description that merely stated who won and who lost.

---

973 See especially Havelock (1973a) 362, (1973b) 386–389; Ong (2002²) 39.

5. CONCLUSION

One of the main findings of this study is the fact that hoplites throughout their long period of existence were characterised by far more similarities than differences. A crucial factor in this is the interdependence between weapons and fighting style. Earlier consensus was largely that hoplite weapons antedated the introduction of the hoplite phalanx. This was largely because archaeological finds of weapons can be dated to before 700, whereas the earliest phalanx representations are datable to *c.* 650. At one point, this supposition even led to an attempt at establishing a link between the rise of the Greek tyrants and the apparently emerging phalanx, since the appearance of tyranny in many Greek *poleis* can be dated to C7m. Although the causal link between these phenomena has since been dismantled, it provides an indicator of the pervasiveness of the previously universally established notion that hoplite weapons came before the phalanx.

This perception was turned upside down with Hanson's important article "Hoplite Technology in Phalanx Battle" from 1991, in which he rightly underlined the normal sequence of events in the technology of weapons development, namely that weapons are almost always developed as a precise response to very specific needs arising out of already existing combat conditions. If this notion is correct, as it no doubt is, it has far-reaching consequences for the perceived causality. In this light, the phalanx, in some form or other, must have existed prior to hoplite weapons, since they were invented to fulfil the very specific needs of that type of combat and were difficult if not impossible to employ in other types of fighting. Accordingly, the phalanx must be datable to C8, since we have weapons finds of a decidedly proto-hoplite character from this period.

So why was there felt to be a link between hoplite phalanxes and tyrants anyway, and why, before Hanson, was 650 a seemingly acceptable *terminus post quem?* This probably has to do with the observation that, contrary to the case of hoplite *weapons,* no good representations of phalanxes can be dated earlier than C7m: thus, the weapons must have been introduced first, paving the way for a gradual introduction of phalanx tactics. This is, however, a glaring *argumentum e silentio*: we cannot claim that something did not exist simply because it is not represented in contemporary art. In fact, the daunting task of representing a phalanx even moderately accurately, with all that it entails of perspective, depth and crowd scenes, probably made it a less than appealing prospect to begin with. This, however, we cannot know: suffice it to say that it is unrealistic to demand artistic representations of a phenomenon in order to vouch for its existence at a given time.

If the hoplite phalanx can be shown to have existed quite some time before the rise of the tyrants, however, there is no good reason to suppose that the two phenomena are in any way causally related; or at least only accidentally so. The inter-

5. Conclusion

dependence of hoplite weapons and tactics actually allows us to date the phalanx at least as early as the earliest weapons finds – in C8l – and accordingly we find that the phalanx must be dated back to at least C8. On this interpretation, any remaining putative links between the emergence of the hoplite phalanx and the rise of the Greek tyrants are therefore severed.

On the other hand, on the new dating of the phalanx to C8, the *terminus ante quem* of the hoplite-style phalanx is more or less exactly contemporary with the emerging *polis,* as recent research has shown that the earliest indications of *polis* activity can be dated to this time. This study corroborates that view and opens up important new perspectives for further research in the formation of the early *polis.* How, for example, are *polis* and hoplite activity interconnected, if at all, and what are the consequences for the understanding of social and military structures of the nascent *polis?* What does the notion of an early phalanx mean for our perception of Greek mercenary activity in the Near East in the Archaic period?

The phalanx quite possibly evolved over a long period of time, but what can be established about the formation that appeared in C8? Recent theories about the early phalanx have primarily focused on pictorial representations, where, it is argued, hoplites fought in an open order, singly or in small groups, and essentially as soloists, possibly even moving forth from the ranks to fight duels and falling back as they saw fit. These theories are mainly based on a combination of archaeological and literary evidence. Early iconographical representations of decidedly hoplite warfare seemingly show mêlée fighting between soloists in no discernible order, and even such depictions of phalanxes as there are can, it is argued, be interpreted as representing a more open order than the classical phalanx. This again squares with the battle scenes in the Homeric poems, which are dated to C8. Here, it is argued, the mixture of closed order formations, evoked in many vivid similes, and the (narratologically preponderant) soloist duels between the main characters, moving about as they deem fit, to the point of riding chariots around on the battlefield, are a reasonably faithful 'snapshot' portrait of warfare contemporary with the poems.

There are, however, serious problems inherent in this. Firstly, iconography of phalanx fighting is in fact not so unambiguous. There are certainly quite early representations of 'canonical', closed-order hoplite phalanxes; perhaps most clearly on the Amathus bowl, datable to *c.* 710–675, showing that Greek hoplites had by this time apparently existed for some time – time enough, at any rate, to acquire a reputation sufficient to get them employment with Middle Eastern rulers. The Chigi vase also, despite allegations to the contrary, is a masterful impression of two closed-order hoplite phalanxes; and the fact that other representations of fighting focus on other aspects of battle than the orderly pre-battle ranks cannot be cited in support of the view that early hoplite fighting did not involve phalanxes.

Secondly, the Homeric battle scenes are not realistic portrayals of how battles were fought in C8. Much like other societal and cultural elements in the poems, or indeed the language itself, the representation of warfare is a 'synthetic' amalgam of close-order infantry battle and the narratologically necessary duels between heroes who move uninterrupted in battle as they please, often on chariots. The hero duels and chariots most likely represent an older stratum, the close-order fighting a comparatively recent one. The two cannot easily be reconciled: on the contrary, they serve to further corroborate the notion that the poems are an amalgam on a societal as well as an historical level. The poems as historical sources should accordingly be approached with the utmost diligence, if at all. In my opinion, historians of early Greek warfare would do better to omit Homer as a historical source altogether, since it is exceedingly difficult, if not downright impossible, to extricate the several intertwined layers from each other, let alone date them.

Early lyric poetry, not much later than the supposed date of the Homeric poems, also reveals a great many similarities with hoplite fighting and some notable references to hoplites fighting shield to shield and toe to toe in densely packed phalanxes: above all, the constant references to the all-importance of keeping order and not yielding in the least, and to the embittered nature of fighting at extremely compressed and close quarters. We also find the accompanying creed of winning glory for one's *polis,* rather than oneself – in fact, the individual is erased from their ethics to the point that the loss of one's own life counts as nothing.

Some adherents of the late phalanx development go even further; and there is a tendency to assume that the phalanx was never truly a close-order formation, not even in Classical times. The basis for this supposition is the claim that the phenomenon of *othismos* is to be understood only in a metaphorical sense ('pushing back' or 'driving back', much like in modern military parlance). Central to this idea is the notion that the great pressure generated by mass-shoving cannot be endured for very long and would impede weapons use. In support of this interpretation of *othismos,* it is argued that the sources describe several hoplite battles as lasting a long time. Consequently, there seems to be a conflict between the traditional conception of *othismos* and the sources seemingly testifying to long-fought battles: if *othismos* was in fact close ranks pushing each other forward, it would seem that battles should be fought to their conclusion fairly quickly.

But as I have pointed out, hoplite weapons are developed with the very specific needs of phalanx fighting in mind, and they cannot be used effectively outside this context. It follows from this that Classical hoplites must have fought in closed-order phalanxes – a notion that is corroborated by many sources testifying directly to pushing and to tightly closed phalanxes, fighting shield by shield. These findings help dismantle the so-called 'heretic' view, according to which hoplite battles were rather more neat affairs, fought as sequences of duels between pairs or small groups of hoplites. This, while theoretically possible, would not be using the heavy and clumsy hoplite equipment to its best advantage and would require a far greater amount of skill than was normal in an average citizen militia, comprised of farmers, who typically did not have the leisure to drill on a regular basis. *Othismos,* understood as a collective effort, undertaken by amateur soldiers desperate to get the grisly work done as quickly and decisively as possible, makes perfect sense. It was the ideal way to make most of the highly compact formation, concentrating a maximum of force forward, while simultaneously negating any drawbacks that might arise from lack of weapons prowess. Moreover, the movement patterns of large, agitated crowds give a very plausible idea of how the shoving may have undulated back and forth as individuals felt they were making headway.

This also helps settle the question of battle duration. The phraseology in the sources is riddled with vagueness and 'empty' designations. Despite frequent allegations to the contrary, phrases such as 'for a long time' or 'until darkness' mean nothing without explicit reference. Furthermore, in most cases it is impossible to determine whether the unspecified amount of time is understood to include other phases of battle, such as marching up, arraying, pursuing, erecting trophies, picking up and exchanging the dead: Xenophon's account of Kunaxa indicates that all these corollary phases could take very long time. Some battles were apparently decided before the phalanxes even met, and such factors may have played a role in determining what was a 'long' battle.

Certainly the oppressive weather conditions in the campaigning period and the heaviness and discomfort of hoplite armour influenced the duration of the average battle. In this connexion, *othismos* also was important, as the collective bodily push was a major factor in quickly obtaining a breakthrough and causing the enemy phalanx to disintegrate – and *vice versa*: since the actual fighting was probably normally relatively quickly over, the physical strain on hoplites does not appear to be too great to bear. Furthermore, accounts of battles lasting all day may be ignored, since they are found in Diodoros, who is demonstrably unreliable in this particular respect, or (in one case) in Herodotos, whose account of the battle of Himera is suspect for a number of reasons.

Finally, it is possible that the concept of long battles was simply culturally ingrained in Greek thought. They were an important part of the superhuman (and, indeed, divine) feats performed in epic poetry, which in turn influenced early historiography. It may be that battles, unless expressly stated otherwise, were simply thought of as long-lasting and hard.

In view of this, we may dismiss the notion that battles lasted 'long' simply because some (late) sources say so. A closer inspection of the battles actually described as 'long' do not allow us to say how long battles normally lasted; and in the absence of this, we must make do with the external criteria: the burden of arms and armour, the necessary and time-consuming other phases before and after the actual fighting (which may very well also be comprised in the notion 'battle'), and the extraordinary conditions involved in fighting evidently lasting a very long time, all militate against an assessment of hoplite battles as drawn-out, all-day affairs. Future studies of hoplite battle should therefore stand on much firmer ground with regard to the notoriously difficult question of battle duration.

In summation, there is no good reason to assume the existence of a fundamentally different early hoplite phalanx. The Homeric poems cannot be assessed as portraying an actual, historical society because of the incongruencies, and this applies especially to the warfare. The evidence from Archaic times, both iconographical and textual, is equally uncompelling, showing distinct traces of hoplite warfare. The qualities of hoplite weapons from the earliest times make it unlikely that they should have been developed for open style combat, only to find their logical application centuries later. This is not a realistic conception of early hoplite combat. Furthermore, the fullness of sources from Archaic and Classical times makes it clear that closed-order phalanx tactics was unquestionably the order of the day.

Based on the overall consistency of the sources as well as the typology of weapons, it therefore seems logical to assume that hoplite fighting in 750 and 338 was *essentially* the same. The massed fighting was in all probability a legacy from the earliest times, to which the hoplite's equipment, despite its many considerable drawbacks in solo fighting, was the logical answer. It came about once technology was sufficiently refined and became the norm throughout the Greek world, largely supplanting other warrior types because of its devastating efficiency when employed in densely packed formations. It remained largely unaltered throughout the hoplite era because the type of fighting did not change significantly, allowing only for minor alterations. Hoplites in C4 fulfilled much the same role as their forbears in C8: that of fighting, pushing, killing and dying on the major battlefields of Greece.

At no point were hoplite arms and armour decidedly light or easily manoeuvrable; rendering the hoplite very poorly qualified to act as a soloist outside the phalanx. As we have seen, the central element of the armament, the hoplite shield, was characteristically large, heavy and unwieldy, while other elements of defensive armour were equally problematic in this respect: the Corinthian helmet offered a maximum of protection while seriously impairing vision and hearing, and bronze cuirasses made movement awkward and generated much extra heat inside. While it may be tempting to assess these elements singularly, we should always remember that they worked best in combination; and this in turn indicates that they were developed and continued to be made for a battle environment in which they made sense. Accordingly, it is a reasonable inference that the conditions for hoplite battle predated the weapons, which in turn may be dated back to *c.* 750.

If this analysis is accepted, it has considerable consequences. Above all, the clear link between weapons and fighting style mean that 750 must therefore be the *terminus ante quem* for fighting of the hoplite variety, which has repercussions for the dating of the early *polis,* with which the emergence of hoplite fighting is usually connected. If this connexion holds true, the notion of early *polis* activity at approximately this time is also corroborated.

Attempting to understand the panoply in terms of *pyrrhiche* and *hoplitodromos* is a roundabout way to assess them: the weapons were not employed in these non-military activities because they were easy to handle and wear – quite the opposite, in fact. What was required, and indeed admired, in the hoplite was above all the ability to grit one's teeth and dig in, to shoulder the heavy burden of taking up one's place in the phalanx and slugging it out in the rank and file despite the appalling conditions and the gruesome slaughter in the front ranks; and this is reflected in the cultural trappings surrounding the shield in particular.

The main defining characteristic of the hoplite was the large, concave wooden hoplite shield. Being made of somewhat elastic wood with the grains running horizontally (often criss-crossing in several layers), it was tough and durable yet slightly pliable and thus ideally suited to rather high pressure in combat. It was often covered with a bronze sheathing which increased protection, but regardless of this, a bronze band, wrapped around and encasing the rim, invariably reinforced the shield edge and helped prevent splintering. The twin grip system invented for this purpose, consisting in a bronze armband (*porpax*) and a strap near the shield's edge (*antilabe*), gave the bearer a very secure grip and enabled him to support the shield on three points: hand, elbow and shoulder. The fact that the shield normally weighed 7–8 kg made this necessary, but at the same time limited the frontal range of the shield considerably, since it could only be held at half an arm's length from the body. Unfortunately, this limited range also made the shield's great size – 80–90 cm in diameter – necessary.

The logical basic stance was sideways, thus placing the hoplite behind the shield's centre, allowing him to take full advantage of the whole surface; but the constant movement of the upper body in combat also inevitably exposed the bearer again and again because of the short distance between shield and body. This would break the hoplite's cover every time he leaned forward to deliver a blow, thereby abandoning his 'safe' position behind the shield. This view is corroborated by the few references to shield handling in combat in the sources: the 'Thessalian feint' described by Euripides suggests that the right side was indeed the preferred point of attack.

The fact that the shield was included in sports and dancing does not prove that it was light or easily manoeuvrable. All other items of armour were eventually discarded from the *hoplitodromos* (several hundred metres long), leaving only the shield, which indicates the symbolic nature and underlines the toughness of the race, and the runner would ideally be of a muscular build. The *pyrrhiche* superficially seems very warlike, but a closer analysis shows that it was apparently often danced by women and children and included acrobatic and artistic movements of doubtful actual combat value. The dance was therefore probably only connected with warfare in a rather loose sense in that it contributed to general physical fitness, strength and suppleness. Moreover, it is strange if hoplite techniques could actually be learned by a cultic dance by all lights predating hoplite fighting by a good margin.

The most successful hoplite helmet, the so-called Corinthian helmet, was a type that combined a maximum of protection and aestheticism with restricted hearing and vision and great discomfort, while the ubiquitous crest increased top-heaviness. The presence of a chin-strap indicates that the helmet was not automatically a perfect fit, exacerbating the narrow field of vision. It is significant that later developments of the helmet all had apertures to facilitate hearing or hinged cheek-pieces or both: indeed, the final stages of the hoplite era saw the introduction of the *pilos,* a conical cap of bronze or felt leaving face, neck and ears completely free.

Body armour, either made of bronze, linen or leather, was also constrictive but offered a high degree of protection. Polished bronze cuirasses will have conducted heat and cold, increasing the discomfort. An ill-fitting cuirass (or corslet), according to Xenophon, was next to useless, and this may often have been the case, since costly bronze cuirasses were family heirlooms. Greaves, usually not being secured by thongs, will have been affected by flexing foot and leg muscles, but offered protection for the ankles and lower legs. Hoplites in C6 may have compounded their armour by wearing arm-guards, thigh-guards, foot-guards and the like, virtually encasing themselves in bronze. This must have met with limited success, however, since the practice ceases with C6. Experiments reveal that bronze armour such as that worn by hoplites is extremely secure against penetration from edged weapons. This fact, coupled with the great weight of the armour, signifies that massive frontal protection was the overriding concern for early hoplites, even at the cost of mobility, comfort and freedom of movement.

The hoplite spear, 1.8–2.4 m long and iron-tipped, was an effective weapon if somewhat prone to breaking, which is why it was supplied with a bronze butt-spike that could double as a reserve spear-head. It also improved balance. The fact that additional throwing-spears disappear from vases altogether after C7 shows that, much like the additional body armour in C6, they were less than effective in the long run, just as there is no definitive link between javelins and hoplites in Archaic poetry. A sword was also carried, but the sources seem to indicate that it was normally resorted to only when the spear was broken or lost. The sword could be either the 'traditional' *xiphos* or the slashing *kopis*; but since the latter is only truly effective when used for great, slashing blows, it seems that it was primarily useful when the ranks were broken.

The weight of an entire set of armour is almost impossible to judge accurately, but the lower limit at any given time must have been approximately 15 kg, with an upper limit of no less than 25–30 kg. Armour was certainly both fuller and heavier in the earliest times, and many items were discarded towards C5l. So, if it is accepted that the hoplite phalanx was a close-order formation in the Classical period, it follows *a posteriori* that early Archaic hoplites, with far heavier armour, can by no means have been light or mobile. Furthermore, even if not all Archaic hoplites did wear a full panoply, the file leaders, who would be in the thick of fighting, would be wearing the fullest sets of armour. Therefore, this in itself reveals nothing about the type of fighting, and it certainly does not support the notion of an open Archaic phalanx.

To this should be added the fact that Greek males in antiquity were considerably smaller than average males today: for men some 162 cm tall, the size and total burden of the equipment must have been far greater. This should be taken into consideration in discussions of what was feasible with hoplite arms and armour. The weight and discomfort is further brought out by many references to armour only donned at the last possible moment and to the presence of *hypaspistai* to carry it, if at all possible.

The suggested mobility of hoplites during the Persian wars consists of two references to manoeuvres which can both be found in the supposedly tighter phalanxes of Thucydides and Xenophon; and one of these is a minor detail which is simply inconsistent with the otherwise perfectly logical *technical* details of Herodotos' account of Thermopylai.

The iconographic evidence offered in support of the early open phalanx is essentially ambiguous, since there is no way to determine what the images are supposed to represent, nor what the artist intended, a problem especially with early Greek vase painting, which is often rather crude. It may be a battle which is in some respects similar to what we find in the *Iliad,* and this may be precisely because it is *meant* to reproduce such a scene – or it may be meant to portray contemporary reality, but be influenced by the numerous paintings of mythical battle scenes. Any interpretation ultimately rests on a contemporary reading and can be no more than a guess. One reason for the scarcity of phalanx representations may be the monotony and sheer technical difficulty of painting rows upon rows of men in perspective, coupled with the fact that far more dramatically and emotionally satisfying tableaux can be produced by concentrating the focus on pairs or smaller groups. Moreover, there is again nothing that precludes the view that vase images of smaller hoplite groups fighting do not simply portray later phases of battle such as *trope* and pursuit for the exact reasons stated: the human interest simply is greater.

Further shaking the theory that early hoplites did not fight in close order and that this is the reason for the scarcity of good phalanx representations is the existence of such masterpieces as the Chigi vase and the Amathus bowl in the Cypro-Phoenician style: datable to 710–675, they undoubtedly show Greek hoplites and seemingly even suggest depth perspective in order to represent hoplites behind each other in orderly lines. This, of course, is subject to the same criteria of doubt as other iconographical analysis, but if nothing else, it demonstrates that there are indeed very early representations of tightly packed hoplite phalanxes.

The centrality of the shield in phalanx fighting is borne out by the many references to *rhipsaspides* and the extremely severe punishments visited upon hoplites found guilty of this. Despite allegations to the contrary, Archilochos' admission of throwing away his shield in combat is certainly meant to provoke outrage at his brazen display of indifference, as is especially demonstrated by Kritias' railing against him, in which he lists an entire catalogue of 'timeless' character flaws for which Archilochos was notorious in antiquity.

That forming a closely-packed defensive line is in fact a logical way of exploiting heavy infantry with large double-grip shields is brought out by a comparison with the use of modern police shields. Despite the shield weighing approximately a third of a Greek hoplite shield, it is not used offensively, and individuals are discouraged from venturing forward from the line. This task, if necessary, is left to policemen with lighter equipment and modified (i.e. halved) standard shields, since these can be wielded freely. Shields are employed in a similar, defensive fashion even today by police forces: here, as in the Greek hoplite phalanx, the keywords are cohesion and contiguity; solo action is strongly discouraged. This is an effective way of creating a strong defensive line, blocking any inroads and generating mutual protection.

Likewise, phalanx file width was probably determined by the width of the shield, i.e. some 90 cm. Ideally, shield edges would touch or overlap, as can be seen by Thucydides' remark that phalanxes tended to drift to the right on the march, as every man sought to bring his unprotected side nearer to the safety of his neighbour's shield. Furthermore, a small file width would help protect the phalanx against disintegration as it marched forward: intervals of two metres or more would probably quickly have opened holes in the formation. Macedonian phalanxes seemingly operated with much smaller file widths; and a pike drill manual from the 17th century shows formations of as little as 0.55 m. Phalanx depth was normally eight men (or 'shields'), and exceptions were almost always deeper, either because a massive forward push was sought generated, or because poor terrain dictated it.

The right wing was rightly considered the most dangerous poisition, followed by the left, and generally a feeling of interdependence and mutual trust was cultivated by posting together men who knew each other intimately on an everyday basis: *phylai* and even demes were often the basis of distribution. This had the undesirable side effect of concentrating losses within these same smaller communities, but evidently the teamwork was considered important enough to override such considerations. The general was expected to take his place among the rank and file, leading his men by a good example. Since he was often conspicuously posted on the right wing and in the front rows, this example was infinitely more important than the need to keep the commanding officer out of harm's way, as can be seen from the massive casualty rate among generals. This in turn suggests that hoplite battles were rather uncomplicated affairs from a tactical and strategical viewpoint.

APPENDIX: BATTLE INVENTORY

The following is an inventory of hoplite battle accounts found in the sources. It is important to note that only selected battles have been included: I make no claim to completeness. The historians – and to a certain degree other sources – are strewn with passing references to battles, often only noting who were victorious and who were defeated. Such battle accounts, while important for the sake of statistics, frequency and dating, are of little interest in themselves, and consequently have been left out. Instead, battle accounts have been selected by a set of criteria such as (comparative) narrative fullness and relative contemporaneity (although later sources, such as Plutarch or Polyainos may be cited in support or to verify criteria such as date or names of commanders). Other important factors are the matter of whether they deal with pitched hoplite battles (or at least one hoplite phalanx against different troop types, such as, e.g., Inventory nos. 15, 19, 36 and 37). The information in the Inventory entries has been tabulated under 29 headings as follows:

1. Name and date
2. Time of year
3. Geographic locality
4. Battlefield topography
5. Weather conditions
6. Encampment
7. Combatants
8. Non-combatants
9. Numbers of armed forces
10. Commanders
11. Pre-battle rites/ exhortations
12. Battle array
13. Hoplite indicators:
    ὁπλίτης
    ἐπιβάτης
    φάλαγξ
    ἀσπίς
    ὠθέω / ὠθισμός
    θῶραξ
    κνημίδες
    κόρυς
    δόρυ
    (ὅπλον / -α)
    παραρρήγνυμι/ παράρρηξις
    παράταξις/ παρατάσσω
14. Indicators of other troop types:
    πελτάστης / πέλτη
    τοξότης / τοξεύειν
    ἱππεύς / ἵππος
    ἀκοντιστής
    σφενδονήτης
    γυμνός / ψιλός
15. Mercenaries
16. Description of battle
17. Duration of battle
18. Decisive factors
19. Victor
20. Defeated
21. Pursuit
22. Casualties (victor)
23. Casualties (defeated)
24. Prisoners
25. Trophy
26. Exchange of bodies
27. Miscellaneous
28. Sources
29. Bibliography

Most of these categories should be self-explanatory and are chosen for their immediate interest and relevance for knowledge of the battles. Geographic localities are given with reference to the *Barrington Atlas of the Greek and Roman World* (Princeton 2000) in page and map grid (e.g. Olpai: Barrington 54 D4). Headings 13 and 14 are sublisted with a number of key terms which indicate the participation of hoplites and other troop types respectively: it is assumed that such terms as φάλαγξ, ὠθέω / ὠθισμός or παράρρηξις are sufficient to demonstrate hoplite activity. Other categories have been chosen for their relevance: the question of season and date, for example, may be significant in determining whether the battle in question was long or short in duration. It is reasonable to assume that a battle fought in midsummer, as opposed to one fought in late autumn, will have been limited in duration because of the oppressive heat. More often than not, one or more of these informations are not given by any of the sources for a particular battle. Such missing points are marked with – . Under heading 29 a list of commentaries and studies of the battles in question may be found: again, I make no claim to have listed all articles and references published on the battle in question. The listings are included as a means for the reader of obtaining a quick overview of secondary literature and research on the battles.

(1) AIGITION 426

1. **Name and date:** Aigition 426 (Thuc. 3.97.2)
2. **Time of year:** summer (Thuc. 3.94.1)
3. **Geographic locality:** Aigition in Aitolia (mod. Strouza, Barrington 55 C4)
4. **Battlefield topography:** a forested, rocky mountain terrain (cf. Thuc. 4.30.1) (about Sphakteria)
5. **Weather conditions:** -
6. **Encampment:** the battle was fought mainly as a series of hit-and-run attacks down the slopes towards the town, which the Athenians had taken and held. They were presumably quartered here as well (Thuc. 3.97.2)
7. **Combatants:** Athenians with allies (Akarnanians, Zakynthians, Kephallenians and Korkyraians) and Messenians *vs.* Aitolians (Ophioneans, Eurytanians, Apodotians) (Thuc. 3.94.1–5, 3.95.2)
8. **Non-combatants:** -
9. **Numbers of armed forces:** 120 Athenian ἐπιβάται from the ships are mentioned, Thuc. 3.95.2; these are scarcely identical with the 120 fallen hoplites "in the prime of life, the best men to die from Athens" mentioned at 3.98.4
10. **Commanders:** Athenians: Demosthenes and Prokles (Thuc. 3.94.3, 3.98.4)
11. **Pre-battle rites/exhortations:** –
12. **Battle array:** –
13. **Hoplite indicators**
    ἐπιβάτης: Thuc. 3.95.2
14. **Indicators of other troop types**
    τοξότης / τοξεύειν etc.: Thuc. 3.98.1
    ἀκοντιστής: Thuc. 3.97.2 (ψιλῶν ἀκοντιστῶν), Thuc. 3.97.3 (ἐσηκόντιζον), Thuc. 3.98.1 (id.), Thuc. 3.98.2 (id.)
    γυμνός / ψιλός: Thuc. 3.97.2 (ψιλῶν ἀκοντιστῶν), Thuc. 3.98.2
15. **Mercenaries:** –

16. **Description of battle:** Demosthenes was persuaded by the Messenians accompanying his army to attack the Aitolians on the assumption that they had no fortified strongholds and lived scattered in the countryside. They accordingly conquered Aigition *without* waiting for a much-needed reinforcement of Lokrian light-armed javelin-throwers, and the inhabitants fled to the mountains above the city. From here, the nimble and quick light-armed Aitolians kept harassing the Athenians by charging down the hillsides and pelting them with javelins. When the Athenians charged out to drive them away, they retired, only to attack them once again as they fell back. The Athenians held out for a while, as their corps of archers helped keeping the Aitolians at a distance; but once their captain was killed and they ran out of arrows, panic began to spread (Thuc. 3.96.3–98.4)
17. **Duration of battle:** ἐπὶ πολύ (Thuc. 3.97.3)
18. **Decisive factors:** the suitability of the Aitolians' armament to the terrain and their knowledge of the landscape; the Athenians' lack of sufficient light-armed troops
19. **Victor:** Aitolians
20. **Defeated:** Athenians and allies
21. **Pursuit:** the battle quickly developed into an all-out pursuit. Many hoplites were killed as they turned to run, and the retreat became a disaster: the hoplites fell into ravines and lost their way in the wilderness (their Messenian guide, Chromon, had been killed); many were overtaken by the nimble Aitolians; and the majority got lost in a forest, which the enemy promptly fired. As Thucydides puts it: πᾶσά τε ἰδέα κατέστη τῆς φυγῆς καὶ τοῦ ὀλέθρου τῷ στρατοπέδῳ τῶν Ἀθηναίων (Thuc. 3.98.3)
22. **Casualties (victor):** –
23. **Casualties (defeated):** 120 Athenian hoplites including Prokles; besides "many of the allies" (Thuc. 3.98.4)
24. **Prisoners:** –
25. **Trophy:** –
26. **Exchange of bodies:** the Athenians received their dead from the Aitolians under a truce (Thuc. 3.98.5)
27. **Miscellaneous:** yet again a forewarning about the importance of light-armed troops, particularly in rough terrain. Thucydides notes that the 120 fallen Athenian hoplites were the flower of the army, "in the prime of life"
28. **Sources:** Thuc. 3.97–98
29. **Bibliography**
    Gomme (1956a) 404–408
    Hornblower (1991) 513–514
    Montagu (2000) 65–66
    Pritchett (1991) 47–82

(2) AMPHIPOLIS 422

1. **Name and date:** Amphipolis 422 (Thuc. 5.8.1; Diod. Sic. 12.74.1)
2. **Time of year:** summer, Thuc. 5.1
3. **Geographic locality:** Amphipolis (Barrington 51 B3; detailed map in Gomme [1956b] 654–655)
4. **Battlefield topography:** the banks and nearest environs of the Strymon are called "marshy" by Thuc. 5.7.4, although Amphipolis itself was perched on top of a hill some 152 m.a.s.l. (Gomme [1956b] 648–660)
5. **Weather conditions:** –
6. **Encampment:** Brasidas' army was stationed partly on the hill of Kerdylion, partly in Amphipolis itself (Thuc. 5.6.5.) Polyainos relates that they had negotiated a takeover so as to avoid pitched battle against the desperate garrison (1.38.3)

7. **Combatants:** –
8. **Non-combatants:** –
9. **Numbers of armed forces:** Spartans: 1500 Thracian mercenaries (probably peltasts) and "all the Edonians", peltasts and horsemen (Thuc. 5.6.4); 1000 Myrkinian and Chalkidian peltasts "besides those in Amphipolis" (Thuc. 5.6.4). Further 2000 Greek hoplites and 300 Greek horsemen, from among which Brasidas selected 1500 (only hoplites or all troop types? For discussion, see Gomme [1956b] 636–637). The rest remained inside Amphipolis with Klearidas (Thuc. 5.6.4–5). Later still, he entered Amhipolis and organised a sallying party of only 150 men (for this suspiciously low figure, see Gomme [1956b] ad locc. 5.9.4 and 5.9.7). Kleon's forces: 30 ships with 1200 hoplites and 300 horsemen, and "even more allies" (Thuc. 5.2.1)
10. **Commanders:** Brasidas inside Amphipolis (Thuc. 5.6.3); Kleon for the attacking Athenians (Thuc. 5.6.1)
11. **Pre-battle rites/exhortations:** Brasidas exhorts his motley crew by a speech, emphasising the military superiority of Peloponnesians and Dorians to Ionians, and giving concrete orders (to Klearidas) (Thuc. 5.9)
12. **Battle array:** –
13. **Hoplite indicators**
    ὁπλίτης: Thuc. 5.2.1, 5.6.5, (5.8.3), 5.8.4, 5.10.9,
    (ὅπλον / -α): Thuc. 5.11.1
14. **Indicators of other troop types**
    πελταστής / πέλτη: Thuc. 5.6.4, 5.10.9*bis*, 5.10.10
    ἱππεύς / ἵππος: Thuc. 5.2.1, 5.6.4, 5.6.5, (5.10.2), 5.10.9, 5.10.10,
15. **Mercenaries:** yes: Thracian mercenaries (most likely peltasts) on Brasidas' side (Thuc. 5.6.4)
16. **Description of battle:** Kleon was somewhat over-confident after his and Demosthenes' exploits at Sphakteria, and accordingly ran far too great risks. After Brasidas' withdrawal into Amphipolis, he proceeded northwards along the city wall to the so-called Thracian gate. Here it was reported to him that, contrary to his expectations, ranks of men and horses were visible under the gates, as if prepared for a counterattack. He decided to postpone an attack on the city until such time as reinforcements had arrived, and therefore ordered a very loosely organised retreat, back again along the city walls towards Eion. The left wing retreated in an orderly fashion, whereas the right part exposed their right sides (τὰ γυμνά) to the enemy (on this manoeuvre, cf. Gomme [1956b] 647–648). Brasidas noticed this and seized the opportunity. His 150 men charged out from the "first gate", thus in effect dividing the shocked Athenian force in two. At this point, Klearidas charged out from the Thracian gate with the main force, turning the battle into utter rout. The left wing, apparently farther away from the fray, fled towards Eion without offering any resistance; the right wing, after Kleon was killed by a Myrkinian peltast, retreated to a hilltop and held their ground as long as possible, until they too were forced to beat a hasty retreat by the shower of javelins and the constant Myrkinian cavalry harassment (Thuc. 5.10)
17. **Duration of battle:** -
18. **Decisive factors:** Kleon's fatal misapprehension of the situation and Brasidas' masterly stratagem of charging from two points at once; according to Anderson (1965), Kleon's failure to let the trumpet call for retreat be preceded by clear orders
19. **Victor:** Brasidas, Peloponnesians and Amphipolitans
20. **Defeated:** Athenians and allies
21. **Pursuit:** the Athenian troops retreat along mountain roads; Thucydides' wording seems to indicate that they were pursued by Chalkidian cavalry and peltasts (Thuc. 5.10.10)
22. **Casualties (victor):** about seven (Thuc. 5.11.2)
23. **Casualties (defeated):** about 600 Athenians (Thuc. 5.11.2)
24. **Prisoners:** –
25. **Trophy:** Thuc. 5.10.6, 5.10.12; Diod. Sic. 12.74.2
26. **Exchange of bodies:** the Athenian fallen are returned to them (Thuc. 5.11.2), under a truce (Diod. Sic. 12.74.2)

27. **Miscellaneous:** in Diodoros' trite account of the fighting, there is none of the scorn of Kleon so obvious in Thucydides
28. **Sources:** Thuc. 5.6–11; Diod. Sic. 12.74; Polyaen. 1.38.3
29. **Bibliography**
    Anderson (1965) 1–4
    Ellis (1978) 28–35
    Gomme (1956b) 635–657
    Hornblower (1996) 435–456
    Jones (1977) 71–104
    Mitchell (1991) 170–192
    Montagu (2000) 70
    Nikolaidis (1990) 89–94

(3) CHAIRONEIA 338

1. **Name and date:** Chaironeia 338 (Hyp. fr. 28 Jensen; Plut. *Cam.* 19.8, *Mor.* 259d; Polyaen. 8.40)
2. **Time of year:** Metageitnion 7: either August 2 or september 1, 338 (Plut. *Cam.* 19.8)
3. **Geographic locality:** at Chaironeia in Boiotia, a plateau between 150–300 m.a.s.l. (Diod. Sic. 16.85.2)
4. **Battlefield topography:** a plateau perhaps 300 m.a.s.l.; north-east of and between Chaironeia itself and the Kephissos brook, and somewhat north of Ta Kerata (detailed battle map: Hammond [1964]: 568)
5. **Weather conditions:** -
6. **Encampment:** –
7. **Combatants:** Philip and his Macedon phalanx vs. a coalition of free Greek states, especially Athens and Thebes, but also counting Euboians, Achaians, Corinthians, Megarians, Leukadians and Korkyraians (Diod. Sic. 16.85.1–2; Dem. 18.237; Plut. *Dem.* 17.3)
8. **Non-combatants:** –
9. **Numbers of armed forces:** Coalition army: allies: 15,000 infantry and 2000 horse (Dem. 18.237; Plut. *Dem.* 17.3); Athenians: some 10,000 infantry and 600 horse (Diod. Sic. 16.85.2); Boiotians: 12,000 infantry and 800 horse; mercenaries: 5000, cf. Kromayer (1903) I 188–195. Macedonians: more than 30,000 infantry and 2000 horse (Diod. Sic. 16.85.5)
10. **Commanders:** Athenians: Chares (Diod. Sic. 16.85.7) and Proxenos and Stratokles (Polyaen. 4.2.2, 4.2.8); Thebans: Theagenes (Plut. *Alex.* 12.3, *Mor.* 259d; Polyaen. 8.40). Macedonians: Philip (Diod. Sic. 16.84.2); Alexander (cavalry) (Diod. Sic. 16.86.1; Plut. *Alex.* 9.2)
11. **Pre-battle rites/exhortations:** –
12. **Battle array:** Macedonians faced south-east, Greeks north-west. Philip commanded the right wing of the Macedonian line and entrusted the left to Alexander's cavalry detachment. In between them was the 'general' Macedonian phalanx. The Athenians occupied the left wing, opposite Philip, and the Boiotians their left, with the Sacred Band just opposite Alexander. The light-armed of both sides were probably to the left of Philip, on the slopes (Diod. Sic. 16.86.1–2; Plut. *Alex.* 9.2)
13. **Hoplite indicators**
    φάλαγξ: Polyaen. 4.2.2
    ἀσπίς: Plut. *Dem.* 20.2
    (ὅπλον / -α): Plut. *Pel.* 18.5; *Dem.* 20.1, 20.2; Polyaen. 4.2.2
    παραρρήγνυμι / παράρρηξις: (Diod. Sic. 16.86.3), 16.86.4
    παράταξις / παρατάσσω: –
14. **Indicators of other troop types**
    ἱππεύς / ἵππος: Diod. Sic. 16.85.5

15. **Mercenaries:** employed by the coalition, Kromayer (1903) vol. I 195
16. **Description of battle:** Philip probably moved across the field slantwise, so that his hypaspists reached the Athenians before the rest of his phalanx. He then ordered his hypaspists to withdraw slowly before the Athenian troops "step by step", leading them on. In this process, the phalanx turned at bay and contracted, protecting itself with the sarissai, and marching up on slightly higher ground to give it an edge over the Athenians. Meanwhile, Alexander's cavalry repeatedly charged the Boiotians and particularly the Sacred Band, who kept their ranks. Meanwhile, the rest of the phalanx moved forward and to the left in order to keep up with the pursuing Athenians, and this inevitably opened a gap at the extreme right, through which Alexander's cavalry poured. At this moment, Philip's men countercharged the Athenians, who had become reckless and disorganised in their ill-advised pursuit. They were trapped in a pocket in the hills below the acropolis, and 1000 were killed and another 2000 taken prisoner. The Sacred Band was cut down where they stood by Alexander's men, and the right wing behind them poured through the gap and began rolling up the entire Greek line, which crumbled and fell apart (Diod. Sic. 16.86; Polyaen. 4.2.2, 4.2.7; Plut. *Alex.* 9.2–3, *Dem.* 20.2, *Pel.* 18.5; Paus. 7.6.5; Just. *Epit.* 9.3)
17. **Duration of battle:** it seems like a quick and effective battle – although Diodoros, as usual, claims that it lasted a long time (Diod. Sic. 16.86.2)
18. **Decisive factors:** Philip's superlative strategy, and the fact that the Athenians at least are completely unseasoned for war
19. **Victor:** Philip, Alexander and Macedonians
20. **Defeated:** Greek coalition army
21. **Pursuit:** Hammond (1938) 214 claims that Philip ordered his cavalry not to pursuit, but gives no source
22. **Casualties (victor):** Diod. Sic. (16.86.2) claims that many died on both sides, probably untrue (although *some* are implied by Plut. *Alex.* 9.2)
23. **Casualties (defeated):** at least 1000 Athenians, and "many Boiotians" (Diod. Sic. 16.86.5; Dem. 18.264); furthermore, almost the entire Sacred Band (Plut. *Pel.* 18.5)
24. **Prisoners:** 2000 Athenians and (among these Demades), again, "many" Boiotians (Diod. Sic. 16.86.6–87.2)
25. **Trophy:** erected by Philip (Diod. Sic. 16.86.6)
26. **Exchange of bodies:** Philip returned the bodies after the battle (Diod. Sic. 16.86.6)
27. **Miscellaneous:** there was a tomb on the spot for all who met their death fighting the Macedonians (Paus. 9.9.10), adorned with a stone lion (Paus. 9.40.10, cf. Strabo 9.2.37); and also one for the fallen Macedonians (Plut. *Alex.* 9.2). The Theban commander, Theagenes, when asked, "how far are you going to pursue?" answered, "to Macedonia!" (Plut. *Mor.* 259d; cf. Polyaen. 8.40)
28. **Sources:** Dem. 18.237, 18.264; Diod. Sic. 16.85–86; Plut. *Alex.* 9.2, 12.3, *Cam.* 19.8, *Dem.* 18–20, *Mor.* 259d; Polyaen. 4.2.2, 4.2.7–8; Paus. 7.6.5, 9.9.10, 9.40.10; Strabo 9.2.37; Just. *Epit.* 9.3.9–10
29. **Bibliography**
    Hammond (1938) 186–218
    Kromayer (1903) vol. I 127–195
    Mazzucco (1973) 671–675
    Montagu (2000) 98–99
    Rahe (1981) 84–87
    Roberts (1982) 367–371
    Wankel (1984) 45–53

(4) CORINTH 392

1. **Name and date:** Corinth 392 (Diod. Sic. 14.86.6; Corinthian War)
2. **Time of year:** –
3. **Geographic locality:** between the Long Walls running from Corinth in the south to the harbour Lechaion in the north (Xen. *Hell.* 4.4.7–10; Barrington 58 D2–E2)
4. **Battlefield topography:** the opposing forces were hemmed in by the long walls. Apparently there were in fact trees inside the walls (Xen. *Hell.* 4.4.10). Besides, the Spartans constructed a palisade by the west wall and threw up a ditch in front of it (Xen. *Hell.* 4.4.9). Otherwise, the terrain is very flat and unimpeded
5. **Weather conditions:** –
6. **Encampment:** the Spartan invaders actually construct a palisade around what must be their encampment (Xen. *Hell.* 4.4.9)
7. **Combatants:** a Spartan *mora* with allies from Sikyon and Corinthian exiles *vs*. The Corinthians themselves and their allies from Argos. In Lechaion itself, to the north of Corinth, there is also a Boiotian garrison (Xen. *Hell.* 4.4.8–10)
8. **Non-combatants:** –
9. **Numbers of armed forces:** *C*. 600 Spartans (usu. = one *mora*) (Xen. *Hell.* 4.4.7); the exiled Corinthians number about 150 Xen. *Hell.* 4.4.9. For the rest, we are left in the dark
10. **Commanders:** Spartans: Praxitas (Xen. *Hell.* 4.4.7) and Pasimachos (cavalry, Xen. *Hell.* 4.4.10); Athenian mercenaries: Iphikrates (Xen. *Hell.* 4.4.9)
11. **Pre-battle rites/exhortations:** –
12. **Battle array:** Spartans to the right, Sikyonians centre, and Corinthian exiles left, by the western wall. Opposed to them Corinthians to the left (opposite the Spartans), Argives opposite Sikyonians (centre), and Iphikrates' mercenaries opposite the Corinthian exiles (Xen. *Hell.* 4.4.9)
13. **Hoplite indicators**
ἀσπίς: Xen. *Hell.* 4.4.10
ὠθέω / ὠθισμός: Xen. *Hell.* 4.4.11; here very concreet indeed: the Argives are *crushed to death* in the throng
14. **Indicators of other troop types**
ἱππεύς / ἵππος: Xen. *Hell.* 4.4.10
15. **Mercenaries:** those of Iphikrates (Xen. *Hell.* 4.4.9)
16. **Description of battle:** the Argives easily routed the opposing Sikyonians, chasing them to the harbour and killing a lot in the process; but the Corinthian exiles meanwhile beat back the enemy opposite them – i.e. Iphikrates' mercenaries, all the way to the city walls of Corinth. Of the Spartans we hear nothing, but they must have beaten back the Corinthians to the city itself, since they did not leave the stockade until afterwards. The Spartan hipparmost, Pasimachos, with a few volunteers, rushed to the aid of the hard-pressed Sikyonians, picking up their shields. They were killed. When the victorious Argives returned from the north, the Spartans had left the stockade (broken anyway in the centre by the Argives' onslaught) to aid the Sikyonians, and reformed by the eastern wall. They now charged into the Argives' unprotected right side; and at the same time, the exiles returned from their pursuit southwards. The Argives were trapped between the western wall, the Spartans to the east, the exiles to the south, and the stockade to the north. They were slaughtered by the Spartans and in the general panic also trampled and shoved themselves to death. The Boiotian garrison in the Lechaion was also disposed of in various kinds of street fighting (Xen. *Hell.* 4.4.10–12). Diodoros' account differs wildly from Xenophon's and includes a raid by night on the Lechaion, resulting in its capture by the Spartans (Diod. Sic. 14.86.2–3)
17. **Duration of battle:** strangely, the invaders waited for a whole day, having entered the walls by night, before the Argives rush out to meet them (Xen. *Hell.* 4.4.9). Of the duration of the battle itself, little information is given

18. **Decisive factors:** the fact that the Spartans *waited* for the Argives to return, rather than charge them head on, placed the Argives in a hopeless situation (Xen. *Hell.* 4.4.11)
19. **Victor:** Spartans and Corinthian exiles
20. **Defeated:** Corinthians, Boiotians and Argives
21. **Pursuit:** apart from the Argives' interrupted pursuit before the decisive Spartan flank attack, there was no pursuit as such, but rather a confused jumble of panicking troops and dead bodies around the stairs to the walls (Xen. *Hell.* 4.4.11–12)
22. **Casualties (victor):** –
23. **Casualties (defeated):** Xenophon mentions dead bodies of Argives in heaps, like stones, corn, or timber: the slaughter by the wall must have been enormous (Xen. *Hell.* 4.4.12). Diodoros (14.86.3) says the Spartans killed "not few", and that in a later attack on the Spartans in Lechaion, the combined forces of Athenians, Argives, Boiotians, and Corinthians lost about 1000 soldiers (14.86.4)
24. **Prisoners etc.:** –
25. **Trophy:** –
26. **Exchange of bodies:** the Corinthians and Argives were allowed to bring their dead home from the field (Xen. *Hell.* 4.4.13)
27. **Miscellaneous:** the Argives presented their γυμνά to the enemy, thus "assisting in their own slaughter" (Xen. *Hell.* 4.4.11–12). The Sikyonians could be identified by the sigmas on their shields (Xen. *Hell.* 4.4.10)
28. **Sources:** Xen. *Hell.* 4.4.7–13; Diod. Sic. 14.86; Andoc. 3.18–20
29. **Bibliography**
    Montagu (2000) 86–87
    Thompson (1985) 51–57
    Szemler (1996) 95–104
    Underhill (1900) 136–142

(5) DELION 424

1. **Name and date:** Delion 424 (Thuc. 4.90.1)
2. **Time of year:** winter; possibly November (Thuc. 4.89.1)
3. **Geographic locality:** near Oropos by the Boiotian-Attic border (Barrington 55 F4)
4. **Battlefield topography:** the site of battle is *c.* 10 stades (1.8 km) from Delion (Thuc. 4.90.4) and bordering on Oropia and Boiotia (Thuc. 4.91, 4.99). The battlefield is commanded by a hill which blocks the line of vision between the two armies (Thuc. 4.93.1, 4.96.5). Furthermore, watercourses largely prevent the cavalry contingents from coming to blows (Thuc. 4.96.2)
5. **Weather conditions:** –
6. **Encampment:** the Athenian army is encamped hard by the sanctuary of Apollon at Delion, Thuc. 4.90.1–4; the Boiotians muster at Tanagra (Thuc. 4.91)
7. **Combatants:** Boiotians *vs.* Athenians
8. **Non-combatants:** –
9. **Numbers of armed forces:** Thebans: about 7000 hoplites, 10,000 ψιλοί, 1000 horse and 500 peltasts; Athenians apparently about 7000 hoplites, uncertain cavalry (Thuc. 4.94.1–2). According to Diodoros (12.69.3), Pagondas commanded no less than 20,000 hoplites and 1000 horse
10. **Commanders:** Boiotians: the Theban *boiotarch* Pagondas, son of Aioladas in supreme command; joint commander: Arianthides, the son of Lysimachidas (Thuc. 4.91). Athenians: Hippokrates (Thuc. 4.90)
11. **Pre-battle rites/exhortations:** exhortations were delivered by both Pagondas to the Boiotians (Thuc. 4.92), and by Hippokrates to the Athenians (Thuc. 4.95); somewhat more briefly and with less convincing arguments

12. **Battle array:** Boiotians: right: Thebans and ξύμμοροι (?); Haliartos, Korone, Kopaïs and other peoples from the lake; centre: Thespians, Tanagreans, and Orchomenians. At the extreme end of either wing cavalry and ψιλοί. The Thebans are 25 *shields deep;* the others ὡς ἕκαστοι ἔτυχον (Thuc. 4.93.3–5; Diod. Sic. 12.70.1). Athenians: hoplites eight deep and "equal to the enemy front" (apparently numerically almost equal to the Boiotians) and cavalry at both ends of the line. 300 horse left by Hippokrates at Delion for guarding and ambush purposes (Thuc. 4.93.2, 4.94; Diod. Sic. 12.70.1–2)
13. **Hoplite indicators**
    ὁπλίτης: Thuc. 4.90.4, 4.91.1, 4.93.3
    ἀσπίς: Thuc. 4.96.2
    ὠθέω / ὠθισμός: Thuc. 4.96.2, 4.96.4 (ὠθισμῷ ἀσπίδων)
    (ὅπλον / -α): Thuc. 4.90.4, 4.93.3; Pl. *Symp.* 221a
    παραρρήγνυμι / παράρρηξις: Thuc. 4.96.6
14. **Indicators of other troop types**
    πελτάστης / πέλτη: Thuc. 4.93.3
    ἱππεύς / ἵππος: Thuc. 4.93.2, 4.93.3, 4.93.4, 4.94.1, 4.95.2, 4.96.5; Diod. Sic. 12.70.3; Pl. *Symp.* 221a
    γυμνός / ψιλός: 4.93.3, 4.93.4, 4.94.1
15. **Mercenaries:** –
16. **Description of battle:** the extreme ends of both armies (mainly cavalry and light-armed) never connected due to watercourses (Thuc. 4.96.2). For the hoplites, it was a tough pitched battle with ὠθισμὸς ἀσπίδων, after the Athenians closed with their foes δρόμῳ (Thuc. 4.96.1–2). The Athenian right conquered the Boiotian left as far as the centre. The Thespians were hard pressed and finally encircled and cut down. In the confusion of the encirclement some of the Athenians even killed each other. The Boiotian right likewise gradually pushed back the opposing Athenians. Meanwhile, Pagondas had sent two cavalry τέλη to assist his left. These appeared suddenly and unexpectedly around the hill and spread panic among the Athenians. Hard pressed on both wings, the Athenians fled, some towards Mt. Parnes, others to the sea and Delion and yet others to Oropos (Thuc. 4.96.1–9; Diod. Sic. 12.70.1–3)
17. **Duration of battle:** uncertain. The battle commenced "late in the day" (Thuc. 4.93.1); and later the Athenians were saved by nightfall, which prevented further pursuit (Thuc. 4.96.8; Diod. Sic. 12.70.4)
18. **Decisive factors:** according to Thucydides, the unexpected appearance of Boiotian cavalry around the hill caused a panic in the entire Athenian army (Thuc. 4.96.5–7)
19. **Victor:** Pagondas and Boiotians
20. **Defeated:** Hippokrates and Athenians
21. **Pursuit:** The Boiotian cavalry and particularly the recently arrived Lokrians pursued the Athenians, slaughtering as many as possible, but were ultimately prevented by nightfall (Thuc. 4.96.8; cf. Pl. *Symp.* 221a–c)
22. **Casualties (victor):** not quite 500 (Thuc. 4.101.2)
23. **Casualties (defeated):** not quite 1000 (Thuc. 4.101.2); "many times 500" (Diod. Sic. 12.70.4); including Hippokrates (Thuc. 4.101.2)
24. **Prisoners:** none. According to Diod. Sic. 12.70.5, the booty taken from the fallen Athenians by the Boiotians was sufficient for them to institute the Delia-festival and also to erect public buildings
25. **Trophy:** erected by the Boiotians (Thuc. 4.97.1)
26. **Exchange of bodies:** given back to the Athenians on request seventeen days after the battle (Thuc. 4.101.1). The reason for the delay was a dispute over the Athenians' occupation of sacred ground (Thuc. 4.97–101)
27. **Miscellaneous:** Sokrates distinguished himself at Delion (Pl. *Ap.* 28e, *Lach.* 181b, *Symp.* 221a–c). The Athenians were so confused in the mêlée that they accidentally killed each other; another proof of the fundamental seclusion of any hoplite: as soon as the relative protection and order of the phalanx was lost (here because of the encircling), confusion ensued. Hip-

pokrates was addressing his army in *successive stages,* repeating the speech, and was interrupted by the Boiotians advancing down the hill when he had reached the centre (Thuc. 4.96.1). For the *othismos,* cf. Liv. 30.34.3 and Tac. *Hist.* 2.42

28. **Sources:** Thuc. 4.90–96, 4.101; Diod. Sic. 12.69–70; Pl. *Ap.* 28e, *Lach.* 181b, *Symp.* 221a–c; Plut. *Alc.* 7.6, *Mor.* 581; Cic. *Div.* 1.54; Strabo 9.2.7; Ath. 5.215c–e; Paus. 1.29.13; *IG* I² 68–69, 70; *IG* VII 1888
29. **Bibliography**
Montagu (2000) 69
Gomme (1956b) 558–572
Hornblower (1996) 286–317 (map of Amphipolis 321)
Pritchett (1969) 24–36

(6) EPHESOS 409

1. **Name and date:** Ephesos 409 (Xen. *Hell.* 1.2.6)
2. **Time of year:** perhaps a month after ἀκμάζοντος τοῦ σίτου; about late June or early July (Xen. *Hell.* 1.2.4–7)
3. **Geographic locality:** between the harbour Koressos some 7 km NE of Ephesos, in Ephesian territory (*Hell. Oxy.* 1.1) and the marshes on the other side of the city (Xen. *Hell.* 1.2.7; Barrington 61 E2)
4. **Battlefield topography:** Xenophon mentions the marshes near which the cavalry, *epibatai* and peltasts disembarked and were later defeated (*Hell.* 1.2.7–9).
5. **Weather conditions:** –
6. **Encampment:** –
7. **Combatants:** Ephesians with allied forces of Syracusans and Selinuntines and Tissaphernes' mercenaries (Xen. *Hell.* 1.2.6–8), possibly native Kilbians (*Hell. Oxy.* 1.2 l. 15); according to the Oxyrhynchus historian, Spartans were also present (*Hell. Oxy.* 1.1) vs. Thrasyllos' Athenians (Xen. *Hell.* 1.1.34)
8. **Encampment:** –
9. **Numbers of armed forces:** Athenians: 1000 hoplites, 100 horse, and 5000 sailors equipped as peltasts; all theses on 50 triremes according to Xenophon (*Hell.* 1.1.34; Diod. Sic. 13.64.1); Diodoros says 30. Ephesians and allies: Tissaphernes with στρατιὰν πολλήν (?) and cavalry (*Hell.* 1.2.6); the Ephesians πανδημεί (Diod. Sic. 13.64.1); besides the *epibatai* from 20 Syracusan and 2 Selinuntine ships and from yet another 5 Syracusan ships, recently arrived (Thuc. 8.26; Xen. *Hell.* 1.2.8).
10. **Commanders:** for the Athenians Thrasyllos, according to Xenophon and the Oxyrhynchus historian (Xen. *Hell.* 1.2.6; *Hell. Oxy.* 1.2); according to Diodoros, Thrasybulos (13.64.1). For the Ephesians Timarchos and Possikrates, *Hell. Oxy.* 1–2; however the newly arrived Syracusan ships were commanded by Eukles, the son of Hippon, and Herakleides, the son of Aristogenes (Xen. *Hell.* 1.2.8).
11. **Pre-battle rites/exhortations:** –
12. **Battle array:** Thrasyllos divided his army: hoplites at Koressos, the others at the marshes; probably with the intention of making a a two-pronged attack or a pincer manoeuvre.
13. **Hoplite indicators**
ὁπλίτης: Xen *Hell.* 1.1.34, 1.2.7, 1.2.9; *Hell. Oxy.* 2.1; Diod. Sic. 13.64.1
ἐπιβάτης: Xen. *Hell.* 1.2.7
14. **Indicators of other troop types**
πελτάστης / πέλτη: Xen. *Hell.* 1.2.1, 1.2.7
ἱππεύς / ἵππος: Xen. *Hell.* 1.2.6, 1.2.7; Diod. Sic. 13.64.1
γυμνός / ψιλός: *Hell. Oxy.* 2.1
15. **Mercenaries:** –

16. **Description of battle:** the Ephesian forces routed first the Athenian hoplites, killing about 100 of them in the process, then turned towards the other troops placed near the marshes and put these too to flight (*Hell.* 1.2.9–10)
17. **Duration of battle:** -
18. **Decisive factors:** possibly the failure of Thrasyllos' pincer tactics. Besides, Tissaphernes was on to their plan, and the attack was no surprise to the Ephesians
19. **Victor:** Ephesians
20. **Defeated:** Athenians
21. **Pursuit:** some Athenians, apparently, were able to withdraw along the coastal road in some sort of order, but those who followed the "higher roads" were destroyed (*Hell. Oxy.* 2.2)
22. **Casualties (victor):** –
23. **Casualties (defeated):** about 1300 according to Xenophon (*Hell.* 1.2.9); to Diodoros: 400 in all (13.64.1)
24. **Prisoners etc.:** –
25. **Trophy:** two trophies erected by Ephesians: one near the marshes, and one at the Koressos.
26. **Exchange of bodies:** the Athenians received their dead back under truce (Xen. *Hell.* 1.2.11)
27. **Miscellaneous:** the Ephesians award the Syracusans and Selinuntines the ἀριστεῖα for their valour, *Hell.* 1.2.10
28. **Sources:** Xen. *Hell.* 1.2.6–11; Diod. Sic. 13.64.1; *Hell. Oxy.* 1–3
29. **Bibliography**
    Montagu (2000) 77

(7) HALIARTOS 395

1. **Name and date**: Haliartos 395 (Boiotian War, Xen. *Hell.* 3.5.17)
2. **Time of year:** –
3. **Geographic locality:** hard by Haliartos, a city in Boiotia on lake Kopaïs (Xen. *Hell.* 3.5.17; Barrington 55 E4)
4. **Battlefield topography:** the precipitous slopes of Mt. Helikon are not far south of Haliartos
5. **Weather conditions:** –
6. **Encampment:** –
7. **Combatants:** Spartans and allies (Phokians, Diod. Sic. 14.81.2) *vs.* Haliartians and their allies, Thebans (Xen. *Hell.* 3.5.17–19; Diod. Sic. 14.81.2)
8. **Non-combatants:** –
9. **Numbers of armed forces:** Pausanias' army was 6000 strong (Diod. Sic. 14.81.1); Lysandros had "many men" (Plut. *Lys.* 28.2)
10. **Commanders:** Spartans: Lysander and (later) Pausanias (Xen. *Hell.* 3.5.17); Haliartians/Thebans: not related
11. **Pre-battle rites/exhortations:** –
12. **Battle array:** Plutarch mentions a φάλαγξ ὀρθία on the part of the Spartans (*Lys.* 28.6); otherwise no information
13. **Hoplite indicators**
    ὁπλίτης: Xen. *Hell.* 3.5.19, 3.5.20
    φάλαγξ: Diod. Sic. 14.81.2, Plut. *Lys.* 28.6, 28.10
    (ὅπλον / -α): Plut. *Lys.* 28.6
14. **Indicators of other troop types**
    ἱππεύς / ἵππος: Xen. *Hell.* 3.5.19, 3.5.23
15. **Mercenaries:** –
16. **Description of battle:** in all probability, Lysander was taken by surprise by the swift attack from the Thebans; according to Xenophon, they came from the outside, but Plutarch maintains that they had snuck in just in time before Lysander arrived. At any rate, the Spartans were

driven south towards Mt. Helikon, with the victorious Thebans in hot pursuit. When they began to ascend the slopes, the tables were turned: the Spartan hoplites turned about and routed the Thebans by virtue of their higher position, actually using their spears as javelins. Nevertheless, 1000 Spartans were killed in the battle and pursuit, including Lysander (Xen. *Hell.* 3.5.19–20; Diod. Sic. 14.81.2; Plut. *Lys.* 28.6)

17. **Duration of battle:** –
18. **Decisive factors:** the unexpected attack of the Thebans
19. **Victor:** Thebans (and Haliartians)
20. **Defeated:** Spartans and allies
21. **Pursuit:** a very long pursuit took place, during which some 1000 Spartans and allies were killed – although some of these certainly fell during the actual battle, Plut. *Lys.* 28.6. Nonetheless, the Spartans faced their pursuers once they reached steep ground where the going was difficult and managed to kill more than 200 of the Thebans, who had ventured too far from their own lines (Xen. *Hell.* 3.5.20; Plut. *Lys.* 28.5–6; Diod. Sic. 14.81.2)
22. **Casualties (victor):** more than 200 (Xen. *Hell.* 3.5.20; Diod. Sic. 14.81.2. 300: Plut. *Lys.* 28.6)
23. **Casualties (defeated):** 1000 (Plut. *Lys.* 28.6)
24. **Prisoners:** –
25. **Trophy:** Xen. *Hell.* 3.5.19 (erected by the Haliartians)
26. **Exchange of bodies:** Pausanias was forced to ask the bodies back under truce, rather than risk another battle with the Theban army *and* the recently arrived Athenian relief force. Besides, the bodies lay directly under the wall, making their recovery exceedingly difficult, even if they should win a battle. The request was granted – naturally – but on condition that the Spartans leave Boiotia immediately (Xen. *Hell.* 3.5.23; Diod. Sic. 14.81.3; Plut. *Lys.* 29.1–2)
27. **Miscellaneous:** it is noteworthy that the Spartan hoplites were not above throwing rocks and using their spears as javelins (Xen. *Hell.* 3.5.20)
28. **Sources:** Xen. *Hell.* 3.5.17–25; Diod. Sic. 14.81.1–3; Plut. *Lys.* 28
29. **Bibliography**
    Montagu (2000) 84
    Underhill (1900) 115–117
    Westlake (1985) 119–133

(8) IDOMENE 426/5

1. **Name and date:** Idomene 426/5 (Thuc. 3.112.1)
2. **Time of year:** Thucydidean "winter", Thuc. 3.105.1 (Gomme suggests November for the battle of Olpai: Gomme [1956] 418)
3. **Geographic locality:** around Idomene in Amphilochia (Thuc. 3.112.1; Barrington 54 D3), some 300 m.a.s.l.
4. **Battlefield topography:** Idomene is situated on two hilltops on the slopes of the Amphilochian range. To the east and north there is very rocky mountainous terrain full of χαράδραι (Thuc. 3.112.6). The terrain is entirely unsuited to hoplites
5. **Weather conditions:** –
6. **Encampment:** Demosthenes' advance party camped (ηὐλίσαντο) on the greater of the two hills of Idomene; apparently the Ambrakiots camped on the other (Thuc. 3.112.1)
7. **Combatants:** Ambrakiots vs. Akarnanians and Amphilochians, backed by allied Athenians (Thuc. 3.105)
8. **Non-combatants:** –
9. **Numbers of armed forces:** 3000 Ambrakian hoplites (Thuc. 3.105.1); Demosthenes commanded 200 Messenian hoplites and 60 Athenian archers (Thuc. 3.107.1), apart from the Akarnanians and Amphilochians (Thuc. 3.105.2). Meanwhile, 20 Athenian ships under the command of Aristoteles and Hierophon blockaded Olpai from the sea (Thuc. 3.105.3, 3.107.1, 3.109.1)

10. **Commanders:** Demosthenes has supreme command of the allied forces (Thuc. 3.107.1)
11. **Pre-battle rites/exhortations:** –
12. **Battle array:** no 'array' as such, as the entire operation is a sneak attack by dawn on the Ambrakiots' camp
13. **Hoplite indicators**
    ὁπλίτης: Thuc. 3.112.6
14. **Indicators of other troop types**
    γυμνός / ψίλος: Thuc. 3.112.6
15. **Mercenaries:** Thuc. 3.109.2 (with the Ambrakiots)
16. **Description of battle:** during the night, Demosthenes lead the allied army out to a surprise attack on the camp of the Ambrakiot relief force on the hill of Idomene Elasson under cover of the dark: the Messenians were accordingly sent forward from the allies' camp on the hill of Idomene Meizon to lure the Ambrakiots' sentinels (they also spoke Dorian). The entire Ambrakiot army was then caught unawares in their sleep. Most were killed here, but many fled into the mountains, where Demosthenes had arranged roadblocks and ambush parties (Thuc. 3.110–112)
17. **Duration of battle:** –
18. **Decisive factors:** Demosthenes' cunning stratagem, the intelligence of the Ambrakiot relief force's arrival and the use of Amphilochians, trained in 'guerilla' warfare and with good knowledge of the terrain
19. **Victor:** Amhilochians, Akarnanians, Messenians and Athenians
20. **Defeated:** Ambrakiots
21. **Pursuit:** the Ambrakiots were pursued ruthlessly and according to an elaborate plan: ambushes and traps had been prepared all over the mountains. The nimble Amphilochians, mountain-dwellers, knew the territory well and had no difficulty in overtaking and slaughtering the panicking Ambrakiots in their clumsy hoplite equipment, or else simply chasing them into χάραδραι, where they perished. A few reached the coast on the other side, only to see the Athenian ships cruising off the beach. If Thucydides can be believed, some of them actually swam toward the ships, preferring death at the hands of "fellow Greeks" rather than τῶν βαρβάρων καὶ ἐχθίστων Ἀμφιλόχων (Thuc. 3.112.6–8)
22. **Casualties (victor):** –
23. **Casualties (defeated):** evidently enormous: Thucydides refuses to reveal the number, because it would be "incredible" (Thuc. 3.113.6): indeed, he calls it the greatest disaster to befall a single city in so many days and claims that the Akarnanians and Ambrakiots – had they wished to do so – could have taken Ambrakia with impunity (Thuc. 3.113.6)
24. **Prisoners:** –
25. **Trophy:** Thuc. 3.112.8
26. **Exchange of bodies:** the Ambrakiots actually sent forth a κῆρυξ to ask for a truce to receive their dead; but Thucydides relates the surreal conversation between him and the Ampihilochian representative: he knew nothing of the second battle and therefore would not the believe the magnitude of the disaster. When he finally realised the truth of the matter, he went away in shock, ἄπρακτος καὶ οὐκέτι ἀπῄτει τοὺς νέκρους (Thuc. 3.113.1–5). Presumably, they were subsequently interred by the Amphilochians
27. **Miscellaneous:** –
28. **Sources:** Thuc. 3.105–113; Diod. Sic. 12.60.2–6
29. **Bibliography**
    Montagu (2000) 66–67
    Gomme (1956a) 415–430
    Hornblower (1991) 531–535
    Pritchett (1992) 1–78

(9) KORONEIA 394

1. **Name and date:** Koroneia 394 (Xen. *Hell.* 4.3.16; Polyaen. 2.1.3)
2. **Time of year:** this battle can be fairly certainly dated because of Xenophon's reference to a partial eclipse of the sun which occurred on August 14, 394 (Xen. *Hell.* 4.3.10)
3. **Geographic locality:** the plain of Koroneia between Mt. Helikon and the Kephisos, some 2 km west of lake Kopais (Xen. *Hell.* 4.3.16; Plut. *Ages.* 18.1; Barrington 55 D4)
4. **Battlefield topography:** –
5. **Weather conditions:** *likely* oppressively hot (due to the season), but the sources are silent
6. **Encampment:** Xenophon states that Agesilaos remained (ἔμεινε) somewhere between Pras and Narthakion after his skirmish with the Thessalian cavalry (*Hell.* 4.3.4–9), thus implying a camp. The Thebans and their allies moved out from the Kephisos (Xen. *Hell.* 4.3.16)
7. **Combatants:** allies: Boiotians, Athenians, Argives, Corinthians, Ainianians, Euboians, both Lokrian peoples. Spartans: themselves (at least 900 hoplites, Xen. *Hell.* 4.3.15: one *mora* from Corinth, and half a *mora* from Orchomenos); 2000 *neodamodeis* (cf. Xen. *Hell.* 3.4.2) and "Herippidas' mercenaries" (probably Kyros' mercenaries, cf. *Ages.* 2.11); furthermore local people (Orchomenian and Phokean hoplites; those from Asia Minor, and some that Agesialos picked up on his way through Europe)
8. **Non-combatants:** –
9. **Numbers of armed forces:** not explicitly mentioned; but Xen. *Ages.* 2.7.9 speaks of them as equal. Pritchett (1969) 93 puts them at about 20,000 each
10. **Commanders:** Spartans: Agesilaos; allies: possibly joint commandership (Xen. *Hell.* 4.3.1; Polyaen. 2.1.3)
11. **Pre-battle rites/exhortations:** Agesilaos cunningly concealed the intelligence of the Spartan defeat in the naval battle off Knidos at the hands of Konon and instead reported a victory to his troops. He therefore sacrificed oxen (ἐβουθύτει) and let the army partake in the meal immediately prior to the battle (Xen. *Hell.* 4.3.13–14; Polyaen. 2.1.3)
12. **Battle array:** Agesilaos' Spartan army on the right, the Orchomenians on the left. Allies: Thebans on the right; Argives to the left (Xen. *Hell.* 4.3.16, Plut. *Ages.* 18.1)
13. **Hoplite indicators**
    ὁπλίτης: Xen. *Hell.* 4.3.15
    φάλαγξ: Xen. *Hell.* 4.3.17, 4.3.18, 4.3.20; Plut. *Ages.* 18.4, 19.1
    ἀσπίς: Xen. *Hell.* 4.3.19
    ὠθέω / ὠθισμός: Xen. *Hell.* 4.3.19 (also συμβαλόντες τὰς ἀσπίδας); Plut. *Ages.* 18.2, 18.4
    δόρυ: Xen. *Hell.* 4.3.17; Plut. *Ages.* 18.3
    (ὅπλον / -α): Xen. *Hell.* 4.3.20; Plut. *Ages.* 18.3
    παράταξις / παρατάσσω: Polyaen. 2.1.3
14. **Indicators of other troop types**
    πελταστής / πέλτη: Xen. *Hell.* 4.3.15
    ἱππεύς / ἵππος: Xen. *Hell.* 4.3.15, 4.3.20
15. **Mercenaries:** Herippidas' force (Kyreans) (Xen. *Hell.* 4.3.15, 4.3.17)
16. **Description of battle:** the armies marched against each other; the allies from Mt. Helikon, the Spartans from Kephisos. When they were about 180 m from each other, the Thebans charged. The Kyreans, Ionians, Aiolians, and Hellespontians countercharged at 30 m distance lest they lose impetus. These parties came to blows, and the Argives were routed; indeed, they did not even wait for Agesilaos, but fled beforehand. Just as the Spartans thought themselves victors, they received news that the Thebans had cut through the Orchomenians and were plundering their baggage. Agesilaos at once wheeled about and faced the Thebans (who in their turn were marching back to join their routed allies on Mt. Helikon). Agesilaos might have let them pass to attack their rear but chose to charge them head on. The outcome was a particularly nasty battle; and although some Thebans cut their way through to their own lines, many were killed.

Losses were heavy on both sides. Agesilaos himself was severely wounded (Xen. *Hell.* 4.3.16–20; *Ages.* 2.9–16)
17. **Duration of battle:** after the Spartan 'victory', they withdrew to prepare supper: καὶ γὰρ ἦν ἤδη ὀψέ (Xen. *Hell.* 4.3.20)
18. **Decisive factors:** not really. The outcome was not clear on either side, although the Spartans erected a trophy
19. **Victor:** [Spartans (Polyaen. 2.1.5)]
20. **Defeated:** [Thebans, allies]
21. **Pursuit:** the phrase πολλοὶ [of the Thebans] δ' ἀποχωροῦντες ἀπέθανον (Xen. *Hell.* 4.3.20) seems to point to a pursuit, but it is not certain
22. **Casualties (victor):** 350 (Diod. Sic. 14.84.2)
23. **Casualties (defeated):** more than 600 (Diod. Sic. 14.84.2)
24. **Prisoners:** –
25. **Trophy:** Xen. *Hell.* 4.3.21; Diod. Sic. 14.84.2; Plut. *Ages.* 19.2
26. **Exchange of bodies:** the Thebans asked permission to pick up their dead, and so a truce was concluded (Xen. *Hell.* 4.3.21)
27. **Miscellaneous:** Plutarch (*Ages.* 18.3) mentions 50 body-guards who helped save Agesilaos' life. Xenophon (*Hell.* 4.3.16) says that this battle was unlike any other in his time, perhaps due to the unnatural silence from the hoplites on both sides (4.3.17)
28. **Sources:** Xen *Hell.* 4.3.15–21, *Ages.* 2.9–16; Diod. Sic. 14.84; Plut. *Ages.* 18; Polyaen. 2.1.3
29. **Bibliography**
    Buckler (1996) 59–72
    Montagu (2000) 86
    Pritchett (1969) 85–95
    Underhill (1900) 131–133

(10) KROMNOS 365

1. **Name and date:** Kromnos 365 (Xen. *Hell.* 7.4.20)
2. **Time of year:** –
3. **Geographic locality:** in and outside Kromnos in Arkadia (Xen. *Hell.* 7.4.20; Barrington 58 C3)
4. **Battlefield topography:** like much of Arkadia, rocky, hilly mountain area, between 1500 and 2700 m.a.s.l. Indeed, a hill is mentioned in Xenophon's narrative (Xen. *Hell.* 7.4.22)
5. **Weather conditions:** –
6. **Encampment:** the Spartan garrison was besieged by the Arkadians inside the city of Kromnos (Xen. *Hell.* 7.4.21)
7. **Combatants:** Arkadians *vs.* Spartans (Xen. *Hell.* 7.4.20–21)
8. **Non-combatants:** –
9. **Numbers of armed forces:** the Spartans inside were three *lochoi*, but these apparently never saw action (Xen. *Hell.* 7.4.20)
10. **Commanders:** Archidamos for the Spartans (Xen. *Hell.* 7.4.21); for the Arkadians: not related
11. **Pre-battle rites/exhortations:** –
12. **Battle array:** the Spartans charged down the road leading to Kromnos, and therefore were forced to march in double file, rather than in battle formation. The Arkadians, on the other hand, were fully deployed (Xen. *Hell.* 7.4.22)
13. **Hoplite indicators** –
14. **Indicators of other troop types**
    πελτάστης / πέλτη: Xen. *Hell.* 7.4.22
    ἱππεύς / ἵππος: Xen. *Hell.* 7.4.22
15. **Mercenaries:** –

16. **Description of battle:** Archidamos initially sent the peltasts round to capture a hill outside the stockade, from which they should be able to oust the besiegers, supported by Spartan cavalry. The Arkadians, however, kept their formation and were not in the least perturbed. Archidamos then attacked himself – down the road. The battle was much as could be expected: the Spartans were forced to keep a double file, and accordingly were easily repelled by the otherwise numerically inferior Arkadians with quite heavy losses. The Spartans withdrew and redeployed, preparing for the next assault, when an older man pointed out the pointlessness of the whole business. The parties agreed and parted for the time being. The Arkadian siege, however, was not lifted (Xen. *Hell.* 7.4.22–25)
17. **Duration of battle:** not related, but hardly very long
18. **Decisive factors:** the Spartans faulty deployment in the first place; the Arkadians' steadfastness
19. **Victor:** Arkadians (if any)
20. **Defeated:** Spartans (likewise)
21. **Pursuit:** –
22. **Casualties (victor):** –
23. **Casualties (defeated):** about 30, albeit noble and brave men (Xen. *Hell.* 7.4.23–24)
24. **Prisoners:** –
25. **Trophy:** erected by the Arkadians (Xen. *Hell.* 7.4.25)
26. **Exchange of bodies:** not necessary, since the two parties agreed to leave the field peacefully
27. **Miscellaneous:** this, if anything, demonstrates the effectiveness of a phalanx: the Arkadians were properly deployed and συνασπιδοῦντες (Xen. *Hell.* 7.4.23); and against this, the Spartans could do nothing with their too narrow front line: they were simply unable to sustain the pressure of the oncoming Arkadians
28. **Sources:** Xen. *Hell.* 7.4.20–25
29. **Bibliography**
    Montagu (2000) 92

(11) KUNAXA 401

1. **Name and date:** Kunaxa 401 (Plut. *Artax.* 8.2)
2. **Time of year:** presumably summer
3. **Geographic locality:** the village of Kunaxa, 8.9 km (500 stades) from Babylon (Barrington 91 E5 ?)
4. **Battlefield topography:** very flat and dry; typical plains land
5. **Weather conditions:** probably warm and dry
6. **Encampment:** –
7. **Combatants:** Kyros with indigenous forces and Greek mercenaries vs. Artaxerxes all-Persian army (Xen. *An.* 1.7.10–13)
8. **Non-combatants:** a multitude of camp-followers (merchants, slaves and prostitutes) accompanied at least Kyros' army (Xen. *An.* 1.10.1–6). Diod. Sic. 14.22.4 mentions an ἀχρεῖος ὄχλος in the King's army, left behind in the camp
9. **Numbers of armed forces:** 10,400 hoplites and 2500 peltasts; furthermore, Kyros' Persian forces; some 100,000 soldiers (and about 20 scythe chariots) Diodoros gives 13,000 Greek mercenaries and 70,000 Persians, of which 3000 were cavalry (Diod. Sic. 14.19.7). Artaxerxes' army Xenophon gives as 1,200,000 strong (apart from 200 scythe chariots), of which only 75% or 900,000 and 150 chariots made it in time for the battle – which is of course greatly exaggerated. Furthermore, 6000 cavalry under Artagerses (Xen. *An.* 1.7.10–13). Diodoros (quoting Ephoros) says 400,000 (Diod. Sic. 14.22.2). Cf. especially Plut. *Artax.* 13.3
10. **Commanders:** Kyros (and the several Greek mercenary generals) *vs.* Artaxerxes with his subordinate generals Abrokomas, Tissaphernes, Gobryas and Arbakes (Xen. *An.* 1.2.1–9, 1.7.12)

11. **Pre-battle rites/exhortations:** different types of sacrifices were performed (Xen. *An.* 1.8.15). Furthermore, a password being agreed upon, it passed through the ranks two times (Xen. *An.* 1.8.16); and the Greeks sang a paian and shouted the war-cry ("eleleu") (Xen. *An.* 1.8.17–18; Diod. Sic. 14.23.1)
12. **Battle array:** Greeks on the extreme right (Klearchos to the right, Menon to the left; Paphlagonian cavalry along with the Greek peltasts; Kyros and his bodyguard (600 strong) in the middle; and Ariaios with the other Persians to the left. Of the King's battle order, not much can be gleaned, except for the fact that Tissaphernes cavalry was posted immediately opposite the Greeks (Xen. *An.* 1.8.4–9; cf. Diod. Sic. 14.22.5–6)
13. **Hoplite indicators**
    ὁπλίτης: Xen. *An.* 1.8.9
    φάλαγξ: Xen. *An.* 1.8.17, 1.8.18, 1.10.10; Diod. Sic. 14.22.7, 14.24.1; Plut. *Artax.* 7.4; Polyaen. 2.2.3
    ἀσπίς: Xen. *An.* 1.8.9, 1.8.18
    ὠθέω / ὠθισμός: Plut. *Artax.* 8.5
    θώραξ: Xen. *An.* 1.8.3, 1.8.6, 1.8.26; Diod. Sic. 14.22.6; Plut. *Artax.* 9.3, 11.2
    δόρυ: Xen. *An.* 1.8.18, Plut. *Artax.* 10.3, 11.1
    (ὅπλον / -α): Xen. *An.* 1.10.3, 1.10.16; Diod. Sic. 14.23.3, 14.23.4; Plut. *Artax.* 8.4
14. **Indicators of other troop types**
    πελτάστης / πέλτη: Xen. *An.* 1.8.5, 1.10.7
    τοξότης / τοξεύειν etc.: Xen. *An.* 1.8.9, (1.8.20); Diod. Sic. 14.23.1
    ἱππεύς / ἵππος: Xen. *An.* 1.8.5, 1.8.6, 1.8.9, 1.8.21, 1.10.12, 1.10.13; Diod. Sic. 14.22.2, 14.22.5, 14.22.6
    γυμνός / ψιλός: (Diod. Sic. 14.23.4)
15. **Mercenaries:** the entire Greek army of Kyros is mercenary
16. **Description of battle:** when the two vast armies were about to engage, the Persians opposite the Greeks, turned and fled "at a bowshot's distance", except for Tissaphernes' cavalry, which passed right through the Greek peltasts. The Greeks, elated, took up pursuit and were carried far away from the actual action. The scythe chariots driving in between them could also be avoided by sidestepping (loose formation?). Kyros, however, saw Artaxerxes and rode forth to engage him with his bodyguard. Artaxerxes' bodyguard counted 6000 men, so naturally the attempt went awry, and Kyros was killed, although he wounded Artaxerxes in the chest. Meanwhile, Greek sentinels successfully drove out Persian marauders (and Tissaphernes) from their camp. The Greek main force wheeled about to face the King, but the two armies awkwardly passed by each other, assuming their original positions. The Greeks charged again, driving the King's troops away and were masters of the field without having struck a single blow – but Kyros was dead, and their camp plundered (Xen. *An.* 1.8.8–29, 1.10.1–19)
17. **Duration of battle:** until about nightfall (Xen. *An.* 1.10.16)
18. **Decisive factors:** the death of Kyros determined the outcome; tactically, however, the Greeks kept the field
19. **Victor:** difficult to decide
20. **Defeated:** –
21. **Pursuit:** the Greeks pursued the main Persian army all over the plains (Xen. *An.* 1.8.8–29, 1.10.11–15)
22. **Casualties (victor):** of the Persian troops, Diodoros (14.24.6) says that some 3000 were killed; of the Greeks – one wounded by an arrow (Xen. *An.* 1.8.20; cf. Diod. Sic. 14.24.6); but since there was no actual contact between the armies as such, it cannot have been much
23. **Casualties (defeated):** Diodoros (14.23.4, 14.24.5) says that more than 15,000 were killed in the pursuit; but if so, Xenophon is oddly silent about this
24. **Prisoners:** –
25. **Trophy:** erected by Klearchos' men (Diod. Sic. 14.24.4)
26. **Exchange of bodies:** –
27. **Miscellaneous:** –

28. **Sources:** Xen. *An.* 1.8, 1.10; Diod. Sic. 14.22–24; Plut. *Artax.* 7–13; Polyaen. 2.2.3
29. **Bibliography**
    Bigwood (1983) 340–347
    Ehrhardt (1994) 1–4
    Gugel (1971) 241–243
    Lendle (1966) 429–452
    Montagu (2000) 82
    Wylie (1992) 119–134

(12) KYNOSKEPHALAI 364

1. **Name and date:** Kynoskephalai 364 (Plut. *Pel.* 32.2)
2. **Time of year:** shortly after July 13, 364 when there was a eclipse of the sun (Diod. Sic. 15.80.2; Plut. *Pel.* 31.2)
3. **Geographic locality:** in the Kynoskephalai mountains (or hills) in Thessaly (Diod. Sic. 15.80.4; Plut. *Pel.* 32.1–2; Barrington 55 D2), between 610 and 1500 m.a.s.l.
4. **Battlefield topography:** terrain with very steep hills (Plut. *Pel.* 32.2)
5. **Weather conditions:** –
6. **Encampment:** Pelopidas' men camped opposite Alexanders' positions, Diod. Sic. 15.80.5
7. **Combatants:** Pelopidas and Thebans and allied Thessalians vs. Alexander of Pherai and his mercenaries (Diod. Sic. 15.80.1; Plut. *Pel.* 32.5)
8. **Non-combatants:** –
9. **Numbers of armed forces:** Pelopidas commanded 7000 Thebans (Diod. Sic. 15.80.2), although Plutarch says he abandoned the Thebans (except for 300 foreign horse) on account of the eclipse, and took command of the Thessalian league, Plut. *Pel.* 31.2–3. Alexander is said to have had 20,000 men (Diod. Sic. 15.80.4); Plutarch merely says "more than twice as many [as Pelopidas' Thessalians]", Plut. *Pel.* 32.1
10. **Commanders:** Pelopidas (Thebans/Thessalians) (Diod. Sic. 15.80.1; Plut. *Pel.* 31.1). Alexander (of Pherai; mercenary army) (Diod. Sic. 15.80.1; Plut. *Pel.* 31.1)
11. **Pre-battle rites/exhortations:** –
12. **Battle array:** Alexander had seized the higher key positions (= hills), and Pelopidas had to make do with such positions opposite as he could (Diod. Sic. 15.80.4; Plut. *Pel.* 32.2–3)
13. **Hoplite indicators**
    ὁπλίτης: Plut. *Pel.* 32.3
    φάλαγξ: Plut. *Pel.* 32.7
    ἀσπίς: Plut. *Pel.* 32.3
    ὠθέω / ὠθισμός: (Plut. *Pel.* 32.4)
    δόρυ: (Plut. *Pel.* 32.6), 32.7
    (ὅπλον / -α): Plut. *Pel.* 32.7
14. **Indicators of other troop types**
    ἱππεύς / ἵππος: Plut. *Pel.* 31.3, 32.2, 32.3, 32.4, 32.7
15. **Mercenaries:** Alexander's soldiers are mercenaries (Plut. *Pel.* 32.5–6)
16. **Description of battle:** Pelopidas' cavalry first conquered the plain, but his infantry's attempts to storm Alexander's positions on the hilltops were repealed with comparative ease, many of his men being harassed with missiles and killed in the process. He therefore called back the cavalry and ordered them to attack the stronger parts of the enemy infantry, while he himself ran out to join the fray on the hillsides. He inspired his men to further attacks, and backed by the cavalry, returning from pursuit, forced the enemy to retreat after successfully repealing two-three attacks. Their collapse was imminent, when Pelopidas spotted Alexander and recklessly charged him, being killed by his guards in the process before the reinforcements could keep up with him (Diod. Sic. 15.80.4–5; Plut. *Pel.* 32.2–7)

17. **Duration of battle:** –
18. **Decisive factors:** according to both Diodoros, the inspiring generalship of Pelopidas (Diod. Sic. 15.80.5; Plut. *Pel.* 32.5–6)
19. **Victor:** Pelopidas and Thebans/Thessalians
20. **Defeated:** Alexander
21. **Pursuit:** the Thessalian infantry and cavalry routed the remains of Alexander's phalanx and pursued them over a great distance, strewing the ground with bodies (Diod. Sic. 15.80.6; Plut. *Pel.* 32.7)
22. **Casualties (victor):** –
23. **Casualties (defeated):** Plutarch says 3000 fallen, but this may be in the rout and pursuit alone (Plut. *Pel.* 32.7)
24. **Prisoners:** –
25. **Trophy:** -
26. **Exchange of bodies:** –
27. **Miscellaneous:** another case of *hopla* meaning defensive 'armour' (Plut. *Pel.* 32.7)
28. **Sources:** Diod. Sic. 15.80; Plut. *Pel.* 31–32
29. **Bibliography**
    Montagu (2000) 93
    Pritchett (1969) 133–144

(13) KYZIKOS 410

1. **Name and date:** Kyzikos 410 (Xen. *Hell.* 1.1.14; Polyaen. 1.40.9)
2. **Time of year:** in the early spring (ἤδη τοῦ χειμῶνος λήγοντος, Diod. Sic. 13.49.2)
3. **Geographic locality:** the city of Kyzikos on the shore of Arktonnesos in Propontis (Diod. Sic. 13.49.4; Barrington 52 B3–4)
4. **Battlefield topography:** such as may be expected from a battle actually fought on a beach
5. **Weather conditions:** according to Xenophon and Plutarch, the battle began at a bright dawn after a night of thunder and heavy rainfall (Xen. *Hell.* 1.1.16; Plut. *Alc.* 28.3)
6. **Encampment:** the Spartans, under command of Mindaros, pitched camp in a besieging effort around Kyzikos (Diod. Sic. 13.49.4). The Athenians spent a night on Prokonnesos, then the generals disembarked the soldiers in Kyzikenian territory (Diod. Sic. 13.49.6)
7. **Combatants:** Spartans and their mercenary allies, financed with Persian money (Xen. *Hell.* 1.1.14; Diod. Sic. 13.50.4), and also Syracusans (Xen. *Hell.* 1.1.18) *vs.* Athenians (Xen. *Hell.* 1.1.11–13; Diod. Sic. 13.49.2)
8. **Non-combatants:** –
9. **Numbers of armed forces:** impossible to say, since the sources do not reveal the number of men, but only of ships: Xenophon puts it at 86 ships (Xen. *Hell.* 1.1.13). We may tentatively, then, suggest something like 86 × 26 = 2236 hoplites (or epibatai) from the Athenian ships (cf. Sybota island, inv. no 35.)
10. **Commanders:** for the Spartans: Mindaros and Pharnabazos (Xen. *Hell.* 1.1.14; Diod. Sic. 13.49.2, 13.49.4; Polyaen. 1.40.9), and also Klearchos (Diod. Sic. 13.51.1); Athenians: Alkibiades, Theramenes and Thrasybulos (Xen. *Hell.* 1.1.11–12; Diod. Sic. 13.50.1; Polyaen. 1.40.9). Also, a Chaireas is mentioned (Diod. Sic. 13.51.2)
11. **Pre-battle rites/exhortations:** –
12. **Battle array:** difficult to extricate from Diodoros' implausible account; but it appears at least that the Athenian squadron was divided into three separate wings, each assigned a different task: Alkibiades, with the first landing party, commanded 20 ships (Xen. *Hell.* 1.1.18; Diod. Sic. 13.50.5); Trasybulos and Theramenes joining the fighting later.
13. **Hoplite indicators**
    ἐπιβάτης: Diod. Sic. 13.50.7, 13.51.2

παραρρήγνυμι / παράρρηξις: Diod. Sic. 13.51.4
14. **Indicators of other troop types**
τοξότης / τοξεύειν etc.: Diod. Sic. 13.51.2
ἱππεύς / ἵππος: Diod. Sic. 13.51.7
15. **Mercenaries:** yes: those of Pharnabazos (Diod. Sic. 13.51.1, 13.51.4)
16. **Description of battle:** Alkibiades with the men of his twenty ships forced his way through the Spartan lines, pursuing the enemy ships ashore. Here, he disembarked the troops and engaged Mindaros' men in a land battle fought on the beach. However, more Peloponnesians and Pharnabazos' mercenary soldiers under Klearchos came to the aid of Mindaros' troops, which in turn made Thrasybulos disembark the *remaining* epibatai (Diod. Sic. 13.50.7), urging Theramenes to join in as well. Thrasybulos' troops – epibatai and archers – aided those of Alkibiades, but are surrounded by Pharnabazos' men. As they were about to despair of rescue, Theramenes appeared with his own men and the infantry of Chaireas (?). After a hard-fought battle, Pharnabazos' mercenaries retreated, and the Peloponnesians followed suit; and the combined forces of Theramenes and Thrasybulos came to aid Alkibiades. Mindaros, however, divided his forces and continued fighting. Finally, after Mindaros was killed, the Peloponnesians and mercenaries were routed (Diod. Sic. 13.50.4–51.6). Polyainos (1.40.9) says that Theramenes cut off Mindaros' route of retreat. He then tried to escape towards Kleroi (?) on Kyzikenian territory, but was prevented by Pharnabazos' camp, apparently a tough obstacle. Meanwhile, Alkibiades was engaged in towing off the deserted ships on the shore; and those who attempted to disembark were cut down by Pharnabazos' cavalry. Finally, Mindaros fell, and Alkibiades was victorious (Diod. Sic. 13.49.2–51.8; Xen. *Hell.* 1.1.11–18; Plut. *Alc.* 28; Polyaen. 1.40.9)
17. **Duration of battle:** Diodoros says it lasts for a long time (13.51.4)
18. **Decisive factors:** possibly the Spartans had lost courage because of the disaster with the ships, a masterstroke on the part of Alkibiades
19. **Victor:** Athenians and allies
20. **Defeated:** Spartans and allies
21. **Pursuit:** the Athenians pursued the beaten Spartans "for a distance" (μέχρι ... τινος), but return to the ships due to the threat from Pharnabazos' cavalry (Diod. Sic. 13.51.7)
22. **Casualties (victor):** –
23. **Casualties (defeated):** many, according to Plutarch (*Alc.* 28.6). Diodoros simply says that there was "great slaughter" (Diod. Sic. 13.50.6)
24. **Prisoners:** many prisoners were taken by the Athenians (Diod. Sic. 13.51.8)
25. **Trophy:** the Athenians erected two trophies, one for the naval cation (on the island of Polydoros), and one for the land battle, "where they first routed the enemy" (Diod. Sic. 13.51.7)
26. **Exchange of bodies:** –
27. **Miscellaneous:** –
28. **Sources:** Diod. Sic. 13.49.2–51.8; Xen. *Hell.* 1.1.11–18; Plut. *Alc.* 28; Polyaen. 1.40.9
29. **Bibliography**
Andrewes (1982) 19–25
Montagu (2000) 77
Underhill (1900) 310–311

(14) LAODOKEION 423

1. **Name and date:** Laodokeion 423
2. **Time of year:** winter (Thuc. 4.134.1)
3. **Geographic locality:** it is not known, but the names given by Thucydides seem to fit in with the vicinity of Megalopolis (Barrington 58 C3)
4. **Battlefield topography:** –
5. **Weather conditions:** –

6. **Encampment:** the Tegeans camped near the battlefield *after* the battle (Thuc. 4.134.2)
7. **Combatants:** Tegeans against Mantineans and their allies (Thuc. 4.134.1)
8. **Non-combatants:** –
9. **Numbers of armed forces:** –
10. **Commanders:** –
11. **Pre-battle rites/exhortations:** –
12. **Battle array:** –
13. **Hoplite indicators** –
14. **Indicators of other troop types** –
15. **Mercenaries:** –
16. **Description of battle:** the usual occurrence in hoplite battles: both defeat the wing in front of them and rout them (probably left defeated by right), and both therefore claim the victory and erect trophies (Thuc. 4.134.1–2)
17. **Duration of battle:** apparently until nightfall (Thuc. 4.134.2)
18. **Decisive factors:** –
19. **Victor:** –
20. **Defeated:** –
21. **Pursuit:** –
22. **Casualties (victor):** quite a few, Thuc. 4.134.2
23. **Casualties (defeated):** as above
24. **Prisoners:** –
25. **Trophy:** Thuc. 4.134.1, 4.134.2 – erected by *both* Mantineans and Tegeans
26. **Exchange of bodies:** –
27. **Miscellaneous:** not only do both parties erect trophies; they also send spoils to Delphi, Thuc. 4.134.2
28. **Sources:** Thuc. 4.134.1–2
29. **Bibliography**
    Montagu (2000) 70
    Hornblower (1996) 415–418

(15) LECHAION 390

1. **Name and date:** Lechaion 390 (Xen. *Hell.* 4.5.10)
2. **Time of year:** early summer (Konecny [2001] 95)
3. **Geographic locality:** in the coastal area between Corinth, Lechaion, and the Corinthian Gulf (Xen. *Hell.* 4.5.14–17; Barrington 58 D2)
4. **Battlefield topography:** somewhat flat, offering a good overview from Corinth (Xen. *Hell.* 4.5.13)
5. **Weather conditions:** apparently clear weather, as might be expected from early summer in Greece. A few days earlier, Agesilaos' troops had been surprised by a thunderstorm (Xen. *Hell.* 4.5.4). Consequently, the sky has probably been bright and the weather fine
6. **Encampment:** –
7. **Combatants:** a spartan *mora* with a cavalry *mora* vs. Iphikrates' mercenary peltasts and Kallias' Athenian hoplites (Xen. *Hell.* 4.5.12–13)
8. **Non-combatants:** –
9. **Numbers of armed forces:** Spartan *mora* of hoplites: 600; cavalry *mora*: not related, probably 100–120 strong (Konecny [2001] 93). The number of Iphikrates' peltasts is hard to define, but it may have been anything between 1000 and 4000 strong. Kallias' hoplites: unknown; may have been about 1000. Xenophon himself gives the number of hoplites and peltasts in Corinth as "many" (Xen. *Hell.* 4.5.12–13)

10. **Commanders:** unnamed Spartan polemarch and hipparmost *vs*. Iphikrates and Kallias (Xen. *Hell.* 4.5.11, 4.5.13)
11. **Pre-battle rites: exhortations:** –
12. **Battle array:** none as such; no pitched battle
13. **Hoplite indicators**
    ὁπλίτης: Xen. *Hell.* 4.5.11, 4.5.12, 4.5.13, 4.5.14, 4.5.15, 4.5.17; Plut. *Ages.* 22
14. **Indicators of other troop types**
    πελτάστης / πέλτη: Xen. *Hell.* 4.5.12, 4.5.13, 4.5.14, 4.5.15, 4.5.16; Diod. Sic. 14.91.3; Plut. *Ages.* 22.3
    ἱππεύς / ἵππος: Xen. *Hell.* 4.5.12, 4.5.16, 4.5.17
15. **Mercenaries:** Iphikrates' peltasts were mercenaries (Plut. *Ages.* 22.3), although not called that by Xenophon
16. **Description of battle:** Iphikrates arranged an ambush for the Spartans returning from their escort service for the Amyklaians: they charged the Spartans, wounding some and killing others, then quickly retreating again. The pursuing Spartan hoplites were roughly handled on *their* flight back to their own lines. The polemarch was powerless to prevent the peltasts' repeated attacks; and the arrival of their cavalry was not much comfort, since they were unable to organise a concerted defence against the attacks, but merely followed the hoplites charging out. The Spartans made a last stand on a hillock some 350 m from the sea and 3 km west of Lechaion, when they suddenly saw Kallias' hoplites approaching. Panic set in, and they fled to the sea. Those who were escorted by cavalry managed to escape over land, some made it to the ships sent out from Lechaion, but most were killed (Xen. *Hell.* 4.5.14–17)
17. **Duration of battle:** hard to tell, but probably not long
18. **Decisive factors:** the Spartans' negligence of the superior peltast forces; also, the poor performance of the cavalry
19. **Victor:** Iphikrates and Kallias
20. **Defeated:** Spartans
21. **Casualties (victor):** probably next to nothing; at any rate, Xenophon does not mention any
22. **Casualties (defeated):** 250 (Xen. *Hell.* 4.5.17); but this seems too low (cf. 4.5.7–8 and Diod. Sic. 14.91.2; Plut. *Ages.* 22.2; Dem. 13.22; Din. 1.75; Nep. *Iph.* 11.2.3)
23. **Prisoners etc.:** –
24. **Pursuit:** almost entirely a pursuing action after the initial skirmishes; many Spartans were killed in the progress (Xen. *Hell.* 4.5.15–17)
25. **Trophy:** –
26. **Exchange of bodies:** –
27. **Miscellaneous:** something is related of the peltasts' tactics against hoplites: they attack at an oblique angle, trying to get at the hoplites' γυμνά (Xen. *Hell.* 4.5.16)
28. **Sources:** Xen. *Hell.* 4.5.11–17; Diod. Sic. 14.91.2; Plut. *Ages.* 22.1–4; Dem. 13.22; Din. 1.75; Nep. *Iph.* 11.2.3.
29. **Bibliography:**
    Konecny (2001) 97–127
    Montagu (2000) 87
    Underhill (1900) 146–147

(16) LEUKTRA 371

1. **Name and date:** Leuktra 371 (Xen. *Hell.* 6.4.4; Din. 1.73; Polyaen. 2.3.2)
2. **Time of year:** Plutarch gives the date as the fifth of Hekatombaion (*Ages.* 28.5). This corresponds roughly to July 8, 371
3. **Geographic locality:** the plain by Leuktra, near Thespiai (Xen. *Hell.* 6.4.4; Diod. Sic. 15.53.2; Barrington 55 E4)
4. **Battlefield topography:** the Spartans and the Thebans were camped on hills opposite each

other, οὐ πολὺ διαλείποντες (Xen. *Hell.* 6.4.4). Between the two armies, the ground was suitable for cavalry fighting on the πεδίον (Xen. *Hell.* 6.4.10)

5. **Weather conditions:** *probably* warm, with a clear sky, cf. pt. 2 above
6. **Encampment:** Kleombrotos marched inland from the coast and pitched camp at Leuktra; the Thebans on top of a hill "just opposite" (Xen. *Hell.* 6.4.4)
7. **Combatants:** Thebans and allied Boiotians (among them Thespians: Polyaen. 2.3.3); Spartans and allies (Xen. *Hell.* 6.1.1, 6.4.4)
8. **Non-combatants:** a number of Boiotian non-combatants (τῶν τὴν ἀγορὰν παρεσκευακότων καὶ σκευοφόρων τινῶν καὶ τῶν οὐ βουλομένων μάχεσθαι) were leaving the ranks just as the battle was about to commence, but were driven back on to the Boiotian phalanx (Xen. *Hell.* 6.4.9). Polyainos 2.3.3 mentions Thespians being allowed to leave their battle stations and weapons prior to the battle
9. **Numbers of armed forces:** four Spartan *morai* were with Kleombrotos, amounting to *c.* 2400 hoplites (Xen. *Hell.* 6.1.1), and only 700 of these actually Spartiates (Xen. *Hell.* 6.4.15). Furthermore, an equal amuont of unspecified allied troops (Xen. *Hell.* 6.1.1). Polyainos, however, says some 40,000 in all (2.3.8). Plutarch gives 10,000 hoplites and 1000 cavalry (*Pel.* 20.1), probably correctly. The Thebans probably amounted to some 3000 of the 6000 Boiotian hoplites (?) present (Diod. Sic. 15.52.2). Diodoros however, is not clear as to the nature of the 6000 mentioned. After the battle, 1500 Thessalian infantry and 500 horse joined the Thebans (Diod. Sic. 15.54.5; Xen. *Hell.* 6.4.20)
10. **Commanders:** Spartans: Kleombrotos; Thebans: Epameinondas, Pelopidas (Sacred Band of 300: Diod. Sic. 15.51.2, 15.52.2) (Xen. *Hell.* 6.4.3; Polyaen. 2.3.2)
11. **Pre-battle rites/exhortations:** apparently, sacrifices were performed: the Thebans received favourable omens from their offerings, the Spartans, on the other hand, did not (Xen. *Hell.* 6.4.7). Plutarch has a strange story about a mare being sacrificed on the eve of battle (*Pel.* 22)
12. **Battle array:** the Spartans occupied their own right, each *enomotia* three abreast, making the depth 12 (Xen. *Hell.* 6.4.12, 6.4.14). On the other side, the main Theban force are on the left, as deep as 50 (Xen. *Hell.* 6.4.12). Stylianou suggests that the Thebans actually formed up a mere 25 deep, placing the Sacred Band right behind the main force (the intention being to charge in two successive waves: the Sacred Band were to pin down the Spartans and lock their position, the main force then determining the outcome). This would be enough to ensure a massive overweight and still leave plenty of hoplites free to use their weapons and agrees well enough with former Theban 25-deep-formations (Xen. *Hell.* 4.2.18; Thuc. 4.93.4). This would produce an impression of a phalanx 50 deep to the Spartans. Now, 2300 or 2400 hoplites arranged 12 deep would present a front 190 wide. Assuming the Theban forces to consist of 3000 men, a formation 25 deep would produce a front line of 120 men, thus not making too great a difference between the two phalanxes (Stylianou [1997] 401–403).
13. **Hoplite indicators**
    ὁπλίτης: Xen. *Hell.* 6.4.13; Plut. *Pel.* 20.1; Polyaen. 1.10, 2.3.3
    φάλαγξ: Xen. *Hell.* 6.4.10, 6.4.12; Diod. Sic. 15.55.2; Plut. *Pel.* 23.1, 23.4; Polyaen. 2.3.3
    ἀσπίς: Xen. *Hell.* 6.4.12; Diod. Sic. 55.5.5 (συνασπίζοντας)
    ὠθέω / ὠθισμός: Xen. *Hell.* 6.4.14
    (ὅπλον / -α): Xen. *Hell.* 6.4.14; Polyaen. 2.3.3
14. **Indicators of other troop types**
    πελτάστης / πέλτη: Xen. *Hell.* 6.4.9
    ἱππεύς / ἵππος / ἱππικόν: Xen. *Hell.* 6.4.9, 6.4.10, 6.4.12, 6.4.13; Plut. *Pel.* 20.1
15. **Mercenaries:** of the Spartans, Xen. *Hell.* 6.4.9
16. **Description of battle:** the Spartan king Kleombrotos charged out before his army was ready; and at first the two cavalry forces skirmished before the lines. The vastly superior Theban cavalry routed the Spartans who were sent floundering back onto their own lines. This may be what resulted in the Spartan μηνοειδὲς σχῆμα of Diodoros (15.53.3). Kleombrotos then advanced into the cavalry before his army was aware of his move, and apparently tried to let his right perform a *kyklosis* (Plut. *Pel.* 23.1–2), but was struck by Pelopidas' (Nep. *Pel.* 4.2) Sacred

Band rushing forward to prevent this very move. For a time the battle was even as proved by the fact that the Spartans succeeded in picking up their dead king, even though a fierce fight ensued around his body (Diod. Sic. 15.55.5). Finally the overwhelming weight of the Thebans (led diagonally across the battlefield under the command of Epameinondas) began to tell. It became increasingly difficult for the Spartans to keep their lines intact and ultimately they were swept off the field, fleeing to their camp. The Spartan left and their allies saw the fate of their right wing and the king, and fled also. According to Polyainos (2.1.13), many Spartans even threw their shields. Apparently the other Boiotians and the Spartans' allies hardly came to blows at all; indeed Diodoros 15.55.2 claims that they were instructed to φυγομαχεῖν (Xen. Hell. 6.4.8–14; Diod. Sic. 15.55.1–56.4; Plut. Pel. 23)

17. **Duration of battle:** no sources tell of it; but it appears to have been an extremely quick victory.
18. **Decisive factors:** ostensibly the λόξη φάλαγξ of Plutarch and Diodoros, whatever it may have been, and the sudden strike of the Sacred Band, which completely took the Spartans by surprise. The flawed cavalry skirmish did not help
19. **Victor:** Thebans (Epameinondas)
20. **Defeated:** Spartans (Kleombrotos)
21. **Pursuit:** –
22. **Casualties (victor):** 300 according to Diodoros (15.56.4)
23. **Casualties (defeated):** 1000 Spartans (including 400 of the 700 Spartiates present (Xen. Hell. 6.4.15). According to Diodoros, however, the number was no less than 4000; and Pausanias claims a mere 47 Thebans against more than 1000 Spartans (9.13.12)
24. **Prisoners:** –
25. **Trophy:** Xen. Hell. 6.4.15
26. **Exchange of bodies:** after the battle, the Spartans discussed what to do, but as their allies were unwilling to fight anew, they acknowledged their defeat by sending a herald to request a truce, which was granted (Xen. Hell. 6.4.15; Diod. Sic. 15.56.4). Pausanias relates that Epameinondas allowed the Spartans to pick up their dead only after the allies had done so; in this way, they would not be able to cover up their actual losses (9.13.11–12)
27. **Miscellaneous:** Plutarch Ages. 28.6 mentions Kleonymos who was, it appears, struck down three times before being finally killed. If this is true, then it might be possible to get up again, even in the mêlée. This is the first direct mention of the famous oblique (λόξη) phalanx: possibly also the first instance of combined cavalry/infantry fighting on the part of the Thebans. Polyainos claims that this was the first pitched battle the Spartans lost and that this was due to their marching into battle not accompanied by *auloi*, 1.10. If nothing else, this may corroborate the view that Kleombrotos was completely unprepared for something like this. Xenophon claims (Hell. 6.4.8) that Kleombrotos and his staff had had some wine to drink and that this perhaps had impaired their judgement. Polyainos (2.3.2) provides the famous cry of Epameinondas: ἓν βῆμα χαρίσασθέ μοι, καὶ τὴν νίκην ἕξομεν!
28. **Sources:** Xen. Hell. 6.4.4–16; Diod. Sic. 15.53–56; Plut. Ages. 28.3–6, Pel. 23; Paus. 9.13.9–12; Nep. Pel. 4.2; Polyaen. 1.10, 2.3.2–3, 2.3.8; Din. 1.72–73; Polyb. 12.25f 3–4; Frontin. Str. 4.2.6.
29. **Bibliography**
    Beister (1973) 47–84
    Buckler (1977) 76–79
    Cawkwell (1972) 254–278
    DeVoto (1989) 115–117
    Hanson (1988) 197–207
    Montagu (2000) 90–91
    Stylianou (1997) 398–407
    Tuplin (1987) 72–107
    Underhill (1900) 244–249

(17) MANTINEIA 418

1. **Name and date:** Mantineia 418 (Thuc. 5.64.4)
2. **Time of year:** uncertain; appears to be early summer (Thuc. 5.57.1)
3. **Geographic locality:** at the plain around Mantineia, just north of Mýtikas and south-west of Mt. Alesio (Thuc. 5.64.5; Barrington 57 A4, 58 C2)
4. **Battlefield topography:** to the south there may have been a wood at the time. The plain tilts slightly downwards to the north (Gomme, Andrewes & Dover [1970] 95). The Spartans march out to battle from the Herakleion (?), near the water sources they have fouled (Thuc. 5.66.1)
5. **Weather conditions:** –
6. **Encampment:** the Spartans and Arkadians invaded Mantineian territory, where they pitched camp and ravaged the ground (Thuc. 5.64.5). Later, the Argives pitched their camp on a hill not far from the Spartans' new vantage point (Thuc. 5.65.6). The Argives camped around Tegea to besiege it (Diod. Sic. 12.79.2)
7. **Combatants:** Spartans and allies: Skiritai, Brasidas' veterans from Thrace, and with these the enfranchised helots *vs.* Argives, Mantineians, Arkadians, Kleonaians and Orneaians, and a few Athenians (Thuc. 5.67)
8. **Non-combatants:** –
9. **Numbers of armed forces:** difficult to assess; according to Thucydides due to the "secrecy of affairs at Sparta" (Thuc. 5.68.2). However, seven *lochoi* were present. The Spartan front line was $4 \times 4 \times 4 \times 7 = 448$ (four hoplites to the front of each *enomotia* – four *enomotiai* to each *pentekostys* – four *pentekostyes* to each *lochos* – seven *lochoi*). These were "generally" deployed to a depth of eight; $448 \times 8 = 3584$. Besides, the Skiritai made out 600, making a total of 4184. One *lochos* consisted of Brasidas' veterans and the freedmen; so the remaining six *lochoi*, the bulk of the Spartan muster, made up *c.* 3072, or about five sixths of the maximum mobilisation (Thuc. 5.68.2–3, cf. 64.3). Gomme puts the number of Argives and allies at 8000 (including, presumably, Diodoros' 3000 Eleans and 3000 Mantineians, and 1000 picked Athenians, Diod. Sic. 12.28.4, 12.79.1) – so if the Spartan army was "visibly larger" (Thuc. 5.68.1), the Spartans' allies must have been considerably more numerous than the Spartans themselves
10. **Commanders:** Sparta: king Agis (Thuc. 5.65.2) with the polemarchs Hipponoidas and Aristokles (Thuc. 5.72.1); Argives – not given
11. **Battle array:** the Skiritai, according to custom, take up the leftmost position (Thuc. 5.67.1), then the Brasideioi, with them the enfranchised helots, then the Spartans themselves, the Arkadians from Heraia, the Mainalians, and the Tegeates. At the extreme left and right some cavalry is posted. On the other side, the Mantineians occupied the rightmost position, then the allied Arkadians, 1000 picked Argive troops, the other Argives, the Kleonaioi and Orneatai, and the Athenians on the left, followed by some cavalry (Thuc. 5.67.1–2)
12. **Pre-battle rites/exhortations:** both armies received a harangue from their respective leaders immediately before battle (Thuc. 5.69.1)
13. **Hoplite indicators**
    ὁπλίτης: Diod. Sic. 12.79.1
    ἀσπίς: Thuc. 5.71.1
    ὠθέω / ὠθισμός: Thuc. 5.72.3
    (ὅπλον / -α): Thuc. 5.74.2
    παραρρήγνυμι / παράρρηξις: Thuc. 5.73.1
14. **Indicators of other troop types**
    ἱππεύς / ἵππος: Thuc. 5.67.1, 5.67.2, 5.73.1; Diod. Sic. 12.79.1
15. **Mercenaries:** -
16. **Description of battle:** according to Thucydides, the Spartans were somewhat baffled to see the Argives appear on a slope, ready for battle and too close for comfort (Thuc. 5.66.1). They hastily prepared for battle, and the two armies moved against each other; the Argives angrily and in great haste; the Spartans steadily, keeping their pace to the sound af many *auloi*. Agis realised

that the Mantineians on his left extended beyond the Skiritai. To remedy this, he ordered the Skiritai and Brasideioi further to the left, and instructed his polemarchs Hipponoidas and Aristokles to fill up the gap with their contingents from the right wing. They refused to alter the array on such short notice, and Agis hurriedly had to recall the Skiritai, but they did not have time to carry out this manoeuvre before the enemy closed. The Mantineians' right then drove back the Skiritai and Brasideioi. The Mantineians with their allies and the 1000 picked Argives then poured through the gap. They encircled the Spartans in this sector, routed them and drove them back to their wagons, where they killed some of the older Spartan men on guard. On the other wing and in the middle (where king Agis and his 'body-guard' were), however, the Spartans routed the older Argives (the so–called five *lochoi* (Thuc. 5.71.4)?), the Kleonaioi, Orneatai, and Athenians, most of these not even waiting for them to come to grips. The Argives' line now fell apart simultaneously on both sides, and the Spartans' and Tegeans' right wings began to encircle the Athenians, now threatened from two sides. These were saved mainly by their cavalry. Agis now came to the rescue of his left, giving the enemy an opportunity to escape, which they did. However, most of the Mantineians were killed (Thuc. 5.70–73)

17. **Duration of battle:** –
18. **Decisive factors:** the allies were not quite prepared to wait for the Spartan charge; consequently their line broke apart in two places at once
19. **Victor:** Agis and Spartans
20. **Defeated:** allies
21. **Pursuit:** the pursuit was not long or particularly vigorous, as was the Spartan custom (Thuc. 5.73); according to Diodoros because a certain Pharax urged the king not to fight desperate men (12.79.6–7). Diodoros also says that the picked Argive corps pursued their beaten enemy, making a great slaughter (Diod. Sic. 12.79.4)
22. **Casualties (victor):** "difficult to get information on", but rumoured to be about 300; for the allies hardly worthy of mentioning (Thuc. 5.74.3). According to Diodoros, the 1000 picked Argives caused πολὺν φόνον (12.79.4)
23. **Casualties (defeated):** 700 Argives, Orneaians, and Kleonaians, 200 Mantineians, 200 Athenaians and Aiginetans – and both generals (Laches and Nikostratos: Thuc. 5.61.1; Diod. Sic. 12.79.1, Thuc. 5.74.3)
24. **Prisoners:** -
25. **Trophy:** Thuc. 5.74.2; Diod. Sic. 12.79.7; Paus. 8.10.5
26. **Exchange of bodies:** the Argives are given their dead under a truce as per usual; nonetheless, the Spartans first plunder the fallen hoplites (Thuc. 5.74.2)
27. **Miscellaneous:** Some of the Kleonaians, Athenians, and Orneaians trampled each other in their attempt to get away from the advancing Spartans (Thuc. 5.72.4). It is also noteworthy that in the first marching up of the troops the day before, king Agis was warned by τῶν πρεσβυτέρων τις (presumably a veteran) that he was about to commit a grave tactical error. He followed the advice and called off the attack at the last possible moment: there was already skirmishing between the light-armed troops (Thuc. 5.65.2)
28. **Sources:** Thuc. 5.64–74; Diod. Sic. 12.78–79; Paus. 8.10.5

29. **Bibliography**
Cooper, III (1978) 35–40
Gomme, Andrewes & Dover (1970) 89–127
Montagu (2000) 70–71
Pritchett (1969) 37–72
Singor (2002) 235–284
Woodhouse (1933)

(18) MANTINEIA 362

1. **Name and date:** Mantineia 362 (Xen. *Hell.* 7.5.18; Diog. Laert. 2.54)
2. **Time of year:** –
3. **Geographic locality:** on the plateau south of Mantineia (Xen. *Hell.* 7.5.7; Diod. Sic. 15.84.2; Barrington 58 C2)
4. **Battlefield topography:** a tableland some 600–1500 m.a.s.l., but level and suited for infantry fighting (see map in Gomme, Andrewes & Dover [1970] facing 96)
5. **Weather conditions:** –
6. **Encampment:** –
7. **Combatants:** Tegeans, with their allies: Thebans (and other Boiotians), Arkadians (Tegeans, Megalopolites, Aseans, Pallantians and others: Lokrians, Sikyonians, Messenians, Malians, Ainianians and Thessalians, Diod. Sic. 15.85.2), Euboians and Argives (Xen. *Hell.* 7.5.4–5; Diod. Sic. 15.84.4); *vs.* Mantineians with their allies: Spartans, Athenians, other Arkadians, Achaians and Eleans (Xen. *Hell.* 7.5.3, 7.5.18; Diod. Sic. 15.82.4, 15.84.4)
8. **Non-combatants:** –
9. **Numbers of armed forces:** with the Mantineians: 6000 Athenians (Diod. Sic. 15.84.2), and the entire coalition army made up some 20,000 infantry and 2000 horse; whereas with the Tegeans, Epameinondas' forces numbered some 30,000 infantry and "no less than" 3000 horse (Diod. Sic. 15.84.4). Polyainos (2.3.14) mentions 1600 horse employed in diversionary tactics
10. **Commanders:** Tegean alliance: Epameinondas (Xen. *Hell.* 7.5.4; Diod. Sic. 15.82.3), with Daiphantos and Iolaidas (Plut. *Mor.* 194c); Mantineian alliance: none related, apart from the Athenian general, Hegesileos, and his colleague, the hipparch Kephisodoros (Diod. Sic. 15.84.2; Diog. Laert. 2.45), though it is scarcely credible that there should not have been one
11. **Pre-battle rites/exhortations:** sacrifices were performed, but deemed unfavourable by the μάντεις on both sides (Diod. Sic. 15.85.1); the signal to attack was given by trumpets (σαλπίγγες, Diod. Sic. 15.85.3)
12. **Battle array:** Epameinondas first lead his men west, towards Mýtikas, and renewed his battle order, so that they faced due north. Xenophon describes the Theban formation as shaped like an ἔμβολον (*Hell.* 7.5.22), but is not very clear as to the actual formation. The Boiotian contingents were led around on the left, together with the Arkadians, to form a much deeper front than the enemy's at this point, Epameinondas' idea being to strike at the enemy line at its strongest, thereby making the entire formation crumble. The Argives commanded the right wing, and the remainder filled the middle: Euboians, Lokrians, Sikyonians, Messenians, Malians, Ainianians and Thessalians (Diod. Sic. 15.85.2). On the other side, the Mantineians and other Arkadians occupied the right wing, then Spartans, Eleians and Achaians, and the Athenians on the right wing. The centre was filled by weaker troops. The Mantineian alliance's cavalry was posted six deep, whereas Epameinondas mirrored the infantry formation with his; the intention being to punch a hole in the enemy formation; but both posted their cavalry on the wings (except for the Eleians, who were posted to the rear of the Mantineian alliance, Diod. Sic. 15.85.7). Epameinondas also had infantry ready to follow up the cavalry shock troops, and posted some infantry and horse on a hill opposite the Athenians to prevent them coming to the aid of those on their right (Xen. *Hell.* 7.5.21–24; Diod. Sic. 15.85.1–2)
13. **Hoplite indicators**
    ὁπλίτης: Xen. *Hell.* 7.5.23, 7.5.24, 7.5.25
    φάλαγξ: Xen. *Hell.* 7.5.22, 7.5.23, 7.5.25; Diod. Sic. 15.85.6, 15.85.7, 15.86.5; Polyaen. 2.3.14
    ἀσπίς: Diod. Sic. 15.87.6
    θῶραξ: Xen. *Hell.* 7.5.22; Diod. Sic. 15.87.1, 15.87.5
    δόρυ: Diod. Sic. 15.86.2, 15.87.1, 15.87.5, 15.87.6
    (ὅπλον / -α): Xen. *Hell.* 7.5.22

14. **Indicators of other troop types**
    πελτάστης / πέλτη: Xen. *Hell.* 7.5.25
    ἱππεύς / ἵππος: Xen. *Hell.* 7.5.23, 7.5.24; Diod. Sic. 15.84.1, 15.84.4, 15.85.2, 15.85.4, 15.85.7, 15.85.8, 15.86.1; Polyaen. 2.3.14
    ἀκοντιστής: Diod. Sic. 15.85.4
    σφενδονήτης: Diod. Sic. 15.85.4
    γυμνός / ψιλός: Diod. Sic. 15.85.4, (15.85.5),
15. **Mercenaries:** some mercenaries are employed, at least by the Thebans (Diod. Sic. 15.85.6, 15.87.3)
16. **Description of battle:** after having made the enemy believe that he would not attack, Epameinondas did just that. The actual battle was preceded by a cavalry skirmish on the flanks, which the Thebans won, since they were backed by three times as many slingers and javelin-throwers. The Athenians finally fled behind the flanks. However, they managed to keep their formation (φάλαγξ) and killed all the Euboians and some mercenaries, sent to conquer some vantage points. The Theban horse, on the other hand, attacked the infantry phalanx and attempted to outflank it; and as the exhausted Athenians turned to flee, the Eleians rode round the flank and repulsed the Theban cavalry. On the other wing, the Thebans and Thessalians beat back their cavalry counterpart with great losses and forced them to seek shelter behind the phalanx. Next, the two phalanxes collided. Most of the spears in the front ranks broke due to the violent action, and the hoplites resorted to swords in a grueling, tightly-locked battle. At some point, Epameinondas grouped his best men in close formation and went directly for the Spartan commander, whom he wounded. The Boiotians broke the enemy line (as at Leuktra) and began to push the Spartans back, constantly killing the hindmost troops. Epameinondas, however, pursued too eagerly and found himself under attack from the Spartans, and in the thick of the fighting, he was fatally struck in the chest by a javelin. The Thebans managed to secure the body, driving the Spartans away and pursuing, but not for long. Some of the infantry supporting the Boiotian cavalry did indeed pursue and kill the enemy, but most of these were killed by the Athenians. As both sides were in possession of many enemy dead, for a time none of them sent an envoy to sue for peace and get their dead back, until finally the Spartans did so (Xen. *Hell.* 7.5.24–25; Diod. Sic. 15.85.3–87; Diog. Laert. 2.54)
17. **Duration of battle:** Diodoros says of the central part of the battle that it lasted a long time (Diod. Sic. 15.86.3–4)
18. **Decisive factors:** probably once again Epameinondas' superior generalship and his unorthodox array, coupled with the much stronger cavalry
19. **Victor:** Tegean alliance, Epameinondas (Polyb. 9.9.7)
20. **Defeated:** Mantineian alliance
21. **Pursuit:** the Thebans followed hard on the heels of the Spartans, cutting them down (Diod. Sic. 15.86.5); and likewise in the endgame, though not for long (Xen. *Hell.* 7.5.25; Diod. Sic. 15.87.2)
22. **Casualties (victor):** the bloodshed on both sides appears to have been terrible, though no exact account is given (Xen. *Hell.* 7.5.25; Diod. Sic. 15.85.6, 15.85.8, 15.86.2–5, 15.87.1)
23. **Casualties (defeated):** –
24. **Prisoners:** –
25. **Trophy:** the outcome is so uncertain that both sides in fact erected trophies and claimed the victory (Xen. *Hell.* 7.5.26; Diod. Sic. 15.87.2)
26. **Exchange of bodies:** the same applies to this point: both, apparently, received their dead back from each other under truce (Xen. *Hell.* 7.5.26–27; Diod. Sic. 15.87.3–4)
27. **Miscellaneous:** trumpets were also used to call back the soldiers from the battle, i.e. to sound the retreat (Diod. Sic. 15.87.2). Also mentioned are the ἄμιπποι, or cavalry-support light troops (Xen. *Hell.* 7.5.23). The Athenians managed to stave off defeat by keeping their line unbroken in the retreat (κατὰ τὴν ἀποχώρησιν οὐκ ἐτάραξαν τὴν ἰδίαν φάλαγγα, Diod. Sic. 15.85.6)
28. **Sources:** Xen. *Hell.* 7.5.21–27; Diod. Sic. 15.84–87; Polyb. 9.8–9; Plut. *Mor.* 194c; Polyaen. 2.3.14

29. **Bibliography**
Kromayer (1903) 27–123
Montagu (2000) 93–94
Underhill (1900) 258–261

(19) MARATHON 490

1. **Name and date:** Marathon 490 (Hdt. 6.102–103.1; 6.117.1)
2. **Time of year:** probably around September, since the Athenian runner sent to Sparta was refused help due to the sacred law forbidding the Spartans to go to war before the fifteenth day of the month. Herodotos is not quite clear as to whether it means any month or this particular month – Karneios, named after the Doric festival of Apollon Karneios. At any rate, the likeliest date is between Metageitnion 17 (= September 12) and Metageitnion 22 (= September 17) (Hdt. 6.106.3)
3. **Geographic locality:** the plain of Marathon, some 40 km north-east of Athens (Hdt. 6.102–3; Barrington 59 C2–D2)
4. **Battlefield topography:** the plain (near modern Vrana) is very wide and drops smoothly down to the sea where the Persian ships lay moored. It is possible that some vegetation (olive trees) was in the way, but generally the plain is very flat and unimpeded. At the ends of the plain it is limited by two marshes, a small one to the south-west and an extensive one to the north-east, which reaches almost to the sea and is virtually impassable. The whole of the plain is divided by a torrent, Charadra, rushing down from Mt. Pentelikos. According to Herodotos, it was chosen because it afforded the Persian cavalry a good possibility of action (Hdt. 6.102)
5. **Weather conditions:** probably good, since the battle took place in mid-September
6. **Encampment:** the Athenian army is drawn up at a precinct sacred to Herakles when they are joined by the Plataians (Hdt. 6.108.1); *idoneo castra fecerunt* (Nep. *Milt.* 5.2)
7. **Combatants:** Persians *vs*. Athenians and their allies, the Plataians (Hdt. 6.108.1)
8. **Non-combatants:** –
9. **Numbers of armed forces:** the numbers of the Persians is not related by Herodotos, but all ancient estimates are wildly exaggerated, cf. Xen. *An.* 3.2.12; Paus. 4.25.5; Plut. *Mor.* 862B. However, they may have outnumbered the Athenians by two – four to one. The Athenians are given as 10,000 (Just. *Epit.* 2.9) or 9000 (Nep. *Milt.* 5.1; cf. Paus. 4.25.5, 10.20.2). Both put the Plataians at 1000, probably exaggerated (cf. Hdt. 9.28.6).
10. **Commanders:** Athenians: Miltiades (formally as one of the ten strategoi, Hdt. 6.103.1) (Hdt. 6.104.1) (polemarch: Kallimachos, Hdt. 6.109.2). Persians: Datis and Artaphernes (Hdt. 6.94.2)
11. **Pre-battle rites/exhortations:** the customary sacrifices were performed (Hdt. 6.112.1)
12. **Battle array:** the Persians and the Sakai took up the centre (Hdt. 6.113.1), their various allies at other places. The Athenians themselves to the right, commanded by Kallimachos, ordered by tribes, and the Plataians at the extreme left. The centre was attenuated, perhaps in an attempt to stretch the line, but more probably this was part of a conscious tactic, Hdt. 6.111.
13. **Hoplite indicators**
ὁπλίτης: Hdt. 6.117.3
ἀσπίς: Hdt. 6.115, 6.117.3
παραρρήγνυμι / παράρρηξις: (Hdt. 6.113.1–2)
14. **Indicators of other troop types**
τοξότης / τοξεύειν etc.: Hdt. 6.112.2
ἱππεύς / ἵππος: Hdt. 6.112.2
15. **Mercenaries:** –
16. **Description of battle:** the Athenians suddenly charged the Persians in a run (perhaps equivalent to modern 'double quick-time'). The Athenian line was broken through in the centre, but

the wings prevailed and sent the Persians scrambling back towards their ships. The Athenians then let their routed enemies flee, wheeled around and attacked the immediately victorious Persian centre in the rear. The result was a complete annihilation of the Persian army, Hdt. 6.112–115

17. **Duration of battle:** μαχομένων χρόνος ἐγίνετο πολλός (Hdt. 6.113.1)
18. **Decisive factors:** the Athenians' body armour, the fact that the Persian cavalry was not employed; the superior tactics of a massed phalanx
19. **Victor:** Athenians
20. **Defeated:** Persians
21. **Pursuit:** the remainder of the Persian army were routed and chased back to the ships, being harassed on the way. A fierce fight about the ships ensued: the Persians managed to escape, but the Athenians captured seven ships (Hdt. 6.113.2–115)
22. **Casualties (victor):** 192, exceptionally buried on the battlefield in the Soros (Hdt. 6.117.1)
23. **Casualties (defeated):** 6400 (Hdt. 6.117.1)
24. **Prisoners:** –
25. **Trophy:** –
26. **Exchange of bodies:** due to the circumstances, no such thing took place: indeed, the Spartan relief force, which arrived *post festum,* asked to see the bodies of the Persians (Hdt. 6.120), and by this time, the remains of the invasion force was long gone (Hdt. 6.116)
27. **Miscellaneous:** the polemarch was chosen by lot (Hdt. 6.109.2). The Athenians charged δρόμῳ, the first Greeks Herodotos knows of to do this (Hdt. 6.112), across a distance of 8 stadia (= 1.440 km). This, however, is highly unlikely: the Athenians may have charged on the double within the last 200 m (within bowshot). The Athenians are "let loose", like runners (Hdt. 6.112.1)
28. **Sources:** Hdt. 6.102–117; Nep. *Milt.* 5; Just. *Epit.* 2.9; Plut. *Mor.* 862b; Paus. 4.25.5, 10.20.2
29. **Bibliography**
    Berthold (1976–77) 84–94
    Bicknell (1970) 427–442
    Doenges (1998) 1–17
    Francis & Vickers (1985) 99–113
    Evans (1993) 279–307
    Hammond (1968) 13–57
    Holoka (1997) 329–353
    How & Wells (1928²) vol. II 106–114
    Kertész (1991) 155–160
    Montagu (2000) 52–54
    Munro (1899) 185–197
    Schreiner (2004) 9–68
    Shrimpton (1980) 20–37
    Storch (2001) 381–394
    van der Veer (1982) 290–321
    Welwei (1979) 101–106
    Whatley (1968), esp. 131–139
    Wyatt Jr. (1976) 483–484

(20) MEGARA 458

1. **Name and date:** Megara 458 (Thuc. 1.105.4)
2. **Time of year:** –
3. **Geographic locality:** Megara (Thuc. 1.105.4), on the Kimolian plain (Diod. Sic. 11.79.4; Barrington 58 E2)

4. **Battlefield topography:** –
5. **Weather conditions:** –
6. **Encampment:** –
7. **Combatants:** Athenians and Corinthians (Thuc. 1.105; Diod. Sic. 11.79.2)
8. **Non-combatants:** –
9. **Numbers of armed forces:** –
10. **Commanders:** Athenians: Myronides (Thuc. 1.105.4; Diod. Sic. 11.79.3)
11. **Pre-battle rites/exhortations:** –
12. **Battle array:** –
13. **Hoplite indicators**
    ὁπλίτης: Thuc. 1.105.3, 1.106.2
14. **Indicators of other troop types**
    γυμνός / ψιλός: Thuc. 1.106.2
15. **Mercenaries:** –
16. **Description of battle:** according to Thucydides, there were actually two separate battles, fought at an interval of twelve days. Myronides led out an army made up of the youngest and the oldest men left in Athens: the rest were besiegeing Aigina or in Egypt. They claimed to have defeated the Corinthians, but apparently the matter was disputed. The Athenians nonetheless put up a trophy. When the Corinthians tried to put up their own trophy, the Athenians overran them completely and defeated the rest of the Corinthian army decisively. On their retreat a large section of the army was separated from the rest and ended pent up in a walled orchard. The Athenian ψιλοί then finished them off with stones (Thuc. 1.105.4–106)
17. **Duration of battle:** ἐπὶ πολὺν χρόνον (Diod. Sic. 11.79.3)
18. **Decisive factors:** –
19. **Victor:** Athenians
20. **Defeated:** Corinthians
21. **Pursuit:** the retreating Corinthians found themselves trapped on a tract of land surrounded by pickets (probably), and were killed to a man by missiles from the light-armed (Thuc. 1.106.2). The situation resembles that of the trapped Athenian hoplites in Sicily in 413 (Thuc. 7.81.4–5)
22. **Casualties (victor):** –
23. **Casualties (defeated):** apparently quite a lot (αὐτῶν μέρος οὐκ ὀλίγον); Diodoros claims the Athenians killed "many" in both battles (Diod. Sic. 11.79.3–4)
24. **Prisoners:** –
25. **Trophy:** Thuc. 1.105.6*bis*
26. **Exchange of bodies:** –
27. **Miscellaneous:** it is interesting to note that ψιλοί played a crucial part in the pursuit and destruction of enemy hoplites as early as this
28. **Sources:** Thuc. 1.105.2–106; Diod. Sic. 11.79.1–4
29. **Bibliography**
    Gomme (1945) 308–311
    Montagu (2000) 60

(21) MILETOS 412

1. **Name and date:** Miletos 412 (Thuc. 8.25.1)
2. **Time of year:** "end of summer", i.e. early autumn (Thuc. 8.25.1)
3. **Geographic locality:** outside Miletos (Thuc. 8.25.1; Barrington 61 E2 and inserted card)
4. **Battlefield topography:** apparently the battle was fought on an isthmus (Thuc. 8.25.5), though from the map it is not clear which one is meant
5. **Weather conditions:** –
6. **Encampment:** the Athenians pitch camp "having crossed over to Miletos" (Thuc. 8.25.2)

7. **Combatants:** 1000 Athenians, 1500 Argives, and 1000 allies (?), conveyed across from Samos on 48 troop-carriers (Thuc. 8.25.1). Against them 800 Milesians, Chalkideus' Peloponnesians, and Tissaphernes' mercenaries and cavalry (Thuc. 8.25.1–2)
8. **Non-combatants:** –
9. **Numbers of armed forces:** 3500 Athenians and allies; 800 Milesians and an unspecified number of allies (Thuc. 8.25.1–2)
10. **Commanders:** Athenians: Phrynichos, Onomakles, and Skironides (Thuc. 8.25.1); Milesians: only Tissaphernes and Chalkideus are mentioned by name (Thuc. 8.25.2)
11. **Battle array:** not clear. At any rate, however, the Argives on one wing (the left?) are stationed opposite the Milesians, and the Athenians opposite the Peloponnesians and barbarians (Thuc. 8.25.3–4)
12. **Pre-battle rites/exhortations:** –
13. **Hoplite indicators**
    ὁπλίτης: Thuc. 8.25.1–2
    ὠθέω / ὠθισμός: Thuc. 8.25.4
    (ὅπλον / -α): Thuc. 8.25.4
14. **Indicators of other troop types**
    ἱππεύς / ἵππος: Thuc. 8.25.2
15. **Mercenaries:** Thuc. 8.25.2
16. **Description of battle:** the Milesians and their barbarian mercenaries attacked first. The Argives, in contempt of the Ionians, broke ranks in pursuit and were defeated by them, losing about 300 in the process. The Athenians, however, pushed back the Peloponnesians and barbarians and any other troops. The Milesians retired to the city after having routed the Argives, not wanting to share their allies' fate, and the Athenians grounded arms just under the walls of Miletos (Thuc. 8.25.3–5)
17. **Duration of battle:** after the battle, a sighting of enemy ships was reported "about sunset" (Thuc. 8.25.6), but as we do not know when hostilities commenced, the duration is uncertain
18. **Decisive factors:** among the decisive factors – according to Thucydides – are the Dorians' disdain of their Ionian enemies on both sides (Thuc. 8.25.3, 8.25.5)
19. **Victor:** Athenians
20. **Defeated:** Milesians
21. **Pursuit:** –
22. **Casualties (victor):** –
23. **Casualties (defeated):** a little less than 300 Argives lost to the Milesians (Thuc. 8.25.3)
24. **Prisoners:** –
25. **Trophy:** Thuc. 8.25.4
26. **Exchange of bodies:** –
27. **Miscellaneous:** –
28. **Sources:** Thuc. 8.25
29. **Bibliography**
    Montagu (2000) 75
    Thompson (1965) 294–297

(22) MUNYCHIA 403

1. **Name and date:** Munychia 403 (Xen. Hell. 2.4.11)
2. **Time of year:** apparently in winter, since an expedition on the part of the Thirty is hampered by heavy snowfall (Xen. Hell. 2.4.3)
3. **Geographic locality:** the hill of Munychia in Piraeus (Xen. Hell. 2.4.11; Barrington 59 B3)
4. **Battlefield topography:** the battle itself was fought on a road going up the Munychia, a very steep hill (Xen. Hell. 2.4.11–12; Diod. Sic. 14.33.2)

5. **Weather conditions:** possibly very cold and snow-filled (cf. Time of year), but it is really impossible to say
6. **Encampment:** -
7. **Combatants:** the followers of the Thirty and the Spartan garrison *vs.* Thrasybulos' rebel force from Phyle (Xen. *Hell.* 2.4.10)
8. **Non-combatants:** –
9. **Numbers of armed forces:** Thrasybulos' force numbered about 1000 men (Xen. *Hell.* 2.4.10; Diod. Sic. 14.33.2: 1200 men)
10. **Commanders:** Thrasybulos on one side (Xen. *Hell.* 2.4.10); for the Thirty presumably most of the tyrants; at any rate, Kritias and Hippomachos were killed (Xen. *Hell.* 2.4.19); two Spartan polemarchs, Chairon and Thibrachos (*IG* II² 11678)
11. **Pre-battle rites/exhortations:** Thrasybulos delivered an exhortatory speech to his men, pointing out the technical advantages of fighting downhill (Xen. *Hell.* 2.4.13–17)
12. **Battle array:** the force with the Thirty was deployed so as to "fill up the road", and they ended up no less than 50 ranks deep (Xen. *Hell.* 2.4.11). Thrasybulos' men, on the other hand, also filled up the road, but no more than 10 ranks deep (Xen. *Hell.* 2.4.12)
13. **Hoplite indicators**
    ὁπλίτης: Xen. *Hell.* 2.4.10, 2.4.12
    ἀσπίς: Xen. *Hell.* 2.4.11, 2.4.12, 2.4.16
    δόρυ: Xen. *Hell.* 2.4.15
    (ὅπλον / -α): Xen. *Hell.* 2.4.12, 2.4.14, 2.4.19*bis*
14. **Indicators of other troop types**
    ἱππεύς / ἵππος: Xen. *Hell.* 2.4.10
    ἀκοντιστής: Xen. *Hell.* (2.4.15)
    γυμνός / ψιλός: Xen. *Hell.* 2.4.12 (incl. πελτοφόροι and πετροβόλοι)
15. **Mercenaries:** –
16. **Description of battle:** very perfunctory. It merely states that Thrasybulos men, although he himself was killed, were victorious (Xen. *Hell.* 2.4.19). This may reflect Xenophon's view of the obviousness of such a battle, but is perhaps excluded to avoid repetition (cf. Xen. *Hell.* 2.4.15–16)
17. **Duration of battle:** a long time (Diod. Sic. 14.33.2)
18. **Decisive factors:** no doubt we have to reckon the unevenly balanced battle, giving Thrasybulos' men all advantage (cf. Diod. Sic. 14.33.2)
19. **Victor:** Thrasybulos' rebel force
20. **Defeated:** the Thirty and their Spartan garrison
21. **Pursuit:** the victors pursued their beaten enemy down to level ground (Xen. *Hell.* 2.4.19); this, at any rate, cannot be a very long distance. Diodoros maintains that they dared not attack on level ground, due to the enemy's numerior superiority (14.33.2–3)
22. **Casualties (victor):** –
23. **Casualties (defeated):** apart from four of the leaders, about 70 (Xen. *Hell.* 2.4.19)
24. **Prisoners:** –
25. **Trophy:** [Xen. *Hell.* 2.4.14: *mentioned* in Thrasybulos' exhortation]
26. **Exchange of bodies:** given back to the defeated under truce (Xen. *Hell.* 2.4.19)
27. **Miscellaneous:** –
28. **Sources:** Xen. *Hell.* 2.4.9–20; Lys. 12.53; [Arist.] *Ath. pol.* 38.1; Diod. Sic. 14.33.2–3; Nep. *Thras.* 2.6–7; *IG* II² 11678
29. **Bibliography**
    Montagu (2000) 82

(23) MYKALE 479

1. **Name and date:** Mykale 479 (Hdt. 9.97, 9.101.1; Polyaen. 7.45.1)
2. **Time of year:** on the same day as the battle of Plataiai (Hdt. 9.90.1; Diod. Sic. 11.34.1); i.e. on Boedromion 4, or about August 1 (Hdt. 9.3.2; Plut. *Arist.* 19.7)
3. **Geographic locality:** on the shore below the slopes of Mt. Mykale (present Samsun), directly east of Samos (Hdt. 9.96.3); near the temple of Demeter (Hdt. 9.101.1; Barrington 61 E2, 57 E4)
4. **Battlefield topography:** varied; at some points level and without obstacles (near the beach); at others rocky and run through by watercourses (Hdt. 9.102.1)
5. **Weather conditions:** –
6. **Encampment:** the Persians pitched their defense works (probably around a camp) near the coast at Mykale (Hdt. 9.97, 9.98.2)
7. **Combatants:** Persians and other barbarians and their force-marched Greek allies (Milesians, Samians and other Ionians) *vs.* allied Greeks: Athenians, Spartans, Corinthians, Sikyonians and Troizenians (Hdt. 9.102.3). According to Diodoros, even Aiolians join in (Diod. Sic. 11.36.5)
8. **Non-combatants:** both Herodotos and Diodoros state that the Samians (Diodoros: all the Persians' Greek allies) were disarmed and thus unable to take part in the fighting (Hdt. 9.99.1; Diod. Sic. 11.35.4)
9. **Numbers of armed forces:** Persian land forces: 60,000 (Hdt. 9.96.2) or 100,000 (Diod. Sic. 11.34.3); Greeks are much fewer, according to Diodoros (11.36.1)
10. **Commanders:** Persians: Tigranes and Mardontes (Hdt. 9.96.2; 9.102.4); Greeks: Leotychides (Hdt. 9.98.2); the Sikyonians' general Perilaos (Hdt. 9.103.1)
11. **Pre-battle rites:** –
12. **Battle array:** the Athenians with the Corinthians, Sikyonians and Troizenians take up position covering about half the line, the Spartans taking the rest (Hdt. 9.102.1). The Persians had prepared for the assault by securing the beached ships behind a palisade and by forming a "wall" of shields (γέρρα ἕρκος εἶναι σφίσι) (Hdt. 9.100.1)
13. **Hoplite indicators**
    ὠθέω / ὠθισμός: Hdt. 9.102.3 (διωσάμενοι)
14. **Indicators of other troop types**
15. **Mercenaries:** –
16. **Description of battle:** as the Greeks advanced towards the Persian stronghold, the Spartan half of the line was deterred by rough terrain, and the Athenians and their Ionian allies reached the Persian front line long before they did. For a while the outcome hung in the air, but finally the Athenians encouraged each other to an extra effort, and they pushed through the Persian shield wall. For a time the Persians fought back, but the Greeks kept pouring through the hole in the line, and they finally turned and fled. Their unwilling Greek allies turned against them and aided the rest in the slaughter. The Milesians, who had orders to guard the escape route to the crags of Mt. Mykale (cf. Polyaen. 7.45.1), led them down the wrong paths and into the thick of the fighting (Hdt. 9.101–104). Diodoros' account differs considerably: here, the Samians and Milesians are not disarmed or posted out of the way, but rush forward to join their liberators, thus actually scaring them. Also, the Persians counter-attack, and the battle is only decided by the appearance of the Milesians and Samians (Diod. Sic. 11.36)
17. **Duration of battle:** difficult to tell, but it seems to have been quickly determined (Hdt. 9.102–103)
18. **Decisive factors:** the breaking of the Persian shield wall and the "treachery" of their Ionian allies (Herodotos); the appearance of the Samians and Milesians (Diodoros)
19. **Victor:** Greeks
20. **Defeated:** Persians
21. **Pursuit:** –
22. **Casualties (victor):** συχνοί (Hdt. 9.103.1)

23. **Casualties (defeated):** "more than 40,000" (Diod. Sic. 11.36.6)
24. **Prisoners:** –
25. **Trophy:** –
26. **Exchange of bodies:** –
27. **Miscellaneous:** –
28. **Sources:** Hdt. 9.96–104; Diod. Sic. 11.35–36; Polyaen. 7.45.1
29. **Bibliography**
    How & Wells (1928²) vol. II 329–333
    McDougall (1990) 143–149
    Montagu (2000) 58

(24) NARYX 395/4

1. **Name and date:** Naryx 395/4 (Diod. Sic. 14.82.8)
2. **Time of year:** –
3. **Geographic locality:** at Naryx in Lokris; possibly at the foothills of Mt. Knemis (Diod Sic. 14.82.8; Barrington 55 D3 (?))
4. **Battlefield topography:** mountainous area; Naryx lies some 300 m.a.s.l. in Lokris Epiknemidia (?)
5. **Weather conditions:** –
6. **Encampment:** Ismenias' coalition army camps in Naryx in Lokris (Diod. Sic. 14.82.8)
7. **Combatants:** Ainianians and Athamanians, persuaded to revolt from Sparta by the Boiotian Ismenias (Diod. Sic. 14.82.7). Against them, the Phokians, commanded by Alkisthenes of Sparta (Diod. Sic. 14.82.8)
8. **Non-combatants:** –
9. **Numbers of armed forces:** Ismenias' army numbered *ca.* 6000 men (Diod. Sic. 14.82.7). Apparently all the Phokians came against them, but how many?
10. **Commanders:** Ismenias for the Ainianians and Athamanians; the Spartan Alkisthenes for the Phokians (Diod. Sic. 14.82.7–8)
11. **Pre-battle rites/exhortations:** –
12. **Battle array:** –
13. **Hoplite indicators**
    (ὅπλον / -α): Diod. Sic. 14.82.8.
    παράταξις / παρατάσσω: Diod. Sic. 14.82.10
14. **Indicators of other troop types**
15. **Mercenaries:** –
16. **Description of battle:** scant. The battle was hard and protracted; but finally the Phokians were routed by their enemies and pursued until nightfall (Diod. Sic. 14.82.9)
17. **Duration of battle:** Diodoros is very explicit in stating that it is a long battle which lasts until nightfall (Diod. Sic. 14.82.9)
18. **Decisive factors:** –
19. **Victor:** Ismenias, Ainianians, and Athamanians
20. **Defeated:** Alkisthenes and Phokians
21. **Pursuit:** the pursuit lasted until nightfall (but Diodoros does not relate when the battle commenced). Diodoros' phrasing *might* indicate that the 1000 casualties on the Phokian side occurred during the pursuit: μέχρι νυκτὸς διώξαντες τοὺς φεύγοντας ἀνεῖλον οὐ πολὺ λείποντας τῶν χιλίων κτλ. (Diod. Sic. 14.82.9)
22. **Casualties (victor):** *c.* 500 (Diod. Sic. 14.82.9)
23. **Casualties (defeated):** a few less than 1000 (Diod. Sic. 14.82.9)
24. **Prisoners:** –
25. **Trophy:** –

26. **Exchange of bodies:** –
27. **Miscellaneous:** most of the Phokians appear to have been killed during the pursuit, Diod. Sic. 14.82.9
28. **Sources:** Diod. Sic. 14.82.7–10
29. **Bibliography**
    Montagu (2000) 85

(25) NEMEA 394

1. **Name and date:** Nemea 394 (Xen. *Hell.* 4.2.14; Diod. Sic. 14.83.2)
2. **Time of year:** –
3. **Geographic locality:** in the dried-up river bed of the Nemea (Xen. *Hell.* 4.2.15, Aeschin. 2.168) near Corinth (Barrington 58 D2). It formed the border between Sikyonia and Corinthia (Strabo 8.6.25)
4. **Battlefield topography:** most likely flat and without obstacles; perfect for a hoplite army. However, Xenophon states that the ground is λάσιον – to such an extent as to obscure the Spartans' seeing the enemy charge until they heard their paian (Xen. *Hell.* 4.2.19)
5. **Weather conditions:** –
6. **Encampment:** the Spartans are mentioned as pitching camp "they, *too*" (κἀκεῖνοι, Xen. *Hell.* 4.2.15); although no mention has been made about the coalition army's camp. The camp was located some 2 km (10 stades) away. After the battle, the Corinthians pitched their tents "at their old camp site" (Xen. *Hell.* 4.2.23)
7. **Combatants:** Sparta and Mantinea (6000 hoplites); Triphylia, Elis, Akroria, and Lasion (3000), Sikyon (1500), Epidauros, Troizen, Hermione, and Halieis (3000). Pellenian hoplites are mentioned (Xen. *Hell.* 4.2.18). Also Kretan archers (300), and 400 Marganian, Letrinian and Amphidolian slingers. Allied forces: Athens (6000 hoplites), Argos (7000), Boiotia (5000), Corinth (3000), and Euboia (3000); besides cavalry from Boiotia (800), Athens (600), Chalkis (100), and about 50 from the Opuntian Lokrians. Again, at Xen. *Hell.* 4.2.18, Thespian hoplites are mentioned, but do not figure in the sum total of the allied forces. Besides, an unknown number of ψιλοί from the Ozolian Lokrians, the Melians, and the Akarnanians (or, more likely, Ainianians) serving with the Corinthians (Xen. *Hell.* 4.2.16–17)
8. **Non-combatants:** –
9. **Numbers of armed forces:** 13,500 hoplites for the Spartans (Xen. *Hell.* 4.2.16–17). Diodoros gives the the size of the army as 23,000 (14.83.1). The allies: 24,000 hoplites in all – but again Diodoros notes only 15,000 hoplites and 500 horse (14.82.10)
10. **Commanders:** for the Spartans: Aristodemos; for the allies, joint commandership (Xen. *Hell.* 4.2.9, 4.2.18)
11. **Pre-battle rites/exhortations:** the Athenians declared the sacrifices favourable and asked the others to prepare for battle (Xen. *Hell.* 4.2.18). Also, a paian was sung, which revealed the coalition army's march on the Spartans (Xen. *Hell.* 4.2.19). They, in turn, quickly sacrificed "the customary she-goat" to Artemis with the enemy as close as 180 m (Xen. *Hell.* 4.2.20)
12. **Battle array:** the Boiotians finally faced the Achaians (not previously mentioned) to the left, and the Athenians the Spartans. The Boiotians furthermore neglected the array agreed upon by the allies and themselves formed up *deeper* than 16. The Spartans were at the extreme right of their own line (facing the Athenians), and next to them the Tegeans (Xen. *Hell.* 4.2.18–21)
13. **Hoplite indicators**
    ὁπλίτης: Xen. *Hell.* 4.2.16, 4.2.17
    φάλαγξ: Xen. *Hell.* 4.2.13, 4.2.18
14. **Indicators of other troop types**
    τοξότης / τοξεύειν etc.: Xen. *Hell.* 4.2.14, 4.2.16,
    ἱππεύς / ἵππος: Xen. *Hell.* 4.2.16, 4.2.17; Diod. Sic. 14.83.1

σφενδονήτης: Xen. *Hell.* 4.2.16
γυμνός / ψίλος: Xen. *Hell.* 4.12.14, 4.2.17
15. **Mercenaries:** –
16. **Description of battle:** the Boiotians on the right wing of the allies' front line charged, leading the phalanx slantwise to the right, probably so as to make an encircling movement. The Athenians followed suit, although realising the danger of becoming encircled themselves. The Spartans were not aware of the charge until they heard the enemy's paian sung; then they, too, moved into battle. The Spartan right extended far beyond the Athenians' left; so much so that only six Athenian *phylai* came to grips with the Spartans; the rest clashed with the Tegeans. The Spartans sacrificed their usual goat at a little more than 180 m distance from the enemy lines, then charged, wheeling their right inwards in order to encircle the Athenians. All Spartan allies were routed except the Pellenians, who stood their ground against the Thespians. Here the battle was static. The Spartans routed the Athenians and encircled them with their right extension; they then marched past them and charged the Argives, the Corinthians and some Thebans returning from pursuit, catching them in their right (γυμνά). The four Athenian *phylai* opposite the Tegeans lost only few, since they had passed the Spartan encircling movement. The remainder of the allied army escaped to Corinth (Xen. *Hell.* 4.2.18–23)
17. **Duration of battle:** according to Diodoros (14.83.2), the battle lasted "until nightfall"
18. **Decisive factors:** the Spartan encircling movement. Probably the allies' breaking ranks to pursue the Spartans' allies aids the Spartan victory or 'rolling up' of the allied lines
19. **Victor:** –
20. **Defeated:** –
21. **Pursuit:** –
22. **Casualties (victor):** Derkylidas reports to Agesilaos that they have lost only 8 Spartans (Xen. *Hell.* 4.3.1) – probably their allies' losses are rather higher. Diodoros (14.83.2) says 1100 Spartans and allies
23. **Casualties (defeated):** Agesilaos again says that enemy casualties are παμπληθεῖς (Xen. *Hell.* 4.3.1); Diodoros (14.83.2) says 2800
24. **Prisoners:** –
25. **Trophy:** Xen. *Hell.* 4.2.23
26. **Exchange of bodies:** –
27. **Miscellaneous:** when the encircling Spartans were about to charge the Argives frontally (– how?), someone shouted out to the polemarch (Xen. *Hell.* 4.2.22) to let their front ranks pass and take them in the flank, τὰ γυμνά
28. **Sources:** Xen. *Hell.* 4.2.9–4.3.1; Diod. Sic. 14.83.1–2; Dem. 20.52–53; Plut. *Ages.* 18; Polyaen. 2.1.19; Frontin. *Str.* 2.6.5; Paus. 3.9.13
29. **Bibliography**
    Montagu (2000) 85
    Pritchett (1969) 73–84
    Roy (1971) 439–441
    Underhill (1900) 124–127

(26) OINOPHYTA 457

1. **Name and date:** Oinophyta 457 (Thuc. 1.108.3; Diod. Sic. 11.83.1)
2. **Time of year:** late summer (two months after the battle of Tanagra [inv. no. 40], Thuc. 1.108.2)
3. **Geographic locality:** at Oinophyta, some 5–6 km west of Tanagra (Thuc. 1.108.2; Barrington 58 F1)
4. **Battlefield topography:** plain
5. **Weather conditions:** –

6. **Encampment:** –
7. **Combatants:** Boiotians vs. Athenians. The Thebans (as usual) had quite a lot of cavalry (Polyaen. 1.35.2)
8. **Non-combatants:** –
9. **Numbers of armed forces:** the Boiotians numbered quite many (Diod. Sic. 11.83.1)
10. **Commanders:** Myronides (Athenians) (Thuc. 1.108.2)
11. **Pre-battle rites/exhortations:** –
12. **Battle array:** –
13. **Hoplite indicators**
    (ὅπλον / -α): Polyaen. 1.35.2 (θεῖναι τὰ ὅπλα)
    παράταξις / παρατάσσω: Polyaen. 1.35.1
14. **Indicators of other troop types:** –
15. **Mercenaries:** –
16. **Description of battle:** according to Polyainos (1.35.1), Myronides employed a stratagem, charging first with the left wing δρόμῳ and shouting out "we are winning on the left!", thereby frightening the Thebans
17. **Duration of battle:** according to Diorodos, the battle lasted *all day* (διημέρευσαν ἐν τῇ μάχῃ) (Diod. Sic. 11.83.1)
18. **Decisive factors:** –
19. **Victor:** Athenians
20. **Defeated:** Boiotians
21. **Pursuit:** –
22. **Casualties (victor):** –
23. **Casualties (defeated):** –
24. **Prisoners:** –
25. **Trophy:** –
26. **Exchange of bodies:** –
27. **Miscellaneous:** –
28. **Sources:** Thuc. 1.108.2–3; Diod. Sic. 11.83.1; Pl. *Menex.* 242b; Polyaen. 1.35
29. **Bibliography**
    Gomme (1945) 317–319
    Montagu (2000) 60

(27) OLPAI 426/5

1. **Name and date:** Olpai 426/5 (Thuc. 3.105.1)
2. **Time of year:** Thucydidean "winter" (Thuc. 3.105.1). Gomme suggests November: Gomme (1956a) 418
3. **Geographic locality:** at the hill of Olpai in the middle of the Gulf of Ambrakia about 4,5 km from Amphilochian Argos (Thuc. 3.105.1; Barrington 54 D4)
4. **Battlefield topography:** the enemies struck camp on each side of a ravine (χαράδρα) running east-west (Thuc. 3.106.3), and Demosthenes placed some troops in ambush in a hollow road which is λοχμώδης (Thuc. 3.107.3)
5. **Weather conditions:** –
6. **Encampment:** Eurylochos' Peloponnesians broke camp and marched forth from Proschion (Thuc. 3.106.1), and later they pitched camp at (ἐπί) the so-called Metropolis (Thuc. 3.107.1). The Ambrakiots with Demosthenes pitched camp near a water-course near Olpai where they remained inactive for five days (Thuc. 3.107.3)
7. **Combatants:** Akarnanians, Amphilochians, Messenians, and Athenians *vs.* Ambrakiots and Peloponnesians (Thuc. 3.106.1, 3.107.1–2)
8. **Non-combatants:** –
9. **Numbers of armed forces:** 3000 Ambrakian hoplites (Thuc. 3.105.1); 3000 Peloponnesian

hoplites already with Eurylochos, Makarios, and Menedaios (Thuc. 3.100.2). Demosthenes commanded 200 Messenian hoplites and 60 Athenian archers (Thuc. 3.107.1), apart from the Akarnanians and Amphilochians (Thuc. 3.105.2). Meanwhile, 20 Athenian ships under the command of Aristotle and Hierophon blockaded Olpai from the sea (Thuc. 3.105.3, 3.107.1, 3.109.1)

10. **Commanders:** Demosthenes for the Akarnanians with allies; for the Ambrakiots and their forces, Eurylochos (with Makarios and Menedaios, cf. Thuc. 3.100.2) (Thuc. 3.107.2; Polyaen. 3.1.2)
11. **Pre-battle rites/exhortations:** –
12. **Battle array:** Demosthenes commanded the right wing with his few Messenians and Athenians; the rest of the line being composed of Akarnanians and Amphilochian javelin-throwers. Besides, Demosthenes had left about 400 hoplites and light troops in an ambush to attack the Peloponnesians' wing, reaching beyond the Akarnanians. The Peloponnesians were drawn up ἀναμίξ the Ambrakiots, save the Mantineans who held the left wing, though not the extreme left: this was held by Eurylochos and his men (?), opposite Demosthenes and his Messenians (Thuc. 3.107.4; Polyaen. 3.1.2)
13. **Hoplite indicators**
    ὁπλίτης: Thuc. 3.100.2, 3.105.1, 3.107.1, 3.107.3; Polyaen. 3.1.2
14. **Indicators of other troop types**
    τοξότης / τοξεύειν etc.: Thuc. 3.107.1
    ἀκοντιστής: Thuc. 3.107.4
    γυμνός / ψίλος: Thuc. 3.107.3; Polyaen. 3.1.2
15. **Mercenaries:** Thuc. 3.109.2
16. **Description of battle:** after the commencement of hostilities, the Peloponnesians began to encircle their enemies' right wing. The Akarnanians then charged out of the ambush and straight into the rear of the encircling part of the Peloponnesian phalanx, putting them to flight. The rest of the army saw that their crack troops and their general were routed and panicked. Now Demosthenes and his Messenians advanced, doing "most of the work". Meanwhile, the Ambrakiots were victorious on the right wing, driving back their adversaries towards Argos. When they saw most of their army beaten they retired, suffering greatly at the hands of the Akarnanians. They lost many men during their retreat to the fort at Olpai because they hurried on without any discipline, except for the Mantineans (Thuc. 3.108.1–3; Polyaen. 3.1.2)
17. **Duration of battle:** it lasted ἐς ὀψέ (Thuc. 3.108.3)
18. **Decisive factors:** the ambush placed by Demosthenes spread panic through the enemy army (Thuc. 3.108.1)
19. **Victor:** Demosthenes and Akarnanians
20. **Defeated:** Peloponnesians and Eurylochos
21. **Pursuit:** Thucydides states that the Ambrakiots at first defeated their immediate adversaries and pursued them towards Argos; but upon returning and seeing the rest of their army defeated, they too turned to flee and just barely reached safety in Olpai, Thuc. 3.108.2–3: this seems to indicate pursuit
22. **Casualties (victor):** c. 300 (Thuc. 3.109.2)
23. **Casualties (defeated):** Diodoros (12.60.4) mentions a force of 1000 Ambrakiots which Demosthenes comes across and very nearly annihilates
24. **Prisoners:** –
25. **Trophy:** put up by Demosthenes and the Akarnanians (Thuc. 3.109.2)
26. **Exchange of bodies:** the Peloponnesians received their dead back under truce; the Akarnanians merely collect their own (Thuc. 3.109.2)
27. **Miscellaneous:** the two armies camped opposite each other for five days before joining battle (Thuc. 3.107.3). The Peloponnesian general Eurylochos was killed in action (Thuc. 3.109.1)
28. **Sources:** Thuc. 3.105–109; Diod. Sic. 12.60; Polyaen. 3.1.2
29. **Bibliography**

Gomme (1956a) 415–422
Hornblower (1991) 531–535
Montagu (2000) 66

(28) OLYNTHOS 381

1. **Name and date:** Olynthos 381 (Xen. Hell. 5.3.3)
2. **Time of year:** at the break of spring (Xen. Hell. 5.3.1)
3. **Geographic locality:** just outside Olynthos in Chalkidike (Xen. Hell. 5.3.3; Barrington 51 A1)
4. **Battlefield topography:** –
5. **Weather conditions:** –
6. **Encampment:** –
7. **Combatants:** Teleutias' Spartan force (with allies ?) vs. Olynthians (Xen. Hell. 5.3.3)
8. **Non-combatants:** –
9. **Numbers of armed forces:** –
10. **Commanders:** Teleutias (Spartans) (Xen. Hell. 5.3.3); Olynthians: not related
11. **Pre-battle rites/exhortations:** –
12. **Battle array:** none, which was exactly the problem: Teleutias' men attacked bit by bit and in no order whatsoever (Xen. Hell. 5.3.3–6)
13. **Hoplite indicators**
    ὁπλίτης: Xen. Hell. 5.3.5, 5.3.6
    φάλαγξ: Xen. Hell. 5.3.6
    (ὅπλον / -α): Xen. Hell. 5.3.5
14. **Indicators of other troop types**
    πελτάστης / πέλτη: Xen. Hell. 5.3.3, 5.3.4, 5.3.5, 5.3.6
    ἱππεύς / ἵππος: Xen. Hell. 5.3.3, 5.3.4, 5.3.5, 5.3.6
15. **Mercenaries:** –
16. **Description of battle:** Teleutias was incensed by the insolence of a company of Olynthian cavalry, which inspected the enemy army at close quarters; and he accordingly ordered the peltasts to charge them. The horsemen re-crossed the river and charged the peltasts, who displayed no caution at all, and caught them unawares, killing more than 100 of them, including their officer, Tlemonidas. Teleutias next lead his hoplites in an ill-advised attack and ordered the remaining peltasts and horsemen to press the attack. Driven back by enemy fire from the walls and towers, they soon became confused and disordered, and an easy prey to first peltasts, then hoplites, taking advantage of their broken order. Teleutias himself is killed fighting (Xen. Hell. 5.3.3–6)
17. **Duration of battle:** –
18. **Decisive factors:** Teleutias' poor generalship; the Spartan peltasts' over-confidence
19. **Victor:** Olynthians
20. **Defeated:** Spartans and Teleutias
21. **Pursuit:** the beaten Spartans retreated in all directions, not keeping any order; and their pursuers slaughtered them in great numbers (Xen. Hell. 5.3.6)
22. **Casualties (victor):** –
23. **Casualties (defeated):** many (Xen. Hell. 5.3.6)
24. **Prisoners:** –
25. **Trophy:** –
26. **Exchange of bodies:** –
27. **Miscellaneous:** –
28. **Sources:** Xen. Hell. 5.3.3–6
29. **Bibliography**
    Montagu (2000) 89
    Underhill (1900) 189

(29) PLATAIAI 479

1. **Name and date:** Plataiai 479 (Pind. *Pyth.* 1.75; called τᾶν πρὸ Κιθαιρῶνος μαχᾶν)
2. **Time of year:** on Boedromion 4, or about August 1 (Hdt. 9.3.2; Plut. *Arist.* 19.7)
3. **Geographic locality:** outside Plataiai, at the foothills of Mt. Kithairon
4. **Battlefield topography:** a level plain by the Asopos river north of Plataiai, chosen by the Persians for its vicinity to Thebes (main provision centre) and its good accommodation for cavalry
5. **Weather conditions:** –
6. **Encampment:** the Persians constructed a palisade with fruit trees from Theban territory: with this, they reinforced their apparently enormous camp (Hdt. 9.15.1–3; Diod. Sic. 11.30.1). The Greeks initially camped on the slopes of Mt. Kithairon (Hdt. 9.19.3), but later moved down towards Plataiai, where the access to water was easier (Hdt. 9.25.2)
7. **Combatants:** Greeks (right to left): 10,000 Spartans (whereof 5000 Spartiates), 1500 Tegeates, 5000 Corinthians, 300 Poteidaians (in Pallene), 600 Orchomenians, 3000 Sikyonians, 800 Epidaurians, 1000 Troizenians, 200 Lepreonians, 400 from Mykene and Tiryns, 1000 Phleiasians, 300 from Hermione, 600 from Eretria and Styra, 400 Chalkidians, 500 Ambrakiots, 800 from Leukas and Anaktor, 200 from Pale (on Kephallenia), 500 Aiginetans, 3000 Megarians, 600 Plataians, 8000 Athenians, commanded by Aristeides, son of Lysimachos (Hdt. 9.28.2–9.30). All of these are hoplites (Hdt. 9.29.1). **Enemy:** Persians, Medians, Baktrians, Indians, Sakai, Phrygians, Mysians, Thracians, Paionians, Aithiopians, two Egyptian tribes: Hermotybians and Kalasirians (both μαχαιροφόροι and marines), and their medising Greek allies: Boiotians, Lokrians, Melians, Thessalians, and 1000 Phokian hoplites, led by Harmokydes; Makedonians and οἱ περὶ Θεσσαλίην οἰκημένοι (Hdt. 9.31–32.2)
8. **Non-combatants:** –
9. **Numbers of armed forces:** 38,700 hoplites, 35,000 light-armed helots with the 5000 Spartiates (seven each: Hdt. 9.10.1, 9.28.2, 9.29.1); 34,500 light-armed from the other Greeks. Total: 38,700 hoplites and 69,500 light-armed: 108,200 Greek soldiers. Apart from these, 1800 Thespian survivors from Thermopylai, without hoplite arms. Against these, 300,000 barbarians and 50,000 medising Greeks (Hdt. 9.32.2, 8.113.3). The cavalry is not calculated. Cf. Diod. Sic. 11.30.1
10. **Commanders:** Pausanias of Sparta for the Greeks (Hdt. 9.10.1–2); Mardonios (and to a certain extent Masistios) for the Persians (Hdt. 9.20, Hdt. 9.1)
11. **Pre-battle rites/exhortations:** the Greeks received favourable omens for fighting defensively, but not for crossing the Asopos (Hdt. 9.36). Thereafter, however, the omens were altogether unfavourable for fighting either Persians or medising Greeks (Hdt. 9.38). The Spartans, under attack from Mardonios' troops much later, sacrificed but still could not get favourable omens (Hdt. 9.61.3); but when Pausanias looked towards the Heraion and prayed for help, the Tegeans suddenly countercharged; and in the same moment the omens became favourable for the Spartans (Hdt. 9.62.1)
12. **Battle array:** Persians opp. Spartans and Tegeans; Medians opp. Corinth, Poteidaia, Orchomenoi, and Sikyon; Baktrians opp. Epidauros, Troizen, Lepreon, Tiryns, Mykene, and Phleius; Indians, opp. Hermione, Eretria, Styra, and Chalkis; Sakai opp. Ambrakia, Leukas, Pale, and Aigina; medising Boiotians, Lokrians, Melians, Thessalians, Phokians, and Makedonians (and "those around Thessaly") opp. Athenians (Hdt. 9.28.2–32.2)
13. **Hoplite indicators**
    ὁπλίτης: Hdt. 9.28.3, 9.29.1, 9.30.1; Plut. *Arist.* 11.2
    φάλαγξ: Diod. Sic. 11.31.1; Plut. *Arist.* 11.7, 14.2, 18.2
    ἀσπίς: Plut. *Arist.* 17.6
    ὠθέω / ὠθισμός: (Hdt. 9.26.1), Hdt. 9.62.2
    θώραξ: Hdt. 9.22.2
    δόρυ: Hdt. 9.62.2; Plut. *Arist.* 14.5 (στύραξ), 18.3
    (ὅπλον / -α): Hdt. 9.25.3, 9.30, 9.52, 9.53.1
    θώραξ: (Hdt. 9.22.2)

14. **Indicators of other troop types**
    τοξότης / τοξεύειν etc.: Hdt. 9.22.1, 9.49.2, 9.60.3; Plut. *Arist.* 14.2, 16.4
    ἱππεύς / ἵππος: Hdt. 9.20, 9.21.1, 9.21.2, 9.23.1, 9.23.2, 9.24, 9.25.1, 9.39.1, 9.40, 9.49.2, 9.51.3, 9.52, 9.54.2, 9.57.3, 9.60.1, 9.60.2, 9.69.2; Plut. *Arist.* 14.1, 16.6; Diod. Sic. 11.30.2
    γυμνός / ψιλός: Hdt. 9.28.2, 9.29.1, 9.29.2, 9.30.1, 9.61.2
15. **Mercenaries:** –
16. **Description of battle:** the Greeks first took up position on the slopes of Mt. Kithairon. Here, the Megarians were attacked by the Persian cavalry at a vulnerable point in the Greek line. The Athenians came to their aid, and they succeeded in killing the cavalry commander Masistios. The Persian cavalry withdrew. The Greeks then moved to lower ground, nearer to a spring. Here, the two armies eyed each other for days, neither making the first move, except for Mardonios' cavalry raid on a Greek supply caravan in the pass over Kithairon. The Persian cavalry again harassed the Greeks and somehow choked the source. Pausanias accordingly withdrew at night to some higher area called the Island. However, some of the Greek contingents moved directly towards Plataiai and never saw action at all; and a Spartan commander, Amompharetos, refused to follow the others but was forced to do so or face certain annihilation at daybreak. The Athenians took a different route than the Spartans, who were sighted (along with the Tegeates) by the Persians at dawn. Mardonios ordered an all-out assault, and Pausanias sent a distress signal to the Athenians. On their way, they were intercepted by the Boiotians, slowing the Athenians' progress. Meanwhile, the Spartans and Tegeates, pinned down by enemy missiles, finally charged the wicker shields covering the Persian archers, breaking them down and forcing the Persians to retreat. This was enough to send Artabazos running with his 40,000 men towards the Hellespont. In the mêlée, Mardonios was killed, and panic ensued. The Spartans charged the enemy camp but were unable to scale the walls. This was solved by the arrival of the Athenians who were apparently more skilled in such matters. They finally made a breach in the wall, and the Greeks poured in to massacre the terrified Persians (Hdt. 9.61–70)
17. **Duration of battle:** probably a long time, though it is hard to tell how long the Spartans and Tegeates cowered under their shields (Plut. *Arist.* 17.6). The sheer workload of routing and slaughtering so many men must have taken some time. Herodotos does say that the battle lasted χρόνον ἐπὶ πολλόν until it came to *othismos* (Persians *vs.* Lakedaimonians, Hdt. 9.62.2), and χρόνον ἐπὶ συχνόν (Athenians *vs.* Boiotians, Hdt. 9.67)
18. **Decisive factors:** the Greeks' body armour and hoplite tactics seem to have decided the outcome once again; also, Mardonios' death
19. **Victor:** allied Greeks
20. **Defeated:** Persians and allies
21. **Pursuit:** the fleeing Persians were naturally pursued by the victorious Greeks (Hdt. 9.68). Those of the Greek contingents that had not participated, now also took up pursuit, but the Megarians and Phleiasians were spotted by Theban cavalry in their disorganised charge, and were driven into the hills, losing some 600 men (Hdt. 9.69.1–2). Diodoros describes the division of labour among the pursuers (Diod. Sic. 11.32.1)
22. **Casualties (victor):** 1,360 (Plut. *Arist.* 19.4; cf. Hdt. 9.70.5: 91 Spartans, 18 Tegeates, and 52 Athenians)
23. **Casualties (defeated):** at least 257,000 (Hdt. 9.66, 70.5; Plut. *Arist.* 19.4.). Diodoros gives at least 100,000 (11.32.5)
24. **Prisoners:** 3000 (by inference, Hdt. 9.70.5). Diodoros says they were more than the Greeks, and butchered to the ludicrous number of 100,000 in order to reduce the risk of unpleasant surprises (11.32.5)
25. **Trophy:** –
26. **Exchange of bodies:** –
27. **Miscellaneous:** –
28. **Sources:** Pind. *Pyth.* 1.75–80; Hdt. 9.19–70; Plut. *Arist.* 11–19; Diod. Sic. 11.30–32
29. **Bibliography**
    Bradford (1992) 27–33

How & Wells (1928²) vol. II 294–317
Hunt (1997) 129–144
Montagu (2000) 57–58
Munro (1899) 185–197
Pritchett (1985b) 92–137
Wallace (1982) 183–192

(30) POTEIDAIA 432

1. **Name and date:** Poteidaia 432 (Diod. Sic. 12.34.3–4)
2. **Time of year:** between early summer and autumn; cf. Gomme (1945) 223–224
3. **Geographic locality:** just north of Pallene – perhaps 3 km – on the isthmus between Poteidaia and Olynthos (Thuc. 1.62.1; Barrington 51 A1)
4. **Battlefield topography:** Olynthos offers a view of the area all the way to Poteidaia, due to its high position (Thuc. 1.63.1)
5. **Weather conditions:** –
6. **Encampment:** the Poteidaians and their allies pitched camp just outside Olynthos, guarding the entrance point to the isthmus. Here they were provided with supplies from an *agora* (Thuc. 1.62.1)
7. **Combatants:** Poteidaians and allied Corinthian and Chalkidian forces as well, Macedonian cavalry (Thuc. 1.60, 1.62.2–4); Athenians and their allies (i.a. Macedonian cavalry) (Thuc. 1.61.4)
8. **Non-combatants:** –
9. **Numbers of armed forces:** 3000 Athenian hoplites, 600 of Philip's and Pausanias' Macedonian horsemen, a lot of allies (ψιλοί?), and 70 triremes (Thuc. 1.61.4). Poteidaians: themselves (?) and 1000 Corinthian hoplites, 600 other Peloponnesians μίσθῳ πειθόμενοι, and 400 ψιλοί (Thuc. 1.60.1–2, 1.61.4; Diod. Sic. 12.34.4: 2000 Corinthians)
10. **Commanders:** for the Poteidaians, Aristeus of Corinth and prince Perdikkas of Macedonia as cavalry commander; for the Athenians Kallias (and four colleagues) (Thuc. 1.62.2–4)
11. **Pre-battle rites/exhortations:** –
12. **Battle array:** Aristeus had laid an ambush consisting of Chalkidians and allies and Philip's 200 horsemen near or at Olynthos. They were to attack the Athenians in the rear when they advanced on to the isthmus. Kallias, in anticipation of this, sent out his 600 Macedonian horse and some of their allies to prevent the attack (Thuc. 1.62.3–4). Other than this, it seems that Aristeus and his Corinthians and picked men were stationed on one wing (Thuc. 1.62.6)
13. **Hoplite indicators**
    ὁπλίτης: Thuc. 1.57.6, 1.60.1, 1.61.1, 1.61.4
    (ὅπλον / -α): Pl. *Symp.* 220e
14. **Indicators of other troop types**
    ἱππεύς / ἵππος: Thuc. 1.61.4, 1.62.2, 1.62.4, 1.63.2
    γυμνός / ψιλός: Thuc. 1.60.1
15. **Mercenaries:** the use of Peloponnesian mercenaries on the part of the Corinthians is implied in Thuc. 1.60.1 (μισθῷ πείσαντες)
16. **Description of battle:** Aristeus and his picked troops drove back the troops opposed to them and pursued them ἐπὶ πολύ, but the Athenians quickly defeated the Poteidaians and the other Peloponnesians, driving them behind the walls of Poteidaia. Aristeus thus was cut off from rescue: his ambush had been signalled but was cut off by the Athenians' allies. He then decided to ξυναγεῖν his troops, brave the rain of missiles δρόμῳ and join the now besieged Poteidaians. This was accomplished by retreating along the breakwater back to the city, though not without a certain number of his men being lost to enemy fire (Thuc. 1.62.4 – 63.2)
17. **Duration of battle:** the Athenians' victory comes about "fast" (Thuc. 1.63.2)

18. **Decisive factors:** probably Aristeus' ambush, which misfired
19. **Victor:** Athenians
20. **Defeated:** Poteidaians and allies
21. **Pursuit:** Aristeus' troops scored a partial victory in their sector and pursued the beaten enemy ἐπὶ πολύ, Thuc. 1.62.6. Returning from the pursuit, they were forced to brave an Athenian barrage of missiles, but succeeded in reaching the city (though not without losses) (Thuc. 1.63.1)
22. **Casualties (victor):** 150, including Kallias (Thuc. 1.63.3)
23. **Casualties (defeated):** somewhat less than 300 (Thuc. 1.63.3; Diod. Sic. 12.34.4): more than 300
24. **Prisoners:** –
25. **Trophy:** erected by the Athenians (Thuc. 1.63.3)
26. **Exchange of bodies:** the Athenians gave back the Poteidaian fallen under truce (Thuc. 1.63.3)
27. **Miscellaneous:** according to Gomme, the isthmus north of Poteidaia is wide enough to accommodate 3000 men drawn up four deep if each man needs 1 m (3 ft.). The ambush troops were signalled for (σημεῖα ἤρθη, Thuc. 1.63.2). According to Plato, Alkibiades was awarded the ἀριστεῖα in connection with the battle (*Symp.* 220d, cf. Plut. *Alc.* 7.3): the prize suggested by Sokrates was a wreath and a panoply
28. **Sources:** Thuc. 1.56–65; Diod. Sic. 12.34, 12.37; *IG* I^2 945 (= Tod 59); Pl. *Chrm.* 153a–c, *Symp.* 219e – 220e; Plut. *Alc.* 7.3
29. **Bibliography**
    Gomme (1945) 199–224
    Hornblower (1991) 97–107
    Montagu (2000) 63
    Shrimpton, G.S. (1984) 7–12
    Tomlinson, R. (2000) 529–532

(31) PYLOS 425

1. **Name and date:** Pylos 425 (Thuc. 4.3.1–2; Diod. Sic. 12.61.1)
2. **Time of year:** early to mid-May (Thuc. 4.2.1): πρὶν τὸν σῖτον ἐν ἀκμῇ εἶναι. About this time, Demosthenes' troops began to fortify Pylos; and the whole complicated chain of events is plausibly calculated by Wilson to last some 18 days (Wilson [1979] 67–72)
3. **Geographic locality:** at the promontory of Pylos in Messene on the south-west tip of the Peloponnese (Barrington 58 B4)
4. **Battlefield topography:** rough and rocky (χαλεπὴ καὶ πετρώδη) all the way to the water (Thuc. 4.9.2, Thuc. 4.12.2)
5. **Weather conditions:** –
6. **Encampment:** Demosthenes' crews and soldiers quickly occupied and fortified Pylos, thus in effect establishing a fortress, where their quarters also were (Thuc. 4.3.2–4.4.3)
7. **Combatants:** Spartans and allied Peloponnesians vs. Athenians and 40 allied Messenian hoplites
8. **Non-combatants:** –
9. **Numbers of armed forces:** difficult to tell: the Spartans quickly mustered great numbers of troops from the entire Peloponnese (Thuc. 4.8.1–2). As for the forces at Demosthenes' disposal, there seems to have been a mere 90 hoplites (*epibatai*) on board the five ships left him by Eurymedon and Sophokles; however, as Wilson argues, the navy bound for Sicily most likely left Demosthenes with more troops than this (Thuc. 4.9.1–2; Wilson [1979] 65–67), since the defence of the fort at Pylos required considerably more than 90 hoplites (and the 60 hoplites under Demosthenes' personal command could never be described as a minority). Possibly some 110 hoplites were left by the other commanders; 50 came from Demosthenes' own five ships, and 40 Messenian hoplites complemented the force (Thuc. 4.9.1)

10. **Commanders:** Demosthenes (Athenians) vs. Thrasymelidas, son of Kratesikles (Spartan admiral), and Brasidas (general) (Thuc. 4.2.4, 4.11.2, 4.11.4)
11. **Pre-battle rites/exhortations:** Demosthenes delivers a *parakeleusis* to his troops (Thuc. 4.10)
12. **Battle array:** the Athenians (perhaps some 150 men, see above) defended the fortress from the landward side, whilst Demosthenes and his 60 picked hoplites kept the Spartans from landing
13. **Hoplite indicators**
    ὁπλίτης: Thuc. 4.9.1, 4.9.2, 4.13.3
    ἀσπίς: Thuc. 4.9.1, 4.12.1; Diod. Sic. 12.62.4–5 (Brasidas')
    ὠθέω / ὠθισμός: Thuc. 4.11.3
    (ὅπλον / -α): Thuc. 4.9.1, 4.14.2
14. **Indicators of other troop types**
    τοξότης / τοξεύειν etc.: Thuc. 4.9.2
15. **Mercenaries:** –
16. **Description of battle:** the Spartan navy detachment from Korkyra arrived early and blockaded the Athenian stronghold from the sea. Despite the efforts of Brasidas and his men, a landing could not be forced from any of their 43 ships, because the Spartans tried to land at the exact spot anticipated by Demosthenes, and because of the roughness of the shoreline. The Spartans next neglected to cut off the entrance at the northern end of Sphakteria as they had originally planned, and the Athenian navy detachment (some 50 triremes with their allies) sailing from Zakynthos defeats their Spartan opponents by attacking from both ends of the strait, sinking some, ramming others on the shore and towing some away, which their crews had left in a panic. The Spartans ran out into the water, taking hold of their ships and putting up fierce resistance. After a bitter struggle with many casualties the Athenians were only able to keep those ships which they took at the outset – but 420 crack Spartan hoplites were now trapped on Sphakteria (Thuc. 4.1–15, 4.23).
17. **Duration of battle:** several days (it is, after all, a blockade); but the first battle (the Spartan ships attacking) lasts "that day and part of the next" (Thuc. 4.13.1)
18. **Decisive factors:** possibly the Spartans' failure to block the ἔσπλους
19. **Victor:** Athenians and Demosthenes and allied Messenians
20. **Defeated:** Spartans and allied Peloponnesians
21. **Pursuit:** –
22. **Casualties (victor):** –
23. **Casualties (defeated):** the Spartans did not spare themselves and lost "many men" (Diod. Sic. 12.61.5, 12.62.6)
24. **Prisoners:** –
25. **Trophy:** Thuc. 4.12.1, 4.14.5
26. **Exchange of bodies:** the Spartan fallen were given back, Thuc. 4.14.5 (strangely, no mention of a truce)
27. **Miscellaneous:** –
28. **Sources:** Thuc. 4.2–23; Diod. Sic. 12.61–62
29. **Bibliography**
    Beardsworth & Wilson (1970) 112–118
    Burrows (1908) 148–150
    Compton & Awdry (1907) 274–283
    Gomme (1956b) 437–463
    Hornblower (1996) 149–180
    Montagu (2000) 68
    Strassler (1990) 110–125
    Westlake (1974) 211–226
    Wilson (1979)
    Wilson & Beardsworth (1970) 42–52

(32) SOLYGEIA 425

1. **Name and date:** Solygeia 425 (Thuc. 4.42.2; Polyaen. 1.39.1)
2. **Time of year:** –
3. **Geographic locality:** near the village of Solygeia, perched on a hill of the same name, 10.5 km (= 60 stades) from Corinth and 3.5 km (= 20 stades) from the Isthmus (Barrington 58 D2)
4. **Battlefield topography:** much of the battle was fought uphill as the Athenians tried to advance into the country from the shore (prob. modern Galatáki), so the coast must be rising steeply at this point (cf. Thuc. 4.43.3)
5. **Weather conditions:** –
6. **Encampment:** –
7. **Combatants:** Athenians with allies: Milesians, Andrians, and Karystians *vs.* Corinthians, Thuc. 4.42.1, 4.42.3
8. **Non-combatants:** –
9. **Numbers of armed forces:** 2000 Athenian hoplites and 200 cavalry on board 20 horse-carrier ships and an uncertain number of allies (Thuc. 4.42.1). Corinthians: πανδημεί (Thuc. 4.42.3)
10. **Commanders:** Nikias (with two colleagues) for the Athenians (Thuc. 4.42.1); Lykophron and Battos for the Corinthians (Thuc. 4.43.1; Polyaen. 1.39.1)
11. **Pre-battle rites/exhortations:** –
12. **Battle array:** Athenians and Karystians apparently on the right wing, Thuc. 4.43.3. Not much else is clear
13. **Hoplite indicators**
    ὁπλίτης: Thuc. 4.42.1, Polyaen. 1.39.1
    ὠθέω / ὠθισμός: Thuc. 4.43.3
    (ὅπλον / -α): Thuc. 4.44.1
14. **Indicators of other troop types**
    ἱππεύς / ἵππος: Thuc. 4.42.1, 4.44.1
15. **Mercenaries:** –
16. **Description of battle:** the Corinthians were pushed back with difficulty by the Athenians and Karystians on the right wing, towards a picket, where they regrouped and charged downhill. They pelted the Athenians with stones, struck up a paian, and charged them again. The Athenians kept them at bay until the enemy was reinforced. The Athenians' right wing was then driven back to the sea, from whence they countercharged. The outcome was uncertain for a long time along the Corinthian right, but the Athenians finally, aided by their cavalry, routed the Corinthians, driving them back up the hill, from whence they do not again come down to fight. The Athenians quickly retire to the ships when the older men from Corinth and half the army, posted at Kenchreai, move towards the cloud of dust that they have churned up (Thuc. 4.43.2–44.5). Polyainos (1.39.1) says that Nikias placed hoplites in ambush first, then surreptitiously sailing away, only to land again very conspicuously. The ambush parties then aided in routing the surprised Corinthians
17. **Duration of battle:** it apparently lasted very long indeed (Thuc. 4.44.1: συνεχῶς, χρόνον πολύν)
18. **Decisive factors:** the Athenians' cavalry helped win the day: the Corinthians had none (Thuc. 4.44.1; Ar. *Eq.* 595–610)
19. **Victor:** Athenians
20. **Defeated:** Corinthians
21. **Pursuit:** there was a partial defeat of the Athenians' right wing, which the Corinthians pursued right down to the edge of the water (Thuc. 4.43.4). Later, most of the Corinthian casualties occurred during the τροπή, not the actual pursuit, since all but the right wing withdrew in good order (Thuc. 4.44.2)
22. **Casualties (victor):** "not quite fifty" (Thuc. 4.44.5–6)
23. **Casualties (defeated):** 212 (Thuc. 4.44.6)

24. **Prisoners:** –
25. **Trophy:** Thuc. 4.44.3
26. **Exchange of bodies:** although the Athenians erected a trophy, they sailed out to some islands just off the coast with their own fallen to avoid a Corinthian relief force, except for two bodies that they could not find. Therefore, although having won the battle, they were forced to ask their two dead back under a truce (Thuc. 4.44.6)
27. **Miscellaneous:** Thucydides is very explicit that the entire battle was tough and ἐν χερσί (Thuc. 4.43.2–4). It is noticeable that, again, the majority of casualties occurred during the rout – including Lykophron (Thuc. 4.44.2)
28. **Sources:** Thuc. 4.42–44; Ar. *Eq.* 595–610; Polyaen. 1.39.1
29. **Bibliography**
    Gomme (1956a) 489–495
    Hornblower (1996) 197–204
    Montagu (2000) 69

(33) SPARTOLOS 429

1. **Name and date:** Spartolos 429 (Thuc. 2.79.2–3)
2. **Time of year:** mid- to late May (Thuc. 2.79.1: τοῦ σίτου ἀκμάζοντος)
3. **Geographic locality:** outside the city of Spartolos in Chalkidike (Thuc. 2.79.2; Barrington 50 D4)
4. **Battlefield topography:** approx. 250 m.a.s.l.
5. **Weather conditions:** –
6. **Encampment:** –
7. **Combatants:** Athenians vs. Chalkidians and Bottiaians and their allies from Krousis and Olynthos (Thuc. 2.79.1, 2.79.4)
8. **Non-combatants:** Thucydides mentions Athenian σκευοφόροι (2.79.5)
9. **Numbers of armed forces:** 2000 Athenian hoplites and 200 horse (Thuc. 2.79.1)
10. **Commanders:** Athenians: Xenophon, son of Euripides, "with two others" (Thuc. 2.79.1): Hestiodoros and Phanomachos (Thuc. 2.70.1)
11. **Pre-battle rites/exhortations:** –
12. **Battle array:** –
13. **Hoplite indicators**
    ὁπλίτης: Thuc. 2.79.1, 2.79.2, 2.79.4
14. **Indicators of other troop types**
    πελτάστης / πέλτη: Thuc. 2.79.4
    ἱππεύς / ἵππος: Thuc. 2.79.1, 2.79.3, 2.79.5, 2.79.6
    ἀκοντιστής: Thuc. 2.79.6 (ἐσηκόντιζον)
    γυμνός / ψιλός: Thuc. 2.79.3, 2.79.5
15. **Mercenaries:** –
16. **Description of battle:** the Chalkidian hoplites (reinforced by hoplites from Olynthos) were defeated by the Athenians and withdrew into the city. However, their horse and ψίλοι were routed by the local force aided by peltasts from Krousis. The relief force from Olynthos, also bringing a number of peltasts, arrived just as the battle had begun. The Chalkidians plucked up courage and rallied, attacking again with the cavalry. A familiar scene ensued: the Athenians were forced to retreat, making sorties every now and then to drive off the pursuers. The ψίλοι and riders then fell back, only to pursue and pelt the Athenians with javelins once again when they attempted to retreat to their main force. The cavalry in particular turned the retreat into a rout by constantly frightening the Athenian hoplites. The Athenians escaped to Poteidaia and next day received their 430 fallen, including all three generals, under truce, Thuc. 2.79.3–6
17. **Duration of battle:** –

18. **Decisive factors:** the appearance of a relief force from Olynthos; the superiority of the local cavalry and light infantry
19. **Victor:** Chalkidians and Bottiaians
20. **Defeated:** Athenians
21. **Pursuit:** the Chalkidian cavalry terrorised the Athenian hoplites and routed them, pursuing them ἐπὶ πολύ towards Poteidaia (Thuc. 2.79.6)
22. **Casualties (victor):** –
23. **Casualties (defeated):** 430 and all three generals (Thuc. 2.79.7)
24. **Prisoners:** –
25. **Trophy:** –
26. **Exchange of bodies:** the Athenians received their dead back under a truce (Thuc. 2.79.7)
27. **Miscellaneous:** the peltasts are specifically named as such to distinguish them from the common ψίλοι. Apparently Thucydides himself was aware of a qualitative difference in armament and employment of the troop types
28. **Sources:** Thuc. 2.79
29. **Bibliography**
    Gomme (1956a) 212–214
    Hornblower (1991) 361
    Montagu (2000) 63

(34) SPHAKTERIA 425

1. **Name and date:** Sphakteria 425 (Diod. Sic. 12.63.3)
2. **Time of year:** Thucydides states that the siege, from the battle of Pylos to the capture of Sphakteria, lasted 72 days of which ca. 20 days were spent in negotiating a truce (Thuc. 4.39.1–2). In all, a not quite three months must have passed since the first encounter, which was probably sometime in late June (Thuc. 4.2.1; cf. Wilson [1979] 67–72); accordingly, the battle and capture of Sphakteria probably occurred in late September to mid-October, given some extra time in both ends. This squares fairly well with Thucydides' assertion that the Athenians were worried about their siege being interrupted by winter storms, Thuc. 4.27.1
3. **Geographic locality:** the island Sphakteria immediately off Pylos (Barrington 58 B4)
4. **Battlefield topography:** Sphakteria is rocky with a summit in the western end (Mt. Elias). The island is described as ὑλώδης (Thuc. 4.8.6), i.e. covered with shrubbery and thickets; and as ἀτρίβης (without paths) διὰ τὴν ἐρημίαν (Thuc. 4.29.3). This is true of the island today as well
5. **Weather conditions:** hot and sunny (Thuc. 4.35.4)
6. **Encampment:** as with Pylos (Thuc. 4.3.2–4.4.3): the Athenians still occupy Pylos and circle the island with the triremes
7. **Combatants:** trapped Spartans vs. Athenians (and some Messenians) and other allies: Imbrians, Lemnians and Ainians, plus possibly a few other allies, summoned by Nikias (Thuc. 4.28.4, 4.30.3)
8. **Non-combatants:** helots and others were persuaded (by promises of freedom or other rewards) to smuggle supplies across to the trapped hoplites on Sphakteria (Thuc. 4.26.5–6)
9. **Numbers of armed forces:** Spartans: 420 hoplites with "their" helots (Thuc. 4.8.9); some 800 hoplites from the joint forces of Kleon and Demosthenes are shipped across (Thuc. 4.31.1–2); also 800 archers and "at least as many light-armed troops" (Thuc. 4.32.2). Further, Demosthenes called out all the rowers except the θαλαμιοί; some 7700 oarsmen in all, equipped "as best they could" (Thuc. 4.32.2); cf. Gomme ad loc.
10. **Commanders:** Spartans: Epitadas, son of Molobros (Thuc. 4.8.9) (replaced by Styphax, son of Pharax, on the latter's death), and his second-in-command, Epitadas (Thuc. 4.38.1); Athenians: Demosthenes and Kleon (Thuc. 4.29.1–2); Komon, the leader of the Messenian troops (Paus. 4.26.2; Thuc. 4.36.1)

11. **Pre-battle rites/exhortations:** –
12. **Battle array:** the Spartans had posted 30 men in a guardpost, and another handful to guard the northern end of the island, towards Pylos. Epitadas and the main force occupied the central part of the island, where the terrain is relatively even, and where there was a source of water (Thuc. 4.31.2). The Athenians were divided into groups of 200 or less, intended to occupy high territory and harass the Spartan hoplites (Thuc. 4.32.3–4). The Athenian hoplites formed a line opposite the Spartans but refused to meet the Spartans, who were threatened from the rear and flanks by the ψιλοί (Thuc. 4.33.1)
13. **Hoplite indicators**
    ὁπλίτης: Thuc. 4.8.9, 4.26.7, 4.31.1, 4.31.2, 4.33.1, 4.33.2, 4.38.5
    ἀσπίς: Thuc. 4.38.1; Diod. Sic. 12.62.4
    ὠθέω / ὠθισμός: Thuc. 4.35.3 NB! *not* a proper hoplite *othismos*
    (ὅπλον / -α): Thuc. 4.30.4, 4.32.1, 4.33.2, 4.37.1, 4.37.2, 4.38.3, 4.40.1
14. **Indicators of other troop types**
    πελταστής / πέλτη: Thuc. 4.28.4
    τοξότης / τοξεύειν etc.: Thuc. 4.28.4, (4.32.4), 4.36.1
    ἀκοντιστής: (Thuc. 4.32.4)
    σφενδονήτης: Thuc. 4.32.4; Paus. 4.26.2
    γυμνός / ψιλός: Thuc. 4.32.4, 4.33.1, 4.33.2, 4.34.1, 4.35.2, 4.36.1
15. **Mercenaries:** –
16. **Description of battle:** Demosthenes' troops overran the guard post near the shore and proceeded further inland. The Spartans, seeing them approaching, formed a battle line. The Athenian hoplites did likewise, but refused to give battle. When the Spartans approached, they were pelted with javelins, stones and arrows, and the well-known pattern ensued: as soon as they sallied, the light-armed retreated, only to harass them on their way back to own lines. Exacerbating the situation further is the accidentally (?) fired forest, which had left a thick layer of ashes that produced huge dust clouds, preventing the Spartans from seeing anything. The war-cry of the Athenians' allies also deafened them and drowned any commands and orders. Finally, the Spartans retreated to their stronghold at the northern end of Sphakteria, in higher terrain, all the while being harassed by the light-armed troops. Here, they managed to hold out for quite a long time, until the leader of the Messenian detachment lead his troops around the Spartan position *via* the steep cliffs facing Pylos. Their appearance behind and above the Spartans was the beginning to the end: the Spartans were trapped between two enemy lines and finally accepted the terms for surrender after consulting with their peers on the mainland (Thuc. 4.30–38)
17. **Duration of battle:** "most of the day" (Thuc. 4.35.4)
18. **Decisive factors:** –
19. **Victor:** Athenians and allies
20. **Defeated:** Spartans
21. **Pursuit:** –
22. **Casualties (victor):** very few: ἡ γὰρ μάχη οὐ σταδαία ἦν (Thuc. 4.38.5)
23. **Casualties (defeated):** 128 out of 420 (Thuc. 4.38.5), including Epitadas and Hippagretas (left for dead), Thuc. 4.38.1
24. **Prisoners:** 292 out of 420; of these 120 Spartiates (Thuc. 4.38.5); Diod. Sic. (12.63.3) says 180 allies
25. **Trophy:** Thuc. 4.38.4
26. **Exchange of bodies:** the Spartans receive their dead back under a truce (Thuc. 4.38.4)
27. **Miscellaneous:** –
28. **Sources:** Thuc. 4.26–41; Diod. Sic. 12.63–65; Paus. 4.26.2; *IG* I³ 522 (inscription on a shield: ΑΘΕΝΑΙΟΙ ΑΠΟ ΛΑΚΕΔΑΙΜΟΝΙΟΝ ΕΚ ΠΥΛΟ)
29. **Bibliography**
    Burrows (1908) 148–150
    Compton & Awdry (1907) 274–283

Gomme (1956b) 466–489
Hornblower (1996) 184–197
Montagu (2000) 69
Nikolaidis (1990) 89–94
Wilson (1979)

(35) SYBOTA ISLAND 433

1. **Name and date:** Sybota island 433 (Thuc. 1.47.1)
2. **Time of year:** almost certainly in summer, as it is actually a battle fought at sea
3. **Geographic locality:** somewhere between the Sybota islands and the Cheimerion promontory (Thuc. 1.47.1–48.1; Barrington 54 B3)
4. **Battlefield topography:** at sea
5. **Weather conditions:** probably fine (as with Time of year)
6. **Encampment:** the Corinthians moored their ships and pitched camp at the promontory of Cheimerion (Thuc. 1.46.4–5); the Korkyraians on the island of Sybota (Thuc. 1.47.1). The Corinthians land army was encamped at the Sybota *harbour,* and so the ships withdrew there (Thuc. 1.50.3)
7. **Combatants:** Corinthians and their allies (Eleans, Megarians, Leukadians, Ambrakiots, Anaktorians) *vs.* Korkyraians and their allies, the Athenians
8. **Non-combatants:** –
9. **Numbers of armed forces:** Korkyraians: 110 ships (Thuc. 1.47.1; Diod. Sic. 12.33.4). Corinthians and allies: 150 ships (90 + 60) (Thuc. 1.46.1; Diod. Sic. 12.33.3). Athenians: ten ships, expected to fight after the "new" fashion, i.e. by ramming, cf. Thuc. 1.49.2–3 (Thuc. 1.45.1; Diod. Sic. 12.33.2). Both Korkyraians and Corinthians are expressly said to have "many hoplites" (Thuc. 1.49.1). In Thuc. 1.29.1 the Corinthians have equipped 75 ships with 2000 hoplites – this *might* indicate an average of approx. 26 hoplites per trireme. If this is so, then the Korkyraians had 2860 hoplites at their disposal; the Corinthians as many as 3900
10. **Commanders:** Athenian squadron: Lakedaimonios, son of Kimon, Diotimos, son of Strombichos and Proteas, son of Epikles (Thuc. 1.45.2); Corinthians: Xenokleides "with four others" (the allies were autonomous) (Thuc. 1.46.2). Korkyraians: Mikiades, Aisimides, Eurybatos (Thuc. 1.47.1)
11. **Pre-battle rites/exhortations:** –
12. **Battle array:** Korkyraians on the left flank in three squadrons, one under each admiral, ten Athenian ships on the right (Thuc. 1.48.3). Corinthians themselves on the left, Megarians and Ambrakiots on the right, and the other allies in the middle (Thuc. 1.48.4)
13. **Hoplite indicators**
    ὁπλίτης: Thuc. 1.49.1, 1.49.3
    (ὅπλον / -α): (Thuc. 1.53.2)
14. **Indicators of other troop types**
    τοξότης / τοξεύειν etc.: Thuc. 1.49.1
    ἀκοντιστής: Thuc. 1.49.1
15. **Mercenaries:** –
16. **Description of battle:** the two lines of ships crashed into each other and quickly became entangled because of the two backward navies' lack of skill. All in all, it resembled more a πεζομαχία than a ναυμαχία because of the reliance on hoplites stationed on the ships (so-called ἐπιβάται, placed on fighting decks, καταστρώματα), a practice which Thucydides deems "old-fashioned" (τῇ μὲν τέχνῃ οὐκ ὁμοίως). These hoplites fought it out with each other ἡσυχαζουσῶν τῶν νεῶν; probably grappling-hooks and suchlike were used to cling to the enemy ship's side. The battle was fierce (Thuc. 1.49.2–3; Diod. Sic. 12.33.4), and fortunes of war were divided: the Corinthian right flank was routed by 20 enemy ships and chased out to sea, where-

upon the pursuers sacked their camp on the shore, burning the tents. Yet the Corinthians' left flank was victorious and put the Korkyraians to flight despite the Athenians' aid (Thuc. 1.48–51.3)

17. **Duration of battle:** until nightfall, when an Athenian relief force of 20 ships appeared, scaring away the Corinthians (Thuc. 1.50.5, 1.51.5)
18. **Decisive factors:** if anything, the appearance of a relief force from Athens, Thuc. 1.50.5; Diod. Sic. 12.33.4
19. **Victor:** Korkyraians (but cf. Thuc. 1.54)
20. **Defeated:** Corinthians (as above)
21. **Pursuit:** –
22. **Casualties (victor):** 70 ships and with corresponding loss of life (Thuc. 1.54.2)
23. **Casualties (defeated):** 30 ships, Thuc. 1.54.2
24. **Prisoners:** 1000 Korkyraian prisoners taken by the Corinthians (Thuc. 1.52.2, 1.54.2)
25. **Trophy:** Thuc. 1.54.1*bis*, 1.54.2*bis*; erected by both sides due to a disagreement as to the outcome of the battle
26. **Exchange of bodies:** the Korkyraians were able to pick up their own dead, because the wind and waves carried them to their side during the night (Thuc. 1.54.1)
27. **Miscellaneous:** this account of a naval battle is included chiefly for its peculiar, heavy reliance on sea-borne hoplite troops and its expressly stated similarity to a land battle (Thuc. 1.49.1–2). Diodoros' account relies heavily on that of Thucydides
28. **Sources:** Thuc. 1.45–55; Diod. Sic. 12.33
29. **Bibliography**
    Gomme (1945) 177–199
    Hornblower (1991) 88–97
    Montagu (2000) 62–63

(36) SYRACUSE 415

1. **Name and date:** Syracuse 415 (Thuc. 6.65.2–3)
2. **Time of year:** late summer, the only time of year to venture a crossing of the Tyrrhenian sea
3. **Geographic locality:** near the temple of Zeus Olympieios (the Olympieion) outside Syracuse (Barrington 47 G4)
4. **Battlefield topography:** most likely flat and relatively unimpeded, as the battle was fought near the shore. Probably the ground was somewhat moist and soft, due to the proximity of the swamps near the Olympieion
5. **Weather conditions:** the Syracusans were intimidated by claps of thunder and heavy showers of rain (Thuc. 6.70.1), so we may assume that it was overcast, perhaps windy (thundersqualls)
6. **Encampment:** the Athenians make a fortification around their beached ships; here, presumably, they spend the night (Thuc. 6.66.1–2)
7. **Combatants:** Syracusans *vs.* Athenians and allies
8. **Non-combatants:** –
9. **Numbers of armed forces:** Athens: 5100 hoplites (1500 Athenians, 700 θῆτες equipped as ἐπιβάται, 500 Argives, 250 Mantineians and mercenaries; the remainder (= 2150) from tribute-paying islands. Besides 480 archers, 80 of these Cretans; 700 Rhodian slingers; 120 ψίλοι in exile from Megara (Thuc. 6.43). Syracusans: πανδημεί with allies (see below). In all likelihood their front line was equal to the enemy's. It is therefore not unreasonable to estimate *c.* 5000 hoplites, along with their allies from Selinus, 200 cavalry and 50 archers from Gela, and 80 cavalry and 50 archers from Kamarine (Thuc. 6.67.1–3)
10. **Commanders:** Athenians: Nikias (Thuc. 6.67.1); Syracusans: Ekphantos (hipparch) (Polyaen. 1.39.2)

11. **Pre-battle rites/exhortations:** Nikias delivers a speech to the Athenians and their allies (Thuc. 6.67.3–69)
12. **Battle array:** from right to left: Argives, Mantineians, Athenians in the middle, other allies for the rest, drawn up eight deep. Only half the army was drawn up; the rest are arranged in the camp nearby in a hollow square to protect the baggage and supply reserves (Thuc. 6.67.1). The Syracusans' hoplites – πανδημεί – were sixteen deep. The cavalry, no fewer than 1200, were placed on their right wing
13. **Hoplite indicators**
    ὁπλίτης: Thuc. 6.43, 6.67.2, 6.69.2, 6.70.3
    ἐπιβάτης: Thuc. 6.43
    ὠθέω / ὠθισμός: Thuc. 6.70.2
    (ὅπλον / -α): Thuc. 6.69.1
    παραρρήγνυμι / παράρρηξις: Thuc. 6.70.2
14. **Indicators of other troop types**
    πελταστής / πέλτη: Polyaen. 1.39.2
    τοξότης / τοξεύειν etc.: Thuc. 6.43, 6.69.2
    ἱππεύς / ἵππος: Thuc. 6.43, 6.67.2, 6.68.3, 6.70.3; Polyaen. 1.39.2
    σφενδονήτης: Thuc. 6.43, 6.69.2
    γυμνός / ψίλος: Thuc. 6.43
15. **Mercenaries:** employed by Athenians (Thuc. 6.43)
16. **Description of battle:** the Syracusans were taken by surprise, not having expected the Athenians so soon. Some had to take up their battle-stations at random ὑστεριζόντες. First, the light-armed troops skirmished and routed each other in turns, inflicting small defeats, as εἰκὸς ψιλοὺς ἀλλήλων ἐποιοῦν. The battle lasted quite long (Thuc. 6.70.1); but finally, perhaps aided by the Syracusans' superstitious fear of a thunderstorm, the Argives push back their opponents, opening a gap for the Athenians to pour through, effectively breaking the army in two and putting it to flight. The powerful Syracusan cavalry prevented an effective pursuit, however, charging any pursuing group darting out too far (Thuc. 6.69–71). Polyainos (1.39.2) relates that Nikias ordered caltrops (τρίβολοι) to be spread out in front of the army, so that the initial cavalry action was turned into a rout. Many of them (grammatical ambiguity: horses? riders?) were unable even to retreat, and so destroyed by the peltasts, who wore shoes with tough soles
17. **Duration of battle:** rather long (ἐπὶ πολύ) (Thuc. 6.70.1)
18. **Decisive factors:** the Syracusans' lacking combat experience, and superstition
19. **Victor:** Athenians
20. **Defeated:** Syracusans
21. **Pursuit:** the Athenians were unable to follow up their success with pursuit: the Syracusan cavalry prevented an all-out pursuit. Instead, they followed the retreating Syracusans as far as possible, retaining their battle order (Thuc. 6.70.3)
22. **Casualties (victor):** 50 (Thuc. 6.71.1)
23. **Casualties (defeated):** c. 260 (Thuc. 6.71.1); 400 (Diod. Sic. 13.6.5)
24. **Prisoners:** –
25. **Trophy:** Thuc. 6.70.3
26. **Exchange of bodies:** the Syracusans asked and received their dead back under truce (Thuc. 6.71.1)
27. **Miscellaneous:** the battle is ἐν χερσί (Thuc. 6.70.1). It appears from Polyainos (1.39.2) that the peltasts wore shoes or boots with tough (στέρεοι) soles; perhaps the Thracian boot
28. **Sources:** Thuc. 6.43, 67–71; Plut. *Nic.* 16; Diod. Sic. 13.6.5; Polyaen. 1.39.2
29. **Bibliography**
    Gomme, Andrewes & Dover (1970) 308–310, 341–346
    Montagu (2000) 71–72

(37) SYRACUSE 414

1. **Name and date:** Syracuse 414 (Thuc. 6.97.2)
2. **Time of year:** summer (Thuc. 6.96.1)
3. **Geographic locality:** Epipolai heights immediately outside Syracuse (Barrington 47 G4)
4. **Battlefield topography:** as the name suggests, Epipolai are indeed flat, but with very steep, rocky precipices at all points
5. **Weather conditions:** the weather was probably fair, since the Syracusans were having an ἐξέτασις on the meadows below (Thuc. 6.96.3)
6. **Encampment:** –
7. **Combatants:** Athenians and allies; Syracusans
8. **Non-combatants:** -
9. **Numbers of armed forces:** not related this time, but at Syracuse 415 (q.v.), some 5100 hoplites fought on the Athenian side; the Syracusans πανδημεί. The conditions here were probably somewhat similar; the Athenians wishing to make sure that they could take and hold Epipolai, in order to commence the investing wall, and the Syracusans were having a general (πανδημεί) drill below (Thuc. 6.96.3)
10. **Commanders:** for the Athenians, most likely Nikias and Lamachos; in Syracuse, the office of general had just been transferred to οἱ περὶ Ἑρμοκράτη στρατηγοί. Otherwise a certain Diomilos is here mentioned; an exile from Andros, who was to have commanded a guard force of 600 hoplites at the entrances to Epipolai (Thuc. 6.96.3)
11. **Pre-battle rites/exhortations:** –
12. **Battle array:** –
13. **Hoplite indicators**
    ὁπλίτης: Thuc. 6.96.3
14. **Indicators of other troop types**
    ἱππεύς / ἵππος: Plut. *Nic.* 17
15. **Mercenaries:** –
16. **Description of battle:** the Athenians succeeded in reaching the heights of Epipolai before the Syracusans realised their presence. Accordingly, they could deploy unmolested and prepare to receive the Syracusan onslaught from below. The Syracusan countercharge was naturally weak, since everybody scrambled up the slopes ὡς ἕκαστος τάχους εἶχε, and there was as much as 25 stades (*c.* 5 km) between the meadow and the Athenians (Thuc. 6.97.2–5). According to Plut. (*Nic.* 17), the Syracusan cavalry was also successfully repelled
17. **Duration of battle:** apparently rather short (Thuc. 6.97.3–4)
18. **Decisive factors:** the Syracusans' exhaustion and, above all, lack of formation
19. **Victor:** Athenians
20. **Defeated:** Syracusans
21. **Pursuit:** –
22. **Casualties (victor):** –
23. **Casualties (defeated):** some 300, including Diomilos (Thuc. 6.97.4, Diod. Sic. 13.7.3; Plut. *Nic.* 17)
24. **Prisoners:** –
25. **Trophy:** Thuc. 6.97.5
26. **Exchange of bodies:** the Athenians gave the Syracusan fallen back, Thuc. 6.97.5
27. **Miscellaneous:** –
28. **Sources:** Thuc. 6.96–97; Diod. Sic. 13.7.3; Plut. *Nic.* 17
29. **Bibliography**
    Gomme, Andrewes & Dover (1970) 370–372
    Montagu (2000) 72

(38) SYRACUSE 413

1. **Name and date:** Syracuse 413 (Thuc. 7.43.2: Epipolai)
2. **Time of year:** Gomme suggests August, Gomme (1956a) 423
3. **Geographic locality:** Epipolai plateau outside Syracuse (Barrington 47 G4)
4. **Battlefield topography:** Epipolai is a plateau with few, steep access points. The battle is fought at night with only moonlight to illuminate the action
5. **Weather conditions:** a clear sky seems to be indicated (Thuc. 7.44.2); at any rate there was quite bright light from the full moon
6. **Encampment:** the Syracusan garrison had a camp hard by the three forts on Epipolai (Thuc. 7.43.4)
7. **Combatants:** Athenians and allies: at least Korkyraians and Argives vs. Syracusans and allies (Boiotians are mentioned) (Thuc. 7.43–44)
8. **Non-combatants:** –
9. **Numbers of armed forces:** the total of the Athenian forces now totalled some 10,000 hoplites and an equal number of light-armed (Thuc. 7.42.1) – but it is by no means certain that all these participated in the battle, in spite of Thuc. 7.43.2. For the Syracusans, at least a garrison of some 300
10. **Commanders:** Athenians: Demosthenes, Eurymedon, and Menander (Thuc. 7.43.2); Syracusans: Gylippos (Thuc. 7.43.6)
11. **Pre-battle rites/exhortations:** –
12. **Battle array:** –
13. **Hoplite indicators**
    ὁπλίτης: Thuc. 7.42.1, 44.2; Plut. *Nic.* 21.1
    ἀσπίς: Plut. *Nic.* 21.8, 21.9
    δόρυ: Plut. *Nic.* 21.5
    (ὅπλον / -α): Thuc. 7.45.2; Plut. *Nic.* 21.7, 21.8
14. **Indicators of other troop types**
    τοξότης / τοξεύειν etc.: Thuc. 7.42.1; Plut. *Nic.* 21.1
    ἱππεύς / ἵππος: Thuc. 7.44.8
    ἀκοντιστής: Thuc. 7.42.1
    σφενδονήτης: Thuc. 7.42.1; Plut. *Nic.* 21.1
15. **Mercenaries:** –
16. **Description of battle:** the Athenians climbed the Epipolai and conquered the outposts, killing the garrison. Some escaped and alerted the main force (600 strong) in the three forts. Meanwhile, Gylippos and his troops from the outworks rushed out to aid them. With the impetus of their first success, the Athenians routed them and pressed their attack. However, they now fell into disorder in their exhilaration, and the Boiotians were able to make a stand and repel their attack, routing them. Now everything fell to pieces, for the Athenians were almost completely unable to communicate, and, exacerbating the situation, their Dorian allies were mistaken for the enemy. Furthermore, the Syracusans knew the ground much better and so had the advantage. In the end, many Athenians were killed in action or throwing themselves down from the plateau; and next day, many who were lost, were rounded up and killed by the Syracusan cavalry (Thuc. 7.43–44)
17. **Duration of battle:** apparently the actual fighting was quickly over (Thuc. 7.44.8)
18. **Decisive factors:** the Athenians' lacking knowledge of the terrain; the confusion of the paians and passwords
19. **Victor:** Syracusans
20. **Defeated:** Athenians
21. **Pursuit:** the victorious Syracusans naturally pursued the Athenians as the circumstances allowed, and apparently enough to drive some of them over the edge and down the slopes (Thuc. 7.44.8; Plut. *Nic.* 21.9)

22. **Casualties (victor):** –
23. **Casualties (defeated):** "a great many" (Thuc. 7.45.2); 2000 (Plut. *Nic.* 21.9)
24. **Prisoners:** –
25. **Trophy:** two erected by the Syracusans (Thuc. 7.45.1)
26. **Exchange of bodies:** the Athenians are allowed to collect their dead under truce (Thuc. 7.45.1)
27. **Miscellaneous:** many of the Athenian hoplites threw their shields away: more weapons were found than the number of Athenian dead warranted (Thuc. 7.45.2). Important observation: even *in daytime* a hoplite hardly knows what is going on around him, possibly due to the helmet (Thuc. 7.44.1)
28. **Sources:** Thuc. 7.42–44; Plut. *Nic.* 21
29. **Bibliography**
    Gomme, Andrewes & Dover (1970) 419–424
    Montagu (2000) 72–73

(39) TA KERATA 409

1. **Name and date:** Ta Kerata 409 (Diod. Sic. 13.65.1)
2. **Time of year:** –
3. **Geographic locality:** outside Nisaia, the port of Megara, on a hill called Ta Kerata, maybe because of its shape (Diod. Sic. 13.65.1; cf. Strabo 9.1.11, Barrington 58 E1–2)
4. **Battlefield topography:** not much to tell; but at least the battlefield apparently is commanded by a hill (= Ta Kerata, Diod. Sic. 13.65.1.)
5. **Weather conditions:** –
6. **Encampment:** –
7. **Combatants:** Megarians and their Sikelian allies *vs.* the Athenians (Diod. Sic. 13.65.1). The text as it is mentions also Lakedaimonians, but this is probably a transmission error. Accordingly, Vogel proposes to read Σικελιῶται rather than Λακεδαιμόνιοι twice
8. **Non-combatants:** –
9. **Numbers of armed forces:** the Megarians πανδημεί (Diod. Sic. 13.65.1), and accordingly said to be many times more than the Athenians (Diod. Sic. 13.65.2), Sikelians not related; the Athenians 1000 πέζοι (hoplites?) and 400 cavalry (Diod. Sic. 13.65.1)
10. **Commanders:** Athenians: Leotrophides and Timarchos (Diod. Sic. 13.65.1)
11. **Pre-battle rites/exhortations:** –
12. **Battle array:** –
13. **Hoplite indicators**
    (ὅπλον / -α): Diod. Sic. 13.65.1
14. **Indicators of other troop types**
    ἱππεύς / ἵππος: Diod. Sic. 13.65.1
15. **Mercenaries:** –
16. **Description of battle:** almost none. The Athenians routed their adversaries, fighting brilliantly, and eschewed pursuing the [Sikelians], concentrating instead on the Megarians, with whom they were angry. Of these, they killed many (Diod. Sic. 13.65.2; *Hell. Oxy.* 4.1). According to the *Hell. Oxy.* (4.1), the Megarians [and Sikelians] were driven headlong (προτροπάδην) back along the road towards Megara, although the Spartans made good their escape by keeping their formation
17. **Duration of battle:** –
18. **Decisive factors:** the Athenians' combat prowess (Diod. Sic. 13.65.2)
19. **Victor:** Athenians
20. **Defeated:** Megarians and Sikelians
21. **Pursuit:** Diodoros is explicit that the Athenians purposely did not pursue the Spartans, but

concentrated instead on the Megarians, the object of their wrath (Diod. Sic. 13.65.2, cf. *Hell. Oxy.* 4.1)
22. **Casualties (victor):** –
23. **Casualties (defeated):** παμπληθεῖς, including 20 Spartans (Diod. Sic. 13.65.2)
24. **Prisoners:** –
25. **Trophy:** erected by the Athenians (*Hell. Oxy.* 4.1)
26. **Exchange of bodies:** Athenians gave back the dead Spartans and Megarians (*Hell. Oxy.* 4.1)
27. **Miscellaneous:** the Athenians were angry with the commanders, believing that they had "gambled with the city" (*Hell. Oxy.* 4.2)
28. **Sources:** Diod. Sic. 13.65.1–2; Strabo 9.1.11; *Hell. Oxy.* 4.2
29. **Bibliography**
Montagu (2000) 77–78

(40) TANAGRA 457

1. **Name and date:** Tanagra 457 (Tod 35.1–2, 36.1; Thuc. 1.108.1)
2. **Time of year:** –
3. **Geographic locality:** on the plain of Tanagra, near the Asopos river (Thuc. 1.108.1; Barrington 58 F1)
4. **Battlefield topography:** very flat; well suited to hoplite battle
5. **Weather conditions:** –
6. **Encampment:** –
7. **Combatants:** Athenians with allies: Argives and others (?); besides a contingent of Thessalian cavalry – which, however, turns traitor (Thuc. 1.107.7; Diod. Sic. 11.80.2)
8. **Non-combatants:** –
9. **Numbers of armed forces:** Athenians πανδημεί, with 1000 Argives, the rest of their allies, and the Thessalians; making up 14,000 in all (Thuc. 1.107.5; Diod. Sic. 11.80.1) Both authors also include Athenian ships cruising off the coast, in the gulf of Krisa
10. **Commanders:** –
11. **Pre-battle rites/exhortations:** –
12. **Battle array:** –
13. **Hoplite indicators**
φάλαγξ: Plut. *Cim.* 17.5
(ὅπλον / -α): Plut. *Cim.* 17.4, 17.7 (πανοπλία)
14. **Indicators of other troop types**
ἱππεύς / ἵππος: Thuc. 1.107.7; Paus. 1.29.6
15. **Mercenaries:** –
16. **Description of battle:** Thucydides on one side and later tradition (Diodoros; Pl. *Menex.* 242a) differ wildly. In short, Thucydides says that the battle was won by the Spartans; Diodoros that it was a draw: at any rate, the Spartans went home unmolested after ravaging in the Megarid. Thucydides is uncharacteristically brief about the battle itself; he may not have had access to proper sources. He does relate, however, that there was great slaughter on both sides (Thuc. 1.108.1). Diodoros claims that the battle was broken off only by nightfall. After this, the Thessalians betrayed the Athenians and attacked a supply caravan from Athens. The Athenians came to the rescue and routed the Thessalians who in turn were aided by the Spartans. A new battle ensued with just as much slaughter, and equally undecided. Agreement was reached through negotiation (Diod. Sic. 11.80.2–6)
17. **Duration of battle:** very long and hard-fought; Diodoros claims that it lasted until nightfall (Diod. Sic. 11.80.2, 11.80.6)
18. **Decisive factors:** –
19. **Victor:** Spartans (if any)

20. **Defeated:** Athenians and allies. An inscription from Olympia (Tod 27) names the "Argives, Athenians, and Ionians" as those defeated, interestingly enough. Cf. Gomme (1945) 315
21. **Pursuit:** –
22. **Casualties (victor):** –
23. **Casualties (defeated):** –
24. **Prisoners:** –
25. **Exchange of bodies:** –
26. **Trophy:** –
27. **Miscellaneous:** Kimon, returning home from an ostracism, was refused to fight due to intrigues. He therefore exhorted his friends (of the Oineis tribe) to fight bravely; and 100 of them were killed (Plut. *Cim.* 17.3–5, *Per.* 10.1.2)
28. **Sources:** Thuc. 1.107.2–108.2; Diod. Sic. 11.80.1–2; Plut. *Cim.* 17.2–6, *Per.* 10.1–2; Paus. 1.29.6–9; Tod 27; Meiggs & Lewis *GHI*² 35, 36
29. **Bibliography**
    Gomme (1945) 313–316
    Hornblower (1991) 167–172
    Meritt (1945) 134–147
    Montagu (2000) 60
    Pritchett (1996) 149–172
    Walters (1978) 188–191

(41) TEGYRA 375

1. **Name and date:** Tegyra 375 (Plut. *Pel.* 16.1)
2. **Time of year:** –
3. **Geographic locality:** near Tegyra (Plut. *Pel.* 17.1; Barrington 55 E3)
4. **Battlefield topography:** very near the northern shore of lake Kopaïs, and apparently with a narrow gorge or pass (Plut. *Pel.* 17.1)
5. **Weather conditions:** –
6. **Encampment:** –
7. **Combatants:** Spartans vs. Thebans (Plut. *Pel.* 17.1)
8. **Non-combatants:** –
9. **Numbers of armed forces:** two *morai* of Spartans (as Plutarch points out, this may signify anything between 500 and 900 soldiers: Plut. *Pel.* 17.2); Pelopidas commanded a mere 300 hoplites (Plut. *Pel.* 17.2). Other *mora* information: 500 strong (Ephoros *FGrH* 70.210); 700 strong (Kallisthenes *FGrH* 124.18); 900 (Polybios fr. 60 Büttner-Wobst)
10. **Commanders:** Pelopidas for the Thebans (Plut. *Pel.* 17.1); Gorgoleon and Theopompos for the Spartans (Plut. *Pel.* 17.3)
11. **Pre-battle rites/exhortations:** –
12. **Battle array:** Pelopidas arranged his 300 men in a closely packed unit facing the centre of the Spartan phalanx and the cavalry on the sides, ready to charge first (Plut. *Pel.* 17.2)
13. **Hoplite indicators** –
14. **Indicators of other troop types** –
15. **Mercenaries:** –
16. **Description of battle:** the Spartans advanced confidently, believing that they would sweep the outnumbered Thebans off the field; but Pelopidas – presumably having let the cavalry wave charge first – attacked the very same point; the centre of the Spartan army, killing the two polemarchs. Dazed, the Spartans divided their ranks to let them pass, but Pelopidas' men marched down this line, killing right and left, until the Spartans fled the field (Plut. *Pel.* 17.3–5)
17. **Duration of battle:** Plutarch says nothing, but it *seems* to have been very short indeed

18. **Decisive factors:** Pelopidas' daring attack, which concentrated all the impact on one point, the polemarchs being right there
19. **Victor:** Pelopidas and Thebans
20. **Defeated:** Spartans
21. **Pursuit:** the Thebans pursued the Spartans, but not for long, since they were worried about the Orchomenians and the Spartan relief force (διαδοχή) (Plut. *Pel.* 17.4)
22. **Casualties (victor):** –
23. **Casualties (defeated):** –
24. **Prisoners:** –
25. **Trophy:** –
26. **Exchange of bodies:** –
27. **Miscellaneous:** this battle apparently did no end of good for Theban morale in the war with Sparta to come, this actually being the first time a numerically inferior force beat a Spartan force in the field (Plut. *Pel.* 17.5–6)
28. **Sources:** Plut. *Pel.* 17
29. **Bibliography**
    Buckler (1995) 43–58
    Montagu (2000) 90

BIBLIOGRAPHY

Alföldi, A. (1967) "Die Herrschaft der Reiterei in Griechenland und Rom nach dem Sturz der Könige", in M. Rohde-Liegle, H.A. Cahn & H.C. Ackermann (edd.) *Gestalt und Geschichte. Festschrift Karl Schefold zu seinem sechzigsten Geburtstag am 26. Januar 1965* (Bern) 13–47
Amandry, P. (1983) "Le bouclier d'Argos" *BCH* 107: 627–634
Ameling, W. (1993) *Karthago. Studien zu Militär, Staat und Gesellschaft* (München)
Anderson, G. (2003) *The Athenian Experiment. Building an Imagined Political Community in Ancient Attica, 508-490 B.C.* (Ann Arbor)
Anderson, J.K. (1963) "The Statue of Chabrias", *AJA* 67: 411–413
– (1965) "Cleon's Orders at Amphipolis", *JHS* 85: 1–4
– (1970) *Military Theory and Practice in the Age of Xenophon* (Berkeley – Los Angeles)
– (1974) *Xenophon* (London)
– (1976) "Shields of Eight Palms' Width", *CSCA* 9: 1–6
– (1982) "Notion and Kyzikos. The Sources Compared", *JHS* 102: 15–25
– (1984) "Hoplites and Heresies: a Note", *JHS* 104: 152
– (1991) "Hoplite Weapons and Offensive Arms" in V.D. Hanson (ed.) 15–37
Andrewes, A. (1956) *The Greek Tyrants* (London)
Angel, J.L. (1944) "A Racial Analysis of the Ancient Greeks: An Essay on the Use of Morphological Types" *American Journal of Physical Anthropology* 2: 329–376
– (1945) "Skeletal Material from Attica", *Hesperia* 14: 279–363

Baer, L. (1994) *Vom Stahlhelm zum Gefechtshelm. Eine Entwicklungsgeschichte von 1915 bis 1993 zusammengestellt in Wort und Bild* vol. I–II (Neu-Anspach)
Barber, G.L. (1935) *The Historian Ephorus* (Cambridge)
Beardsworth, T. & Wilson, J. (1970) "Bad Weather and the Blockade at Pylos", *Phoenix* 24: 112–118
Beister, H. (1973) "Ein thebanisches Tropaion bereits vor Beginn der Schlacht bei Leuktra. Zur Interpretation von *IG* VII 2462 und Paus. 4, 32, 5", *Chiron* 3: 47–84
Berthold, R.M. (1976–77) "Which Way to Marathon?", *REA* 78–79: 84–95
Bicknell, P. (1970) "The Command Structure and Generals of the Marathon Campaign", *AntCl* 39: 427–442
Bigwood, J. M. (1983) "The Ancient Accounts of the Battle of Cunaxa", *AJPh* 104: 340–347
Blyth, H. (1982) "The Structure of a Hoplite Shield in the Museo Gregoriano Etrusco", *BMMP* 3: 5–21
Boardman, J. (1952) "Pottery from Eretria", *BSA* 47: 1–48
– (1980²) *The Greeks Overseas. Their Early Colonies and Trade* (London)
Boegehold, A.L. (1996) "Group and Single Competitions at the Panathenaia", in J. Neils (ed.) *Worshipping Athena. Panathenaia and Parthenon* (Madison, WI) 95–105
Borthwick, E.K. (1967) "Trojan Leap and Pyrrhic Dance in Euripides' *Andromache* 1129–41", *JHS* 87: 18–23
– (1970a) "P. Oxy. 2738: Athena and the Pyrrhic Dance", *Hermes* 98: 318–331
– (1970b) "Two Scenes of Combat in Euripides", *JHS* 90: 15–21
Bowden, H. (1995) "Hoplites and Homer: Warfare, Hero Cult, and the Ideology of the Polis", in J. Rich and G. Shipley (edd.) *War and Society in the Greek World* (London) 45–63
Bowie, E. (1990) "Miles Ludens? The Problem of Martial Exhortation in Early Greek Elegy", in O. Murray (ed.) *Sympotica: A Symposium on the Symposion* (Oxford) 221–229

Bradeen, D.W. (1964) "Athenian Casualty Lists", *Hesp.* 33: 16–62
Bradford, A.S. (1992) "Plataea and the Soothsayer", *AncWorld* 23: 27–33
Briant, P. (1999) "The Achaemenid Empire", in K. Raaflaub & N. Rosenstein (edd.) 105–128
Bryant, J.M. (1990) "Military Technology and Socio-Cultural Change in the Ancient Greek City", *Sociological Review* 38: 484–516
Buckler, J. (1977) "The Thespians at Leuktra", *WS* 11: 76–79
 – (1985) "Epameinondas and the Embolon", *Phoenix* 39: 134–143
 – (1996) "The Battle of Koroneia and its Historiographical Legacy", in J.M. Fossey & P.M. Smith (edd.) *Boeotia Antiqua 6. Proceedings of the 8th International Conference on Boiotian Antiquities, Chicago 24–26 May 1995* (Amsterdam) 59–72
Burn, A.R. (1962) *Persia and the Greeks. The Defence of the West, c. 546–478 B.C.* (Oxford)
Burrows, R.M. (1908) "Pylos and Sphakteria", *JHS* 37: 148–150

Cahn, D. (1989) *Waffen und Zaumzeug. Ausstellung Antikenmuseum Basel und Sammlung Ludwig* (Basel)
Cartledge, P. (1977) "Hoplites and Heroes: Sparta's Contribution to the Technique of Ancient Warfare", *JHS* 97: 11–27
 – (1996) "La nascita degli opliti e l'organizzazione militare", in S. Settis et al. (edd.) *I Greci. Storia Cultura Arte Società* (Torino) vol. 2, 681–714
 – (2001) *Spartan Reflections* (London)
Cary, M. (1949) *Geographical Background of Greek and Roman History* (Oxford)
Cawkwell, G. (1972) "Epaminondas and Thebes", *CQ* 22: 254–278
 – (1978) *Philip of Macedon* (London)
 – (1989) "Orthodoxy and Hoplites", *CQ* 39: 375–389
Chantraine, P. (1968) *Dictionnaire étymologique de la langue grecque* (Paris)
Chrimes, K.M.T. (1949) *Ancient Sparta. A Re-examination of the Evidence* (Manchester)
Christensen, J. (2001) "Misothebaios? Om Xenophons skildring af Theben i *Hellenika*", *AIGIS* 1: 1–54 (http://www.igl.ku.dk/~aigis/2001,1/Vol1.html)
Clairmont, C.W. *Patrios Nomos. Public Burial in Athens during the Fifth and Fourth Centuries B.C. The archaeological, epigraphic-literary and historical evidence* vol. I–II, BAR International Series 161 (Oxford)
Coldstream, J.N. (1968) *Greek Geometric Pottery. A Survey of Ten Local Styles and their Chronology* (London)
Compton, H.C. & Awdry, H. (1907) "Two Notes on Pylos and Sphacteria", *JHS* 37: 274–283
Connolly, P. (1998²) *Greece and Rome at War* (London)
Connor, W. (1988) "Early Greek Land-Warfare as Symbolic Expression", *P&P* 119: 3–29
Cooper III, G.L. (1978) "Thuc. 5.65.3 and the Tactical Obsession of Agis II on the Day Before the Battle of Mantinea", *TAPhA* 108: 35–40
Courbin, P. (1957) "Une tombe géométrique d'Argos", *BCH* 81: 322–386
Craik, E. (1988) *Euripides: Phoenician Women. Edited with Translation and Commentary by* (Warminster)
Crouwel, J.H. *Chariots and Other Wheeled Vehicles in Iron Age Greece* (Amsterdam)
Cunliffe, R.J. (1963²) *A Lexicon of the Homeric Dialect* (Norman, OK)

Dawson, D. (1996) *The Origins of Western Warfare. Militarism and Morality in the Ancient World* (Boulder, CO)
Delbrück, H. (1900) *Geschichte der Kriegskunst im Rahmen der politischen Geschichte. Erster Theil: das Alterthum* (Berlin)
Develin, R. (1989) *Athenian Officials 684–321B.C.* (Cambridge)
Devine, A.M. (1983) "ΕΜΒΟΛΟΝ: A Study in Tactical Terminology", *Phoenix* 37: 201–217
DeVoto, J. (1989) "Pelopidas and Kleombrotos at Leuktra", *AHB* 3: 115–117
Doenges, N.A. (1998) "The Campaign and Battle of Marathon", *Historia* 47: 1–17
Donlan, W. (1970) "Archilochus, Strabo and the Lelantine War", *TAPhA* 101: 131–142

Donlan, W. & Thompson, J. (1976) "The Charge at Marathon: Herodotus 6.112", *CJ* 71: 339–343
–(1979) "The Charge at Marathon Again", *CW* 72: 419–420
Dover, K.J. (1968) *Aristophanes Clouds. Edited With Introduction and Commentary by* (Oxford)
Ducrey, P. (1986) *Warfare in Ancient Greece* (New York)

Edwards, M.W. (1997) "Homeric Style and Oral Poets", in I. Morris & B. Powell (edd.) *A New Companion to Homer* (Leiden) 261–283
Ehrhardt, C.T.H.R. (1994) "Two Notes on Xenophon, *Anabasis*", *AHB* 8: 1–4
Ellis, J.R. (1978) "Thucydides at Amphipolis", *Antichthon* 12: 28–35
Evans, J.A.S. (1964) "The 'Final Problem' at Thermopylae", *GRBS* 4: 231–237
– (1993) "Herodotus and the Battle of Marathon", *Historia* 42: 279–307

Fantar, M.H. (1993) *Carthage. Approche d'une civilisation* vol. I–II (Tunis)
Fehling, D. (1971) *Die Quellenangaben bei Herodot. Studien zur Erzählkunst Herodots* (Berlin)
Fenik, B. (1968) *Typical Battle Scenes in the Iliad. Studies in the Narrative Techniques of Homeric Battle Description. Hermes* Einzelschriften 21 (Wiesbaden)
Figueira, T.J. (2003) "The Demography of the Spartan Helots", in N. Luraghi & S.E. Alcock (edd.) *Helots and Their Masters in Laconia and Messenia. Histories, Ideologies, Structures* (Cambridge, MA) 193–239
Finley, M. (1979²) *The World of Odysseus* (London)
Forrest, W.G. (1966) *The Emergence of Greek Democracy. The Character of Greek Politics 800–400 BC* (London)
Foxhall, L. (1997) "A View from the Top: Evaluating the Solonian Property Classes", in L.G. Mitchell and P.J. Rhodes (edd.) *The Development of the Polis in Archaic Greece* (London) 113–136
Foxhall, L. & Forbes, H.A. (1982) "Σιτομετρεία: The Role of Grain as a Staple Food in Classical Antiquity", *Chiron* 12: 41–90
Francis, E.D. & Vickers, M. (1985) "The Oenoe Painting in the Stoa Poikile, and Herodotus' Account of Marathon", *BSA* 80: 99–113
Francis, P. (1796) *Proceedings in the House of Commons on the Slave Trade, and State of the Negroes in the West India Islands* (London)
Franz, J.P. (2002) *Krieger, Bauern, Bürger. Untersuchungen zu den Hopliten der archaischen und klassischen Zeit* (Frankfurt am Main)
Fraser, A.D. (1942) "The Myth of the Phalanx-Scrimmage", *CW* 36: 15–16
Fuhrmann, F. (1988) *Plutarque. Œuvres morales, tome III. Apophtegmes de rois et de généraux. Apophtegmes laconiens. Texte établi et traduit par* (Paris)
Fuller, J.F.C. (1946) *Armament and History. A Study of the Influence of Armament on History from the Dawn of Classical Warfare to the Second World War* (London)

Gabrici, E. (1912) "Vasi greci arcaici della necropoli di Cuma", *MDAI(R)* 27: 124–147
Gabriel, R.A. & Metz, K.S. (1991) *From Sumer to Rome. The Military Capabilities of Ancient Armies* (Westport, CT)
Gabrielsen, V. (2001) "Socio-economic Classes and Ancient Greek Warfare", in K. Ascani, V. Gabrielsen, K. Kvist & A.H. Rasmussen (edd.) *Ancient History Matters. Studies Presented to Jens Erik Skydsgaard on His Seventieth Birthday* (Rome) 203–220
Gardiner, E.N. (1903) "Notes on the Greek Foot Race", *JHS* 23: 261–291
Garlan, Y. (1975) *War in the Ancient World* (London)
Garnsey, P. (1999) *Food and Society in Classical Antiquity* (Cambridge)
Gat, A. (2006) *War in Human Civilization* (Oxford)
Gehrke, H.-J. & Wirbelauer, E. (2004) "Akarnania and Adjacent Areas" in M.H. Hansen & T.H. Nielsen (edd.) 351–378
Goldsworthy, A.K. (1997) "The *Othismos*, Myths and Heresies. The Nature of Hoplite Battle", *War in History* 4: 1–26
Gomme, A.W. (1945) *A Historical Commentary on Thucydides* vol. I (Oxford)

– (1956a) *A Historical Commentary on Thucydides* vol. II (Oxford)
– (1956b) *A Historical Commentary on Thucydides* vol. III (Oxford)
Gomme, A.W., Andrewes, A. & Dover, K. (1970) *A Historical Commentary on Thucydides* vol. IV (Oxford)
Goulaki-Voutira, A. (1996) "Pyrrhic Dance and Female Pyrrhic Dancers", *Répertoire International d'Iconographie Musicale* 21: 3–12
Grant, J.R. (1961) "Leonidas' Last Stand", *Phoenix* 15: 14–27
Gray, D.H.F. (1947) "Homeric Epithets for Things", *CQ* 41: 109–121
Greenhalgh, P.A.L. (1973) *Early Greek Warfare. Horsemen and Chariots in the Homeric and Archaic Ages* (Cambridge)
Gröschel, S.-G. (1989) *Waffenbesitz und Waffeneinsatz bei den Griechen* (Frankfurt am Main)
Gugel, H. (1971) "Die Aufstellung von Kyros' Heer in der Schlacht von Kunaxa. Zu Xen.An.1,8,5", *Gymnasium* 78: 241–243

Hamel, D. (1998) *Athenian Generals. Military Authority in the Classical Period* (Leiden)
Hammond, M. (1979–80) "A Famous *Exemplum* of Spartan Toughness", *CJ* 75: 97–109
Hammond, N.G.L. (1938) "The Two Battles of Chaeronea (338 B.C. and 86 B.C.)", *Klio* 31: 186–218
– (1973) "The Campaign and Battle of Marathon", in N.G.L. Hammond *Studies in Greek History. A Companion Volume to* A History of Greece to 322 B.C. (Oxford) 170–250
Hansen, M.H. (1973) *Atimistraffen i Athen i klassisk tid* (Odense)
– (1976) Apagoge, Endeixis *and* Ephegesis *against* Kakourgoi, Atimoi *and* Pheugontes. *A Study in the Athenian Administration of Justice in the Fourth Century B.C.* (Odense)
– (1988) *Three Studies in Athenian Demography* (Copenhagen)
– (1999²) *The Athenian Democracy in the Age of Demosthenes* (London)
– (2000) "The Hellenic *Polis*", in M.H. Hansen (ed.) *A Comparative Study of Thirty City-State Cultures* (Copenhagen) 141–187
– (2004) *Polis, den oldgræske bystatskultur* (Copenhagen)
– (2006) *The Shotgun Method Used to Establish the Number of Inhabitants in the Ancient Greek City-States* (Columbia, MO)
Hansen, M.H. & Nielsen, T. (2004, ed.) *An Inventory of Archaic and Classical Poleis* (Oxford)
Hansen, M.H., Spencer, N. & Williams H. (2004) "Lesbos", in M.H. Hansen & T.H. Nielsen (edd.) 1018–1032
Hanson, V.D. (1988) "Epameinondas, the Battle of Leuktra (371 B.C.) and the 'Revolution' in Greek Battle Tactics", *ClAnt* 7: 197–207
– (1991) "Hoplite Technology in Phalanx Battle", in V.D. Hanson (ed.) 63–84
– (1991, ed.) *Hoplites. The Classical Greek Battle Experience* (London)
– (1995) *The Other Greeks. The Family Farm and the Agrarian Roots of Western Civilization* (New York)
– (1998²) *Warfare and Agriculture in Classical Greece* (Berkeley – Los Angeles)
– (1999) "Hoplite Obliteration. The Case of the Town of Thespiai", in J. Carman & A. Harding (edd.) *Ancient Warfare. Archaeological Perspectives* (Gloucestershire) 203–217
– (2000²) *The Western Way of War. Infantry Battle in Classical Greece* (Berkeley – Los Angeles)
Harris, H.A. (1963) "Greek Javelin Throwing", *G&R* 10: 26–36
– (1966) "Nutrition and Physical Performance. The Diet of Greek Athletes", *Proceedings of the Nutrition Society* 25: 87–90
Hatzfeld, J. (1948) *Xénophon: Helléniques* vol. I–II (Paris)
Havelock, E.A. (1973a) "Prologue to Greek Literacy 1", in C.G. Boulter et al. (edd.) *Lectures in Memory of Louise Taft Semple* (Norman, OK) 329–363
– (1973b) "Prologue to Greek Literacy 2", in C.G. Boulter et al. (edd.) *Lectures in memory of Louise Taft Semple* (Norman, OK) 364–391

– (1988) "The Coming of Literate Communication to Western Culture", in E.R. Kintgen, B. Kroll & M. Rose (edd.) *Perspectives on Literacy* (Carbondale) 127–134
Helbig, W. (1911) *Über die Einführungszeit der geschlossenen Phalanx. Sitzungsberichte der Königlich Bayerischen Akademie der Wissenschaften. Philosophisch-philologische und historische Klasse* 1911, 12
Heubeck, A. (1966) *Aus der Welt der frühgriechischen Lineartafeln. Eine kurze Einführung in Grundlagen, Aufgaben und Ergebnisse der Mykenologie* (Göttingen)
Hignett, C. (1963) *Xerxes' Invasion of Greece* (Oxford)
Holladay, A.J. (1982) "Hoplites and Heresies", *JHS* 102: 94–103
Holoka, J.P. (1997) "Marathon and the Myth of the Same-Day March" *GRBS* 38: 329–353
Hope Simpson, R. (1972) "Leonidas' Decision", *Phoenix* 26: 1–11
Hornblower, J. (1981) *Hieronymus of Cardia* (Oxford)
Hornblower, S. (1991) *A Commentary on Thucydides* I (Oxford)
– (1996) *A Commentary on Thucydides* II (Oxford)
– (2002) "Sticks, Stones, and Spartans", in H. van Wees (ed.) *War and Violence in Ancient Greece* (London) 57–82
How, W.W. (1919) "On the Meaning of ΒΑΔΗΝ and ΔΡΟΜΩΙ", *CQ* 13: 40–42
How, W.W. & Wells, J. (1928²) *A Commentary on Herodotus* vol. I–II (Oxford)
Hunt, P. (1997) "Helots at the Battle of Plataea", *Historia* 46: 129–144
– (1998) *Slaves, Warfare and Ideology in the Greek Historians* (Cambridge)
Hurwit, J.M. (2002) "Reading the Chigi Vase", *Hesperia* 71: 1–22
Huß, Werner (1990) *Die Karthager* (München)

Jackson, A.H. (1991) "Hoplites and the Gods: The Dedication of Captured Arms and Armour", in V.D. Hanson (ed.) 228–249
Jacques, J.-M. (1998) *Ménandre. Le bouclier. Texte établi et traduit par* (Paris)
Jaeger, W. (1966) "Tyrtaeus on True Arete", in W. Jaeger *Five Essays* (Montreal) 101–142
Jarva, E. (1995) *Archaiologia on Archaic Greek Body Armour* (Rovaniemi)
Jones, N. (1977) "The Topography and Strategy of the Battle of Amphipolis in 422 B.C.", *CSCA* 10: 71–104
Jones, N.F. (1987) *Public Organization in Ancient Greece. A Documentary Study* (Philadelphia)
de Jong, I.J.F. (1987) "Silent Characters in the *Iliad*", in J.M. Bremer, I.J.F. de Jong & J. Kalff (edd). *Homer: Beyond Oral Poetry. Recent Trends in Homeric Interpretation* (Amsterdam) 105–121
Keegan, J. (1993, Danish 2000) *Krigens historie* (Copenhagen)
Kertész, I. (1991) "Schlacht und Lauf bei Marathon. Legende und Wirklichkeit", *Nikephoros* 4: 155–160
Kirk, G.S. (1962) *The Songs of Homer* (Cambridge)
Köchly, H. & Rüstow, W. (1862) *Geschichte des griechischen Kriegswesens von der ältesten Zeit bis auf Pyrrhos. Nach den Quellen bearbeitet* (Aarau)
Konecny, A. (2001) "κατέκοψεν τὴν μόραν Ἰφικράτης. Das Gefecht bei Lechaion im Frühsommer 390 v. Chr.", *Chiron* 31: 79–127
Kourouniotis, K. (1904) " Ἀγγεῖα Ἐρέτριας", *AE* 1903: 1–38
Krentz, P. (1985a) "Casualties in Hoplite Battles", *GRBS* 26: 13–20
– (1985b) "The Nature of Hoplite Battle", *ClAnt* 4: 50–61
– (1991) "The *Salpinx* in Greek Warfare", in V.D. Hanson (ed.) 110–120
– (1994) "Continuing the *Othismos* on *Othismos*", *AHB* 8: 45–49
– (2002) "Fighting by the Rules: The Invention of the Hoplite *Agôn*", *Hesperia* 71: 23–39
– (2007) "Warfare and Hoplites", in H.A. Shapiro (ed.) *The Cambridge Companion to Archaic Greece* (Cambridge) 61–84
Kromayer, J. (1903) *Antike Schlachtfelder in Griechenland. Bausteine zu einer antiken Kriegsgeschichte* vol. I (Berlin)
Kromayer, J. & Veith, G. (1928²) "Heerwesen und Kriegführung der Griechen und Römer", in W. Otto (ed.) *Handbuch der Altertumswissenschaft* vol. IV, 3 (München)

Kubitschek, W. (1928) *Grundriss der antiken Zeitrechnung* (München)
Kunze, E. & Schleif, H. (1938) *Bericht über die Ausgrabungen in Olympia* vol. II (Berlin)
Kunze, E. (1958) *Bericht über die Ausgrabungen in Olympia* vol. VI (Berlin)
– (1961) *Bericht über die Ausgrabungen in Olympia* vol. VII (Berlin)
– (1967) *Bericht über die Ausgrabungen in Olympia* vol. VIII (Berlin)
Küsters, A. (1939) *Cuneus, Phalanx und Legio. Untersuchungen zur Wehrverfassung, Kampfweise und Kriegführung der Germanen, Griechen und Römer* (Würzburg – Aumühle)
Kyle, D.G. (1992) "The Panathenaic Games: Sacred and Civic Athletics", in J. Neils (ed.) *Goddess and Polis. The Panathenaic Festival in Ancient Athens* (Princeton) 77–101

Latacz, J. (1977) *Kampfparänese, Kampfdarstellung und Kampfwirklichkeit in der Ilias, bei Kallinos und Tyrtaios*. Zetemata 66 (München)
Lazenby, J.F. (1985) *The Spartan Army* (Warminster)
– (1991) "The Killing Zone", in V.D. Hanson (ed.) 87–109
– (1993) *The Defence of Greece 490–479 B.C.* (Warminster)
Lazenby, J.F. & Whitehead, D. (1996) "The Myth of the Hoplite's *Hoplon*", *CQ* 46: 27–33
Lee, H.M. (2001) *The Program and Schedule of the Ancient Olympic Games*, *Nikephoros* Beihefte 6 (Hildesheim)
Lendle, O. (1966): "Der Bericht Xenophons über die Schlacht von Kunaxa", *Gymnasium* 73: 429–452
– (1995) *Kommentar zu Xenophons Anabasis (Bücher 1–7)* (Darmstadt)
Lendon, J.E. (2005) *Soldiers and Ghosts: A History of Battle in Classical Antiquity* (New Haven – London)
Leumann, M. (1950) *Homerische Wörter* (Basel)
Lévêque, P. & Vidal-Naquet, P. (1960) "Épaminondas Pythagoricien ou le problème tactique de la droite et de la gauche", *Historia* 9: 294–308
Lord, A.B. (1960) *The Singer of Tales* (Harvard)
Lorimer, H.L. (1947) "The Hoplite Phalanx with Special Reference to the Poems of Archilochus and Tyrtaeus", *BSA* 42: 76–138
– (1950) *Homer and the Monuments* (London)
Lotze, D. (1959) ΜΕΤΑΞΥ ΕΛΕΥΘΕΡΩΝ ΚΑΙ ΔΟΥΛΩΝ. *Studien zur Rechtsstellung unfreier Landbevölkerungen in Griechenland bis zum 4. Jahrhundert v. Chr.* (Berlin)
Luckenbill, D.D (1924) *The Annals of Sennacherib. Oriental Institute Publications 2* (Chicago)
Luginbill, R.D. (1994) "*Othismos*: the Importance of the Mass-Shove in Hoplite Warfare", *Phoenix* 48: 51–61
– (2002) "Tyrtaeus 12 West: Come Join the Spartan Army", *CQ* 52: 405–414
Luraghi, N. (2006) "Traders, pirates, warriors: The proto-history of Greek mercenary soldiers in the Eastern Mediterranean", *Phoenix* 60: 21–47

McDougall, I. (1990) "The Persian Ships at Mycale", in E.M. Craik (ed.) *Owls to Athens. Essays on Classical Subjects Presented to Sir Kenneth Dover* (Oxford) 143–149
Machiavelli, N. (1520, English 2003) *Arte della guerra / Art of War. Translated, Edited, and with a Commentary by Christopher Lynch* (Chicago)
Maier, F.G. (2004) "Cyprus", in M.H. Hansen & T.H. Nielsen (edd.) 1223–1232
Majno, G. (1975) *The Healing Hand. Man and Wound in the Ancient World* (Cambridge, MA)
Mallwitz, A. & Herrmann, H.-V. (1980) *Die Funde aus Olympia* (Athen)
Markle, III, M.M. (1977) "The Macedonian Sarissa, Spear and Related Armor", *AJA* 81: 323–339
Markoe, G. (1985) *Phoenician Bronze and Silver Bowls from Cyprus and the Mediterranean* (Berkeley)
Mastronarde, D.J. (1998) *Euripides: Phoenissae. Edited with Introduction and Commentary by* (Cambridge)
Mazzucco, C. (1973) "Cleombroto a Cheronea. I precedenti della battaglia di Leuttra", *RIL* 107: 671–675

Meiggs, R. & Lewis, D. (1988²) *A Selection of Greek Historical Inscriptions to the End of the Fifth Century B.C.* (Oxford)
Meritt, B.D. (1945) "The Argives at Tanagra", *Hesperia* 14: 134–147
Miller, S.G. (2004) *Ancient Greek Athletics* (New Haven – London)
Mitchell, B. (1991) "Kleon's Amphipolitan Campaign. Aims and Results", *Historia* 40: 170–192
Momigliano, A.D. (1969²) *Studies in Historiography* (London)
Montagu, J.D. (2000) *Battles of the Greek and Roman World. A Chronological Compendium of 667 Battles to 31 BC, from the Historians of the Ancient World* (London)
Moretti, L. (1953) *Iscrizioni Agonistiche Greche* (Rom)
Morgan, C. (1990) *Athletes and Oracles. The Transformation of Olympia and Delphi in the Eighth Century BC* (Cambridge)
— (2001) "Symbolic and Pragmatic Aspects of Warfare in the Greek World of the 8th to 6th Centuries BC", in T. Bekker-Nielsen & L. Hannestad (edd.) *War as a Cultural and Social Force. Essays on Warfare in Antiquity* (Copenhagen) 20–44
Morris, I. (1986) "The Use and Abuse of Homer", *ClAnt* 5: 81–138
Munro, J.A.R. (1899), "Some Observations on the Persian Wars", *JHS* 19: 185–197
Murray, O. (1993²) *Early Greece* (London)
Myres, J.L. (1933) "The Amathous Bowl", *JHS* 53: 25–39

Nierhaus, R. (1938) "Eine frühgriechische Kampfform", *JdI* 53: 90–113
Nikolaidis, A.G. (1990) "Thucydides 4, 28, 5, or Kleon at Sphakteria and Amphipolis", *BICS* 37: 89–94

Ober, J. (1991) "Hoplites and Obstacles", in V.D. Hanson (ed.) 173–196
O'Connell, R.L. *Of Arms and Men. A History of War, Weapons and Aggression* (Oxford)
Oman, C. (1924²) *A History of the Art of War in the Middle Ages* vol. I–II (London)
Ong, W.J. (2002²) *Orality and Literacy. The Technologizing of the Word* (London – New York)
Osborne, R. (2004) "Homer's Society", in R. Fowler (ed.) *The Cambridge Companion to Homer* (Cambridge) 206–219

Palmer, L. (1963) *The Interpretation of Mycenaean Greek Texts* (Oxford)
Parry, M. (1971) *The Making of Homeric Verse: the Collected Papers of Milman Parry* (Oxford)
Paul, G.M. (1987) "Two Battles in Thucydides", *EMC* 31: 307–312
Poliakoff, M.B. (1987) *Combat Sports in the Ancient World. Competition, Violence and Culture* (New Haven, CT and London)
Poulsen, V. (1962) "Geometrisk Kunst paa Glyptoteket", *MedKøb* 19: 1–17
Powell, J.E. (1960²) *A Lexicon to Herodotus* (Hildesheim)
Poznanski, L. (1992) *Asclépiodote. Traité de tactique. Texte établi et traduit par* (Paris)
Pritchett, W.K. (1969) *Studies in Ancient Greek Topography* vol. II (Berkeley – Los Angeles)
— (1971) *The Greek City-State at War* vol. I (Berkeley – Los Angeles)
— (1974) *The Greek City-State at War* vol. II (Berkeley – Los Angeles)
— (1979) *The Greek City-State at War* vol. III (Berkeley – Los Angeles)
— (1985a) *The Greek City-State at War* vol. IV (Berkeley – Los Angeles)
— (1985b) *Studies in Ancient Greek Topography* vol. V (Berkeley – Los Angeles)
— (1991) *Studies in Ancient Greek Topography* vol. VII (Berkeley – Los Angeles)
— (1992) *Studies in Ancient Greek Topography* vol. VIII (Berkeley – Los Angeles)
— (1994) *Essays in Greek History* (Amsterdam)
— (1996) "Thucydides' Campaign of Tanagra", in: *Greek Archives, Cults, and Topography* (Amsterdam) 149–172
Raaflaub, K.A. (1997) "Homeric Society", in I. Morris & B. Powell (edd.) *A New Companion to Homer* (Leiden) 624–648
— (1998) "A Historian's Headache. How to Read 'Homeric Society'?", in N. Fisher and H. van Wees (edd.) *Archaic Greece. New Approaches and New Evidence* (Chippenham) 169–193

- (1999) "Archaic and Classical Greece", in K. Raaflaub & N. Rosenstein (edd.) 129–161
Raaflaub, K. & Rosenstein, N. (1999, edd.) *War and Society in the Ancient and Medieval Worlds* (Washington)
Rahe, P.A. (1981) "The Annihilation of the Sacred Band at Chaeronea", *AJA* 85: 84–87
Rankin, H.D. (1975) "Archilochus fr. 6 D (13 L/B; 8 T)", *LF* 98: 193–198
Ratti, O. & Westbrook, A. (1973) *Secrets of the Samurai: a Survey of the Martial Arts of Feudal Japan* (Rutland, VT)
Rawlings, L. (2000) "Alternative Agonies. Hoplite Martial and Combat Experiences beyond the Phalanx", in H. van Wees (ed.) *War and Violence in Ancient Greece* (London) 233–250
Rhodes, P.J. (1981) *A Commentary on the Aristotelian* Athenaion Politeia (Oxford)
Rieth, A. (1964) "Ein etruskischer Rundschild", *AA* 1: 101–109
Ringwood Arnold, I. (1937) "The Shield of Argos", *AJA* 41: 436–440
Roberts, J.T. (1982) "Chares, Lysicles and the Battle of Chaeronea", *Klio* 64: 367–371
Robinson, D.M. (1941) *Excavations at Olynthus. Part X: Metal and Minor Miscellaneous Finds. An Original Contribution to Greek Life* (Baltimore)
Rosivach, V.J. (2002) "*Zeugitai* and Hoplites", *AHB* 16: 33–43
Rouché, C. (1993) *Performers and Partisans at Aphrodisias in the Roman and Late Roman Periods*. J.R.S. Monograph no. 6. Society for the Promotion of Roman Studies (London)
Roy, J. (1971) "Tegeans at the Battle Near the River Nemea in 394 B.C.", *PP* 26: 439–441
Rusch, S.M. (2002) "The Plausibility of Diodorus 13.72.3–73.2", in V.B. Gorman & E.W. Robinson (edd.) *Oikistes. Studies in Constitutions, Colonies, and Military Power in the Ancient World Offered in Honor of A.J. Graham* (Leiden) 285–300
Ruschenbusch, E. (1966) ΣΟΛΩΝΟΣ ΝΟΜΟΙ. *Historia* Einzelschriften 9 (Wiesbaden)
Rüstow, W. & Köchly, H. (1852) *Geschichte des griechischen Kriegswesens von der ältesten Zeit bis auf Pyrrhos* (Aarau)

Sacks, K.S. (1990) *Diodorus Siculus and the First Century* (Princeton)
Salazar, C.F. (2000) *The Treatment of War Wounds in Graeco–Roman Antiquity. Studies in Ancient Medicine 21* (Leiden)
Salmon, J. (1977) "Political Hoplites?", *JHS* 97: 84–101
Schneider, R. (1893) *Legion und Phalanx. Taktische Untersuchungen* (Berlin)
Schreiner, J.H. (2004) *Two Battles and Two Bills: Marathon and the Athenian Fleet* (Oslo)
Schuller, W. (1982²) *Griechische Geschichte* (Munich)
Schwartz, A. (2002) "The Early Hoplite Phalanx: Close Order or Disarray?", *C&M* 53: 31–63
- (2004) *Hoplitkrigsførelse i arkaisk og klassisk tid* (Copenhagen)
Schwertfeger, T. (1982) "Der Schild des Archilochos", *Chiron* 12: 253–280
Seaford, R. (1984) *Euripides: Cyclops. With Introduction and Commentary by* (Oxford)
Sekunda, N.V. (1986) *Warriors of Ancient Greece* (Oxford)
- (1994) "Classical Warfare", in J. Boardman (ed.) *CAH²*. *Plates to Volumes V and VI* (Cambridge) 167–194
Shear, T.L. (1937) "The Campaign of 1936", *Hesperia* 6: 333–381
Shrimpton, G. (1980) "The Persian Cavalry at Marathon", *Phoenix* 34: 20–37
- (1984) "Strategy and Tactics in the Preliminaries to the Siege of Potidaea (Thuc. 1.61–65)", *SO* 59: 7–12
Sinclair, R.K. (1966) "Diodorus Siculus and Fighting in Relays", *CQ* 16: 249–255
Singor, H.W. (1988) *Oorsprong en betekenis van de hoplitenphalanx in het archaische Griekenland* (Leiden)
- (2002) "The Spartan army at Mantinea and its organisation in the fifth century B.C.", in: W. Jongman & M. Kleijwegt (edd.) *After the Past. Essays in Ancient History in Honour of H.W. Pleket, Mnemosyne* Supplement 233 (Leiden) 235–284
Skafte-Jensen, M. (1980) *The Homeric Question and the Oral-formulaic Theory* (Copenhagen)
Snell, B. (1969) *Tyrtaios und die Sprache des Epos. Hypomnemata* 22 (Göttingen)

Snodgrass, A.M. (1964a) *Early Greek Armour and Weapons from the end of the Bronze Age to 600 B.C.* (Edinburgh)
- (1964b) "Carian Armourers: the Growth of a Tradition", *JHS* 84: 107–118
- (1965) "The Hoplite Reform and History", *JHS* 85: 110–122
- (1967) *Arms and Armour of the Greeks* (London)
- (1974) "An Historical Homeric Society?", *JHS* 94: 114–125
- (1980) *Archaic Greece. The Age of Experiment* (London)

Solin, H. (1974) "Bemerkungen zu einer mantineischen Namenliste", *ZPE* 14: 270–276
Sommerstein, A. (1982) *The Comedies of Aristophanes vol. 3: Clouds. Edited with Translation and Notes by* (Warminster)
de Ste. Croix, G. (2004) *Athenian Democratic Origins and other Essays* (Oxford)
Stewart, A. (1990) *Greek Sculpture* vol. I–II (New Haven)
- (1997) *Art, Desire, and the Body in Ancient Greece* (Cambridge)

Storch, R.H. (1998) "The Archaic Greek 'Phalanx,' 750–650 BC", *AHB* 12: 1–7
- (2001) "The Silence is Deafening. Persian Arrows Did Not Inspire the Greek Charge at Marathon", *AAntHung* 41: 381–394

Strasburger, H. (1972) "Homer und die Geschichtsschreibung" *Sitzungsberichte der Heidelberger Akademie der Wissenschaften. Philosophisch-historische Klasse* 1972, 1
Strassler, R.B. (1988) "The Harbor at Pylos, 425 BC", *JHS* 108: 198–203
- (1990) "The Opening of the Pylos Campaign", *JHS* 110: 110–125

Stylianou, P.J. (1997): *A Historical Commentary on Diodorus Siculus Book 15* (Oxford)
Szemler, G.J. (1996) "Two Notes on the Corinthian War", *AncW* 27: 95–104

Talbert, R.J.A. (2000, ed.) *Barrington Atlas of the Greek and Roman World* (Princeton)
Thompson, W.E. (1965) "Tissaphernes and the Mercenaries at Miletos", *Philologus* 109: 294–297
- (1985) "Chabrias at Corinth", *GRBS* 26: 51–57

Tölle, R. (1963) "Figürlich bemalte Fragmente vom Kerameikos", *AA* 1963: 642–665
Tomlinson, R. (2000) "From Pydna to Potidaia. Thucydides I 61", in D. Pantermales (ed.) Μύρτος. Μνήμη Ιουλίας Βοκοτοπούλου (Thessaloniki) 529–532
Törnkvist, S. (1969) "Note on Linen Corslets", *ORom* 7: 81–82
Tuplin, C.J. (1987) "The Leuctra campaign. Some Outstanding Problems", *Klio* 69: 72–107

Underhill, G.E. (1900) *A Commentary With Introduction and Appendix on the Hellenica of Xenophon* (Oxford)

Vanderpool, E. (1955) "News Letter from Greece", *AJA* 59: 221–229
Vaughn, P. (1991) "Identification and Retrieval of Hoplite Battle-dead", in V.D. Hanson (ed.) 38–62
Veer, J.A.G. van der (1982) "The Battle of Marathon. A Topographical Survey", *Mnemosyne* 35: 290–321
Ventris, M. & Chadwick, J. (1973²) *Documents in Mycenaean Greek* (Cambridge)
Vial, C. (1977) *Diodore de Sicile. Bibliothèque Historique Livre XV* (Paris)

Wallace, P.W. (1982) "The Final Battle at Plataia", in: *Studies in Attic epigraphy History and Topography Presented to Eugene Vanderpool* (Hesperia suppl. 19) 183–192
Walters, K.R. (1978) "Diodorus 11.82–84 and the Second Battle of Tanagra", *AJAH* 3: 188–191
Wankel, H. (1984) "Die athenischen Strategen der Schlacht bei Chaironeia", *ZPE* 55: 45–53
van Wees, H. (1994) "The Homeric Way of War: the *Iliad* and the Hoplite Phalanx" (I) and (II), *GR* 41: 1–18 and 131–156
- (2000) "The Development of the Hoplite Phalanx. Iconography and Reality in the 7th Century", in H. van Wees (ed.) *War and Violence in Ancient Greece* (London) 125–166
- (2001) "The Myth of the Middle-Class Army: Military and Social Status in Ancient Athens", in

T. Bekker-Nielsen & L. Hannestad (edd.) *War as a Cultural and Social Force. Essays on Warfare in Antiquity* (Copenhagen) 45–71
– (2004) *Greek Warfare. Myths and Realities* (London)
Warner, R. (1949) *Xenophon: The Persian Expedition* (London)
– (1979²) *Xenophon: A History of My Times* (London)
Waterlow, J.C. (1989) "Diet of the Classical Period of Greece and Rome", *European Journal of Clinical Nutrition* 43: 3–12
Webster, T.B.L. (1972) *Potter and Patron in Classical Athens* (London)
Welwei, K.W. (1979) "Das sogenannte Grab der Plataier im Vranatal bei Marathon", *Historia* 28: 101–106
Westlake, H.D. (1974) "The Naval Battle at Pylos and its Consequences", *CQ* 24: 211–226
– (1985) "The Sources for the Spartan Debacle at Haliartus", *Phoenix* 39: 119–133
Whatley, N. (1964) "On Reconstructing Marathon and Other Ancient Battles", *JHS* 84: 119–139
Wheeler, E.L. (1982) "*Hoplomachia* and Greek Dances in Arms", *GRBS* 23: 223–233
– (1991) "The General as Hoplite", in V.D. Hanson (ed.) 121–154
Whitehead, D. (1981) "The Archaic Athenian ΖΕΥΓΙΤΑΙ", *CQ* 31: 282–286
Wilson, J.B (1979) *Pylos 425 BC: a Historical and Topographical Study of Thucydides' Account of the Campaign* (Guildford)
Wilson, J. & Beardsworth, T. (1970) "Pylos 425 B.C. The Spartan Plan to Block the Entrances", *CQ* 20: 42–52
Woodfall, W. (1796) *An Impartial Report of the Debates that Occur in the Two Houses of Parliament in the Course of the Sixth Session of the Seventeenth Parliament of Great Britain, Called to Meet at Westminster, on Thursday the 29th of October, 1795* vol. IV (London)
Woodhouse, W.J. (1933) *King Agis of Sparta and his Campaign in Arkadia in 418 B.C.* (Oxford)
Wyatt Jr., W.F. (1976) "Persian Dead at Marathon", *Historia* 25: 483–484
Wylie, G. (1992) "Cunaxa and Xenophon", *AC* 61: 119–134

Yates, R.D.S. (1999) "Early China", in K. Raaflaub & N. Rosenstein (edd.) 7–45

LIST OF ILLUSTRATIONS

1. Hoplite with shield and spear. Red-figure kylix (the 'centaur kylix') by Euphronios, c. 490–480. Museum antiker Kleinkunst, Munich (2640). Reprinted from Ernst Pfuhl: *Malerei und Zeichnung der Griechen* (Munich 1923) with kind permission of Stiebner Verlag GmbH.
2. Hoplite running with shield. Terracotta plaque by Euthymides, c. 510–500. Acropolis Museum, Athens (1037). Reprinted from Ernst Pfuhl: *Malerei und Zeichnung der Griechen* (Munich 1923) with kind permission of Stiebner Verlag GmbH.
3. Hoplite warrior in departure scene. White-ground lekythos by the Achilles master, C5s. National Museum, Athens (1818). Reprinted from Ernst Pfuhl: *Malerei und Zeichnung der Griechen* (Munich 1923) with kind permission of Stiebner Verlag GmbH.
4. 'Illyrian' helmet, front and side view. Line drawing by Merete Egeskov.
5. Early Corinthian helmet, front and side view. Line drawing by Merete Egeskov.
6. Corinthian helmet, front and side view. Line drawing by Merete Egeskov.
7. Corinthian helmet, front and side view. Line drawing by Merete Egeskov.
8. *Pilos* helmet (front and side view identical). Line drawing by Merete Egeskov.
9. The Argos panoply. Bronze cuirass and helmet with crest, C8l. Photo © École Française d'Athènes/E. Sérafis (cl. 26.335).
10. Hoplite wearing a muscled bronze cuirass. Red-figure kalyx krater by the painter of the Berlin hydria, c. 465. Metropolitan Museum of Art, New York (07.286.86). Reprinted from Ernst Pfuhl: *Malerei und Zeichnung der Griechen* (Munich 1923) with kind permission of Stiebner Verlag GmbH.
11. Departure scene with hoplite wearing a *linothorax*. Red-figure stamnos by the Achilles painter, C5m. British Museum, London (E 448). Reprinted from Ernst Pfuhl: *Malerei und Zeichnung der Griechen* (Munich 1923) with kind permission of Stiebner Verlag GmbH.
12. The sack of Troy (*Iliou persis*) depicting the double-edged straight sword (*xiphos*). Red-figure kylix by the Brygos painter, c. 480. Louvre, Paris (G 152). Reprinted from Ernst Pfuhl: *Malerei und Zeichnung der Griechen* (Munich 1923) with kind permission of Stiebner Verlag GmbH.
13. Apollon killing Tityos with a *kopis* or *machaira*. Red-figure kylix by the Penthesileia painter, C5f. Staatliche Antikensammlung, Munich (2689). Reprinted from Ernst Pfuhl: *Malerei und Zeichnung der Griechen* (Munich 1923) with kind permission of Stiebner Verlag GmbH.
14. The sack of Troy (*Iliou persis*) with detailed depiction of different types of weapons and armour. Red-figure kalpis hydria by the Kleophrades painter, c. 480–475. Museo Nazionale Arceologico, Naples (2422). Reprinted from Ernst Pfuhl: *Malerei und Zeichnung der Griechen* (Munich 1923) with kind permission of Stiebner Verlag GmbH.
15. Hoplite battle scene. Sherd of black-figure Middle Corinthian column krater, attributed to the Cavalcade painter, c. 590–570. The Metropolitan Museum of Art, New York (12.229.9; gift of John Marshall, 1912). Photo © The Metropolitan Museum of Art.
16. Battle between hoplite phalanxes. Middle Protocorinthian olpe (the 'Chigi olpe') by the Macmillan painter, c. 650. Villa Giulia, Rome (22679). Reprinted from Ernst Pfuhl: *Malerei und Zeichnung der Griechen* (Munich 1923) with kind permission of Stiebner Verlag GmbH.
17. Hoplite battle scene. Middle Protocorinthian aryballos (the 'Berlin aryballos') by the Macmillan painter, c. 650. Antikensammlung, Berlin (3773). Reprinted from Ernst Pfuhl: *Malerei und Zeichnung der Griechen* (Munich 1923) with kind permission of Stiebner Verlag GmbH.
18. Hoplite battle scene. Middle Protocorinthian aryballos (the 'Macmillan aryballos') by the Macmillan painter, c. 650. British Museum, London (1889.4–18.1). Drawing after *JHS* 11 (1890) pl. II.5.

19. Siege scene with opposing hoplite phalanxes. The Amathus bowl, possibly Assyrian or Phoenician silverwork, c. 710–675. British Museum, London (ANE 123053). Drawing after *JHS* 53 (1933).

INDICES

by Niels Grotum Sørensen

GENERAL INDEX

Abu Simbel   134 n. 545
Abydos   214–215 n. 930
Achaians   112, 114, 173 n. 717, 223
Achilles painter   33 fig. 3, 71 fig. 11
Achilleus   150
acrobatics   50–51, 53, 231
advance   17, 101, 114, 120, 122, 126, 128, 133, 146, 155, 160, 161, 165, 181, 187, 195–198, 200, 205, 209–210, 229, 233
Aemilius Paullus   204
African society   106
Agamemnon   111 n. 444
Agatharchides   218 n. 944
age   13, 36, 49, 66, 99–101, 119, 122, 138, 142, 197, 216, 219 n. 946
Agesilaos   136, 164 n. 669, 181–182, 199
Agis II   136, 163, 211
Agis III   193
agonal culture   12, 16, 47
Aias   113 n. 450, 148, 223
Aigina   181–182
Aigition   *see* battle inventory: no. 1
Ailian   20, 157
Aineias   223
Aiolic dialect   107
Aischines   153
Aischylos   18, 86–88, 117
Aitolians   136 n. 551
Akarnania   118 n. 477, 178
akontia   84
   see also javelin
akontismata   84
   see also javelin
akrothinion   21
Alexander   27 n. 62
Alkaios   19, 134 n. 545, 149
Alkibiades   215
allies   38, 172–173, 174, 197, 200, 213, 219
   see also coalition
Alypetos   181–182
Amasis (pharaoh)   70
Amathus   130, 134
Amathus bowl   130–135, 227, 233
amazons   94
Ambrakia   178–179
ambush   178, 181, 201
Amphiaria (at Oropos)   49

Amphilochians   178
Amphipolis   *see* battle inventory: no. 2
Anakreon   149
Anaxibios   181
Anaxikrates   181–182
Anchimolios   181–182
anchor   38, 147
Anopaia   101
ankle-guards   22, 78–79, 96
   see also armour
antilabe   21, 29 fig. 1, 32–34, 53–54, 69 fig. 10, 84, 91, 105 n. 413, 122, 148, 194, 230–231
   see also double grip system, shield: size and shape of
Antimenidas   134 n. 545
Antiochos III   136 n. 551
Aphrodisias   49 n. 160
apobates   49
apoboleus   154
Apollo   87 fig. 13, 224
archers   110, 121, 131–133, 138, 160
Archidamos   169
Archilochos   14, 19, 88, 92–93, 122, 147–151, 154 n. 635, 180, 233
archon   141, 152, 181–182
Ares   108 n. 426, 198
Argive games   46–47
Argives   136, 172–173, 196, 197, 200, 211, 213, 214 n. 929
– in Homer (= Achaians)   114
Argos   46–47, 176 n. 732, 197
Argos grave finds   13, 14, 66, 103, 135
Argos panoply   55, 66–67, 103
Argyraspides   *see* Silver Shields
Aristagoras   89
Aristeides   176 n. 734
Aristodemos   139, 151
ariston   209
Aristophanes   18, 36, 57, 62, 154, 180
Aristotle   60, 76 n. 271, 141, 174 n. 722, 175 n. 729, 195, 198
Arkadians   169, 190, 200, 211
Arkadian federation   173 n. 717
Arkado-Cypriotic dialect   107
armband (on shield)   *see porpax*
arm-guards   22, 78, 96, 231
   see also armour

armour 13, 14–15, 20, 21–22, 27, 35, 46, 55–81, 94, 131, 134, 141, 144, 189, 231–232
- body 15, 22, 41, 66–81, 101, 134, 231
- development of 14–15, 22, 91, 96, 97, 101, 102–105, 123, 135, 143–146, 193, 226–232
- effectiveness of 79–81, 92–93, 94, 231–232
- mobility 46, 81, 97–98, 134, 144, 146, 161, 230, 232
- samurai 81
- size of 20, 22
- supplementary 22, 96, 231–232
- weight of 20, 22, 45, 46, 48, 81, 95–98, 100–101, 123, 146, 161, 197, 210, 214, 216–217, 228–232
   see also arm-guards, ankle-guards, breastplate, corslet, cuirass, foot-guards, greaves, helmet, mail shirt, thigh-guards, *thorax*
army size 97–98, 136–137, 142, 168, 169, 170, 178, 195, 210–211, 219, 222
arraying 74, 161, 209–211, 213, 215, 222, 229
   see also deployment
Arrian 20, 157–158, 185, 193
arrows 34, 35, 52, 57, 70–71, 77, 80, 84, 160, 184, 221
Artaxerxes 209–210
Ashkelon 134 n. 545
Ashurbanipal 134
Asklepiodotos 20, 157, 159
Asopios 181–182
aspis 25–27, 28, 32, 34, 121, 124, 158 n. 644, 167, 187, 205
- as a landmark 28 n. 65
- *aspis Argolike* 46
- *aspis omphaloesse* 117, 118
- metonym 167
   see also shield
Assyrian 79, 110, 131, 133–134
Astyanax 59, 128
Athena 51, 52, 55 n. 182, 149
Athenian 28 n. 65, 57, 73, 85, 94, 100, 101 n. 398, 135, 136, 141–143, 147, 149, 152–154, 168, 170, 172, 174, 175, 177, 178, 179 n. 751, 180, 181–182, 183, 185, 189, 190, 191, 195, 197, 199–200, 202, 205, 206–207, 208, 211, 213, 214–215 n. 930, 216, 219, 220–221
Athens 11, 21, 49, 101 n. 398, 106, 141–143, 151–154, 172, 173 n. 717, 174, 175, 217
Attika 31, 98–99, 142, 168
athletics 46–49, 51 n. 167, 231
atimia 27, 151, 152–153, 164
aulos 50, 62–63, 126, 196
   see also piper

auxiliary troops 14, 126, 140 n. 577
axe 80, 94, 131, 186–187 n. 786

Ba'al Hammon 219 n. 946
baden 197 n. 851
Basel shield 21–22, 28, 31
battle inventory:
   no. 1 (Aigition 426) 181–182, 202 n. 869, 236–237
   no. 2 (Amphipolis 422) 136, 170, 172, 181–182, 217 n. 941, 237–239
   no. 3 (Chaironeia 338) 13, 92, 177, 183, 211 n. 914, 239–240
   no. 4 (Corinth 392) 136, 181–182, 185, 189 n. 797, 214 n. 929, 241–242
   no. 5 (Delion 424) 136, 154, 169, 179 n. 752, 181–182, 185, 191, 195, 199, 204 n. 877, 205–206, 207 n. 896, 211 n. 914, 214–215 n. 930, 242–244
   no. 6 (Ephesos 409) 244–245
   no. 7 (Haliartos 395) 139, 181–182, 245–246
   no. 8 (Idomene 426/5) 178–179, 246–247
   no. 9 (Koroneia 394) 136, 181–182, 185, 190, 196 n. 842, 198 n. 854, 199, 205, 207 n. 896, 211 n. 914, 213, 248–249
   no. 10 (Kromnos 365) 169, 190, 249–250
   no. 11 (Kunaxa 401) 172, 180 n. 752, 182 n. 763, 196, 207 n. 896, 209–211, 214–215 n. 930, 222, 229, 250–252
   no. 12 (Kynoskephalai 364) 181–182, 185, 211 n. 914, 252–253
   no. 13 (Kyzikos 410) 181–182, 253–254
   no. 14 (Laodokeion 423) 205–206, 207 n. 896, 254–255
   no. 15 (Lechaion 390) 255–256
   no. 16 (Leuktra 371) 142 n. 587, 169–171, 173–174, 179–180, 181–182, 184, 185, 189–190, 196 n. 843, 211 n. 914, 256–258
   no. 17 (Mantineia 418) 136, 162–164, 176 n. 732, 179 n. 753, 181–182, 185, 211, 213, 259–261
   no. 18 (Mantineia 362) 153, 181–182, 211 n. 914, 261–263
   no. 19 (Marathon 490) 85, 135–136, 176 n. 734, 179 n. 753, 181–182, 197, 202–203, 204–205, 211 n. 914, 216, 263–264
   no. 20 (Megara 458) 264–265
   no. 21 (Miletos 412) 185, 265–266
   no. 22 (Munychia 403) 168–169, 181–182, 191, 266–267
   no. 23 (Mykale 479) 185, 207 n. 896, 268–269

no. 24 (Naryx 395/4) 214–215 n. 930, 269–270
no. 25 (Nemea 394) 169, 172–173, 176 n. 733, 177, 179 n. 752, 179 n. 753, 190 n. 806, 197, 207 n. 896, 211 n. 914, 213–214, 270–271
no. 26 (Oinophyta 457) 217, 271–272
no. 27 (Olpai 426/5) 179 n. 753, 181–182, 205, 207 n. 896, 272–274
no. 28 (Olynthos 381) 181–182, 274
no. 29 (Plataiai 479) 88–89, 91, 139–140, 156, 174–175, 185, 202 n. 869, 205, 207, 211 n. 914, 219, 275–277
no. 30 (Poteidaia 432) 74, 179 n. 753, 181–182, 205, 207, 213, 277–278
no. 31 (Pylos 425) 152, 185, 205–207, 278–279
no. 32 (Solygeia 425) 181–182, 185, 189, 199–200, 202 n. 869, 205–207, 280–81
no. 33 (Spartolos 429) 181–182, 281–282
no. 34 (Sphakteria 425) 57, 181–182, 202 n. 869, 220–222, 282–284
no. 35 (Sybota island 433) 284–285
no. 36 (Syracuse 415) 113, 179 n. 753, 185, 202 n. 869, 205–207, 285–286
no. 37 (Syracuse 414) 181–182, 287
no. 38 (Syracuse 413) 208, 214–215 n. 930, 288–289
no. 39 (Ta Kerata 409) 289–290
no. 40 (Tanagra 457) 175, 207 n. 896, 211 n. 914, 290–291
no. 41 (Tegyra 375) 169, 181–182, 291–292
battle order 157–167, 175, 177, 210
– centre 135, 174, 203
– closed 13–14, 15, 81, 98, 103, 108–115, 119–120, 123, 127–128, 129, 135, 139, 143, 144, 146, 157–161, 166–167, 170, 171, 196, 220–221, 227–230, 232–233
– column 161, 169, 190, 191
– depth 132, 140, 155, 165 n. 670, 166, 167–171, 180–181, 189, 190, 194–195, 195–196, 233
– *embolon* 171
– file 108, 131, 132, 200
– flank 136, 163, 170–175, 220
– front 123, 168, 169, 170, 171, 172, 173, 176, 179, 181, 182, 183–184, 189, 191, 199, 222, 230, 234
– front ranks 15, 17, 90, 91, 96–97, 104 n. 410, 108, 114, 117, 123, 138, 140, 176, 183–184, 186, 191, 193, 194–195, 198, 199, 230
– in the *Iliad* 108–115
– in Tyrtaios 115–123
– *kata phylas* 175–177
– *kata physin* 157, 159
– *keras* see battle order: wings
– line 54, 108, 111, 155-157, 164, 183, 196, 198, 233
– loose-order 15–16, 39, 108–115, 135, 143, 145, 156, 198, 209
– open 14, 115, 116, 157–161, 163, 198, 227, 232
– *promachoi* 15, 108, 114, 117–118
– ranks 15, 97, 114, 126, 129, 132, 140, 143, 155, 161, 167–171, 173, 183–184, 185–186, 188, 192, 199, 220–221, 227–228
– rear 172, 183–184, 220, 222
– rear ranks 97, 104 n. 410, 117, 124, 126–127, 183–184,185–186, 192, 194–195, 1999
– re-grouping 114
– re-spacing 160–161
– second rank 91, 126, 193
– *stix* 15, 108, 120 n. 489, 144
– third rank 91, 193
– wedge 171
– width of file 93–94, 112, 137, 157–167, 233
– width of front 168, 169, 170, 171, 180–181
– wings 135, 136, 162–163, 172–175, 179–180, 195, 199–200, 203, 213, 234
battle station 139, 151, 153, 172–175, 175–176, 179–180, 183–184, 186, 209, 222, 234
Benaki painter 31
Berlin aryballos 127–128
Berlin hydria (painter of) 69 fig. 10
Boiotia 48 n. 153, 49, 99, 181–182
Boiotian 169, 171, 182 n. 763, 205, 212, 214–215 n. 930
see also shield: Boiotian type
Bomarzo shield 21–22, 28–31, 63
bow 70–71, 82, 88, 89, 131
boxing 39, 43 n. 131, 49, 99 n. 384, 156, 215–216
Brasidas 28 n. 67, 92, 136, 152, 170, 172, 181–182, 206
bronze age 106–107, 115
Brygos painter 87 fig. 12
buckler 38, 54
see also shield
butt-spike 22, 82, 83, 89–91, 184, 193, 232
see also spear

Carthage 219
Carthaginians 218–219

Casaubon, Isaac 70
casualties 17, 137, 170, 176–179, 181–182, 183, 190, 202, 206, 211, 219, 234
casualty lists 19, 175–176, 177
Cavalcade painter 125 fig. 15
cavalry 67, 78, 82, 86, 90, 95, 97 n. 378, 128–129, 131–132, 141–142, 156, 166, 180 n. 758, 185, 191 n. 812, 195 n. 836, 198, 204, 211, 214, 219
cavalry armour 67, 78, 180 n. 758
Celts 110 n. 440
census 141–143
centaur 29 fig. 1
Chabrias 101
Chairon 181–182
Chaironeia 89 n. 335, 89 n. 337, 99 n. 384
 see also battle inventory: no. 3
Chalkedon 181–182
Chalkideus 181–182
Chalkis 13, 212
chalkoma 21, 30, 79, 230
 see also shield
charge 29 n. 73, 85, 90–91, 92, 121, 126–127, 128, 138, 165, 186, 191, 195, 196–198, 210, 213, 216–217
chariot 19 n. 40, 49, 110–111, 113–115, 168, 227
Charoiades 181–182
cheir see arm-guard
Chersonese 182 n. 763
Chigi olpe 14, 62–63, 84, 122, 124–127, 128–129, 130, 132–133, 227, 233
China 94
Chinese 50, 110
chiton 132
chora 178 n. 750
citizen-soldier 13, 17, 142, 180, 228
class 99, 104 n. 410, 141–143, 149–150
coalition 13, 137, 171, 173, 174, 179 n. 752, 205, 210, 212, 213, 219 n. 949
 see also allies
cohesion 116, 123, 147, 175, 179, 195–198, 233
collision 92, 121, 122, 129, 136, 155, 193, 197–198, 206, 211–212
comedy 18, 27 n. 62, 36, 154
command 18, 131, 150, 173–174, 180, 181, 200, 234
community ties 175–179
cooperation 16
Corinth 55–56, 73–74, 158, 176, 205
 see also battle inventory: no. 4
Corinthia 99

Corinthian 172–173, 189, 199–200, 205, 213
 see also helmet: Corinthian type
Cornish 57 n. 189
corslet 29 fig. 1, 66, 70–73, 74–75, 97 n. 378, 124, 131–132, 231
– development of 70 n. 244, 73
– Egyptian type 70
– *epomides* 71 fig. 11, 72
– lacing 71
– layers 70, 72, 73
– leather 66, 70–73, 231
– linen 29 fig. 1, 66, 70–73, 88 fig. 14, 95, 231
– materials 70, 72, 95–96
– movement 72, 73, 74–75, 231
– neck-guard 72
– protection 70–73, 79, 231
– *pteryges* 71 fig. 11, 72
– reinforcement 72–73, 95–96
– scale-corslet 70
– shoulder piece 70
– temperature and ventilation 74
– weight of 70, 72, 73, 74–75, 95–96
 see also armour
cowardice 116, 123, 149–152, 154, 172, 176, 184
Crete 49, 99 n. 388, 193 n. 826
cuirass 15, 22, 66–70, 71, 96–97, 131–133, 144, 184, 230–231
– back plate 66, 67, 68
– bell rim 68, 69, 72
– bell type 13, 41, 66–68, 95, 103
– belly-guard 68–70
– breastplate 25–55, 41, 47 n. 148, 55, 66–70, 78, 80, 89 n. 335, 92, 158 n. 644, 187, 216 n. 937
– bronze 13, 14, 41, 89 n. 335, 125, 144, 184, 189, 230–231
– development of 66–70, 230
– *gyala* 66, 68
– hem 67, 68
– iron 70
– lining 67
– movement 41, 67, 69, 75, 144, 230–231
– muscular type 66, 68–70, 73
– omega curve 67
– padding 73
– protection 67, 68, 69, 73, 79–81, 231
– *pteryges* 72
– size of 67
– skirt 67
– stand collar 67, 68

- temperature and ventilation 73–74, 144, 230–231
- weight of 67–68, 73, 75, 95, 144
  see also armour

dagger 193
  see also encheiridion
daimon 223
Daiphantos 181–182
dance 36, 49–53, 231
  see also pyrrhiche
Daton 181–182
dedication 21–22, 28, 30, 31, 96–97, 149–150, 184
deile 209–210, 219
Deinon 173
Delion see battle inventory: no. 5
Delphi 21, 52, 68
Demaratos 147, 164
deme 175, 176 n. 732, 234
demography 142, 178 n. 750
Demosthenes 19, 153, 178, 182–183, 208, 220
Dendra armour 66
deployment 157–183, 205, 209, 212, 213, 222
  see also arraying, battle order, formation
desertion 153
dialect 107, 222
  see also individual dialects
diaulos race 46–47
Dienekes 138 n. 560
diet 99
Dikaiogenes 181–182
Diodoros 19–20, 25–27, 38–39, 153 n. 631, 165, 167, 168, 183, 202–203, 205, 217–220, 222, 229
Diomilos 181–182
Dirphys 212
discipline 20 n. 44, 37–38, 51 n. 166, 138–139, 163
diskos 49
disruption 54, 123, 155, 162–163, 165, 183, 184, 186–187, 195–198, 199, 233
doratismos 92, 199
Dorian 197
Doric dialect 197, 208, 222
dorsal protection 37, 65, 146, 159–161
dory see spear
dorydrepanon 94
double-grip system 32–34, 36, 37, 53–54, 71 fig. 11, 77, 87 fig. 12, 88 fig. 14, 98, 102, 105, 148, 192, 230–231, 233
  see also *antilabe*, *porpax*, shield
drill 17, 155, 166, 228, 233

drill manual 166–167
dromos 85, 197–198
  see also charge
duel 15–16, 17, 19, 24, 41–42, 98, 108, 110, 113, 115, 121, 146, 150, 188, 223, 227–228
  see also *monomachia*, single combat
duration of battle 17, 20, 184, 190, 201–225, 228–229

Egypt 19, 70, 178 n. 747, 181–182, 217
Egyptian 110, 131, 187, 192
  see also corslet: Egyptian type
Egyptian phalanx 158
epigraphy 19
ekecheiria 47
Elam 110 n. 440
Eleans 200
eleleu 198
Eleusis 181–182
Elis 173 n. 717
élite 27 n. 62, 176, 177, 211
encheiridion 93, 193–194
  see also sword
Enyalios 198
Epameinondas 49, 153, 171, 173–174, 181–182, 189–190
Ephesos see battle inventory: no. 6
ephor 140
Ephoros 153 n. 631, 218
Epidauros 176 n. 732
Epipolai 208, 214–215 n. 930
Epitadas 181–182
Epitales 182 n. 763
Epiteles 182 n. 763
Erechtheid *phyle* 177, 182 n. 763
Eretria 31
Esarhadon 134
escalade 131, 133–134
Eteokles 41–43
Euboia 13, 212
Euboians 92–93
Euphronios 29 fig. 1
Eupolis 154
Eurasian society 106
Euripides 18–19, 41–43, 49, 52, 89, 92, 101, 164, 190, 231
Euripos 212
Europe 12, 16, 156
European 59, 99
Eurylochos 181–182
Euthymides 33 fig. 2
exodos 175 n. 728

family 139 n. 570, 178, 216
fencing 203
file leader 199, 232
flanking 110, 136, 163, 171, 172–173, 177, 220
flight 22–23, 37, 111, 116–119, 128, 136–139, 148, 150, 154–155, 165, 172, 184, 186, 206, 211–212, 214–215, 217
– feigned 136–139
foot-guard 79, 96, 231
  see also armour
formation 14, 19, 24, 53, 74, 103–104, 108–114, 123, 126–128, 139, 140, 168, 169, 174, 180, 186, 191, 196, 197, 204, 210, 228, 233
frontal combat 37, 45, 69, 81
frontal protection 45, 57, 81, 144, 146, 159–162, 192, 231–232

Gauls 70 n. 246, 136 n. 551
Gelon 219
general 18, 94, 150, 172, 173, 175, 176, 177, 179–183, 217 n. 941, 222, 234
  see also strategos
generalship 135–136, 201
geography 12, 142
Glabrio, Marcus Acilius 136 n. 551
Glockenpanzer see cuirass: bell type
Gorgoleon 181–182
Gorgopas 181–182
graphe deilias 152
graphe [tou] apobeblekenai ten aspida 27, 152-153
greaves 15, 22, 26, 29 fig. 1, 37, 47, 60, 75–77, 78–79, 88 fig. 14, 96–97, 103, 124, 128, 131, 144, 231
– attachment 76, 231
– decoration 77
– development of 76–77
– effectiveness 79–81
– padding 76
– protection 77, 79, 231
– shape of 75–77
– size of 77, 79
– weight of 47, 77, 95–96, 144, 216 n. 937
  see also armour
grip (on shield) see *antilabe*, double grip system
Gymkhana 48–49 n. 155

Halai 190
halberd 94, 166 n. 676
Haliartos see battle inventory: no. 7
Hamilkar 219

hammer 80–81
Hannibal 133
harmost 181
Hastings 191 n. 812
Hektor 59, 112, 114, 165, 223
hegemonia 173–174
Helenos 114
helmet 22, 26, 33 fig. 2, 47, 48–49 n. 155, 55–66, 66, 94, 96, 112, 131–133, 144, 146, 165, 167, 180, 230–231
– Assyrian type 79, 133
– Attic type 55, 63
– balance 59, 65, 144, 231
– cap 65, 133
  see also helmet: *pilos* type
– Chalcidian type 55, 56 n. 183, 63
– cheek-guards 55, 56, 58 fig. 7, 61 n. 207, 65, 67 fig. 9, 231
– chin plate 131
– chin-strap 26, 62, 63, 65, 231
– closed 144
– conical 13, 14, 55, 57
– Corinthian type 22, 33 fig. 3, 55–66, 97, 124, 128, 131, 230–231
– crest 48–49 n. 155, 55, 56, 59–60, 62, 63, 65, 112, 128, 131, 133, 165, 180, 231
– crest-holder 65
– crest-stilt 13, 55, 59, 67 fig. 9
– development of 55–59, 63, 230, 231
– effectiveness 79
– eye slits 56, 58 fig. 6–7, 59, 61–62, 65
– hearing 62–63, 64–65, 76 n. 271, 98, 144, 230–231
– Illyrian type 55, 56 fig. 4, 56 n. 183, 59, 61 n. 207, 103, 131
– Ionian type 131 n. 535
– *Kegelhelm* type 55, 56 n. 183, 66, 67 fig. 9
– materials 55, 57, 59, 60, 63, 64, 111, 231
– measurements 57, 79
– modern steel 61, 63–64
– Montefortino type 65–66
– neck-guard 55, 56, 65, 133
– nose-guard 56, 57, 58 fig. 5
– padding 57 n. 187, 60–61, 63, 65, 73, 76 n. 271
– *pilos* type 57–59, 97, 231
– pointed 133
– plume see crest
– rounded 55, 56, 57, 65
– temperature and ventilation 56, 57, 59, 64–66, 144
– types of 55–59
– vision 61–62, 64–65, 98, 144, 146, 230–231

– weight of 47, 57, 59, 63–64, 65, 67–68, 95–96, 144, 146, 216 n. 937, 231
   see also armour
helots 139–140
hemiogdoa 179
Hera 108 n. 426
Herakleia 181–182
herald 178–179, 223
Herculaneum 99 n. 382
Herodotos 16, 18, 22, 55, 70, 85, 88–89, 92, 109, 134 n. 545, 135–140, 147, 149, 151, 156, 174–175, 185, 197, 202–203, 209 n. 902, 212, 218–220, 222, 224–225, 229, 232
hetairai 51
Hesiod 18, 116, 224
Hestiodoros 181–182
hieros lochos *see* Sacred Band
Himera 218–220, 229
Hippagretas 181–182
hipparch 182
hippeus 141–142, 211
Hippokrates (Athenian general) 181–182
Hippokrates (Spartan harmost) 181–182
Hippomachos 181–182
historians 12, 18, 19–20, 23, 93, 196, 198, 201–202, 209, 225
   see also individual historians
historiography 19–20, 23, 201–202, 222, 224–225, 229
– Latin 186
Hittites 110
Homer 14, 18, 22, 47, 67 n. 232, 105–106, 109, 115, 116–118, 144, 148, 167, 174, 176, 202–203, 222, 228
Homeric poems 12, 14, 15, 16–17, 19, 105–108, 109–110, 117, 119, 128, 135, 150, 183 n. 767, 224, 227–229
– language 105–107, 112 n. 448, 115, 117, 222, 227
– similes 19, 112 n. 448, 113, 227
Homeric society 16, 18, 105–108, 114–115, 144, 227–229
– architecture 107
– burial 106–107
– marriage 106–107
– metallurgy 106
hoplite race *see* hoplitodromos
hoplitodromos 33 fig. 2, 46–49, 98, 230–231
hoplomachos 94
hoplon 25–27, 109 n. 434
   see also shield
Horace 117, 148 n. 598

hostages 139 n. 570
Hurrians 110
hybris 151
hypaspistes 100, 232

Iberian 38
iconography 14, 15–16, 18, 20–21, 23–24, 28, 31–32, 34–35, 45, 64, 68, 70, 73, 84, 94, 97, 102, 103, 115, 116 n. 462, 122, 123–130, 130–135, 143, 145, 155, 191, 226–227, 229, 232–233
Idomene *see* battle inventory: no. 8
Igel (Swiss) 166 n. 676, 167
Iliad 12, 14, 15, 16, 19, 75, 105, 106, 108–115, 116–120, 128–129, 135, 144, 148, 150, 165, 202–203, 222–224, 232
Iliou persis 87 fig. 12, 88 fig. 14, 128
Immortals 137
insubordination 152
Iolaidas 181–182
Iolaos 101
Ionian 134 n. 545
Ionic dialect 107, 222, 224
iron age 105–107
Iron Man 48
Isaios 19
Isokrates 152
Italy 28
itea 28–29 n. 70
   see also shield
itys 29–31, 32–34, 42, 44, 45, 77, 89, 91, 94, 98, 161, 162, 190, 192, 198, 230
   see also shield

Japan 16
Japanese 50, 59, 81, 94
javelin 13–14, 57, 77, 78, 84–85, 96, 121–122, 126, 128, 139, 143, 145, 160, 221, 232
– effectiveness 80, 85

Kaineus 154
Kalamata 73–74
Kallias 181–182
Kallimachos 136, 181–182
Kallinos 14, 19, 84, 103, 108, 119, 122
karate 43 n. 131
Karians 102, 134 n. 545
Karystians 189, 199–200
kata (Japanese) 50
Kilikia 181–182
kinship 175, 179, 222, 234
Kimon 175
Kerameikos 32

Klearchos 94, 172, 182 n. 763, 210
Klearidas 170
Kleinias (the Kretan) 11
Kleinios 181–182
Kleisthenic reforms 175
Kleombrotos 26, 171, 173, 181–182
Kleomenes 89
Kleon 154, 172, 181–182, 220
Kleonymos (Athenian) 154
Kleonymos (Spartan) 70
Kleonymos (Spartan, son of Sphodrias) 173, 184
kleruch 97 n. 377
Knielauf 124–125
kopis 86, 87–88 fig. 13–14, 94–95, 232
  see also sword
Korkyra 197
Koroneia 181–182
  see also battle inventory: no. 9
korys 55
kranos 55
Kreusis 26
Krimisos 92
Kritias 140, 150–151, 181–182, 233
Kromnos *see* battle inventory: no. 10
Kunaxa *see* battle inventory: no. 11
kydoimos 121
Kyme 128
kyneë 55
Kynoskephalai *see* battle inventory: no. 12
kyrioi 178
Kyros 168, 169, 172, 209–210
Kyzikos *see* battle inventory: no. 13

Labotas 181–182
Laches 94, 181–182
ladder (scaling) 131–133
laiseïon 111–112
Lamachos 180, 181–182
Lamia 181–182
Laodokeion *see* battle inventory: no. 14
lateral protection 37, 44, 45, 63, 98, 144, 146, 159–164, 172, 174, 233
leadership 18, 180–183, 187
  see also hegemonia
Lechaion 32
  see also battle inventory: no. 15
Lee, Bruce 39
legion 204
legionary 53, 65–66, 146
Leonidas (of Rhodes) 47
Leonidas (of Sparta) 118 n. 477, 181–182
Leosthenes 181–182

Lesbian dialect 107
Leuktra *see* battle inventory: no. 16
Levant 134
light-armed troops 14, 54, 121–122, 128, 133, 135, 138, 139–140, 141, 156, 204, 211 n. 914, 214, 219 n. 949, 220–222, 233
Linear B 110
linothorax 70
  see also corslet: linen
lipotaxiou 153
Lipsius 152 n. 622
literary conventions 42, 108, 117, 201, 203, 217–218, 222–225
logistics 14, 177
logographers 19
  see also individual logographers
lovers 176–177
Lukianos 49–50
Lusitanians 38
Lykambes 147
Lykia 181–182
Lykophron 181–182
Lykurgos (of Athens) 183
Lykurgos (of Sparta) 215 n. 932
Lysander 139, 181–182
Lysias 19, 26–27, 152–153, 176
Lysikles (Athenian general, *PA* 9417) 181–182
Lysikles (Athenian general, *PA* 9422) 183

Macedon 158, 165
Macedonia 99
Macedonian 14, 27 n. 62, 67, 82, 186
  see also shield: Macedonian type
Macedonian phalanx 19–20, 158–159, 165–167, 184, 186, 193, 204, 233
machaira 86, 87–88 fig. 13–14, 158 n. 644
  see also kopis, sword
Machanidas 90 n. 338
Machiavelli 167 n. 680
machomai 225
Macmillan aryballos 128
Macmillan painter 124–128
mail shirt 133
  see also armour
Makarios 181–182
Mantineia 176 n. 732
  see also battle inventory: nos. 17–18
Mantitheos 175 n. 728, 176
Marathon *see* battle inventory: no. 19
Marathon race 48 n. 154
Mardonios 219
martial arts 39, 40, 43 n. 131, 50–51, 53

see also boxing, karate, *kata*, *tao lu*, wrestling, *wushu*
massed fighting 15, 19, 21, 98, 104, 108–110, 114–115, 230
meals 207, 209, 211, 223
Medes 137
Megara 26 n. 58, 49 n. 160, 57, 176 n. 732
see also battle inventory: no. 20
Melanippos 149–150
Melea 199, 211
Melesandros 181–182
Menander 27 n. 62
mercenary 19, 27 n. 62, 88, 94, 97 n. 378, 133–135, 145, 150, 166 n. 676, 168, 169, 172, 210, 227
Messenian War, second 14, 116
metaichmion 197–198
metalepsis 90–91
see also spear
Miletos 196 n. 842
see also battle inventory: no. 21
Miltiades 136
Mimnermos 19, 84, 103, 122
Mindaros 181–182
miners 57 n. 189
missiles 34, 52, 53–54, 77, 79, 84, 124, 131, 143, 155, 160, 184, 220
see also arrows, javelin, slingshots, stones
mixed phalanx 96–97, 104 n. 410, 140
Mnaseas 181–182
Mnasippos 211 n. 915
mobility 41, 54, 81, 98, 111, 116, 120, 122, 135–136, 139, 146, 156, 161, 167, 193, 197, 220, 230, 232
momentum 155, 191, 192, 197–198
monomachia 54
see also duel, single combat
morale 116–119, 164–165, 173, 180–183, 194–195, 196–197, 234
Mummius 158
Munychia see battle inventory: no. 22
musket 79
mutiny 139–140
Mycenean 75, 85, 106, 107, 108 n. 426, 110
Mykale see battle inventory: no. 23
Mykalessos 133–134 n. 540
Myrmidons 112 n. 448
Mysian 50–51
Mytilene 149

naginata 94
Napoleonic wars 59, 79
Naqš-i-Rustam 138 n. 560

Naryx 181–182
see also battle inventory: no. 24
naval battle 94, 185, 219
navy 141, 219
Nebuchadnezzar II 134 n. 545
Nemea see battle inventory: no. 25
Neobule 147
Neoptolemos 52, 88 fig. 14, 128
Nereid monument 130
Nerikos 181–182
Nestor 110, 176, 223
Nikostratos 181–182

oblique phalanx 171
Odysseus 148
Odyssey 105, 106, 115, 222
officer 176, 180, 182–183 n. 765, 200, 234
oikiai see family, kinship
Oinokles painter 69 n. 241
Oinophyta see battle inventory: no. 26
oligarchy 149
Olpai see battle inventory: no. 27
Olympia 21, 28, 30, 31, 46, 48–49 n. 155, 55 n. 178, 56 n. 183, 61, 82, 96, 181–182, 185, 200
Olympic games 46, 200
Olynthos 30–31 n. 81
see also battle inventory: no. 28
opse 205–207
orchestris 51
Oropos 49
ostracism 151, 175
othismos 17, 112 n. 448, 119, 183–200, 205, 215–216, 221, 228–230
Oxyrhynchus historian 218

Pagondas 206
paian 155, 196–198, 208
paidotribai 49 n. 160
palta 82
Panathenaia 36, 49, 52
panoply 37, 47, 55 n. 182, 62, 79, 95–96, 100, 103, 126, 144, 161, 230, 232
Panormos 181–182
Pantites 151
Papua New Guinea 15, 113
parade 168
parameridia 78
see also thigh guards
pararrexis 213
Parthians 70 n. 246
partisan 94
Pasimachos 181–182

Pategyas 210
Pausanias (king of Sparta) 38, 190
Pausanias (periegetes) 46, 47, 68
Peace treaties 19
pechys 157–159, 165
Peiraieus 133–134 n. 540, 168, 179 n. 753, 185, 190, 191
Peisistratos 55 n. 182, 106
Pellenians 213–214
Pelopidas 169, 181–182
Peloponnese 55 n. 178, 73–74, 102
Peloponnesian war 18, 19, 97 n. 378, 135–136, 139
Pelousion 217
peltast 25, 121, 141, 160 n. 653
pelte 25, 50, 121
 see also shield
pentakosiomedimnoi 141, 142 n. 589
Penthesileia painter 88 fig. 14
Perseus (king of Macedonia) 204
Persia 168, 169, 180, 195 n. 836
Persians 88–89, 91, 101, 135–138, 156, 185, 187, 189, 203, 205, 209–211, 214–215 n. 930
Persian dance 50
Persian wars 16, 46, 86–88, 104, 115, 135–140, 145, 224, 232
phalanx 15, 144, 205
– etymology 109
– *loxe phalanx* see oblique phalanx
– nomenclature 103–104, 108, 120 n. 489
Phaleron 181–182
Phanomachos 181–182
Pharnabazos 214–215 n. 930
phases of battle 17, 98, 111, 113, 114, 127–128, 130, 184, 188, 191, 198, 199, 203, 205–207, 208–212, 213, 214–215, 220, 222, 229
– combat 202, 205–207, 212, 214–215, 229
– development of 17
– general's harangue 211, 212, 215
– mopping up 86, 212
– preparatory 74, 207, 209, 211, 214–215
 see also arraying, flight, pursuit, retreat, retrieval of corpses, rout, *trope*
Philip II 13, 70, 136 n. 551, 165, 167
Philip V 136 n. 551
Philippi 148 n. 598
Philopoimen 90 n. 338
philosophers 12, 158
philosophy 158, 180
Philostratus 47–48
Phleius 173 n. 717
Phoenician 131, 134, 218

Phoibidas 181–182
Phokian 101, 182 n. 763, 214–215 n. 930
phatrai 179
Phye 55 n. 182
phyle 175–179, 222, 234
Phyle 215 n. 933
physiology 17, 47–48, 54, 95, 98–101, 186–187, 188–189, 190–191, 197, 198, 215–217, 221, 228–229, 231, 232
pike 166–167, 186, 233
pikemen 156, 166–167
pincer manoeuvre 135
Pindar 19, 198 n. 855
piper 63, 126–127
 see also aulos
Pistias 73–74
Pittakos 149
Plataiai 133–134 n. 540, 208
 see also battle inventory: no. 29
Plataians 85, 208
Plataian games 46, 48
Plato 11–12, 51, 94, 153–154, 176 n. 736, 214–215, 219 n. 949
playwrights 18–19
 see also individual playwrights, comedy, tragedy
Pliny 28–29, 70
plunder 96–97, 210, 214
Plutarch 19–20, 38, 91, 92, 93 n. 355, 150, 152, 164, 169, 172 n. 713, 174, 176, 193–194, 196 n. 847, 203 n. 873, 204, 212, 222
poetry
– Archaic 14, 16, 19, 23, 84, 103, 115–123, 135, 143, 149, 228, 229, 232
– elegiac 19, 116
– epic 12, 18, 42, 108–109, 117, 218, 222–225, 229
– iambic 148
– Ionian 148, 224
– lyric 19, 228
– Mycenaean 108 n. 426
– oral 15, 105–107, 112, 117
 see also individual poets
polemarch 173, 181–182
police 51, 53–54, 98, 155–157, 233
politicians 19
 see also individual politicians
Polyainos 190, 219 n. 951
polyandrion 89 n. 335, 212
Polybios 19–20, 90, 157–158 n. 642, 161 n. 656, 165 n. 671, 185–186, 218 n. 944
Polycharmos 181–182
Polygnotos 68

Polyneikes 41–42
Pompeii 99 n. 382
porpax 21, 29 fig. 1, 32, 53–54, 69 fig. 10, 102, 105 n. 413, 140, 147, 148, 192, 230–231
    see also double grip system, shield: size and shape of
Poseidonios 158
Poteidaia see battle inventory: no. 30
Priam 88 fig. 14
prisoners 28 n. 65, 183
Prokles 181–182
prothelymnos 112–113
Psammetichos (= Psamtik) I 134 n. 545
Psamtik II 134 n. 545
psiloi 140, 141, 211 n. 914
Punic wars 133
pursuit 54, 86, 95, 110, 111, 127, 136, 139, 154-155, 204, 205–206, 209, 210, 211 n. 915, 213, 214–215, 220, 222, 229, 233
Pydna 158, 203 n. 873, 204, 222
pyknosis 157–158, 165
pyknotes 165
Pylos 152
    see also battle inventory: no. 31
Pylos shield 21, 28, 31
pyrrhic dance see pyrrhiche
pyrrhiche 46, 49–53, 98, 230–231
Pythagoreanism 174

rally 22, 113, 114, 123, 127, 210, 214
Rameses II 134 n. 545
redeployment 190
religion 12, 46–47, 174, 196, 198, 219
replica 35, 61, 73, 83 n. 314, 85 n. 324, 95–96, 216–217
reserves 170, 190, 194, 204
retreat 37, 51, 95, 100, 114, 116, 120, 122–123, 136, 138, 147, 148, 181, 200, 215, 220
retrieval of corpses 205, 209, 212, 229
revolt 139–140
rhapsodes 108, 115
rhipsaspia 22–23, 26–27, 147–155, 164, 214, 233
rifle 35, 79
rim (on shield) see itys
Roland, Chanson de 138 n. 563
Rome 16
Roman 53, 65–66, 90, 93, 99 n. 382, 146, 158, 186–187 n. 786, 204, 219 n. 946
rout 111, 120, 123, 127–128, 139, 172, 202, 205, 206, 211–212, 213, 214, 217
    see also trope
rugby 216

sabre see kopis
Sacred Band 170–171, 176–177
sacrifice 210, 211, 212, 219
– human 219
– self-sacrifice 117–118, 150, 152, 176, 219
Saian 148
Salamis 219
salpinx 63
samurai 59, 81, 94 n. 362
Sandion 181–182
sarissa 81–82, 158 n. 644, 166
satyr 164
satyr play 48–49 n. 155
sauroter 83
    see also butt-spike
scabbard 85–86, 87 fig. 12, 92, 96
scutum 42 n. 129, 53, 146
    see also shield
seasons 23, 229
– autumn 74
– spring 74
– summer 23, 57, 64–65, 73–74, 144, 197, 208, 216, 221
– winter 23, 74
    see also weather
Semitic 219 n. 945
Sennacherib 110 n. 440
shame 37–38, 123, 148–151, 176, 182–183
sheathing see chalkoma
shield 12, 13, 21–22, 23, 25–54, 55, 62, 63, 69 fig. 10, 81, 84–85, 86, 91, 93–94, 96, 102, 104–105, 112–113, 121, 128, 131–132, 137, 140, 144, 146–157, 158 n. 644, 160, 161, 162, 163, 164, 165, 167, 172, 174, 184, 185, 186, 187, 190, 191–194, 196, 199, 228, 230–231, 233
– blazons 31–32, 33 fig. 3, 125 fig. 16, 128–129, 131–132, 147
– Boiotian type 124
– carrying strap 37, 148
– cover 28
– Cypriote type 133 n. 539
– defensive usage 38–41, 54, 77, 78, 155–157, 164, 233
– effectiveness 28–29, 79, 92
– handling 21–22, 25–26, 32–37, 38–41, 46, 54, 74, 77, 93–94, 98, 121, 144, 146–147, 154, 155–157, 164, 167, 192, 198, 230–231
– Macedonian type 165–166
– materials 21–22, 28–31, 79, 111, 125 fig. 16, 230
– metonym 167
– nomenclature 25–27

- offensive usage 41–45, 233
- padding 21, 30–31
- rimless 165
- size and shape of 21–22, 25–26, 28–32, 33 fig. 2–3, 35, 36–37, 38, 42, 44, 45, 53–54, 77, 79, 100, 111, 121, 146–147, 156–157, 158 n. 644, 161-162, 165, 187, 191, 192, 230
- spiked bosses 131, 133
- weight of 25, 28–29 n. 70, 31, 33 fig. 2–3, 34, 35, 37, 38, 43, 45, 46, 54, 77, 81, 95–96, 98, 133, 144, 146–147, 148, 154, 156, 164, 191, 192, 216, 230–231, 233
- wicker 185, 187

see also antilabe, aspis, buckler, chalkoma, double grip system, hoplon, itea, itys, pelte, porpax, scutum, targe, thyreos

shield apron 34, 69 fig. 10, 96
Sicily 28, 178 n. 747, 181–182, 219
sickle-spear see dorydrepanon
siege 74, 131–134, 168, 169, 223
Sigeion 149–150
Silenos 164
Silver Shields 27 n. 62
Simonides 19
single combat 14, 16, 38–39, 41–42, 54, 98, 129–130, 146, 156, 159, 161, 227, 230, 233
see also duel, monomachia
Siphnian treasury 130 n. 526
Skiritai 174
skirmishers 160
slaves 138–139 n. 563, 139
slingers 121
slings 77
slingshots 60 n. 204, 84, 160
Sokrates 74–75, 150, 215
solidarity 177
Solon 141–143, 152
Solonian census classes 141–143
Solygeia see battle inventory: no. 32
Sophanes 147, 181–182
Sophokles 18
Sosias painter 60 n. 200
Spanish civil war 130 n. 225
Spanish mercenaries 133
Sparta 18, 20, 28, 31, 37–38, 49–50, 51 n. 167, 57, 60, 63, 84, 93, 115–119, 122, 133–134 n. 540, 135–140, 147, 150, 151–152, 154, 156, 163, 164, 168, 169–171, 172–173, 174, 174–175 n. 724, 177, 180, 181–182, 190, 191, 193–194, 196, 198 n. 854, 199, 205, 206, 207, 211, 213, 214–215, 220–221

Spartolos see battle inventory: no. 33
spear 12, 22, 23, 26, 27, 33 fig. 2, 34, 37, 40, 43, 44, 45, 57, 66, 72, 79, 81–85, 93, 109, 112, 120, 124, 126, 129, 131–132, 137, 139, 140, 144, 156, 158 n. 644, 165, 167, 184, 187, 190, 191, 197–198, 232
- effectiveness 80–81, 92, 93, 197–198
- in combat 86–92, 93, 159, 167, 184, 191, 193–194, 197–198
- materials 22, 81–83, 106
- range 83, 88, 90, 93, 159, 164, 193
- secondary 31, 66, 84–85, 91, 122, 126, 135
- size of 22, 83, 88–89, 223
- throwing see javelin
- weight of 80, 83, 95–96, 216

see also butt-spike, dorydrepanon, metalepsis, palta, sarissa, spear-head

spear-head 22, 35, 66, 81, 88, 90 n. 338, 91, 129, 131, 137, 172, 211, 232
- materials 22, 82, 106, 232
- size and shape of 82–83
- weight of 83

see also spear

Sphakteria see battle inventory: no. 34
sphinx 131
Sphodrias 173
spolas see corslet: leather
spoliation 21, 75, 96–97, 209
spongos Achilleios 60, 76 n. 271
squire 129
stages of battle see phases of battle
stance 39–41, 78, 120, 124, 127, 155, 160, 162, 191–193, 198, 231
Steiria 176
Stesileos (PA 12905) 94
Stesileos (son of Thrasyleos, PA 12906) 181–182
Stone Lion monument 99 n. 384
stones 52, 53, 54, 77, 121, 221
stratagems 135–139, 170–171, 172–174, 201
strategos 150, 163, 175, 179, 221 n. 957
see also general
Stratolas 181–182
styrax 83
see also butt-spike
surprise 44, 45, 101, 127, 155, 170, 178, 197, 208, 209, 213, 215 n. 933, 220
surrounding 136, 139, 163, 220
swimming 48 n. 154, 81
sword 22, 26, 30, 31, 37, 40, 42–43, 44, 45, 57, 77, 85–86, 89, 90, 91, 92–95, 106, 144, 158 n. 644, 164 n. 669, 184, 190, 198, 232
- effectiveness 80–81, 93–95

- hilt 66, 85
- in combat 87 fig. 13, 92–95, 193–194
- materials 22, 85–86, 106
- size and shape of 85–87, 89, 93, 193–194
- weight of 67–68, 80, 95–96, 216 n. 937
  see also encheiridion, kopis, machaira, xiphos
Sybota island see battle inventory: no. 35
symposium 50–51
synaspismos 113 n. 451, 119, 157–158, 164, 165–167
systoichiai 174 n. 722
Syracuse 100, 181–182, 197, 219
  see also battle inventory: nos. 36–38
Syrian phalanx 158
Syro-Levantine 110

tacticians 20, 157–159, 165, 171
  see also individual tacticians
tactics 12–14, 20, 22–23, 43–44, 49, 81, 112–113, 120, 130 n. 527, 135–139, 143–146, 155–157, 167, 169–171, 172, 177, 180–181, 184, 187, 191, 198, 200, 204, 233–234
- development of 15–16, 17, 22, 84–85, 96, 102–105, 109, 112, 115–116, 122–123, 135, 139, 143–146, 158–160, 169, 171, 186, 226–230
- in the *Iliad* 108–115, 129, 144, 223 n. 959
- in Tyrtaios 119–123
- position 136–137, 139, 191, 197, 220–221
ta gymna 162–163, 172
ta hoplismena 172
Ta Kerata see battle inventory: no. 39
Tanagra see battle inventory: no. 40
tao lu 50
targe 38, 54
  see also shield
tasso 140
taxiarch 176
taxis 175
Tearless Battle 199, 211
Tegea 174, 176 n. 732
Tegyra see battle inventory: no. 41
telamon see shield: carrying strap
Teleutias 181–182
terrain 136–137, 155, 168–169, 180–181, 191, 195–196, 208, 210, 213, 216, 220, 233
territory 177, 178 n. 750
Theban 18, 89 n. 335, 136, 139, 169–171, 172–174, 176–177, 179–180, 190, 191, 195, 198 n. 854, 199, 205, 206, 212, 213
Thebes 18, 99 n. 384, 176, 205, 217
Themistokles 176 n. 734

Theomnestos 26
Theopompos 181–182
Thermopylai 92, 101, 135–139, 151, 179 n. 752, 181–182, 185, 232
Thersites 150
Thespiai 179 n. 752, 181–182
Thespians 136, 138–139 n. 563, 213
Thessalian feint 41–44, 231
Thessalians 43
Thessaly 41–42, 181–182
thetes 141–143
Thibrachos 181–182
thigh guards 78, 96, 231
thorax 25 n. 55, 57 n. 187, 74–75, 147, 158 n. 644, 187
Thrace 50, 51 n. 167, 121
Thrasybulos 168–169, 191, 215 n. 933
Thucydides 18, 23, 57, 100, 141, 142, 162–164, 174, 178–179, 195, 196, 199–200, 203, 205–206, 208, 215, 220–221, 224, 232, 233
Thymbrara 189
thyreos 42 n. 129, 146, 158 n. 644
  see also shield
Thyrrheion 118 n. 477
time of battle 74, 204, 206, 207–208, 208–210, 215, 218–220, 223
- night 197, 206–208, 214–215 n. 930
timema 141–142
Tiryns 32
titans 224
Tityos 87 fig. 13
Tolmides 181–182
Torquatus, T. Manlius 42 n. 129
Trachian 138 n. 560
tragedy 18–19, 48–49 n. 155, 117, 178 n. 749
training 17, 37, 49–53, 138, 156, 177, 190, 220
trampling 137, 184, 189 n. 797, 211
treachery 92, 149
tremblers see *tresantes*
tresantes 151
triakades 179
triathlon (modern) 48
tribe see *phyle*
Tritogeneia 36
Troad 149
Trojans 112, 114, 119, 165, 223–224
Trojan war 154, 165
trope 111, 215, 233
  see also rout
trophy 152, 183, 184, 209, 212, 229
Troy (fall of) see *Iliou persis*
trumpeter 63

tyrants 102, 226–227
- the thirty 150, 168, 181–182
Tyriaeion 168
Tyrtaios 14, 19, 60, 84, 89, 100, 103, 108, 115–123, 200

Vegetius 93–95, 204, 222
Vergina 67, 70
Vietnam war 130 n. 525

Wallhausen, Johann Jacobi von 166–167
war cry 198
watchword 197, 208, 210
weapons (offensive) 14, 20, 21–22, 23–24, 25, 27, 36, 44, 53, 65, 131, 141, 146, 171, 187–188, 190, 191, 199, 204, 210, 228
- development of 14, 15, 17, 22, 28, 68, 94, 96, 102–105, 123, 135, 144–146, 193, 226–230
- edged 73, 77, 81, 232
- size of 20, 100
- weight of 20, 100, 210, 216–217, 228–230
see also arrows, axe, bow, dagger, *dory-drepanon*, *encheiridion*, halberd, hammer, javelin, missiles, *naginata*, *palta*, partisan, pike, *sarissa*, shield, slingshots, spear, stones, sword
weather 26, 65, 73–74, 154, 197, 216–217, 221, 229
- clouds 208, 221
- rain 74, 208
- storm 26, 113, 208
see also seasons
withdrawal see retreat
whip 131, 137, 138–139 n. 563
William the Conqueror 191 n. 812
World War, First 104
World War, Second 35, 130 n. 225, 136 n. 551
wrestling 49, 190
writing 106, 112, 115
wushu 50–51

Xenares 181–182
Xenophon (Athenian general) 181–182
Xenophon (historian) 18, 25–26, 37–38, 50, 70–71, 74–75, 78, 82, 91–92, 95, 100 n. 393, 110 n. 440, 151, 162, 168, 169–171, 172, 180, 182 n. 763, 183–184, 187, 189, 190, 192, 195, 199, 205, 207, 209–211, 213–214, 215 n. 933, 229, 231, 232
Xerxes 138 n. 563, 219
xiphos 85–86, 87 fig. 12, 94, 232
see also sword
xynklesis 163, 221

yoke 141–143

zeugites 141–143
zeugos 141–143
Zeus 52 n. 169, 224
Zulu 59
zygon 141
zygos 141

INDEX OF SOURCES

I. LITERARY SOURCES

Aelianus

De natura animalium
9.17:   70 n. 243

Varia Historia
10.13:   151 n. 612

Aelianus

Tactica
11.1–5:   157–158 n. 642
14.6:   185 n. 779

Aeneas Tacticus
37.6–7:   28 n. 65, 30 n. 67

Aeschines
1.28–32:   152 n. 623
2.132–133:   136 n. 551
2.168:   270
3.175–176:   152 n. 624
3.176:   153 n. 630

Aeschylus

Agamemnon
60–67:   89 n. 332

Persae
239–240:   86–88, 138 n. 560

Septem
315:   27 n. 61

Fr. 112 Radt: 117

Alcaeus
Fr. 48 Voigt: 134 n. 545
Fr. 140.1–8 Voigt: 62 n. 211
Fr. 140.4–8 Voigt: 59 n. 195
Fr. 140.10 Voigt: 70 n. 243
Fr. 350 Voigt: 134 n. 545

Fr. 401 Ba Voigt: 149
Fr. 401 Bb Voigt: 149

Anacreon
Fr. 85 Gentili: 149

Andocides
1.74:   152 n. 623
1.75:   153 n. 629
3.18–20:   242

Anthologia Latina
104 Shackleton-Bailey: 51 n. 165

Anthologia Palatina
6.122:   82 n. 302
6.123:   82 n. 302
7.253:   119 n. 480
7.255:   119 n. 480
16.26:   212

Antiphanes
16 Kassel-Austin: 100 n. 396

Antiphon
Fr. 61 Thalheim: 141 n. 581

Apollodorus

Bibliotheca
2.2.1:   46 n. 144

Appianus

Syriaca
4.17–20:   136 n. 551

Archilochus
Fr. 2 West: 88
Fr. 3.3–5 West: 92–93

Fr. 5 West: 148
Fr. 114 West: 150, 180
Fr. 139 West: 122
Fr. 190 Bergk: 52 n. 171

Aristophanes

Acharnenses
57:   62 n. 211
88–90:   154 n. 635
277–279:   62 n. 211
573:   121 n. 494
842–844:   154 n. 635
1071–1234:   180 n. 758
1180–1183:   60 n. 199

Aves
289–290:   154 n. 635
291–292:   46 n. 138
364:   198 n. 855
491:   29–30 n. 74
1473–1481:   154 n. 635

Equites
595–610:   280, 281
843–859:   32 n. 93, 140 n. 572
953–958:   154 n. 635
1369–1372:   32 n. 93, 140 n. 572, 154 n. 635

Lysistrata
562:   57

Nubes
351–354:   154 n. 635
400:   154 n. 635
670–680:   154 n. 635
988–989:   36

Pax
444–449:   154 n. 635
673–678:   154 n. 635

1172–1178: 180 n. 758
1295–1298: 154 n. 635

Ranae
726–733: 36 n. 107
1037–1038: 60 n. 199, 62

Thesmaphoriazusae
604–607: 154 n. 635

Vespae
12–20: 154 n. 635
821–825: 154 n. 635
1081–1085: 185 n. 779

Fr. 248 Kassel-Austin: 141 n. 581
Fr. 650 Kassel-Austin: 28–29 n. 70

Aristoteles

[*Athenaion politeia*]
7.3–4: 141–142
22.2: 175 n. 729
38.1: 267
42.1: 175 n. 727
53.4: 100 n. 390

Historia animalium
548a 31 – b 4: 60

Metaphysica
986a 15–21: 174 n. 722

Politica
1297b 20–21: 198
1303b 12–14: 195
1306b 36 – 1307a 2: 116 n. 460

Rhetorica
1411b 6–10: 101 n. 398

Fr. 519 Rose: 52 n. 171
Fr. 532 Rose: 25 n. 55

Arrianus

Anabasis
5.17.7: 164 n. 666

Tactica
1.1: 158 n. 645
3.2: 158
4.1: 78 n. 287
11.1–6: 157–158 n. 642
12.1–2: 186 n. 785

12.3: 185 n. 779
12.4: 195 n. 834
12.10–11: 185 n. 779
12.11: 172 n. 710, 184 n. 771
16.13–14: 185 n. 779, 185, 193
26.3: 171 n. 706
27.3–4: 63 n. 217
37.5: 172 n. 712

Artemidorus
1.63: 47 n. 146

Asclepiodotus
3.2–5: 186 n. 785
4.1–3: 157
5.1: 165–166 n. 674
10.1: 171 n. 706
11.1: 171 n. 706

Athenaeus
5.215c–e: 244
8.347e: 117 n. 471
13.561e–f: 176–177 n. 739
14.628–629: 49 n. 159, 51 n. 166
14.631a: 49 n. 160
15.701d–e: 196 n. 846

Aulus Gellius
1.11.1–10: 63 n. 215
9.13.7–19: 42 n. 129

Callimachus
In lavacrum Palladis (Hymn. 5)
35–41: 46 n. 144

Callinus
Fr. 1.5 West: 84, 122
Fr. 1.8–9 West: 119
Fr. 1.12–13 West: 119
Fr. 1.14–15 West: 122
Fr. 14 West: 84

Callisthenes
FGrH 124.18: 291

Cicero
De divinatione
1.54: 244

Claudius Quadrigarius
Fr. 10b Peter: 42 n. 129

Critias
Fr. 88 B 37 Diels-Kranz: 32 n. 93, 140
Fr. 88 B 44 Diels-Kranz: 151

Curtius Rufus
3.2.13: 165 n. 671
4.13.27: 27 n. 62
8.5.4: 27 n. 62

Demosthenes
1.22: 43 n. 133
4.47: 182–183
13.22: 256*bis*
15.32: 153 n. 630
18.237: 211 n. 914, 239*bis*, 240
18.264: 240*bis*
19.84: 136 n. 551
20.52–53: 271
39.17: 175 n. 730
50.23: 74 n. 262
54.4: 100 n. 396
[61.23–29]: 49 n. 156
[61.25]: 49 n. 156

Dinarchus
1.72–73: 169 n. 695, 258
1.73: 256
1.75: 256*bis*

Dio Chrysostomus
Orationes
33.17: 150 n. 608

Diodorus Siculus
2.18.3: 218 n. 942
3.54.7: 218 n. 942
4.16.2: 218 n. 942
4.66.4: 218 n. 942
5.34.5: 38, 196 n. 845
8.27.1–2: 116 n. 460
9.16: 218 n. 942
11.4.3–5: 553 n. 137
11.5.4: 100–101
11.7.1: 218 n. 942
11.7.2: 218 n. 942
11.9.4: 218 n. 942
11.21.4–22: 219

I. Literary sources 323

11.30–32: 276
11.30.1: 211 n. 914, 219 n. 948, 275*bis*
11.30.2: 218 n. 942, 276
11.31.1: 275
11.32.1: 276
11.32.2: 218 n. 942
11.32.5: 276*bis*
11.34.1: 268
11.34.3: 268
11.35–36: 269
11.35.4: 268
11.36: 268
11.36.1: 268
11.36.5: 268
11.36.6: 269
11.60.1–2: 218 n. 944
11.74.3: 218 n. 942
11.79.1–4: 265
11.79.2: 265
11.79.3: 218 n. 942, 265*bis*
11.79.3–4: 265
11.79.4: 264
11.80.1: 211 n. 914, 290
11.80.1–2: 291
11.80.2: 207 n. 896, 290*bis*
11.80.2–6: 290
11.80.6: 207 n. 896, 290
11.83.1: 217 n. 938, 271, 272*ter*
11.83.4: 218 n. 942
12.3.4: 182 n. 763
12.6.2: 182 n. 763, 218 n. 942
12.28.4: 211 n. 914, 259
12.33: 285
12.33.2: 284
12.33.3: 284
12.33.4: 284*bis*, 285
12.34: 278
12.34.3–4: 277
12.34.4: 277, 278
12.37: 278
12.46.2: 218 n. 942
12.60: 273
12.60.2–6: 247
12.60.4: 273
12.61–62: 278
12.61.1: 278
12.61.5: 278
12.62.4: 28 n. 67, 283
12.62.4–5: 278
12.62.6: 278

12.63–65: 283
12.63.3: 282, 283
12.66.2: 100–101
12.69–70: 244
12.69.3: 242
12.70: 169 n. 691
12.70.1: 243
12.70.1–2: 243
12.70.1–3: 243
12.70.3: 180 n. 757, 190 n. 807, 243
12.70.4: 207 n. 896, 214–215 n. 930, 243*bis*
12.70.5: 243
12.74: 239
12.74.1: 237
12.74.1–2: 217 n. 941
12.74.2: 238*bis*
12.78–79: 260
12.79.1: 182 n. 763, 211 n. 914, 259*ter*, 260
12.79.2: 259
12.79.4: 214 n. 927, 260*bis*
12.79.6–7: 214 n. 927, 260
12.79.7: 260
12.80.8: 218 n. 942*bis*
12.82.6: 218 n. 942
13.6.5: 286*bis*
13.7.3: 287*bis*
13.17.1: 218 n. 942
13.46.2: 218 n. 942
13.49.2: 253*ter*
13.49.2–51.8: 254*bis*
13.49.4: 253*ter*
13.49.6: 253
13.50.1: 253
13.50.4: 253
13.50.4–51.6: 254
13.50.5: 253
13.50.6: 254
13.50.7: 253, 254
13.51.1: 253, 254
13.51.2: 253*bis*, 254
13.51.4: 217 n. 941, 218 n. 942*bis*, 254*ter*
13.51.7: 254*ter*
13.51.8: 254
13.56.6: 218 n. 942
13.59.8: 218 n. 942
13.60.7: 218 n. 942
13.62.3: 218 n. 942
13.64.1: 218 n. 942, 244*quinquies*, 245*bis*

13.65.1: 289*novies*
13.65.1–2: 290
13.65.2: 289*ter*, 290*bis*
13.66.2: 218 n. 942
13.67.5: 218 n. 942
13.72.5–6: 168
13.72.7: 218 n. 942
13.78.2: 218 n. 942
13.79.4: 218 n. 942
13.80.6: 218 n. 942
13.87.1: 218 n. 942
13.99.5: 218 n. 942
13.110.3: 218 n. 942
14.12.7: 218 n. 942
14.19.7: 250
14.22–24: 252
14.22.2: 250, 251
14.22.4: 250
14.22.5: 251
14.22.5–6: 251
14.22.6: 251*bis*
14.22.7: 251
14.23.1: 196 n. 845, 251*bis*
14.23.3: 251
14.23.4: 251*ter*
14.24.1: 251
14.24.4: 251
14.24.5: 251
14.24.6: 251*bis*
14.33.2: 218 n. 942*bis*, 266, 267*ter*
14.33.2–3: 267*bis*
14.43.2–3: 97 n. 378
14.80.3: 218 n. 942
14.81.1: 245
14.81.1–3: 246
14.81.2: 139 n. 564, 245*ter*, 246*ter*
14.81.3: 246
14.82.7: 269*bis*
14.82.7–8: 269
14.82.7–10: 270
14.82.8: 269*quinquies*
14.82.9: 214–215 n. 930, 218 n. 942*bis*, 269*quinquies*, 270
14.82.10: 211 n. 914, 269, 270
14.83.1: 211 n. 914, 270*bis*
14.83.1–2: 271
14.83.2: 207 n. 896, 270, 271*ter*
14.84: 249

14.84.1: 217 n. 941
14.84.2: 249*ter*
14.86: 242
14.86.2–3: 241
14.86.3: 242
14.86.4: 242
14.86.6: 241
14.90.4: 218 n. 942
14.91.2: 256*bis*
14.91.3: 256
14.105.2: 100–101
15.13.3: 218 n. 942
15.32.4: 191 n. 814
15.32.5: 101
15.33.6: 182 n. 763
15.34.2: 218 n. 942*bis*
15.37.1: 218 n. 942
15.39.1: 180 n. 757, 190 n. 807
15.41.5: 218 n. 942
15.44.2–3: 25–27
15.44.3: 121 n. 500
15.51.2: 257
15.52.2: 211 n. 914, 257*bis*
15.53–56: 169 n. 695, 258
15.53.2: 256
15.53.3: 257
15.54.5: 257
15.55.1: 179 n. 753
15.55.1–56.4: 258
15.55.2: 171 n. 706, 257, 258
15.55.5: 257, 258
15.55.5–15.56.2: 181 n. 761
15.56.4: 258*bis*
15.62.2: 218 n. 942
15.72.3: 211 n. 915
15.78.3: 218 n. 942
15.80: 253
15.80.1: 252*ter*
15.80.2: 211 n. 914, 252*bis*
15.80.4: 252*ter*
15.80.4–5: 217 n. 941, 252
15.80.5: 182 n. 763, 218 n. 942, 252, 253
15.80.6: 214 n. 927, 253
15.82.3: 261
15.82.4: 261
15.84–87: 262
15.84.1: 262
15.84.2: 261*ter*
15.84.2–4: 211 n. 914
15.84.4: 261*ter*, 262
15.85.1: 261

15.85.1–2: 261
15.85.2: 174–175 n. 724, 261*bis*, 262
15.85.3: 261
15.85.3–87: 262
15.85.4: 262*quater*
15.85.5: 262
15.85.6: 196 n. 841, 261, 262*ter*
15.85.7: 261*bis*, 262
15.85.8: 262*bis*
15.86.1: 262
15.86.2: 89 n. 332, 261
15.86.2–5: 262
15.86.3–4: 262
15.86.4: 218 n. 942*bis*
15.86.5: 261, 262
15.87.1: 92 n. 348, 182 n. 763, 190 n. 807, 261*bis*, 262
15.87.2: 262*ter*
15.87.3: 262
15.87.3–4: 262
15.87.5: 261*bis*
15.87.6: 153, 261*bis*
16.3.2: 164 n. 666, 165 n. 671
16.4.5–6: 218 n. 942*bis*
16.31.3: 218 n. 942
16.38.1–2: 136 n. 551
16.38.6–7: 182 n. 763
16.39.5: 218 n. 942
16.46.9: 217 n. 938
16.48.5: 182 n. 763, 218 n. 942
16.77.4–80: 74 n. 262, 92 n. 353
16.79.6: 218 n. 942
16.84.2: 239
16.85–86: 240
16.85.1–2: 239
16.85.2: 239*bis*
16.85.5: 239*bis*
16.85.7: 239
16.86: 240
16.86.1: 239
16.86.1–2: 239
16.86.2: 217 n. 941, 218 n. 942*bis*, 240*bis*
16.86.3: 239
16.86.4: 239
16.86.5: 240
16.86.6: 240*bis*

16.86.6–87.2: 240
16.88.1–2: 183
17.11.5: 218 n. 942
17.18.1: 25 n. 54
17.21.2: 25 n. 54
17.63.2: 218 n. 942
17.100.6–7: 89 n. 332
17.103.3: 218 n. 942
18.13.3–5: 182 n. 763
18.14.3: 218 n. 942
18.15.3: 218 n. 942
18.26.4: 100–101
18.34.4: 218 n. 942
18.61.1: 100–101
18.70.6: 218 n. 942
19.6.6: 172 n. 712
19.76.2: 218 n. 942*bis*
19.83.4: 218 n. 942
19.84.1: 218 n. 942
19.89.2: 218 n. 942*bis*
19.108.3: 218 n. 942
20.1.3: 217 n. 940
20.2.1–2: 217 n. 940
20.38.5: 218 n. 942
20.42.5: 100–101
20.87.3: 218 n. 942
20.88.8: 100–101, 218 n. 942*bis*
20.89.2: 218 n. 942
22.10.3: 218 n. 942
31.19.5: 218 n. 942
37.2.13: 218 n. 942
87.1: 180 n. 757

Diogenes Laertius
1.74: 149 n. 606
2.45: 261
2.54: 261, 262
8.1.4–5: 46 n. 144

Dionysius Halicarnassensis
Antiquitates Romanae
7.72.7: 52 n. 171

Ephorus
FGrH 70.191: 218 n. 944
FGrH 70.210: 291

Eupolis
Fr. 100 Austin: 154

Euripides

Andromache
1127–1136: 52

Bacchae
303–305: 211 n. 916
1280–1300: 178 n. 749

Cyclops
5–8: 28–29 n. 70
39: 164
241–243: 86 n. 326

Heraclidae
375–376: 28–29 n. 70
685: 92 n. 348
720–726: 101
738: 92 n. 348
836–837: 120 n. 490

Hercules furens
190–194: 89

Phoenissae
1382–1406: 92 n. 352
1396–1399: 89 n. 332
1404–1415: 41–43

Supplices
476–493: 178 n. 749
650–730: 19 n. 40
694–696: 28–29 n. 70
846–848: 91 n. 346

Troades
1192–1193: 28–29 n. 70
1196–1199: 42 n. 130, 190

Fr. 282.16–21 Nauck: 49, 119 n. 485

Frontinus

Strategemata
2.6.5: 271
4.2.6: 169 n. 695, 258

Galenus

2.303 Kühn: 47 n. 150
2.775 Kühn: 47 n. 150
3.253 Kühn: 47 n. 150
6.155 Kühn: 51 n. 167
[14.724] Kühn: 47 n. 150
18^1.737 Kühn: 47 n. 150

Hellenica Oxyrhynchia

1–2: 244
1–3: 245
1.1: 244*bis*
1.2: 244
1.2 l. 15: 244
2.1: 244*bis*
2.2: 245
4.1: 196 n. 841, 289*bis*, 290*ter*
4.2: 290*bis*

Herodotus

Prooemion: 219–220 n. 952
1.60.4–5: 55
1.62.3: 100–101
1.86: 219 n. 946
1.135: 70 n. 244
2.152: 134 n. 545
2.161: 134 n. 545
2.173.1: 209 n. 905
2.182: 70
3.47: 70
3.104.2: 209 n. 905
4.43.1: 219 n. 945
4.181.3: 209 n. 905
4.195.1: 219 n. 945
4.196.1–2: 219 n. 945
4.200.2–3: 28 n. 65, 30 n. 76
5.49.3–4: 89
5.63: 182 n. 763
5.74.2: 100–101
5.77: 212
5.95: 149 n. 603
5.95.1–2: 149
5.104.1: 134 n. 541
5.111: 100 n. 393
6.80: 100 n. 395
6.81: 100 n. 395
6.94.2: 263
6.102: 263
6.102–103: 263*bis*
6.102–117: 264
6.103.1: 263
6.104.1: 263
6.106.3: 263
6.108.1: 263*bis*
6.109.2: 263, 264
6.111: 179 n. 753, 263
6.111.1: 174–175 n. 724, 175 n. 727
6.112: 264
6.112–115: 263
6.112.1: 264*bis*
6.112.1–3: 85, 197, 197 n. 853
6.112.2: 263*bis*
6.113: 135–136, 202–203
6.113.1: 202 n. 869, 263, 264
6.113.1–2: 263
6.113.2–115: 214 n. 927, 264
6.114: 182 n. 763
6.115: 263
6.116: 264
6.117.1: 263, 264*bis*
6.117.3: 263*bis*
6.120: 264
7.40.1: 100 n. 394
7.56.1: 137 n. 554
7.61.1: 89 n. 331
7.64.1: 89 n. 331
7.72.1: 89 n. 331
7.77: 89 n. 331
7.78: 89 n. 331
7.79: 89 n. 331
7.103.3–4: 137 n. 554
7.144.1: 219 n. 949
7.158.4: 219 n. 949
7.165: 219 n. 948
7.165–166: 219 n. 945
7.166: 219 n. 945, 219 n. 947
7.175: 136 n. 550
7.176.2: 136 n. 550
7.202: 138 n. 558
7.211.2: 137
7.211.3: 136 n. 549, 137
7.218.1–2: 101
7.223.2–4: 553 n. 137
7.223.3: 137
7.224: 182 n. 763
7.224.1: 89 n. 332, 92
7.225: 181 n. 761
7.225.1: 185 n. 779
7.225.2–3: 138 n. 560
7.226: 138 n. 560
7.229.1: 100 n. 393, 100 n. 396
7.231–232: 151 n. 617
8.37.3: 198 n. 855
8.113.3: 275
9.1: 275
9.3.2: 219 n. 947, 268, 275
9.10: 100 n. 396
9.10.1: 139, 140, 211 n. 914, 275

9.10.1–2: 275
9.15.1–3: 275
9.19–70: 276
9.19.3: 275
9.20: 275, 276
9.21.1: 276
9.21.2: 276
9.22.1: 276
9.22.2: 275*bis*
9.23.1: 276
9.23.2: 276
9.24: 276
9.25.1: 276
9.25.2: 275
9.25.3: 275
9.26–28.1: 174–175
9.26.1: 185 n. 779, 275
9.28–29: 100 n. 396
9.28.2: 139, 275
9.28.2–6: 174–175 n. 724
9.28.2–9.30: 275
9.28.2–32.2: 211 n. 914, 275, 276
9.28.3: 275
9.28.6: 263
9.29.1: 139, 140, 275*ter*, 276
9.29.2: 276
9.30: 275
9.30.1: 275, 276
9.31–32.2: 275
9.32.2: 202 n. 869, 219 n. 948, 275
9.36: 275
9.38: 275
9.39.1: 276
9.40: 276
9.49.2: 276*bis*
9.51.3: 276
9.52: 100–101, 275, 276
9.53.1: 275
9.54.2: 276
9.57.3: 276
9.60.1: 276
9.60.2: 276
9.60.3: 276
9.61–70: 276
9.61.1: 205
9.61.2: 276
9.61.3: 275
9.62.1: 275
9.62.2: 88–89, 89 n. 332, 156, 185 n. 779, 275*bis*, 276

9.66: 276
9.67: 205, 276
9.68: 276
9.69.1–2: 276
9.69.2: 276
9.70.5: 276*ter*
9.71.3–4: 139 n. 565
9.74.1–2: 147
9.75: 182 n. 763
9.80: 100 n. 395
9.90.1: 268
9.96–104: 269
9.96.2: 268*bis*
9.96.3: 268
9.97: 268*bis*
9.98.2: 268*bis*
9.99.1: 268
9.100.1: 268
9.101–104: 268
9.101.1: 268*bis*
9.101.2: 207 n. 896
9.102–103: 268
9.102.1: 268*bis*
9.102.1–3: 174–175 n. 724
9.102.2–3: 185
9.102.3: 185 n. 779, 268*bis*
9.102.4: 268
9.103.1: 268*bis*

Hesiodus

Theogonia
629–640: 224
646–648: 224
686: 198 n. 855
711–712: 224

Homerus

Ilias
1.37–42: 107 n. 424
1.271–272: 223 n. 958
2.217–219: 150 n. 609
2.362–363: 175 n. 727, 176–177 n. 739
2.529–530: 70 n. 243
2.828–833: 70 n. 243
3.316: 111 n. 446
3.336: 111 n. 446
3.369: 111 n. 446
4.254: 108 n. 431
4.274–282: 113 n. 450
4.280–282: 108 n. 431
4.293–310: 110

4.298–300: 172 n. 710
4.331–335: 108 n. 431
4.422–426: 108 n. 431
4.422–429: 113 n. 451
4.446–451: 112 n. 448
4.457–459: 111 n. 446
5.4: 111 n. 446
5.12–26: 111 n. 444
5.43–48: 111 n. 444
5.92–94: 108 n. 431
5.159–165: 111 n. 444
5.182: 111 n. 447
5.275–310: 111 n. 444
5.289: 108 n. 426
5.302–304: 223
5.445–446: 107 n. 424
5.452–453: 111 n. 447
5.576–583: 111 n. 444
5.608–609: 111 n. 444
5.703–705: 111 n. 444
5.743: 111 n. 446
6.9: 111 n. 446
6.16–19: 111 n. 444
6.83–85: 108 n. 431
6.102–109: 114
6.117–118: 111 n. 447
6.242–245: 107 n. 424
6.297–300: 107 n. 424
6.318–319: 223
6.467–470: 59
6.470–472: 111 n. 446
6.494: 111 n. 446
7.54–66: 108 n. 431
7.61–62: 113 n. 450
7.61–66: 113 n. 451
7.62: 111 n. 446
7.81–86: 107 n. 424
7.239: 108 n. 426
7.273–282: 223
8.1: 223
8.53–74: 223
8.60–63: 112 n. 448
8.60–65: 112 n. 448
8.97–123: 111 n. 444
8.192–193: 111 n. 447
8.485–502: 223
8.493–494: 223
10.151–154: 83 n. 310
10.257–259: 111 n. 446
10.261–265: 111 n. 446
10.265: 60
10.335: 111 n. 446
10.458: 111 n. 446

I. Literary sources

11.1 – 16.777–780: 223
11.1 – 18.239–245: 223
11.41–42: 111 n. 446
11.86–112: 111 n. 444
11.122–147: 111 n. 444
11.150–153: 110 n. 442
11.264–274: 111 n. 444
11.320–322: 111 n. 444
11.328–335: 111 n. 444
11.351–353:: 111 n. 446
11.357–360: 111 n. 444
11.396–400: 111 n. 444
11.592–595: 112 n. 448
11.735–758: 223
12.105–107: 112 n. 448
12.160: 111 n. 446
12.183–184: 111 n. 446
12.381–383: 223 n. 958
12.384: 111 n. 446
12.415–426: 108–109 n. 431
12.425–426: 111 n. 447
12.443–449: 223 n. 958
13.126–135: 108–109 n. 431
13.128–133: 112, 165
13.131–133: 111 n. 446
13.132–133: 59–60 n. 197
13.145–148: 108–109 n. 431
13.188: 111 n. 446
13.265: 111 n. 446
13.341: 111 n. 446
13.383–401: 111 n. 444
13.567–568: 89 n. 336
13.611: 111 n. 447
13.611–612: 94, 94 n. 358
13.614–615: 111 n. 446
13.652–658: 111 n. 444
13.714: 111 n. 446
13.795–801: 113 n. 451
14.371: 111 n. 447
14.372: 111 n. 446
14.402–406: 111–112 n. 447
15.352–355: 110 n. 442
15.408–409: 108–109 n. 431
15.445–458: 111 n. 444
15.480: 111 n. 446
15.535–538: 111 n. 446
15.615–616: 112 n. 448
15.645–646: 111–112 n. 447
15.711: 94, 94 n. 358
16.70: 111 n. 446
16.137–138: 59–60 n. 197, 111 n. 446
16.155–217: 112 n. 448
16.210–217: 112 n. 448, 165 n. 672
16.214–216: 111 n. 446
16.338: 111 n. 446
16.342–344: 111 n. 444
16.413: 111 n. 446
16.731–743: 111 n. 444
17.262–271: 112 n. 448
17.274–277: 112 n. 448
17.294–295: 111 n. 446
17.364–365: 112 n. 448
18.207–213: 223
18.610–612: 111 n. 446
19.164–170: 223
19.359: 111 n. 446
20.78: 108 n. 426
20.267–272: 111–112 n. 447
20.279–281: 111–112 n. 447
20.385–287: 223 n. 958
20.484–489: 111 n. 444
21.229–232: 224
21.581: 111–112 n. 447
22.71–76: 119 n. 482
22.267: 108 n. 426
22.294: 111–112 n. 447
22.314: 111 n. 446
22.364–366: 119 n. 483

Odyssea
1.365–366: 107 n. 424
3.386–389: 107 n. 424
4.20–25: 107 n. 424
6.7–10: 107 n. 424
7.78–83: 107 n. 424
18.74: 47

Horatius

Carmina
1.28.9–15: 46 n. 144
2.7.9–12: 148 n. 598
3.2.13–16: 118 n. 476

Hyginus

Fabulae
170.9–11: 46 n. 144
273.1–3: 46 n. 144

Hyperides
Fr. 28 Jensen: 239

Iamblichus

Vita Pythagorae
14.63: 46 n. 144

Isaeus
2.42: 175 n. 728, 175 n. 730
5.11: 100 n. 396
5.36: 49 n. 160
5.42: 182 n. 763

Isocrates
8.143: 152, 152 n. 623

Justinus

Epitoma
2.9: 263, 264
2.9.9: 211 n. 914
6.8.11–13: 153 n. 631
8.2.8: 136 n. 551
9.3: 240
9.3.9–10: 240
12.7.5: 27 n. 62
13.5.12: 182 n. 763
18.6.11: 219 n. 946

Lexica Segueriana
217.21–25: 152 n. 623

Libanius
25.63: 32 n. 93, 140 n. 572

Livius
28.7.3: 136 n. 551
30.34.3: 244
30.34.3–4: 186–187 n. 786
36.15.5–19.13: 136 n. 551

[Longinus]

De sublimitate
26: 52 n. 170

Lucianus

De saltatione
8: 49 n. 160
9: 52 n. 171
10: 49 n. 160, 49–50
14: 49 n. 159

Dialogi deorum
13: 52 n. 169, 52 n. 171

Toxaris
55: 89 n. 337

Lycurgus

In Leocratem
106–107: 116 n. 463

Lysias
10.1: 27
10.9: 27*bis*, 626
10.12: 27
10.21: 27
10.23: 27
10.25: 27
11.5: 27
11.7: 27
11.8: 27
12.53: 267
13.79: 175 n. 727, 175 n. 730
14.9: 153 n. 630
16.14: 175 n. 728
16.15: 176, 177 n. 740
16.16: 176 n. 735
[20.23]: 175 n. 728
21.2: 49 n. 160
21.4: 49 n. 160
31.15: 175 n. 728

Maximus Tyrius
10.2: 46 n. 144

Mimnermus
Fr. 14.5–8 West: 122
Fr. 14.8 West: 84

Nepos

Iphicrates
11.2.3: 256*bis*

Miltiades
5: 264
5.1: 211 n. 914, 263
5.2: 263

Pelopidas
4.2: 257, 258

Thrasybulus
2.6–7: 267

Onasander
21.8: 171 n. 706
23: 180 n. 758
24: 176 n. 738
26: 63 n. 217
29.1: 191 n. 814
33.1: 180 n. 758

Pausanias
1.21.7: 70 n. 243
1.29.4: 176 n. 732
1.29.6: 290
1.29.6–9: 291
1.29.13: 244
1.32.3: 175 n. 727, 176 n. 732
2.11.8: 46 n. 138
2.17.3: 46 n. 144
2.25.7: 46 n. 144
3.9.13: 271
3.14.3: 46, 47 n. 146
4.8.2: 185 n. 779
4.18.2–3: 116 n. 460
4.25.5: 211 n. 914, 263*bis*, 264
4.26.2: 282, 283*bis*
5.8.10: 46 n. 142
6.10.4: 47
7.6.5: 240*bis*
8.10.5: 260*bis*
8.50.2: 90 n. 338
9.13.9–12: 258
9.13.11–12: 258
9.13.12: 258
9.9.10: 240*bis*
9.40.10: 240*bis*
10.20.2: 211 n. 914, 263, 264
10.20.3–22.1: 136 n. 551
10.22.8–13: 136 n. 551
10.26.5–6: 68
10.34.5: 46 n. 138

Philostratus

De gymnastica
7: 46 n. 140, 47 n. 146
8: 46 n. 141, 48 n. 152, 48 n. 153
19: 49 n. 159
33: 47

Pindarus

Olympia
7.81–87: 46 n. 143

Pythia
1.75: 275
1.75–80: 276
4.85: 209 n. 905

Nemea
3.60: 198 n. 855
10.19–24: 46 n. 143

Isthmia
7.10: 198 n. 855

Plato

Apologia Socratis
28d: 176 n. 736
28e: 243, 244

Charmides
153a–c: 278

Critias
112d–e: 219 n. 949

Gorgias
469d: 209 n. 905

Laches
181b: 243, 244
183d: 94

Leges
625d: 50 n. 161
625e–626a: 12
629a–b: 116 n. 460
815a: 51
815a–b: 51
816d: 51 n. 167
944a–c: 154
944c: 153–154
944c–945b: 154

Menexenus
242a: 290
242b: 272

Symposium
179a–b: 176–177 n. 739
219e–220b: 74 n. 263
219e–220e: 278
220d: 278
220e: 277
221a: 243*bis*
221a–c: 214 n. 927, 243*bis*, 244
221c: 214–215

I. Literary sources 329

Plinius

Naturalis historia
8.192:   70
16.209: 28–29 n. 70
19.2.11: 70 n. 243

Plutarchus

Aemilius Paullus
22.1:   203 n. 873, 204

Agesilaus
18:      213 n. 923, 249, 271
18.1:   248*bis*
18.2:   248
18.2–4: 185 n. 779
18.3:   248*bis*, 249
18.4:   248*bis*
19.1:   182 n. 763, 248
19.2:   249
22:      256
22.1–4: 256
22.2:   256
22.3:   256*bis*
28.3–6: 258
28.5:   256
28.6:   258
33.3:   211 n. 915
36.2:   182 n. 763

Alcibiades
7.3:   278*bis*
7.6:   244
28:    254*bis*
28.3: 253
28.6: 254

Alexander
9.2:   239*bis*, 240*ter*
9.2–3: 240
12.3:   239, 240
16.6:   89 n. 332
63.5:   92 n. 348

Aristides
5.3:   175 n. 726, 176 n. 734
5.5:   175 n. 727
11–19: 276
11.2:   275
11.7:   275
14.1:   276
14.2:   275, 276
14.5:   90 n. 338, 275
16.4:   276

16.6:   276
17.6:   275, 276
18.2:   91, 275
18.3:   275
19.4:   276*bis*
19.7:   219 n. 947, 268, 275
23.2:   38

Artaxerxes
7–13:   252
7.4:   251
8.2:   250
8.2–7: 172 n. 713
8.4:   251
8.5:   251
9.3:   251
10.3:   251
11.1:   251
11.2:   251
13.3:   250

Camillus
19.8:   239*bis*, 240

Cato Maior
13–14: 136 n. 551

Cimon
17.3–5: 175 n. 731, 291
17.4:   290
17.2–6: 291
17.5:   290
17.7:   290

Cleomenes
2.3:   118 n. 477
17.4–5: 28 n. 65
21.3:   28 n. 65

Demosthenes
17.3:   211 n. 914, 239*bis*
18–20: 240
20.1:   239
20.2:   239*bis*, 240

Eumenes
7.3:   89 n. 332

Flamininus
8.4:   164 n. 666

Lycurgus
19.2:   193–194
21–22: 196 n. 847
22.2–3: 196 n. 843

Lysander
28:     246
28.2:   245
28.5:   182 n. 763
28.5–6: 246
28.6:   139 n. 564, 245*ter*, 246*ter*
28.10: 245
28.11: 214 n. 927
29.1–2: 246

Moralia
190b: 92
191e: 93 n. 355, 194 n. 828
193e 18: 49
194c: 182 n. 763, 261, 262
210f: 164 n. 669
217e: 93 n. 355, 193
219c: 92
220a: 152 n. 620, 164
220a 2: 151 n. 616
225a–d: 553 n. 137
232e: 193
233e: 190 n. 807
234f – 235a: 152 n. 620
238b: 196 n. 847
239b: 150
241f: 62 n. 211, 93 n. 355, 152 n. 620, 194
241f 16: 20 n. 44, 28 n. 65, 164 n. 669
259d: 239*bis*, 240*bis*
282e: 174
581:    244
628d – 629b: 179 n. 753
639d–e: 47 n. 146
639f: 180 n. 757
639f – 640a: 190 n. 807
741b 1: 49 n. 160
788a: 190 n. 807
862b: 263, 264

Nicias
16:     286
17:     287*quater*
18.3:   182 n. 763
21:     289
21.1:   288*ter*
21.5:   288
21.7:   288
21.8:   288*bis*
21.9:   288*bis*, 289

Pelopidas
1.5: 164 n. 669
2.3: 92 n. 348
7.3: 190 n. 807
16.1: 291
17: 292
17.1: 291*quater*
17.1–10: 169 n. 693
17.2: 291*ter*
17.3: 182 n. 763, 291
17.3–5: 291
17.4: 292
17.5–6: 292
18: 176–177
18.5: 177 n. 743, 239, 240*bis*
19.3: 184 n. 771
20.1: 211 n. 914, 257*ter*
22: 257
23: 258*bis*
23.1: 257
23.1–2: 257
23.1–3: 169 n. 695, 179 n. 753
23.4: 142, 257
31–32: 253
31.1: 252*bis*
31.2: 252
31.2–3: 252
31.3: 252
32.1: 211 n. 914, 252
32.1–2: 252
32.2: 252*ter*
32.2–3: 252
32.2–7: 252
32.3: 101 n. 400, 252*ter*
32.4: 185 n. 779, 252*bis*
32.5: 252
32.5–6: 252, 253
32.6: 252
32.7: 182 n. 763, 214 n. 927, 252*quater*, 253*ter*

Pericles
10.1–2: 291*bis*

Philopoemen
9.2: 164 n. 666
10.7–8: 90 n. 338

Phocion
20.1: 49 n. 156

Pyrrhus
32.1–4: 28 n. 65

Timoleon
28.1–4: 74 n. 262, 92

Pollux

Onomasticon
1.135: 78 n. 285
2.188–189: 47 n. 150
7.70: 70 n. 247
8.40: 152 n. 623
8.132: 142

Polyaenus

1.10: 196 n. 843, 257, 258*bis*
1.16.3: 215 n. 932
1.27.2: 219 n. 951
1.35: 272
1.35.1: 272*bis*
1.35.2: 272*bis*
1.38.3: 237, 239
1.39.1: 280*quater*, 281
1.39.2: 285, 286*quinquies*
1.40.9: 253*ter*, 254*ter*
1.45.2: 30 n. 76
2.1.2: 101
2.1.3: 247, 248*ter*, 249
2.1.5: 249
2.1.13: 258
2.1.19: 271
2.2.3: 251, 252
2.3.2: 189–190, 256, 257, 258
2.3.2–3: 169 n. 695, 258
2.3.3: 257*quinquies*
2.3.8: 169 n. 695, 257, 258
2.3.10: 100 n. 393
2.3.14: 261*bis*, 262*bis*
3.1.2: 273*sexies*
3.8.1: 25 n. 54
3.9.8: 91 n. 344
3.9.26: 191 n. 814
3.9.27: 190 n. 801
4.2.2: 164 n. 666, 239*ter*, 240*bis*
4.2.7: 240
4.2.7–8: 240
4.2.8: 239
4.3.8: 190 n. 801
7.8.1: 30 n. 76
7.41: 25 n. 54
7.45.1: 268*bis*, 269
8.40: 239*bis*, 240

Polybius

1.45.9: 141 n. 583
2.30.3: 42 n. 129
3.81.2: 141 n. 583
4.64.6–7: 164 n. 666
6.25.6–9: 90
6.25.9: 90
9.8–9: 262
9.9.7: 262
11.18.4: 90 n. 338
12.19.7–8: 157–158 n. 642
12.21.3: 164 n. 666
12.25f3–4: 258
16.33.2–3: 90 n. 338
18.29–30: 164 n. 666
18.29–30.2: 186
18.29.2: 157–158 n. 642
18.29.5: 141 n. 583
18.29.5–7: 165 n. 671
18.30.4: 185–186
18.31.5: 195 n. 837

Fr. 60 Büttner-Wobst: 291

Porphyrius

Vita Pythagorae
26–27: 46 n. 144

Quintilianus

Institutio oratoria
1.10.14: 196 n. 843

Scholia

in Eur. *Phoen.* 1102: 196 n. 846
in Eur. *Phoen.* 1404–1414: 42 n. 128
in Eur. *Phoen.* 1408: 43 n. 131, 43 n. 133
in Pind. *Ol.* 7.81–87: 46 n. 143
in Pl. *Leg.* 629a–b: 116 n. 460
in Thuc. 2.39.1: 28 n. 65, 164 n. 669
in Thuc. 4.34.3: 57 n. 187

Simonides

Fr. 8 Campbell: 119 n. 480
Fr. 9 Campbell: 119 n. 480
Fr. 87 Diehl: 212

I. Literary sources

Pseudo-Skylax
103: 134

Stobaeus

Florilegium
3.7.30: 28 n. 65, 164 n. 669

Strabo
3.3.6: 32 n. 92
8.6.25: 270
9.1.11: 289, 290
9.2.7: 244
9.2.37: 240*bis*
10.4.16: 49 n. 160, 52 n. 171
13.1.38: 149 n. 602

Suda
s.v. ἀναυμαχίου: 152 n. 623
s.v. τορνευτολυρασπιδοπ-
  ηγοί: 29–30 n. 74
s.v. Τυρταῖος: 116 n. 460

Tacitus

Historiae
2.42: 186–187 n. 786, 244

Theocritus

Idyllia
22.198–204: 119 n. 481

Theophrastus

Characteres
25.3–6: 175 n. 728
25.4: 100 n. 396

Historia plantarum
3.12.2: 82 n. 300

Thucydides
1.29.1: 284
1.45–55: 285
1.45.1: 284
1.45.2: 284
1.46.1: 284
1.46.2: 284
1.46.4–5: 284
1.47.1: 284*quater*
1.47.1–48.1: 284
1.48–51.3: 285
1.48.3: 284
1.48.4: 284

1.49.1: 284*quater*
1.49.1–2: 285
1.49.2–3: 284*bis*
1.49.3: 284
1.50.3: 284
1.50.5: 285*bis*
1.51.5: 285
1.52.2: 285
1.53.2: 284
1.54: 285
1.54.1: 285*bis*
1.54.2: 285*quater*
1.56–65: 278
1.57.6: 277
1.60: 277
1.60.1: 277*ter*
1.60.1–2: 277
1.61.1: 277
1.61.4: 277*quinquies*
1.62.1: 277*bis*
1.62.2: 277
1.62.2–4: 277*bis*
1.62.3–4: 277
1.62.4: 277
1.62.4–63.2: 277
1.62.6: 179 n. 753, 214
  n. 927, 277, 278
1.62.6–63.1: 213 n. 920
1.63.1: 196 n. 844, 277, 278
1.63.2: 205, 277*bis*, 278
1.63.3: 182 n. 763,
  278*quater*
1.105: 265
1.105.2–106: 265
1.105.3: 265
1.105.4: 264*bis*, 265
1.105.4–106: 265
1.105.6: 265
1.106.2: 265*ter*
1.107.2–108.2: 291
1.107.5: 211 n. 914, 290
1.107.7: 290*bis*
1.108.1: 290*ter*
1.108.2: 271*bis*, 272
1.108.2–3: 272
1.108.3: 271
1.110.1: 178 n. 747
2.1: 23 n. 50
2.2.4: 100–101
2.13.6–7: 142
2.69.2: 182 n. 763
2.70.1: 281
2.75–78: 133–134 n. 540

2.79: 282
2.79.1: 281*sexies*
2.79.2: 281*bis*
2.79.2–3: 281
2.79.3: 281*bis*
2.79.3–6: 281
2.79.4: 281*ter*
2.79.5: 100 n. 394, 281*ter*
2.79.6: 214 n. 927, 281*bis*,
  282
2.79.7: 182 n. 763, 282*bis*
3.7.4: 182 n. 763
3.16.1: 141 n. 581
3.17.4: 100 n. 396
3.19.2: 182 n. 763
3.20–23: 74 n. 263
3.20.1–24.2: 208 n. 899
3.22.2: 174 n. 721
3.23.4: 172 n. 712
3.52.1–2: 133–134 n. 540
3.90.2: 182 n. 763
3.94.1: 236
3.94.1–5: 236
3.94.3: 236
3.95.2: 236*ter*
3.96.3–98.4: 237
3.97–98: 237
3.97.2: 236*ter*
3.97.3: 202 n. 869, 236, 237
3.98.1: 236*bis*
3.98.2: 236*bis*
3.98.3: 237
3.98.4: 182 n. 763, 214
  n. 927, 236*bis*, 237
3.98.5: 237
3.100.2: 273*ter*
3.105: 246
3.105–109: 273
3.105–113: 247
3.105.1: 178 n. 746, 246*bis*,
  272*quater*, 273
3.105.2: 246, 273
3.105.3: 246, 273
3.106.1: 272*bis*
3.106.3: 272
3.107.1: 246*bis*, 247, 272,
  273*quater*
3.107.1–2: 272
3.107.3: 273
3.107.3: 272*bis*, 273*ter*
3.107.4: 179 n. 753, 273*bis*
3.108.1: 273
3.108.1–3: 273

3.108.2–3: 273
3.108.3: 196 n. 841, 196 n. 844, 205–206, 207 n. 896, 273
3.109.1: 182 n. 763, 246, 273bis
3.109.2: 247, 273quater
3.110–112: 247
3.111.3–4: 178 n. 746
3.111.3–3.113: 178–179
3.112.1: 246ter
3.112.3–8: 178 n. 748
3.112.6: 246, 247bis
3.112.6–8: 214 n. 927, 247
3.112.8: 247
3.113.1–5: 247
3.113.4: 178 n. 746
3.113.6: 178, 247bis
4.1–15: 279
4.2.1: 278, 282
4.2–23: 279
4.2.4: 279
4.3.1–2: 278
4.3.2–4.4.3: 278, 282
4.8.1–2: 278
4.8.6: 282
4.8.9: 282bis, 283
4.9.1: 278, 279ter
4.9.1–2: 278
4.9.2: 278, 279bis
4.10: 279
4.11.2: 279
4.11.3: 185 n. 779, 279
4.11.4: 279
4.12.1: 28 n. 67, 152 n. 621, 206, 279bis
4.12.2: 278
4.13.1: 206, 279
4.13.3: 279
4.14.2: 279
4.14.5: 279bis
4.16.1: 100 n. 396
4.23: 279
4.26–41: 283
4.26.5–6: 282
4.26.7: 283
4.27.1: 282
4.28.4: 282, 283bis
4.29.1–2: 282
4.29.3: 282
4.30–38: 220–221, 283
4.30.1: 236
4.30.3: 282

4.30.4: 283
4.31.1: 283
4.31.1–2: 282
4.31.2: 283bis
4.32.1: 283
4.32.2: 220 n. 953, 282bis
4.32.3–4: 283
4.32.4: 283quater
4.33–35: 220 n. 954
4.33.1: 283ter
4.33.2: 283ter
4.34.1: 283
4.34.2–35.4: 221
4.34.3: 57
4.35.1–4: 220 n. 955
4.35.2: 283
4.35.3: 283
4.35.4: 202 n. 869, 220 n. 953, 282, 283
4.36.1: 282, 283bis
4.37.1: 283
4.37.2: 283
4.38.1: 182 n. 763, 282, 283bis
4.38.3: 283
4.38.4: 283bis
4.38.5: 283quater
4.39.1–2: 282
4.40.1: 283
4.42–44: 281
4.42.1: 280qui n. quies
4.42.2: 280
4.42.3: 280bis
4.43.1: 280
4.43.2: 206
4.43.2–4: 281
4.43.2–44.5: 280
4.43.3: 185 n. 779, 189, 199–200, 280ter
4.43.4: 280
4.44.1: 100–101, 202 n. 869, 206, 280quater
4.44.2: 182 n. 763, 280, 281
4.44.3: 281
4.44.5–6: 280
4.44.6: 280, 281
4.44.8: 214–215 n. 930
4.54.2: 221 n. 957
4.68.3: 100–101
4.89.1: 242
4.90: 242
4.90–96: 244
4.90.1: 242

4.90.1–4: 242
4.90.4: 100–101, 242, 243bis
4.91: 100–101, 242ter, 243
4.91.1–2: 242
4.92: 242
4.93.1: 206, 214–215 n. 930, 242, 243
4.93.1–4.96.8: 207 n. 896
4.93.2: 243bis
4.93.3: 100–101, 243quinquies
4.93.3–5: 243
4.93.4: 169 n. 691, 243bis, 257
4.94: 243
4.94.1: 100 n. 396, 243bis
4.94.1–2: 211 n. 914
4.95: 242
4.95.2: 243
4.96.1: 191, 244
4.96.1–2: 243
4.96.1–9: 136 n. 548, 243
4.96.2: 185 n. 779, 195, 199, 242, 243ter
4.96.4: 185 n. 779, 243
4.96.5: 242, 243
4.96.5–7: 243
4.96.6: 243
4.96.8: 206, 214 n. 927, 214–215 n. 930, 243bis
4.97–101: 243
4.97.1: 243
4.99: 242
4.101: 244
4.101.1: 243
4.101.2: 100 n. 394, 182 n. 763, 243ter
4.103.1–2: 74 n. 263
4.104.1: 163 n. 663
4.126.5: 196 n. 841
4.134.1: 254, 255bis
4.134.1–2: 255bis
4.134.2: 206, 207 n. 896, 255quinquies
5.1: 237
5.2.1: 238ter
5.6–11: 239
5.6.1: 238
5.6.3: 238
5.6.4: 238quinquies
5.6.4–5: 238
5.6.5: 237, 238bis
5.7.4: 237

5.8.1: 237
5.8.3: 238
5.8.4: 170 n. 700, 238
5.9: 238
5.9.1–8: 170 n. 700
5.9.3: 196 n. 841
5.9.4: 238
5.9.7: 238
5.10: 136 n. 548, 238
5.10.2: 238
5.10.2–12: 172 n. 714
5.10.6: 238
5.10.8–9: 182 n. 763
5.10.9: 238*ter*
5.10.9–10: 214 n. 927
5.10.10: 238*ter*
5.10.12: 238
5.11.1: 238
5.11.2: 238*ter*
5.26: 18 n. 38
5.34.2: 151 n. 615
5.47.7: 173 n. 718
5.51.2: 182 n. 763
5.57.1: 259
5.59.5: 176 n. 732
5.61.1: 260
5.64–74: 260
5.64.3: 211 n. 914, 259
5.64.4: 259
5.64.5: 259*bis*
5.65.2: 259, 260
5.65.6: 259
5.66.1: 259*bis*
5.67: 259
5.67.1: 174 n. 723, 174–175 n. 724, 259*bis*
5.67.1–2: 259
5.67.2: 173 n. 718, 259
5.68.1: 211 n. 914, 259
5.68.1–3: 211 n. 914
5.68.2: 259
5.68.2–3: 259
5.68.3: 141 n. 583
5.69.1: 259
5.69.2–5.70: 196
5.70: 63, 196 n. 843
5.70–73: 136 n. 548, 260
5.71.1: 162–164, 172, 259
5.71.1–2: 174–175 n. 724
5.71.2–72.2: 163 n. 662
5.71.4: 260
5.72.1: 259
5.72.3: 185 n. 779, 259

5.72.3–73.3: 213 n. 921
5.72.4: 176 n. 732, 179 n. 753, 211, 260
5.73: 260
5.73.1: 259*bis*
5.73.4: 214 n. 927
5.74.2: 100–101, 259, 260*bis*
5.74.3: 182 n. 763, 260*bis*
6.43: 141 n. 581, 285, 286*octies*
6.65.2–3: 285
6.66.1–2: 285
6.67–71: 286
6.67.1: 179 n. 753, 285, 286
6.67.1–3: 285
6.67.2: 286*bis*
6.67.3–69: 286
6.68.3: 286
6.69–71: 286
6.69.1: 101 n. 400, 286
6.69.2: 286*ter*
6.70.1: 74 n. 262, 202 n. 869, 206–207, 285, 286*ter*
6.70.2: 185 n. 779, 286*bis*
6.70.3: 286*quater*
6.71.1: 286*ter*
6.96–97: 287
6.96.1: 287
6.96.3: 287*quater*
6.97.2: 287
6.97.2–5: 287
6.97.3–4: 287
6.97.4: 182 n. 763, 196 n. 841, 287
6.97.5: 287*bis*
6.98.4: 175 n. 727
6.101.5: 175 n. 727
6.101.6: 182 n. 763
6.102.2: 100 n. 395
7.3.1: 100–101
7.13.2: 100 n. 396
7.16.1: 141 n. 581
7.29.2–3: 133–134 n. 540
7.42–44: 289
7.42.1: 288*quinquies*
7.43–44: 288*bis*
7.43.2: 288*ter*
7.43.2–44.8: 208 n. 898
7.43.4: 288
7.43.6: 288
7.44.1: 208, 289
7.44.2: 288*bis*
7.44.6: 197 n. 849

7.44.8: 288*ter*
7.45.1: 289*bis*
7.45.2: 288, 289*bis*
7.53.2: 196 n. 841
7.75.5: 25 n. 54, 100
7.78.2: 100 n. 394
7.81.4–5: 265
7.82.3: 28 n. 65
7.83.5: 100–101
7.84.3–85.1: 214 n. 929
7.87.6: 178 n. 747
8.10.4: 196 n. 841
8.24.1: 182 n. 763
8.24.2: 141 n. 581
8.25: 266
8.25.1: 265*ter*, 266*bis*
8.25.1–2: 266*ter*
8.25.2: 265, 266*ter*
8.25.3: 196 n. 841, 196 n. 842, 266*bis*
8.25.3–4: 266
8.25.3–5: 266
8.25.4: 100–101, 185 n. 779, 266*ter*
8.25.5: 265, 266
8.25.6: 266
8.26: 244
8.29.4: 175 n. 730
8.92.2: 209 n. 905
8.92.4: 175 n. 727
8.93.1: 100–101
8.97.1: 97 n. 377

Tyrtaeus
Fr. 10.1–2 West: 118, 153–154 n. 632
Fr. 10.1–14 West: 117
Fr. 10.17–18 West: 117
Fr. 10.21–27 West: 89
Fr. 10.21–30 West: 119 n. 482
Fr. 10.27–30 West: 117, 118 n. 479
Fr. 10.31–32 West: 120 n. 490
Fr. 11.3–6 West: 117, 118, 153–154 n. 632
Fr. 11.11–13 West: 116 n. 465, 116–117
Fr. 11.15–20 West: 119 n. 482
Fr. 11.17 West: 119
Fr. 11.20–34 West: 200

Fr. 11.21–34 West: 120
Fr. 11.23–24 West: 100
Fr. 11.26 West: 60
Fr. 11.28 West: 84, 116 n. 465
Fr. 11.29–34 West: 119–120
Fr. 11.35–38 West: 121
Fr. 12 West: 116 n. 466
Fr. 12.1–14 West: 49 n. 157, 119
Fr. 12.12 West: 120
Fr. 12.15 West: 119
Fr. 12.15–17 West: 116 n. 465
Fr. 12.15–19 West: 120 n. 490
Fr. 12.19–22 West: 120
Fr. 12.20–21 West: 119 n. 480
Fr. 12.22 West: 121
Fr. 12.23–34 West: 118 n. 478
Fr. 12.25 West: 117
Fr. 12.25–26 West: 92 n. 348, 118
Fr. 19.2 West: 121
Fr. 19.7–15 West: 121
Fr. 19.17–20 West: 92 n. 348
Fr. 19.19–20 West: 121

Vegetius

De re militari
1.12: 93–94
3.9.2: 204

Xenophanes

Fr. 11 B 2.14–22 Diels-Kranz: 49 n. 157, 119 n. 485

Xenophon

Agesilaus
2.7.9: 248
2.9–16: 136 n. 548, 249*bis*
2.10: 198 n. 855
2.11: 248
2.12: 185 n. 779, 213 n. 923
2.13: 182 n. 763
2.14: 92, 190 n. 804
2.15: 207 n. 894, 207 n. 895
5.3: 180 n. 756

Anabasis
passim: 18 n. 38
1.2.1–9: 250
1.2.9: 168 n. 686
1.2.15: 168 n. 682, 168
1.2.15–16: 97 n. 378
1.2.16: 168 n. 687
1.2.16–17: 91 n. 344
1.2.17: 198 n. 855
1.5.12: 94
1.5.14: 100–101
1.5.17: 100–101
1.6.4: 100–101
1.7.10–13: 250*bis*
1.7.12: 250
1.8: 209–211, 252
1.8.3: 251
1.8.4–9: 251
1.8.5: 251*bis*
1.8.6: 78 n. 287, 251*bis*
1.8.8: 209
1.8.8–17: 210
1.8.8–29: 251*bis*
1.8.9: 251*quater*
1.8.12: 210
1.8.13: 172 n. 713
1.8.14: 210
1.8.15: 180 n. 758, 209 n. 904, 251
1.8.16: 251
1.8.17: 196 n. 845, 251
1.8.17–18: 196 n. 840, 251
1.8.18: 198 n. 855, 251*ter*
1.8.20: 251*bis*
1.8.21: 251
1.8.26: 251
1.10: 252
1.10.1–6: 250
1.10.1–17: 209–211
1.10.1–19: 251
1.10.3: 251
1.10.4: 210 n. 913
1.10.7: 251
1.10.10: 251
1.10.11–15: 251
1.10.12: 251
1.10.13: 251
1.10.15: 210
1.10.16: 100–101, 207 n. 896, 214 n. 927, 251*bis*
1.10.19: 209
2.1.9: 100 n. 395

2.2.8: 100–101
2.2.21: 100–101
2.6.1: 182 n. 763
3.1.2: 209 n. 904
3.1.4: 180 n. 758
3.1.37: 180 n. 756
3.1.47: 180 n. 758
3.2.12: 211 n. 914, 263
3.2.18–19: 195 n. 836
3.3.8: 214 n. 928
3.3.8–19: 214 n. 927
3.3.20: 97 n. 378*bis*
3.4.3–4: 214 n. 928
3.4.46–49: 180 n. 756, 180 n. 758
4 *passim*: 74 n. 262
4.1.18: 70–71, 92 n. 348
4.2.16: 100–101
4.2.20: 100 n. 393
4.3.17: 100–101
4.3.26: 100–101
4.4.11–12: 180 n. 756
4.4.16: 94 n. 360
4.8.9–19: 169 n. 690
4.8.11: 195
5.2.3–13: 169 n. 690
5.2.8: 100–101
5.2.19: 100–101
5.4.11: 100–101
5.4.24: 214 n. 928
6.1.5: 51 n. 163
6.1.8: 100–101
6.1.9–10: 50 n. 162
6.1.12: 51
6.5.3: 100–101
6.5.14: 180 n. 756
6.5.26: 198 n. 855
6.5.27: 91
7.1.22: 100–101

Cyropaedia
1.4.18
2.1.16: 162
2.1.16–17: 91–92
2.3.9–11: 91–92, 162 n. 660
3.3.60: 110 n. 442
5.3.40: 100 n. 394
5.4.45: 172 n. 712
6.3.4: 100 n. 394
6.3.21–23: 195
6.3.25: 184 n. 771
6.4.1: 78 n. 287
6.4.17: 195

I. Literary sources

7.1.1–2: 78 n. 287
7.1.33–34: 187, 192
7.1.34: 189
7.1.35: 121 n. 494

De equitandi ratione
12.1: 75 n. 266
12.5: 78 n. 285
12.7: 78
12.11: 86, 95
12.12: 82

De republica Lacedaemoniorum
9.6: 151
11.3: 30 n. 76*bis*
11.9: 172 n. 711
12.4: 140 n. 572

Hellenica
1.1.11–12: 253
1.1.11–13: 253
1.1.11–18: 254*bis*
1.1.13: 253
1.1.14: 253*ter*
1.1.16: 253
1.1.18: 182 n. 763, 253*bis*
1.1.34: 244*ter*
1.2.1: 244
1.2.2–3: 214 n. 928
1.2.4–7: 244
1.2.6: 244*ter*
1.2.6ff.: 244
1.2.6–11: 245
1.2.7: 244*quinquies*
1.2.7–9: 244
1.2.8: 244*bis*
1.2.9: 244, 245
1.2.9–10: 245
1.2.10: 245
1.2.11: 245
1.2.16: 214 n. 928, 214–215 n. 930
1.2.18: 182 n. 763
1.3.5–6: 182 n. 763
1.8.8–29: 214–215 n. 930
1.10.11–15: 214–215 n. 930
2.4.3: 266
2.4.5: 100–101
2.4.6: 214 n. 927, 214 n. 928, 215 n. 933
2.4.9–20: 267
2.4.10: 267*quinquies*
2.4.10–12: 169 n. 688
2.4.11: 266*bis*, 267*bis*
2.4.11–12: 266
2.4.11–19: 191 n. 814
2.4.12: 100–101, 267*quinquies*
2.4.13–17: 267
2.4.14: 267*bis*
2.4.15: 267*bis*
2.4.15–16: 267
2.4.16: 267
2.4.19: 182 n. 763, 267*sexies*
2.4.25: 25 n. 54
2.4.30: 179 n. 753
2.4.32: 138 n. 561
2.4.32–34: 190
2.4.33: 182 n. 763
2.4.34: 185 n. 779
3.1.9: 37
3.1.23: 100–101
3.4.2: 248
3.4.14: 82 n. 300, 82 n. 301, 89 n. 332
3.4.16: 190 n. 807
3.4.22: 100 n. 394
3.4.23: 138 n. 561, 191 n. 814, 197 n. 851
3.5.11: 164 n. 666
3.5.17: 245*ter*
3.5.17–19: 245
3.5.17–25: 246
3.5.18–19: 191 n. 814
3.5.19: 182 n. 763, 245*bis*, 246
3.5.19–20: 139 n. 564, 246
3.5.20: 245, 246*ter*
3.5.23: 245, 246
4.2.9: 179 n. 753, 270
4.2.9–4.3.1: 271
4.2.13: 169 n. 692, 190 n. 806, 270
4.2.14: 270*bis*, 271
4.2.15: 270*bis*
4.2.16: 270*ter*, 271
4.2.16–17: 211 n. 914, 270*bis*
4.2.16–23: 174–175 n. 724
4.2.17: 176 n. 733, 177 n. 740, 270*bis*, 271
4.2.18: 169 n. 692, 173 n. 717, 190 n. 806, 257, 270*quinquies*
4.2.18–21: 270
4.2.18–23: 271
4.2.19: 175 n. 727, 179 n. 753, 197 n. 848, 213, 270*bis*
4.2.20: 213–214, 270
4.2.20–22: 173, 213 n. 922
4.2.21: 176 n. 733, 177 n. 740
4.2.22: 271
4.2.23: 270, 271
4.3.1: 248, 271*bis*
4.3.4–9: 248
4.3.8: 182 n. 763
4.3.10: 248
4.3.13–14: 248
4.3.15: 248*quinquies*
4.3.15–21: 249
4.3.16: 179 n. 753, 248*quater*, 249
4.3.16–20: 136 n. 548, 249
4.3.17: 191 n. 814, 196 n. 842, 198 n. 854, 198 n. 855, 211 n. 916, 248*ter*, 249
4.3.17–19: 213 n. 923
4.3.18: 248
4.3.19: 120 n. 490, 185 n. 779, 199, 248*bis*
4.3.20: 182, 207, 207 n. 896, 248*ter*, 249*bis*
4.3.21: 249*bis*
4.4.7: 241*bis*
4.4.7ff.: 241
4.4.7–13: 242
4.4.8ff: 241
4.4.9: 174–175 n. 724, 241*sexies*
4.4.9–11: 213 n. 924
4.4.10: 182 n. 763, 241*quater*, 242
4.4.10–12: 136 n. 548, 241
4.4.11: 185 n. 779, 241, 242
4.4.11–12: 172 n. 712, 214 n. 929, 242*bis*
4.4.12: 242
4.4.13: 242
4.5.3: 74 n. 262
4.5.4: 255
4.5.7–8: 256
4.5.8: 100–101
4.5.10: 255
4.5.11: 256*bis*
4.5.11–17: 256
4.5.12: 256*ter*

4.5.12–13: 255*bis*
4.5.13: 255, 256*ter*
4.5.13–18: 172 n. 712
4.5.14: 100 n. 393, 100 n. 395, 256*bis*
4.5.14ff.: 255
4.5.14–17: 256
4.5.15: 256*bis*
4.5.15–16: 138 n. 561
4.5.15–17: 256
4.5.16: 160 n. 653, 256*ter*
4.5.17: 256*ter*
4.8.9–10: 133–134 n. 540
4.8.35–39: 181 n. 762
4.8.37–39: 101 n. 400
4.8.39: 100 n. 393
5.1.11–12: 182 n. 763
5.1.12: 196 n. 841
5.2.40: 100–101
5.2.40–41: 174–175 n. 724
5.3.1: 274
5.3.3: 274*sexies*
5.3.3–6: 274*ter*
5.3.4: 274*bis*
5.3.5: 274*quater*
5.3.5–6: 214 n. 927
5.3.6: 182 n. 763, 274*sexies*
5.3.18: 100–101
5.4.8: 100–101
5.4.17–18: 26, 28 n. 65, 74 n. 262, 74 n. 263
5.4.18: 26 n. 58
5.4.20–21: 133–134 n. 540
5.4.33: 184
5.4.41: 182 n. 763
5.4.44–45: 182 n. 763

5.4.52: 182 n. 763
6.1.1: 211 n. 914, 257*ter*
6.2.20: 211 n. 915
6.4.1–16: 174–175 n. 724
6.4.3: 257
6.4.4: 256*bis*, 257*ter*
6.4.4–16: 169 n. 695, 258
6.4.7: 257
6.4.8: 258
6.4.8–14: 258
6.4.9: 100 n. 396, 257*quater*
6.4.10: 257*ter*
6.4.12: 170, 171, 257*quinquies*
6.4.13: 181 n. 761, 182 n. 763, 257*bis*
6.4.13–14: 173 n. 719, 179 n. 753, 180 n. 754
6.4.14: 100–101, 171, 185 n. 779, 190, 257*ter*
6.4.15: 211 n. 914, 257, 258*ter*
6.4.17: 100 n. 390
6.4.20: 257
7.1.14: 173 n. 717
7.1.31–32: 199, 211
7.3.9: 100–101
7.4.20: 249*ter*
7.4.20–21: 249
7.4.20–25: 250
7.4.21: 249*bis*
7.4.22: 249*quater*
7.4.22–23: 169 n. 689
7.4.22–25: 250
7.4.23: 89 n. 337, 164 n. 666, 190, 250

7.4.23–24: 250
7.4.25: 250
7.4.31: 182 n. 763, 185 n. 779, 200
7.5.3: 173 n. 717, 261
7.5.4: 261
7.5.4–5: 261
7.5.7: 261
7.5.18: 261*bis*
7.5.21–24: 261
7.5.21–27: 262
7.5.22: 100–101, 261*quater*
7.5.23: 261*bis*, 262*bis*
7.5.24: 261, 262
7.5.24–25: 262
7.5.25: 214 n. 927, 261*bis*, 262*ter*
7.5.26: 262
7.5.26–27: 262

Memorabilia
3.1.8: 172, 183–184
3.10.9–13: 74–75
3.10.12: 32 n. 95
3.10.14–15: 75

Oeconomicus
8.4–7: 195 n. 837
21.4–7: 180 n. 756

Symposium
2.11: 51 n. 165

Zenobius
4.29: 43 n. 133

II. INSCRIPTIONS AND PAPYRI

***CEG* I**
10: 117 n. 472
27: 117 n. 472
112: 117 n. 472

CIG
2758 G: 49 n. 160
2759: 49 n. 160

GVI
I: 212
749: 118 n. 477

***IG* I²**
1.8–10: 97 n. 377
68–69: 244
70: 244
929: 176 n. 732, 177
931: 176 n. 732
932: 176 n. 732
943: 176 n. 732
945: 278
1085: 175 n. 730

***IG* I³**
439.4: 182 n. 763
476.380: 29–30 n. 74
522: 283
1162.4: 182 n. 763

***IG* II²**
112.34–35: 173 n. 717
1155: 175 n. 730
1926: 100 n. 390
2311.72–74: 49 n. 160
6320: 47 n. 147
11678: 182 n. 763, 267*bis*

***IG* IV²**
1.28: 176 n. 732

***IG* V**
2.173: 176 n. 732
2.174: 176 n. 732

***IG* VII**
190: 49 n. 160
1888: 244

Meiggs-Lewis, *GHI*²
7a–g: 134 n. 545
14: 97 n. 377
33: 176 n. 732, 177, 177 n. 742
33.5–6: 182 n. 763
33.63: 182 n. 763
35: 176 n. 732, 291
36: 291
48: 176 n. 732
48.4: 182 n. 763*bis*

P Oxy.
1087: 154 n. 636

Roueché, *Performers*
52.IV: 49 n. 160
53: 49 n. 160

Tod, *GHI*
27: 291*bis*
35.1–2: 290
36.1: 290
59: 278

HISTORIA – EINZELSCHRIFTEN

Herausgegeben von Kai Brodersen, Mortimer Chambers, Martin Jehne, François Paschoud und Aloys Winterling.

Franz Steiner Verlag   ISSN 0341-0056

176. Anthony Francis Natoli
The Letter of Speusippus to Philip II
Introduction, Text, Translation and Commentary. With an Appendix on the Thirty-First Socratic Letter attributed to Plato
2004. 196 S., kt.
ISBN 978-3-515-08396-6

177. Karl-Wilhelm Welwei
Res publica und Imperium
Kleine Schriften zur römischen Geschichte. Hg. von Mischa Meier und Meret Strothmann
2004. 328 S., geb.
ISBN 978-3-515-08333-1

178. Konrad Vössing
Biographie und Prosopographie
Internationales Kolloquium
zum 65. Geburtstag von Anthony R. Birley
2005. 146 S., 2 Taf., kt.
ISBN 978-3-515-08538-0

179. Vera-Elisabeth Hirschmann
Horrenda Secta
Untersuchungen zum frühchristlichen Montanismus und seinen Verbindungen zur paganen Religion Phrygiens
2005. 168 S., kt.
ISBN 978-3-515-08675-2

180. Thomas Heine Nielsen (Hg.)
Once Again: Studies in the Ancient Greek *Polis*
(Papers from the Copenhagen *Polis* Centre 7)
2004. 202 S., kt.
ISBN 978-3-515-08438-3

181. Gideon Maier
Amtsträger und Herrscher in der *Romania Gothica*
Vergleichende Untersuchungen zu den Institutionen der ostgermanischen Völkerwanderungsreiche
2005. 363 S., kt.
ISBN 978-3-515-08505-2

182. David Whitehead / P. H. Blyth
Athenaeus Mechanicus, On Machines (Περὶ μηχανημάτων)
Translated with Introduction and Commentary
2004. 236 S., kt.
ISBN 978-3-515-08532-8

183. Wolfgang Blösel
Themistokles bei Herodot: Spiegel Athens im fünften Jahrhundert
Studien zur Geschichte und historiographischen Konstruktion des griechischen Freiheitskampfes 480 v. Chr.
2004. 422 S. mit 2 Ktn., geb.
ISBN 978-3-515-08533-5

184. Klaus Scherberich
Koinè symmachía
Untersuchungen zum Hellenenbund Antigonos' III. Doson und Philipps V. (224–197 v. Chr.)
2009. 254 S., geb.
ISBN 978-3-515-09406-1

185. Stephan Berrens
Sonnenkult und Kaisertum von den Severern bis zu Constantin I. (193–227 n. Chr.)
2004. 283 S., 2 Taf., kt.
ISBN 978-3-515-08575-5

186. Norbert Geske
Nikias und das Volk von Athen im Archidamischen Krieg
2005. 224 S., kt.
ISBN 978-3-515-08566-3

187. Christopher L. H. Barnes
Images and Insults
Ancient Historiography and the Outbreak of the Tarentine War
2005. 170 S., kt.
ISBN 978-3-515-08689-9

188. Massimiliano Vitiello
Momenti di Roma ostrogota
aduentus, feste, politica
2005. 162 S., kt.
ISBN 978-3-515-08688-2

189. Klaus Freitag / Peter Funke / Matthias Haake (Hg.)
Kult – Politik – *Ethnos*
Überregionale Heiligtümer im Spannungsfeld von Kult und Politik
2006. 287 S., kt.

ISBN 978-3-515-08718-6
190. Jens Uwe Krause / Christian Witschel (Hg.)
**Die Stadt in der Spätantike –
Niedergang oder Wandel?**
Akten des internationalen Kolloquiums
in München am 30. und 31. Mai 2003
2006. 492 S. mit 42 Abb., kt.
ISBN 978-3-515-08810-7
191. Heinz Heinen
**Vom hellenistischen Osten
zum römischen Westen**
Ausgewählte Schriften zur Alten
Geschichte. Hg. von Andrea Binsfeld
und Stefan Pfeiffer
2006. XXVIII, 553 S. mit Frontisp. und 34
Abb., geb.
ISBN 978-3-515-08740-7
192. Itzhak F. Fikhman
**Wirtschaft und Gesellschaft
im spätantiken Ägypten**
Kleine Schriften. Hg. von Andrea Jördens
unter Mitarb. von Walter Sperling
2006. XVIII, 380 S. mit Frontisp. und 2
Farbktn., geb.
ISBN 978-3-515-08876-3
193. Adalberto Giovannini
**Les relations entre états dans
la Grèce antique**
Du temps d'Homère à l'intervention
romaine (ca. 700–200 av. J.-C.)
2007. 445 S., kt.
ISBN 978-3-515-08953-1
194. Michael B. Charles
Vegetius in Context
Establishing the Date of the *Epitoma Rei
Militaris*
2007. 205 S., kt.
ISBN 978-3-515-08989-0
195. Clemens Koehn
Krieg – Diplomatie – Ideologie
Zur Außenpolitik hellenistischer
Mittelmeerstaaten
2007. 248 S., kt.
ISBN 978-3-515-08990-6
196. Kay Ehling
**Untersuchungen zur Geschichte der
späten Seleukiden (164–63 v. Chr.)**
Vom Tode des Antiochos IV. bis zur Einrichtung der Provinz Syria unter Pompeius
2007. 306 S. mit 1 Kte. und 1 Falttaf.
in einer Tasche, kt.
ISBN 978-3-515-09035-3
197. Stephanie L. Larson
Tales of Epic Ancestry
Boiotian Collective Identity in the Late
Archaic and Early Classical Periods
2007. 238 S., kt.
ISBN 978-3-515-09028-5
198. Mogens Herman Hansen (Hg.)
The Return of the *Polis*
The Use and Meanings of the Word *Polis*
in Archaic and Classical Sources
(Papers from the Copenhagen *Polis*
Centre 8)
2007. 276 S., kt.
ISBN 978-3-515-09054-4
199. Volker Grieb
Hellenistische Demokratie
Politische Organisation und Struktur in
freien griechischen *Poleis* nach Alexander
dem Großen
2008. 407 S., geb.
ISBN 978-3-515-09063-6
200. Cristina Rosillo López
**La corruption à la fin
de la République romaine
(IIe–Ier s. av. J.-C.)**
Aspects politiques et financiers
2010. 276 S., geb.
ISBN 978-3-515-09127-5
201. Manuel Tröster
**Themes, Character, and Politics
in Plutarch's *Life of Lucullus***
The Construction of a Roman Aristocrat
2008. 206 S., kt.
ISBN 978-3-515-09124-4
202. David D. Phillips
Avengers of Blood
Homicide in Athenian Law and Custom
from Draco to Demosthenes
2008. 279 S., geb.
ISBN 978-3-515-09123-7
203. Vassiliki Pothou
**La place et le rôle de la digression
dans l'œuvre de Thucydide**
2009. 189 S., 2 Taf., geb.
ISBN 978-3-515-09193-0
204. Iris Samotta
Das Vorbild der Vergangenheit
Geschichtsbild und Reformvorschläge
bei Cicero und Sallust
2009. 506 S., geb.
ISBN 978-3-515-09167-1
205. Efrem Zambon
Tradition and Innovation
Sicily between Hellenism and Rome
2008. 326 S., geb.
ISBN 978-3-515-09194-7
206. Susanne Carlsson
Hellenistic Democracies

Freedom, Independence and Political
Procedure in Some East Greek City-States
2010. 372 S. mit 2 Abb., geb.
ISBN 978-3-515-09265-4

207. Adam Schwartz
Reinstating the Hoplite
Arms, Armour and Phalanx Fighting
in Archaic and Classical Greece
2009. 337 S. mit 19 Abb., geb.
ISBN 978-3-515-09330-9

208. Elizabeth A. Meyer
**Metics and the Athenian
Phialai-Inscriptions**
A Study in Athenian Epigraphy and Law
2010. 167 S. und 47 Taf., geb.
ISBN 978-3-515-09331-6

209. Margret Dissen
**Römische Kollegien und deutsche
Geschichtswissenschaft
im 19. und 20. Jahrhundert**
2009. 337 S., geb.
ISBN 978-3-515-09387-3

210. Joachim Szidat
Usurpator tanti nominis
Kaiser und Usurpator in der Spätantike
(337–476 n. Chr.)
2010. 458 S., geb.
ISBN 978-3-515-09636-2

211. Armin Eich (Hg.)
**Die Verwaltung der kaiserzeitlichen
römischen Armee**
Studien für Hartmut Wolff
2010. 210 S. mit 4 Abb., geb.
ISBN 978-3-515-09420-7

212. Stefan Pfeiffer
**Der römische Kaiser
und das Land am Nil**
Kaiserverehrung und Kaiserkult in
Alexandria und Ägypten von Augustus bis
Caracalla (30 v. Chr. – 217 n. Chr.)
2010. 378 S., geb.
ISBN 978-3-515-09650-8

213. M. A. Robb
Beyond *Populares* and *Optimates*
Political Language in the Late Republic
2010. 225 S., geb.
ISBN 978-3-515-09643-0

214. Kai Brodersen / Jaś Elsner (ed.)
**Images and Texts
on the "Artemidorus Papyrus"**
Working Papers on P. Artemid. (St. John's
College Oxford, 2008)
2009. 169 S. mit 70 Abb., geb.
ISBN 978-3-515-09426-9

215. Eberhard Ruschenbusch (†)
**Solon: Das Gesetzeswerk –
Fragmente**
Übersetzung und Kommentar
Hrsg. v. Klaus Bringmann
2010. 168 S. mit Frontispiz, geb.
ISBN 978-3-515-09709-3

216. David Whitehead
**Apollodorus Mechanicus:
Siege-matters (Πολιορκητικά)**
Translated with Introduction
and Commentary
2010. 162 S. mit 6 Abb., geb.
ISBN 978-3-515-09710-9

217. Bernhard Linke / Mischa Meier /
Meret Strothmann (Hg.)
**Zwischen Monarchie
und Republik**
Gesellschaftliche Stabilisierungsleistungen
und politische Transformationspotentiale
in den antiken Stadtstaaten
2010. 236 S., geb.
ISBN 978-3-515-09782-6

218. Julia Hoffmann-Salz
**Die wirtschaftlichen Auswirkungen
der römischen Eroberung**
Vergleichende Untersuchungen der
Provinzen Hispania Tarraconensis, Africa
Proconsularis und Syria
2011. 561 S. mit 26 Tab. und 3 Ktn., geb.
ISBN 978-3-515-09847-2

219. Dirk Schnurbusch
Convivium
Form und Bedeutung aristokratischer
Geselligkeit in der römischen Antike
2011. 314 S., geb.
ISBN 978-3-515-09860-1

220. Gabriel Herman (ed.)
**Stability and Crisis in the Athenian
Democracy**
2011. 165 S. mit 3 Tab., geb.
ISBN 978-3-515-09867-0

221. Christoph Lundgreen
**Regelkonflikte
in der römischen Republik**
Geltung und Gewichtung von Normen
in politischen Entscheidungsprozessen
2011. 375 S., geb.
ISBN 978-3-515-09901-1

222. James H. Richardson
The Fabii and the Gauls
Studies in historical thought and
historiography in Republican Rome
2012. 186 S., geb.
ISBN 978-3-515-10040-3